W0114879

CITRUS

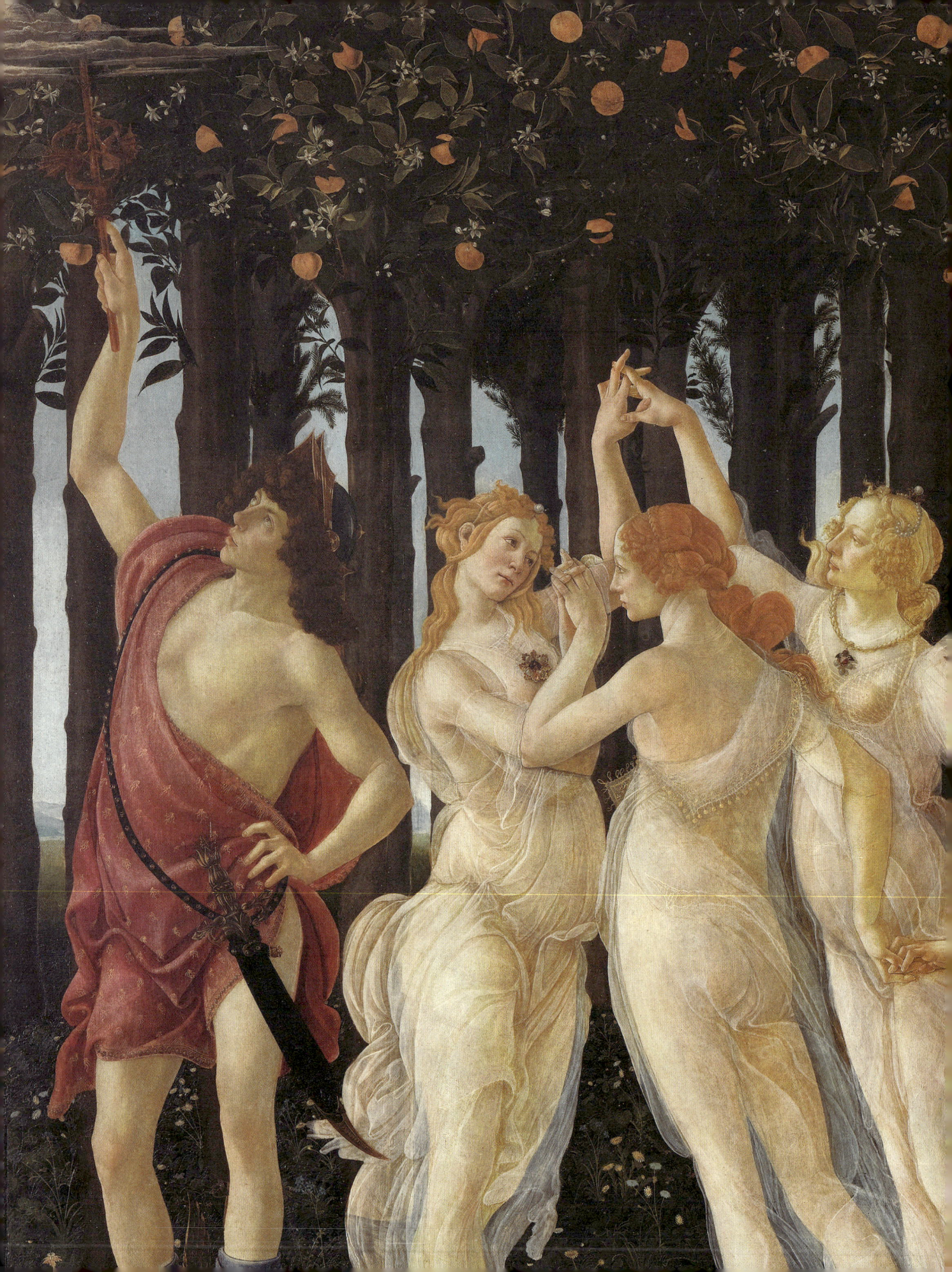

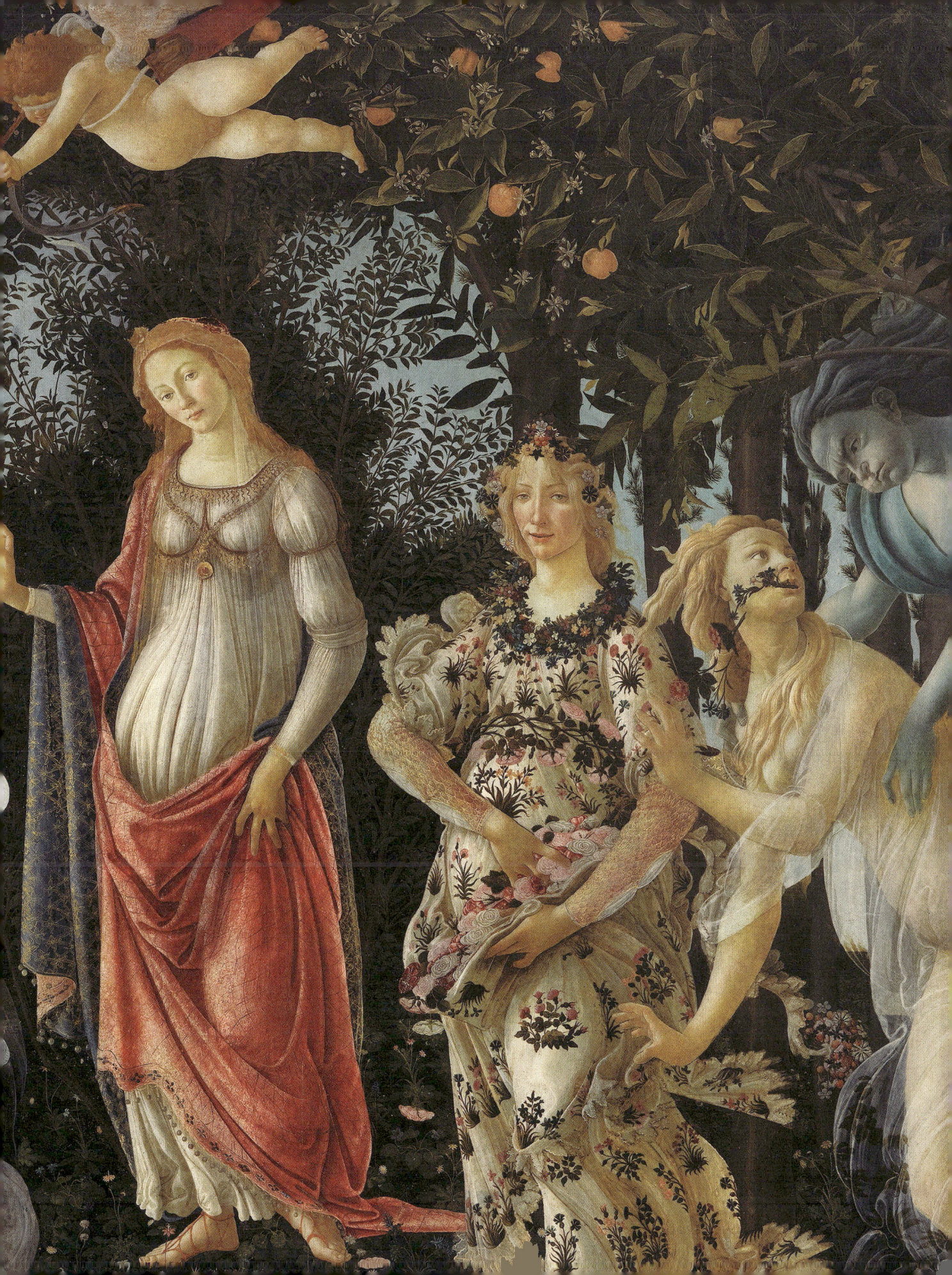

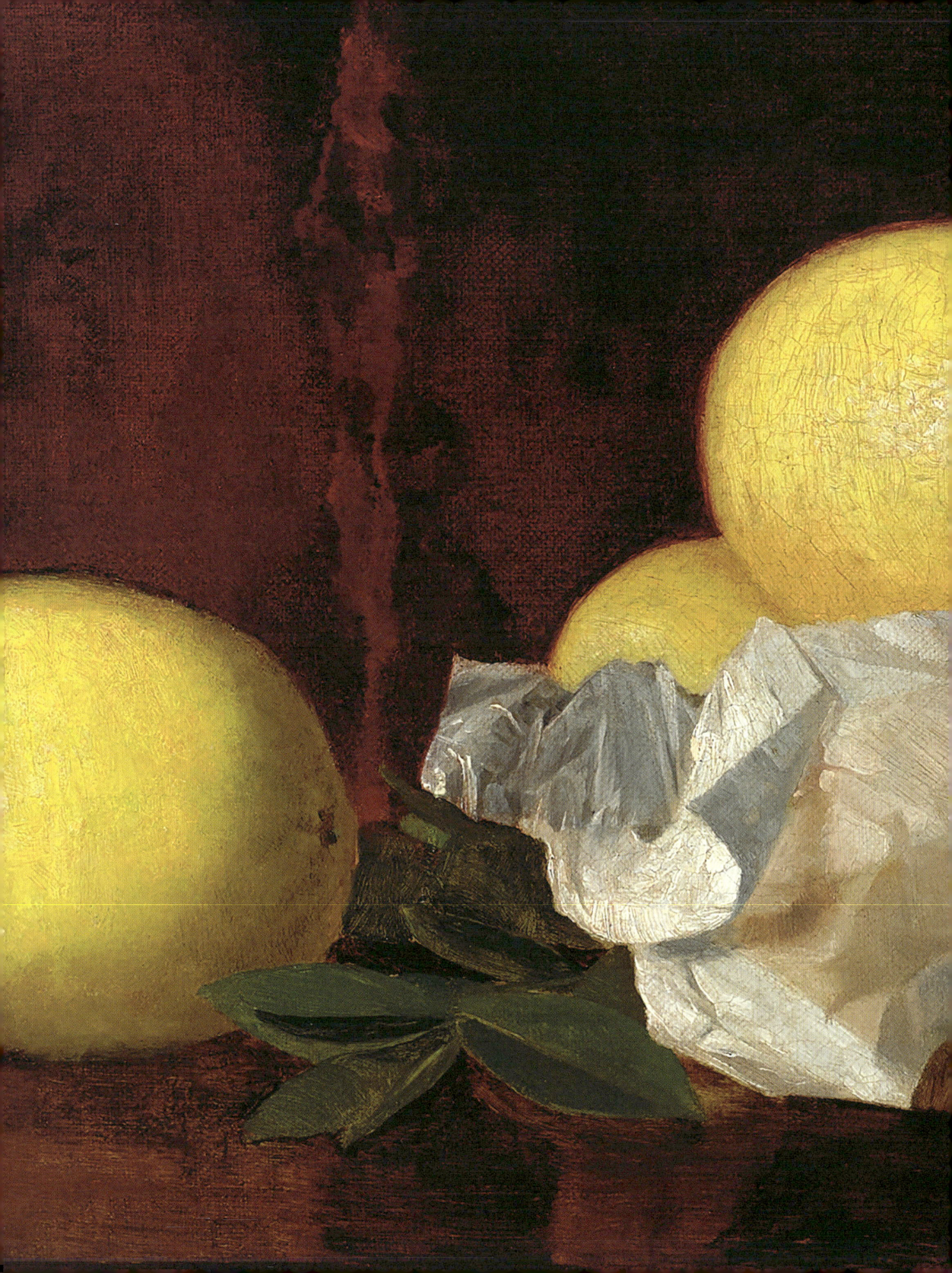

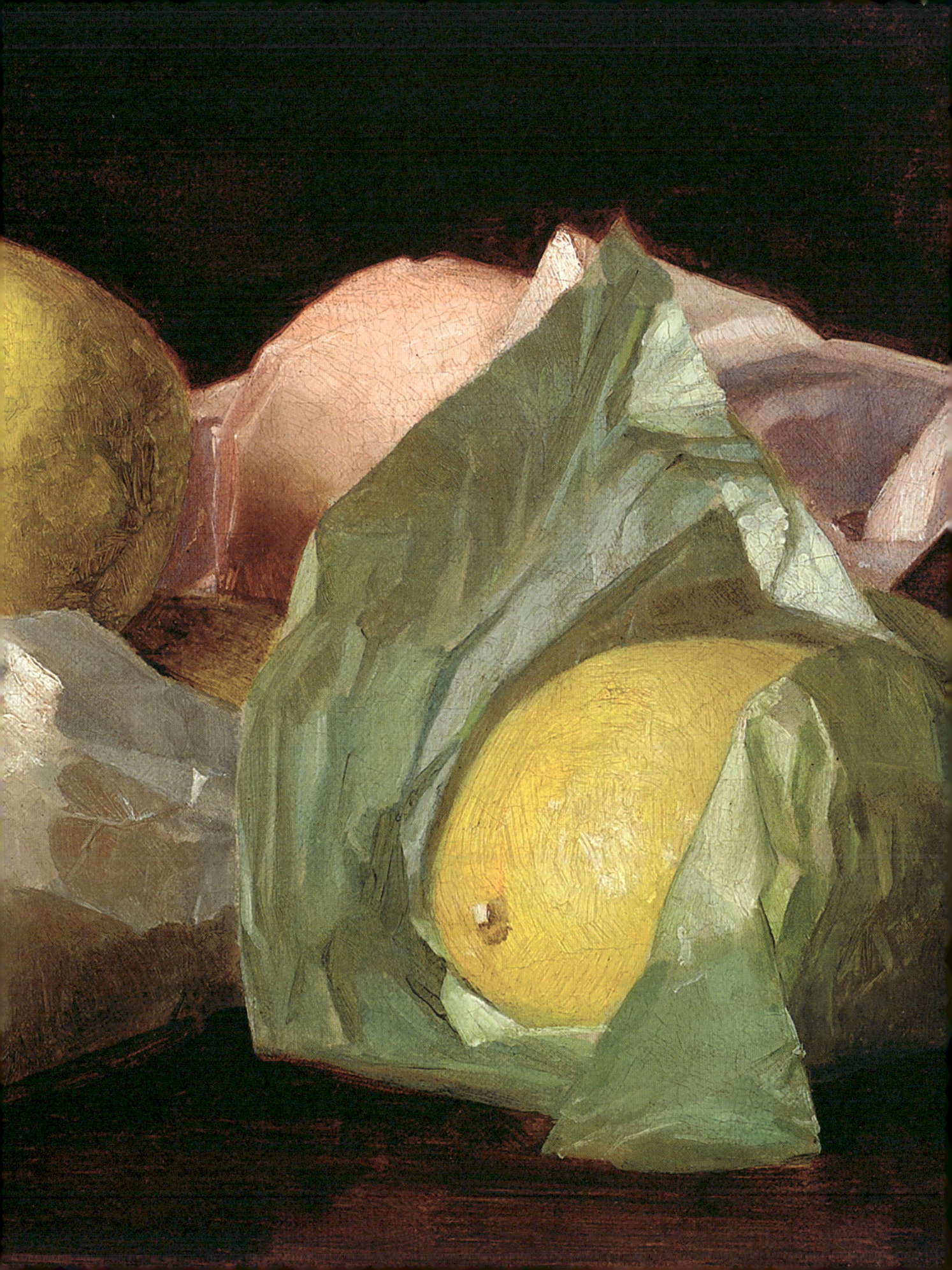

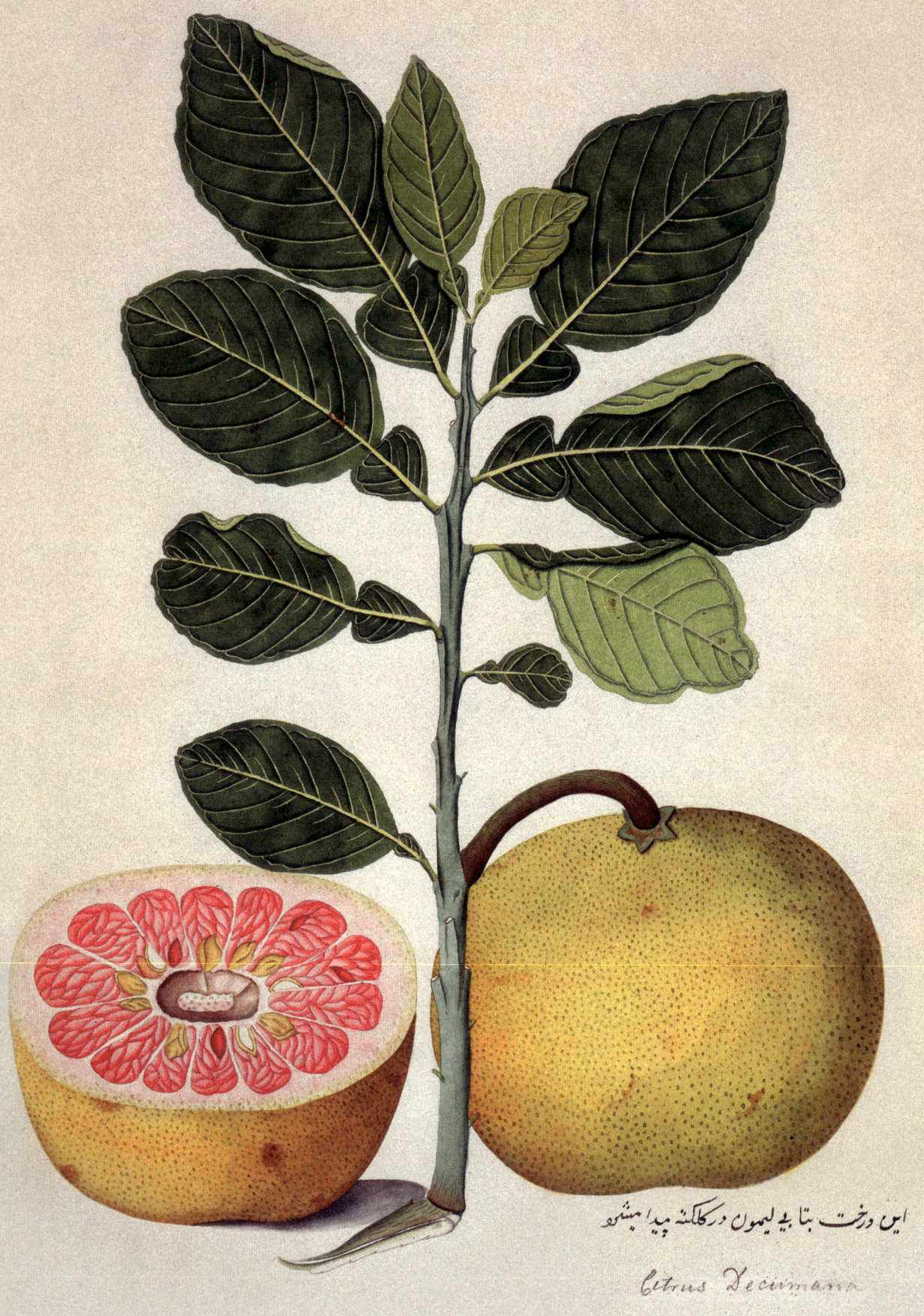

این درخت بتا بید لیمون درکلکته پیدا میشو

Citrus Decumana

Plate I.

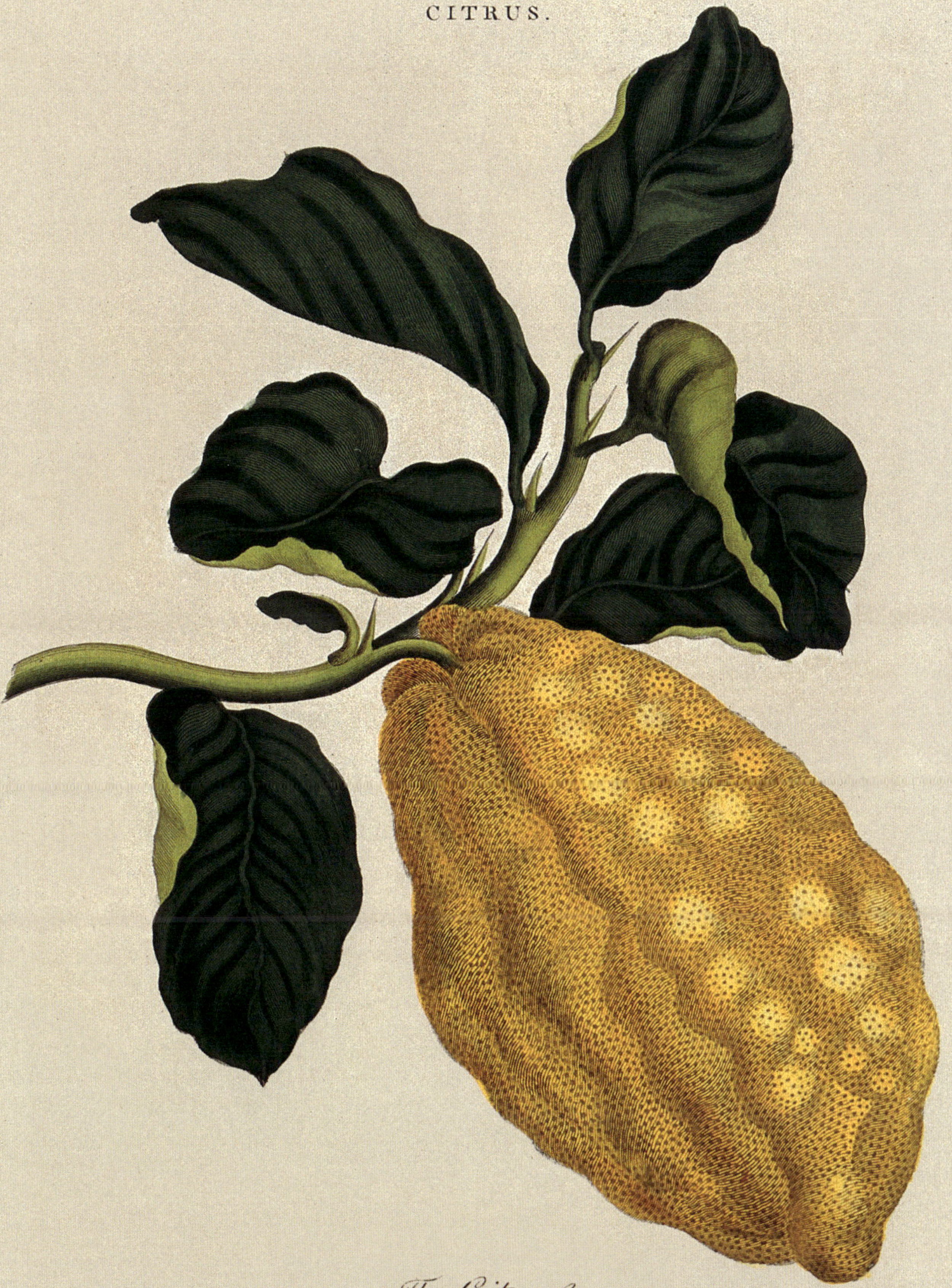

The Citron.

London. Published as the Act directs, Oct.r 1 800 by J. Wilkes.

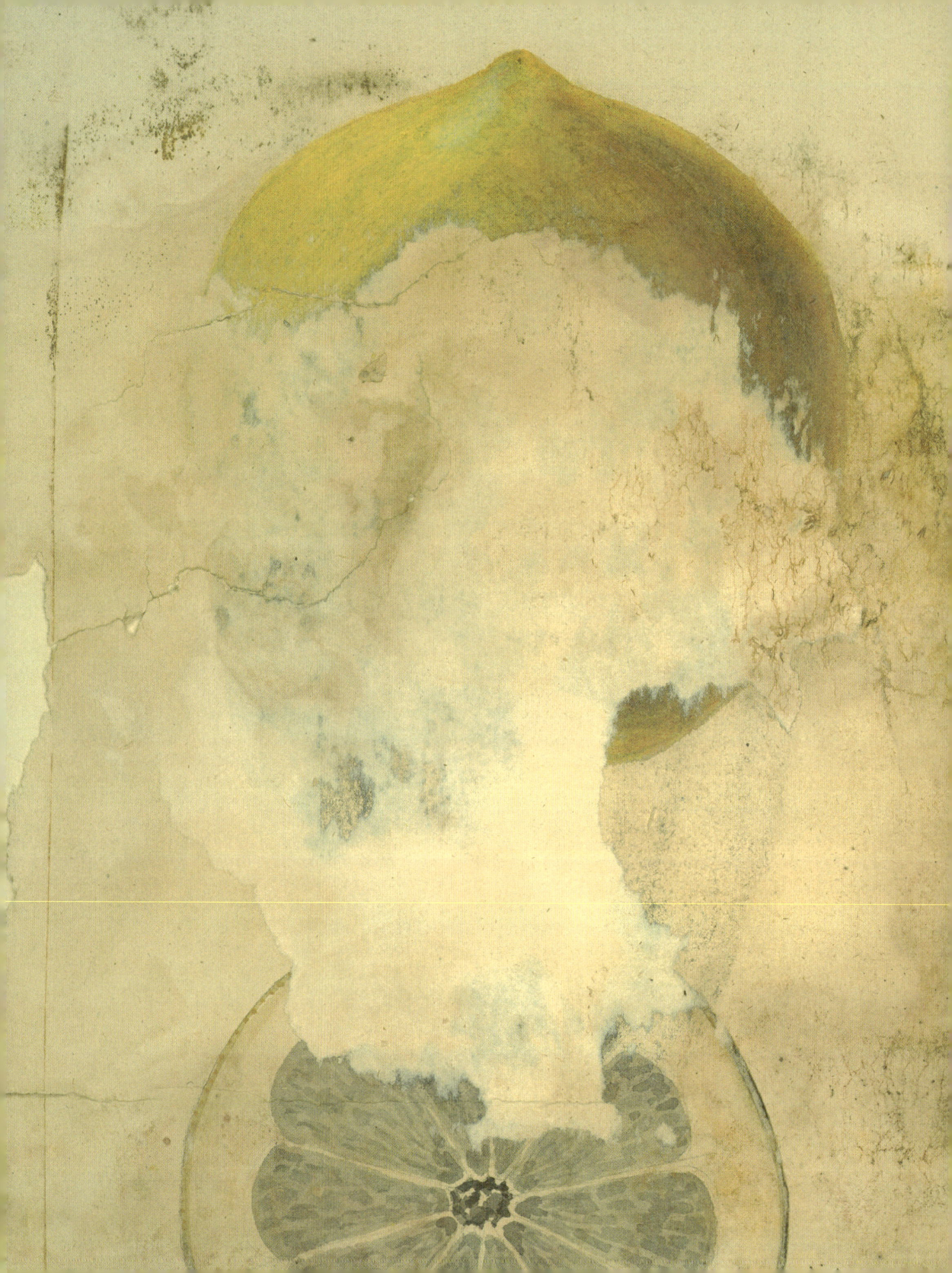

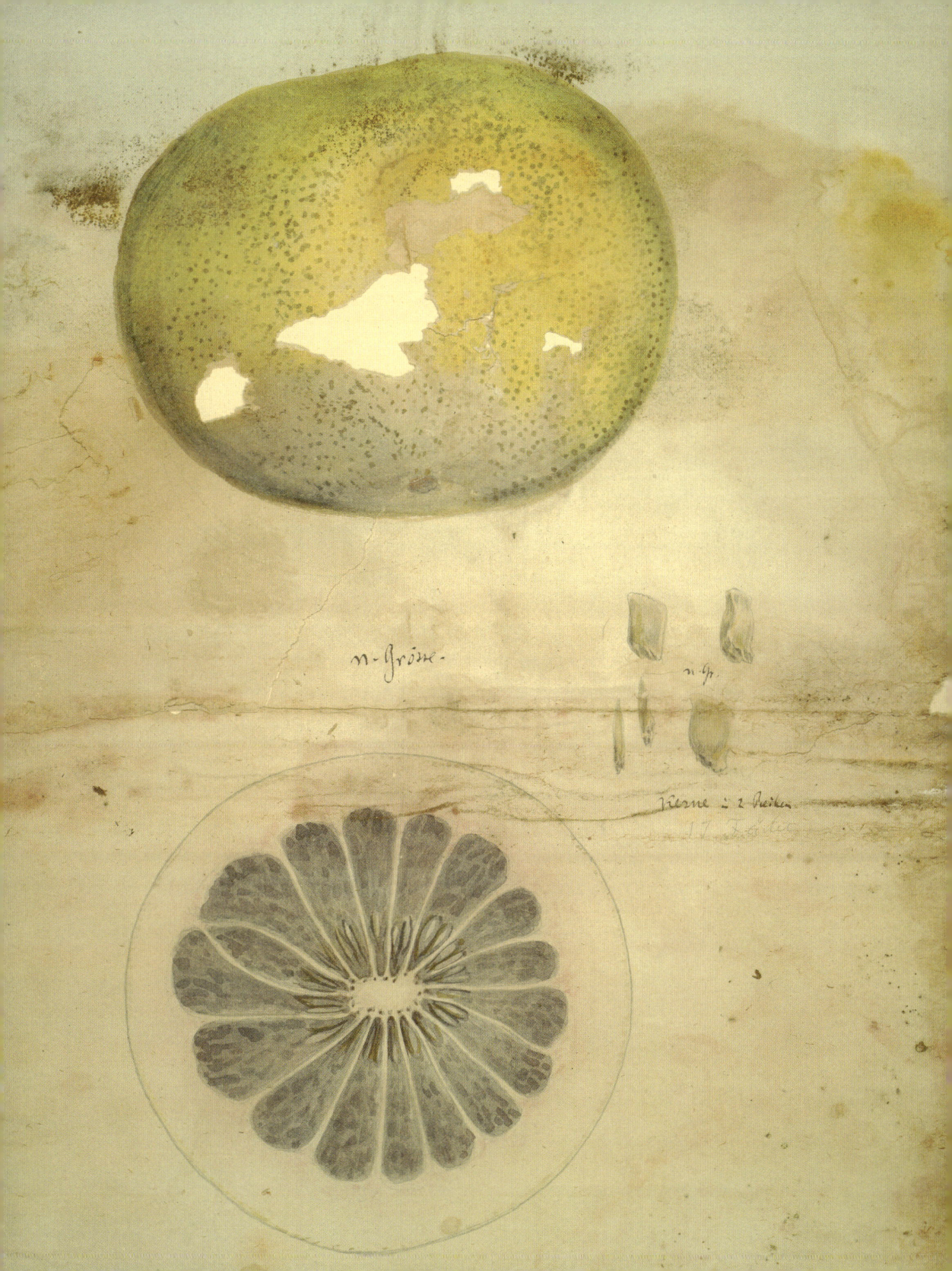

n. Größe.

n. Gr.

Kerne : 2 Reihen

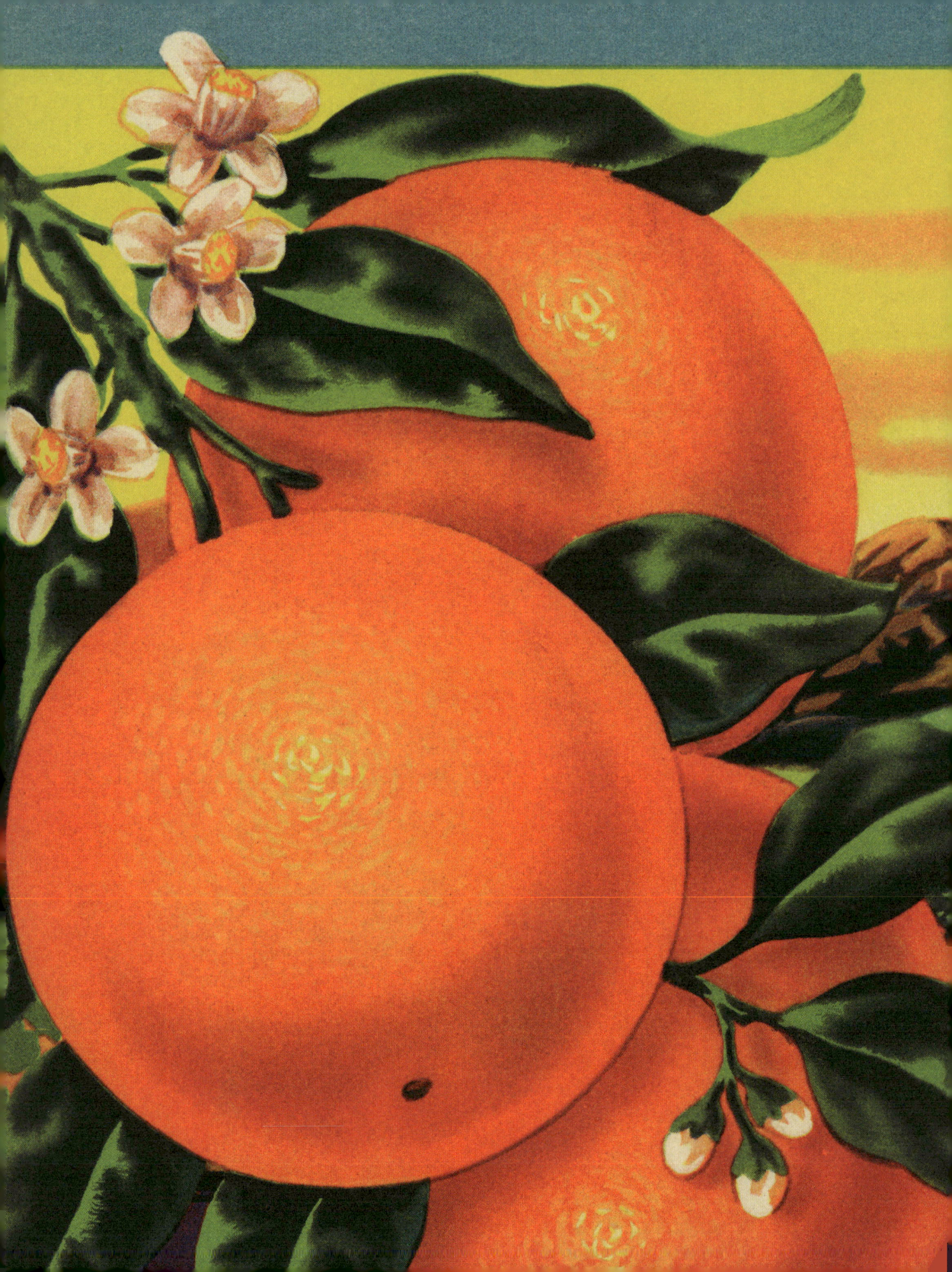

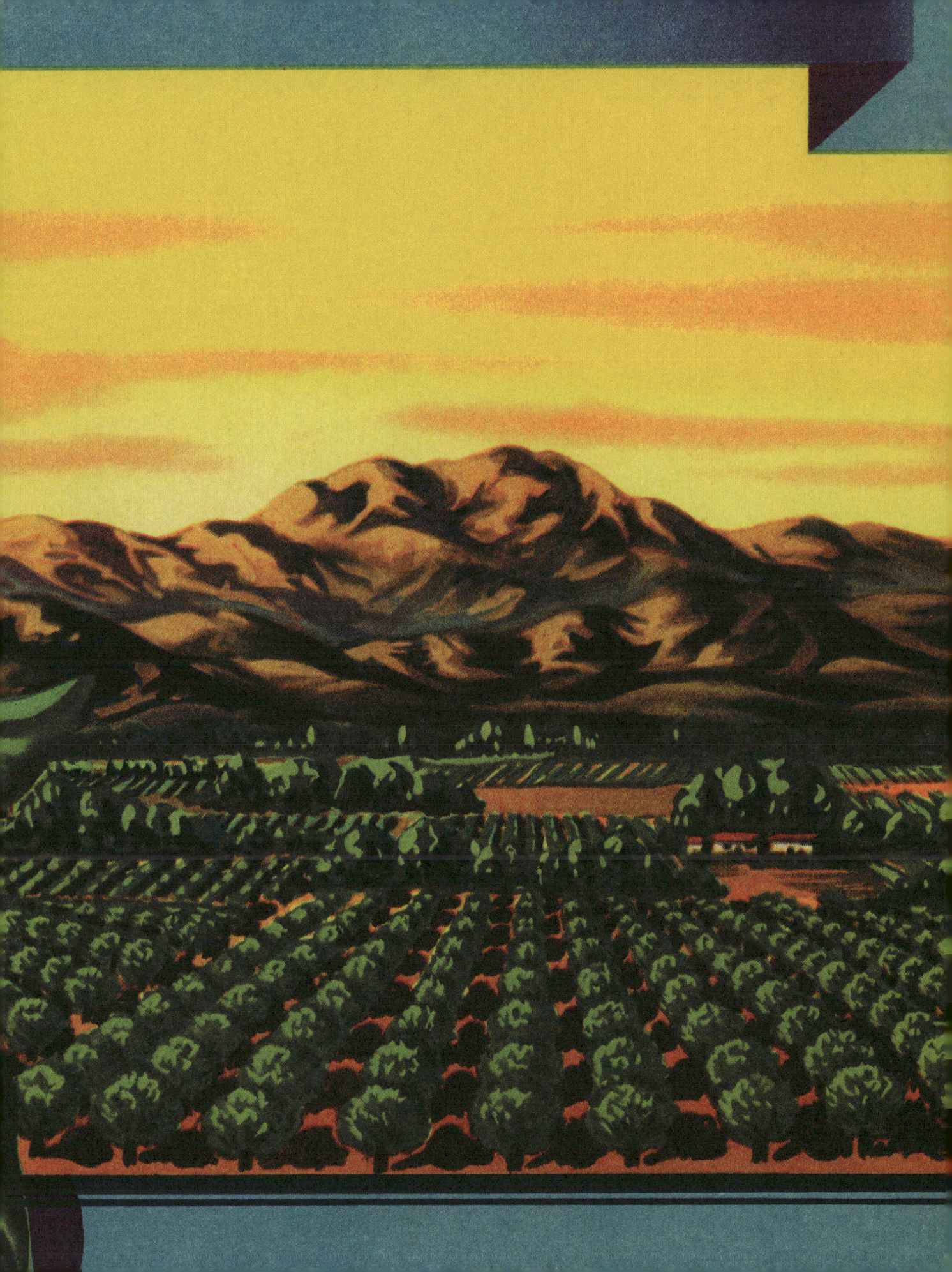

David J. Mabberley

CITRUS

A WORLD HISTORY

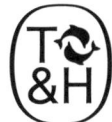

CONTENTS

INTRODUCTION 16

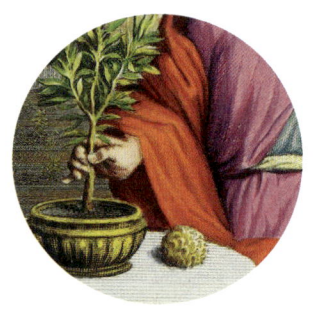

ONE

The Ancient World 26

The origins of *Citrus*; The importance of hybrids;
Yu the Great to Alexander the Great; The Feast
of the Tabernacles

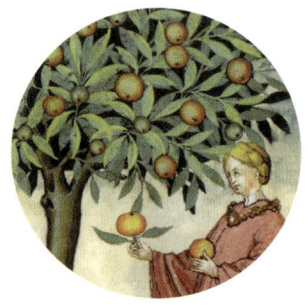

TWO

Herbals to Hesperides 56

Medicine, Medici and the Middle East; Citrus in
art: the Renaissance; Bergamot and the *bizzarria*;
Ferrari's *Hesperides*

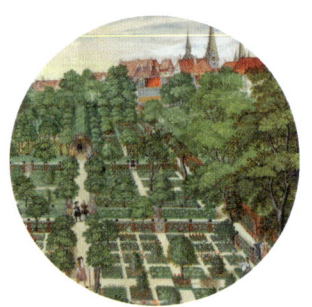

THREE

The House of Orange 114

The Dutch Golden Age and 'citrusmania'; Seville
oranges and marmalade madams; Orangeries and
lemon houses; Still life and botanical drawing

FOUR

Empire, Exports and Vitamin C 150

Scurvy: the plague of the sea; Colonization and
the citrus trade; The lime and the grapefruit;
Australasia and the future of citrus

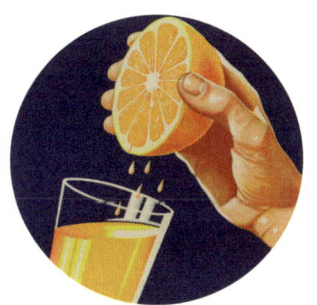

FIVE

Progress and Perils 194

Modern medicine, mass production and
the mafia; Tangerines and satsumas: citrus
goes global; Orange juice and cultural icons;
Huanglongbing: the end?

NOTES 252

BIBLIOGRAPHY 258

SOURCES OF ILLUSTRATIONS 265

ACKNOWLEDGMENTS 267

INDEX 268

INTRODUCTION

FROM ITS ORIGINS in the Indo-Pacific, to its development in Europe, the Americas and beyond, to recent advances in medicine, the story of citrus permeates human history, as recorded in the literature and art of many civilizations from antiquity to the present day. The hitherto obscure origins of the species and their hybrids that are the ultimate basis of the modern citrus industry are explained, as is the possible demise of commercial plantations, perhaps heralding the end of an enormous global enterprise – and the last drops of Everyman's orange juice.

In tracing the intricate history of oranges, lemons, limes, grapefruit and countless other hybrids and cultivars, it is shown how human preoccupation with these amenable vitamin-rich fruits has led, and continues to lead, not only to advances in medicine and agricultural policy and practice, but also to some of the world's most famous works of art, and scientific intrigue. The story of citrus epitomizes the reverential as well as cavalier approach of humans to food, cultivation and the living world.

Growing oranges for juice, as well as fresh fruit, is now the most important fruit industry in warm countries and contributes significantly to the economies of China, the USA (particularly California and Florida), Brazil, Australia, South Africa, Israel and Egypt, as well as several countries in southern Europe (notably Spain and Portugal). In 2021, world commercial citrus fruit production was 161 million tonnes, up from 45 million in 1972.[1] This number greatly exceeds global apple production (93 million tonnes)[2] in 2021, while banana production that year was 124 million tonnes, up from 33 million in 1972.[3]

Although over half of all commercial citrus crops are grown in Asia (China produced 47 million tonnes in 2021, some 28 per cent of world production), more than half of citrus exports are from the Mediterranean region and Africa, accounting for around 80 per cent of their total citrus production: pre-eminent here are Spain, South Africa, Turkey and Egypt (now the world's biggest exporter of oranges). The main importers, on the other hand, are European countries, which imported over 6 million tonnes of citrus in 2021, 42 per cent of the global share. Annual consumption per capita in 2021 was greatest in Scandinavia (14.5 kg [32 lb]), followed by France (13.5 kg [30 lb]) and Germany (12.8 kg [28 lb]).[4]

In Asia, citrus fruits have been used in local and commercial medicinal preparations for hundreds of years. Moreover, the readily transported and easily stored fruits have entered the art, religion and folklore, as well as the cuisine, of countries way beyond.

But for many peoples in Western countries, this mighty industry seems a relatively new one. In the post-war English countryside where I grew up, we squeezed lemon juice on to our fish on Fridays (and on to pancakes on Shrove Tuesday), and we were given concentrated orange juice as a source of vitamin C. It was not always easy to get, and, when there were shortages, the juice was largely replaced by rose-hip syrup prepared from wild roses (we children were paid threepence per pound for the hips). At Christmas we marvelled at boxes of mandarins, with each precious fruit individually wrapped in tissue paper, and – luxury of luxuries – candied citrus peel. Fresh

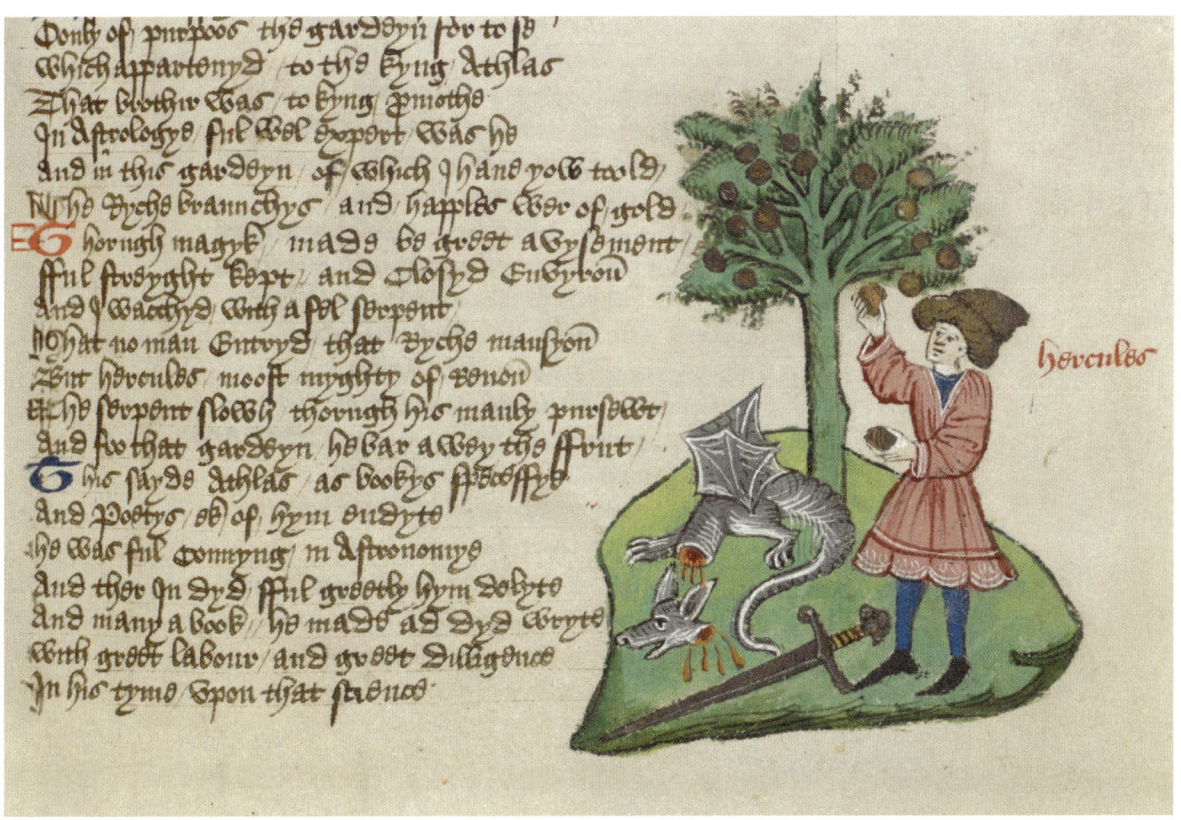

A detail of Heracles from *Fall of Princes* by John Lydgate, *c*. 1450–60. The association of the 'golden apples' stolen by Heracles from the garden of the Hesperides with oranges perpetuates the taxonomic confusion for which the genus *Citrus* is notorious.

orange juice, despite its ubiquity today, especially in the USA, was for most people almost unheard of. Yet, in other ways, citrus has long been with us, adding brightness and zest to our lives: orange blossom and eau de Cologne, Curaçao and Cointreau, Earl Grey and orange pekoe teas, marmalade.

During the Napoleonic wars, British ships carried lemons and limes to protect their sailors against scurvy (caused by vitamin C deficiency): the 'Limeys' ruled the waves. British people are still called 'Limeys' by Americans, and today citrus is associated with good health, freshness and cleanliness. Lemon scents are increasingly encountered in everything from health drinks to household cleaning products.

Oranges, lemons and citrons (*Citrus medica*, the source of candied peel) have been classified in the botanical genus *Citrus* since the time of the eighteenth-century Swedish naturalist Carl Linnaeus (1707–1778), who took up the Classical name once used for the sweet-scented timber of *Tetraclinis articulata*, a conifer known to the ancient Greeks as *thyon* and to the Romans as *citrus*; the Romans used the term *citreum* to define what we now call citrus fruits.[5] *Tetraclinis* wood was used for making insecticidal chests for woollen clothes, and the fruits

of the citron placed among clothes for the same purpose.[6] The Greek word *kedros*, used for other conifers, cedars and (perhaps originally) junipers, was the source of the Latin *citrus*. In his *De materia medica*, Pedanius Dioscorides (see p. 47), a Greek physician and botanist, described the citron as '*kedromela*, which the Romans call *citria*'.

Before Linnaeus's time, the fruits had been known in Europe as *aurantia*, from the Latin for gold – it is possible that this explains *orenge* in Old French, from *or* (gold). The Arabic word *naranj* is apparently derived from Hindi *narangi* via the Persian *narang* and Sanskrit *na ranga*, ultimately from an imported Dravidian or Tamil word meaning fragrant.[7] The word 'orange' probably reached England from Spain via Old French and Old Provençal, from both of which the initial 'n' had already been 'sporadically lost'. From *citrus* has come the English citron

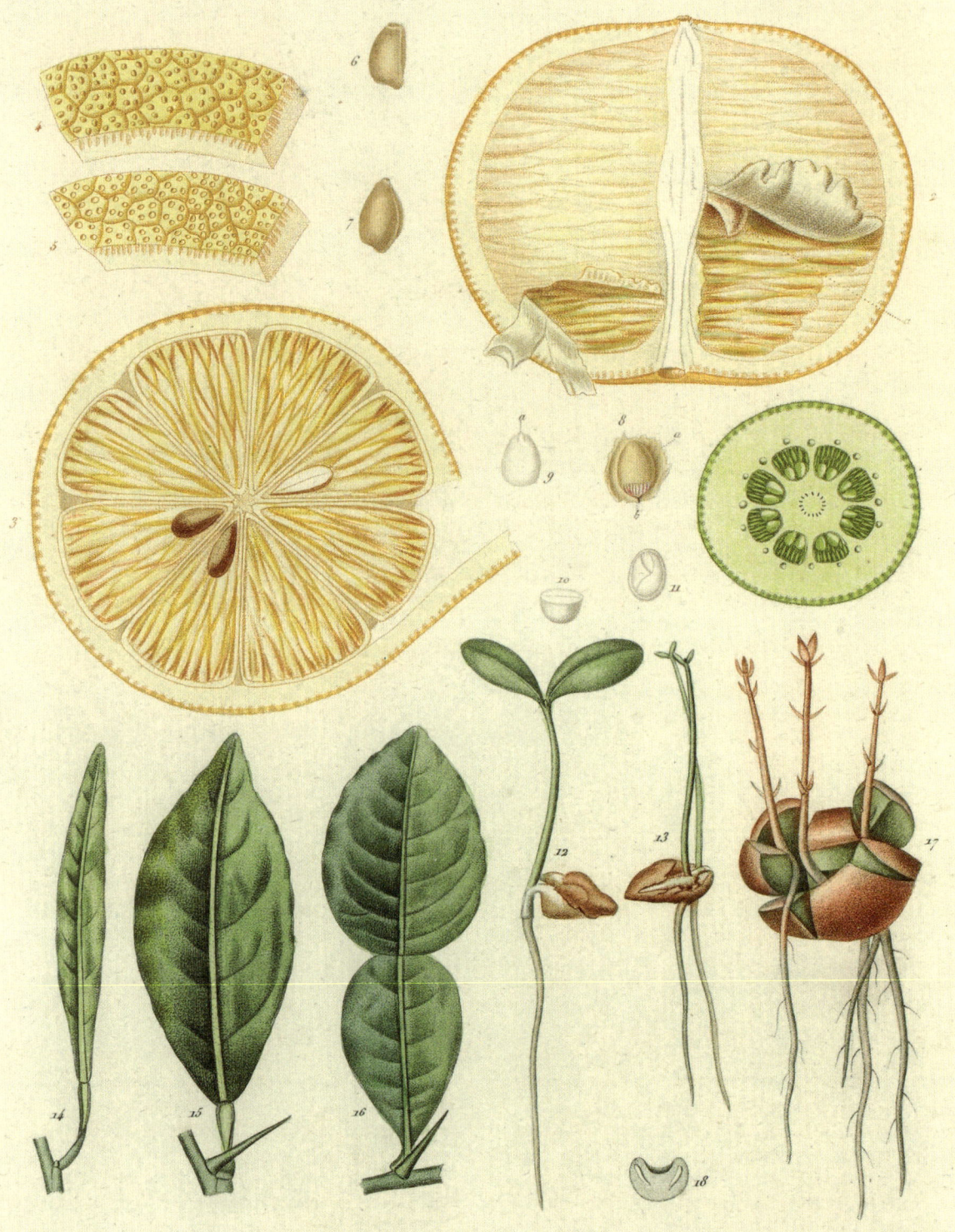

ANATOMIE

Tab. 2.

Poiteau Pinx.^t

Gabriel Sculp.^t

(*Citrus medica*), but the *citron* of France, *citroen* of the Netherlands and *Zitrone* of Germany are all lemons (*Citrus × limon*)*, though originally *lemoenen* in Dutch.[8]

WHAT IS CITRUS?

The genus *Citrus*, as now understood in the light of modern work on DNA on top of classical taxonomic techniques, comprises some twenty-five usually thorny tree species growing naturally in the wild from South Australia northwards as far as China, extending east to New Caledonia and west to north-eastern India. Citrus fruits have been introduced to most of the warm parts of the world, in many of which they have become naturalized so as to appear wild. The genus now includes the formerly segregated genera *Eremocitrus* (desert lime), *Fortunella* (kumquats), *Microcitrus* (Australian limes) and *Poncirus* (trifoliate orange). Contrary to some sources, most species are not native to China and neighbouring territories; the region with the highest number of indigenous *Citrus* species is Australasia. Some Australian species, such as finger limes (*Citrus australasica*, formerly *Microcitrus australasica*), are being hybridized with long-established citrus crops, and may now be of vital importance for the future of the citrus industry (see p. 226).

The origins of the genus *Citrus* have been much debated. Wood fossils of two species of *Citrusoxylon* are known from the Eocene era over 41 million years ago (mya),[9] while a species of *Citroxylon* is recorded from Bavaria in the Miocene (23–5.3 mya). Fossil leaves referred to *Citrus*, which could like those fossils be referred to a number of allied genera if they were alive today, are known from the Pliocene (5.3–2.6 mya) of Italy (*Citrus meletensis*),[10] but also reportedly from the Palaeocene (56–33.7 mya)/Eocene of Guangdong, China (*Citrus nigra*), the Cretaceous (Cenomanian, 100.5–93.9 mya) as well as the Eocene of North America and possibly the Oligocene (33.7–23.8 mya) of the Caucasus (*Citrophyllum* spp.).

Citrus anatomy from *Histoire naturelle des orangers* by Antoine Risso and Pierre-Antoine Poiteau, 1818–22. Citrus fruits are of a unique type known as a hesperidium. The tough peel surrounds the segments containing the juicy pulp vesicles and seeds.

* The '×' between the two terms (genus, species) making up a species binomial indicates that the plant is of hybrid origin (see p. 29); a name in quotes and roman font after a binomial indicates a cultivar (cultivated variety).

Perhaps the most convincing is the late Miocene leaf fossil *Citrus linczangensis* from Yunnan, China.[11] How this overall fossil record relates to the present distribution of modern *Citrus* species is unclear, but it is not uncommon for many genera of plants currently found only in the tropics and the subtropics of south-east Asia to be represented by fossils in Europe.[12] Following the ideas of the Russian botanist Nikolai Vavilov (1887–1943) and earlier authors, it has been argued that *Citrus* originated in the north-east India/ northern Myanmar/south-west China nexus, with a secondary radiation into the Pacific. The American citrologist Walter Tennyson Swingle (1871–1952, see p. 201), however, was convinced that the genus arose in the south-west Pacific. This fits various biogeographical patterns suggesting a Gondwana origin,[13] although other commentators, using known fossils as oldest rather than youngest ancestors, continue to cling to the idea of an Asian origin.

Although the greatest number of wild *Citrus* species today is from the south-west Pacific, the orange and lemon story properly begins in ancient China. The wild progenitors of those two, as well as limes, grew in the forests of southern and eastern Asia – from Assam to the Philippines – in forests that have now been largely cleared.

The genus *Citrus* is classified in the family Rutaceae, found throughout the tropical and other warm parts of the world and very closely related to the mahogany family, Meliaceae. The family Rutaceae is represented by 2,000 or so species in about 155 genera, especially well represented in the tropics.[14] They are predominantly aromatic trees and shrubs, rarely herbs, though the family name is based on the genus *Ruta*, which includes the strongly aromatic Mediterranean herb rue (*Ruta graveolens*). The characteristic scents are due to the presence of oil-producing secretory cavities, which, in the leaves of most species, appear to the naked eye as transparent dots. Their leaves are usually pinnate or trifoliolate, more rarely simple blades, and their fruits take a wide range of forms, from those that split open to fleshy ones such as berries or drupes.

The aromatic nature of these plants and the fleshy fruits of many are the basis of their interest to humans as flavour enhancers. Besides rue, there is curry leaf (*Bergera koenigii*), an essential in certain dishes in south and south-east Asia, while angostura (*Galipea* spp.) from South America is used to make aromatic bitters. The essential oils of *Agathosma* spp. from Africa are the source of *buchu*, used in medicine

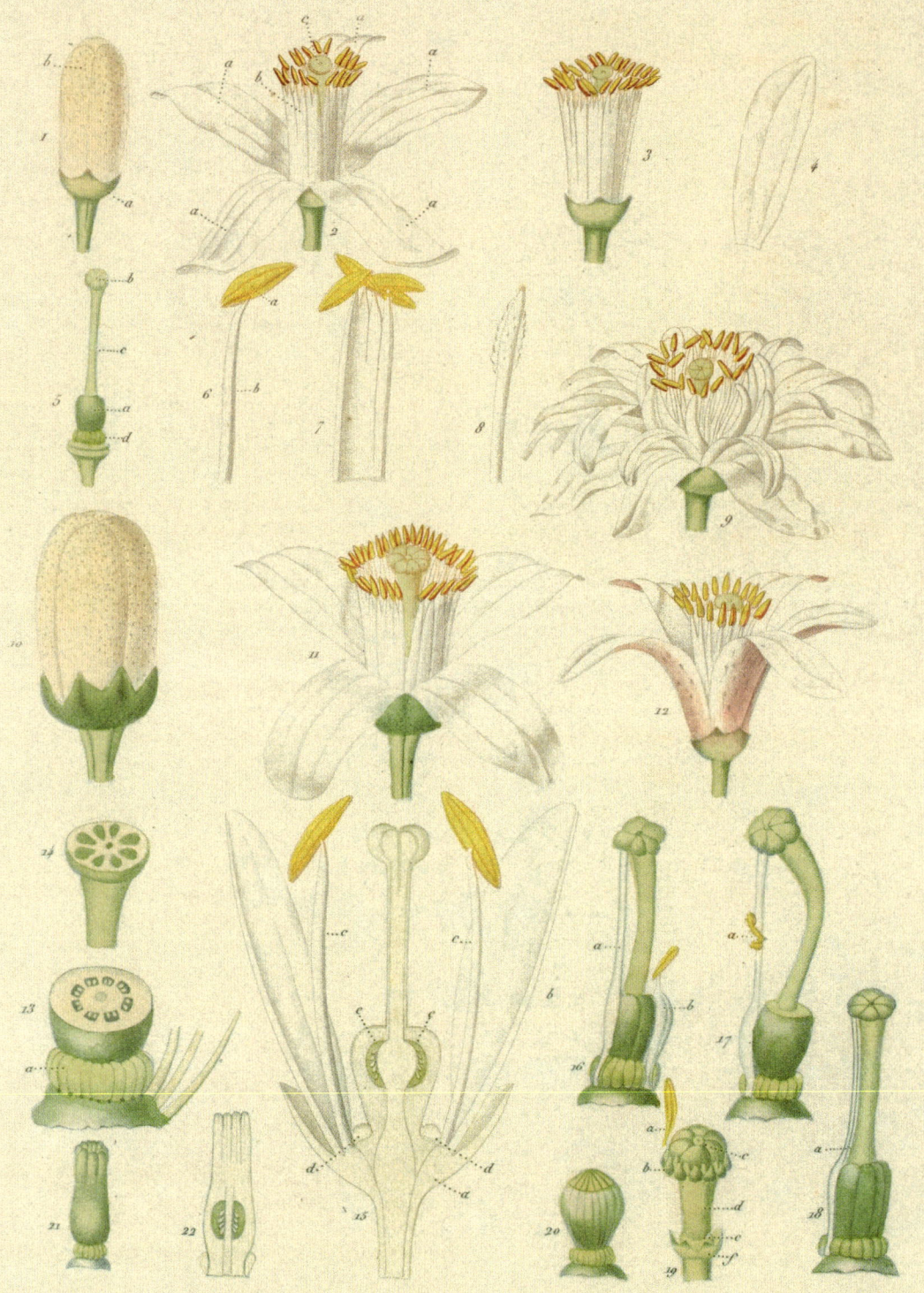

ANATOMIE

Tab. 1.

Poiteau Pinx.t   Gabriel Sculp.t

as well as flavouring, while species of *Zanthoxylum* from Asia yield fruits sold as spices, notably *sansho*, a key ingredient in the Japanese condiment *shichimi togarashi* ('seven spice').

Some Rutaceae provide timber, such as satinwood (*Chloroxylon swietenia* from southern India and Sri Lanka) and sneezewood (*Ptaeroxylon obliquum* from Africa), but also other commercial timbers from tropical American species of *Amyris*, *Balfourodendron* and *Euxylophora* and Australian species of *Bouchardatia*, *Flindersia*, *Geijera* and *Halfordia*. The wood of *Amyris balsamifera* was known as 'citron' – *bois citron* and *citronnier* being French names for this and other timbers – while the name lemonwood was applied not only to *Citrus* species, but also to the unrelated *degame* (*Calycophyllum candidissimum*).[15] Orangewood from citrus was also used as a substitute for boxwood, though the name was used for other orange-coloured woods, especially Brazilwood (*Paubrasilia echinata*). Also from tropical America come the medicinal products known as *jaborandi* (*Pilocarpus* spp.) used in the treatment of glaucoma. Important constituents of commercial scents include extracts from species of *Boronia* and, most importantly, *Citrus*.

Many Rutaceae are beautiful ornamentals. Notable in northern-hemisphere gardens are the burning bush (*Dictamnus albus*), the Mexican *Choisya* species and hybrids, *Ptelea* spp. from North America and *Orixa*, *Skimmia* and *Tetradium* spp. from temperate east Asia. In the tropics and subtropics, widely grown ornamentals include Cape chestnut (*Calodendrum capense*) from Africa, *Correa* spp. from Australia and species of *Atalantia*, *Euodia*, *Melicope*, *Murraya* and *Triphasia* from the Asia-Pacific region. But, in terms of global commerce, it is the fruits of Rutaceae that are the most significant.

GOLDEN APPLES

Citrus are characterized by their fruits being of a unique type known as a hesperidium: leathery-coated berries containing 'pulp vesicles' inside the segments.

OPPOSITE: Citrus flowers from *Histoire naturelle des orangers* by Antoine Risso and Pierre-Antoine Poiteau, 1818–22. In tropical countries, citrus trees flower throughout the year. **ABOVE:** Otaheite orange, ornamental plant catalogue, 1899. The flowers are sweet-scented due to the presence of essential oil glands and are important in commercial fragrance. Some cultivars, like *Citrus* × *otaitensis* 'Otaheite', are also important in the pot-plant trade.

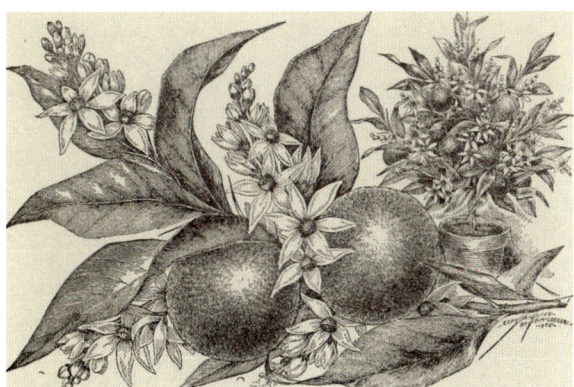

What we have, in effect, is a tough but non-shattering sphere containing juice bathing the seeds. All parts of the hesperidium are of commercial significance. The fruit wall, or rind, contains the 'zest', full of insecticidal materials, many of them of significant medicinal value. The vesicles yield the juice, which also has a number of medicinal qualities. The seeds, twigs and leaves are the source of petitgrain oil used in perfumery.

The word 'hesperidium' alludes to the Greek mythological golden apples of the garden of the Hesperides (meaning 'daughters of the evening' in Greek), said to lie beyond the Atlas Mountains of North Africa.[16] The original golden apples seem in fact to have been quinces (*Cydonia oblonga*), but, with the introduction of citrus fruits to the Mediterranean, the name was soon transferred to them (see p. 47). The botanical term hesperidium perpetuates the confusion.

The fragrance of citrus leaves and fruits is due to the release of essential oils when the tissues are bruised. The biological function of the oil is not fully understood, but it is likely to do with deterrence of pests: indeed, the use of citrons and other citrus against moths and other insects has long been known and exploited by humans. It is notable that it is only the stamens that are free of oil glands, so as not to deter pollinators by the release of any such oil while collecting pollen.[17] Interestingly, the stamens contain concentrations of caffeine (which has insecticidal qualities) almost as high as in coffee beans.[18] It has been shown that bees rewarded with caffeine are three times as likely to remember a floral scent as those rewarded with sucrose, so that it has been argued that there is selective advantage for the plant to produce enough caffeine to ensure pollinator fidelity, but not enough to be repellent.[19] *Citrus* also has the highest nectar concentrations known of the insect neurotransmitters octopamine and tyramine, which at least in bumblebees interact with caffeine to affect floral preferences and sucrose responsiveness besides long-term memory.[20]

The fact that citrus fruits can hang so long on the tree, and can be transported over long distances while remaining fresh, is testament to the efficacy of the peel components in deterring not only insects but also fungi and other pathogens. The peel of both lemons and oranges has even been shown to contain limonoids that are toxic to other plants, inhibiting their root growth.[21] The waste from citrus food-processing in Japan has been mooted for suppression of weed germination in agriculture.[22] Citric acid solutions are already used as spot-treatment herbicides.[23]

Other deterrent compounds in citrus include derivatives of tryptamine, found in the highest levels in the leaves and the seeds, which are most frequently attacked. Under the federal Controlled Substances Act in the USA these compounds are classified in the same category as heroin. Of these, dimethyltryptamine is one of the principal psychoactive compounds found in various shamanistic formulations, such as ayahuasca, long used in South America and now a recreational drug in Europe and North America. The levels in the fruits are extremely low, though according to the legislation 'any material' containing 'any quantity' of a Schedule I drug is legally equivalent to the drug.

Like many kinds of fruit trees, citrus grown from seeds remain sterile for many years, but, when mature, a tree can produce more than 200,000 flowers. Flowering is initiated by low temperatures or water-stress some two to three months before flowering shoots unfurl,[24] as Louis XIV's gardeners discovered when required to provide a continuous supply of orange blossom for the King at Versailles (see p. 134). As few as 1–5 per cent of the flowers mature as fruits.[25] The sweet-scented flowers, the source of the fragrant neroli oil used in eau de Cologne and redolent of the Mediterranean Riviera, have long been associated with the Virgin Mary, and are hence a symbol of virginity when worn by the bride at her wedding.

Citrus fruits mature over a long period – Valencia oranges, for example, taking twelve to fourteen months[26] – and so trees start flowering again before the previous season's fruits are ripe. The vesicles in the fruits hold a very large amount of water, up to 92 per cent by weight, and this can be withdrawn

to other tissues under stress in dry seasons.[27] They also contain vitamins (particularly vitamin C), sugars and acids, principally citric, but also, in grapefruit, tartaric, malic and oxalic acids. The bitter taste of pomelos and grapefruit may well be linked to resisting fungal attack. As the Borneo-based biologist Quentin Phillipps remarked, 'Based on this hypothesis "bitter" fruits are more likely to have evolved in hot wet "rainforest" conditions and "sweet" fruits in drier more seasonal conditions'.[28]

DISPERSAL AND EVOLUTION

The large, heavy, watery, long-lived fruits pose questions with regard to seed dispersal, which is little studied.[29] In Japan the fruits of tachibana (Citrus × tachibana) are taken by monkeys and birds, such as bulbuls and crows, but they also float and are possibly spread by river waters, as recorded for pomelo (Citrus maxima) introduced in Fiji. In Jamaica, citrus seeds are dispersed by birds and, in Madagascar, by lemurs, bush pigs and humans. Indeed, the large, succulent vitamin- and sugar-rich fruits strongly suggest animal dispersal: the smaller species perhaps involving birds (notably the kumquat, Citrus japonica, with edible inseparable peel), while the easily peeled mandarins suggest dextrous manipulation by primates. The very large fruits, like the pomelo and the citron, are still, on the whole, enigmatic, though their very thick rind would support the idea of water dispersal, during which the antifungal and insecticidal compounds in the peel would delay rotting and insect damage.

According to the Maltese doctor Emanuel Bonavia (1826–1908), in South America, 'whole forests of orange trees have been created through the sole agency of parrots', while in India 'certain animals, especially parrots, are fond of the citrus pulp whether sweet or sour, and are quite equal to carrying its seeds to long distances … carried with the pulp to their nests for their young, and dropping it on the way'.[30] Bonavia hypothesized that birds, particularly parrots, selected for increased flesh in citrus (hence the increasing size of the hesperidium), the bitter seeds being avoided but dispersed.[31] He also argued that the

---

Limonia crenulata (now Naringi crenulata) from Plants of the Coast of Coromandel by William Roxburgh, 1798. Citrus is a member of the Rutaceae family, its leaves equivalent to the apical leaflet, rarely the apical three, of pinnate leaves of genera like Naringi, whose name comes from the Hindi word for orange (see p. 17).

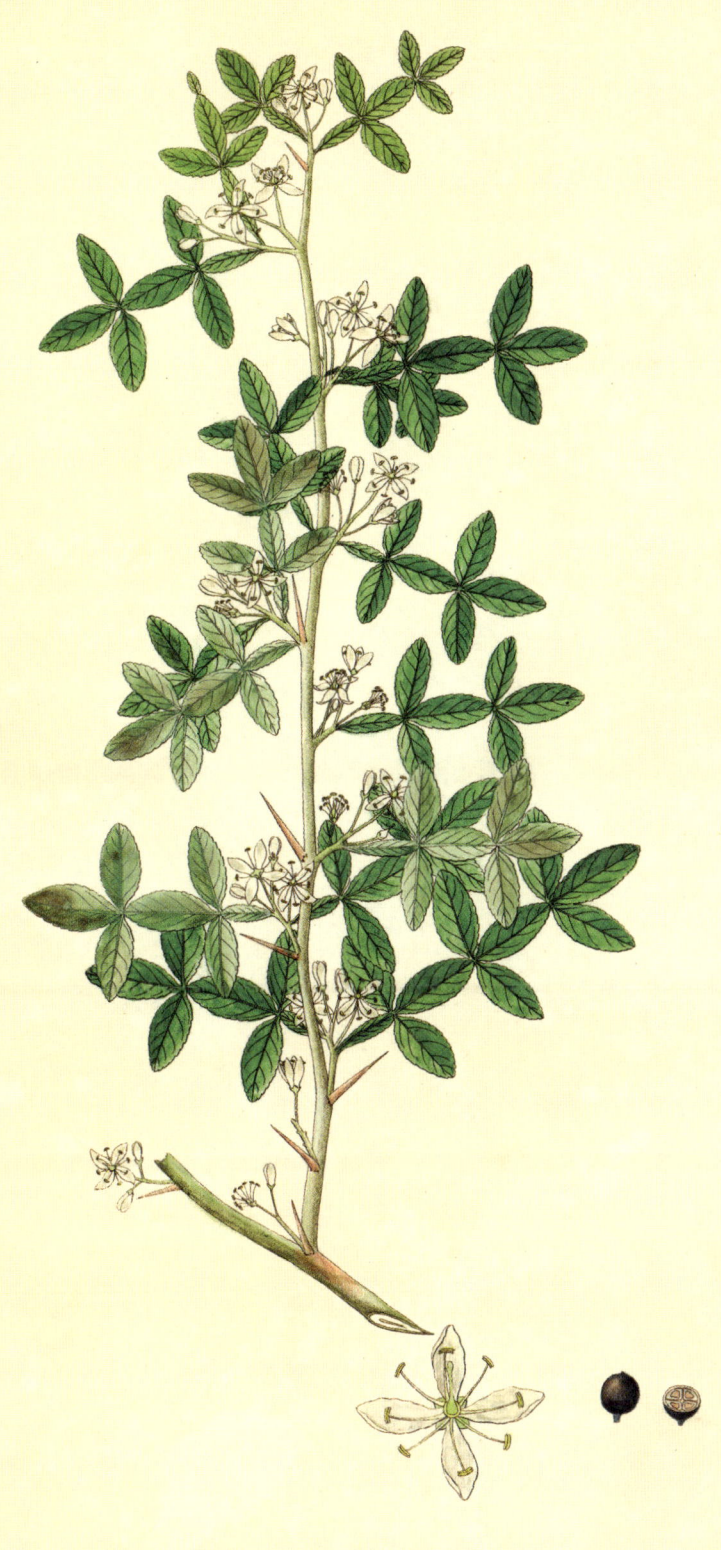

Limonia crenulata

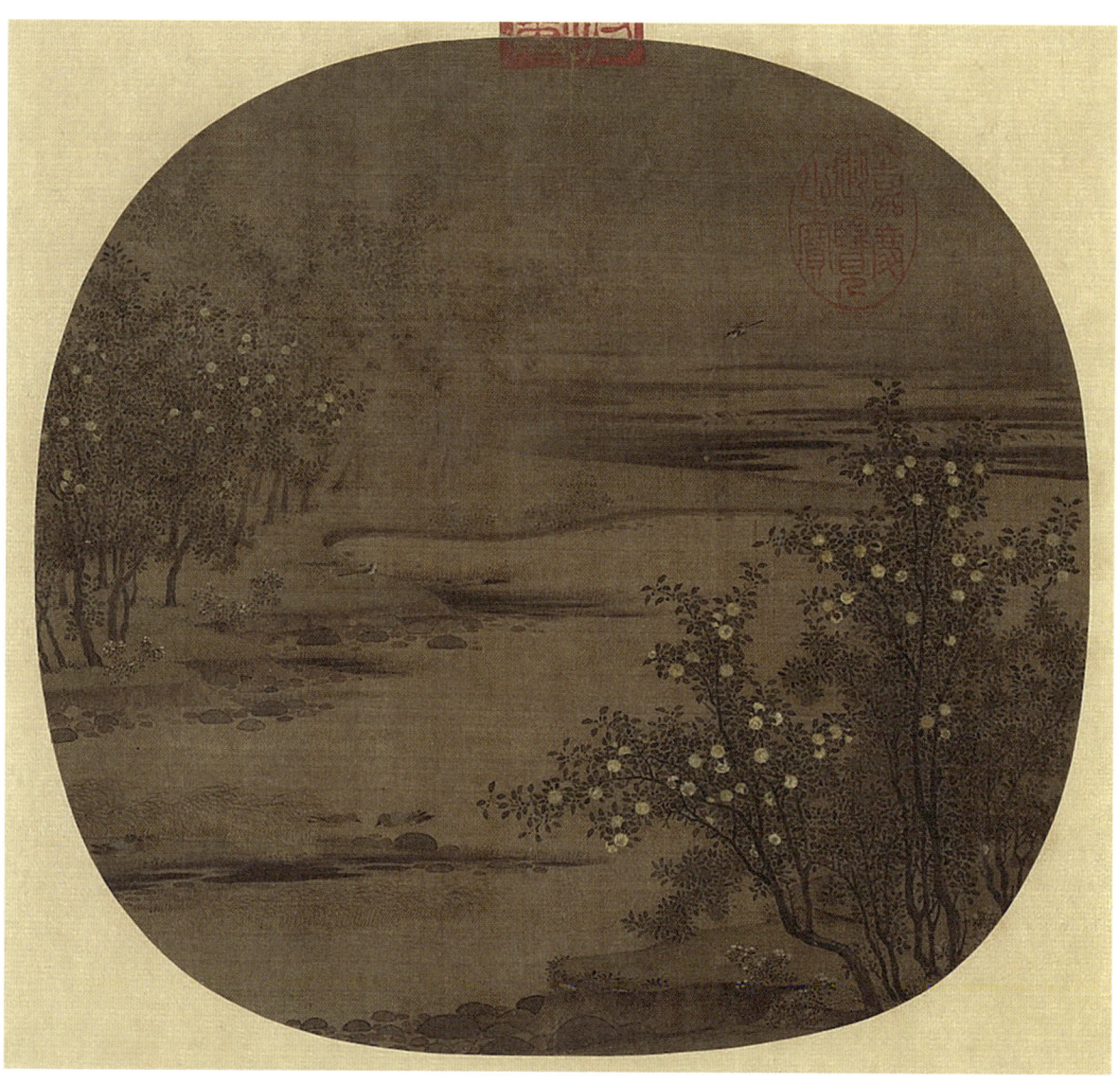

Citrus trees, Zhao Lingrang, *c.* 1070–1100. Citrus fruits have enthralled scientists and inspired artists for thousands of years. This Song-dynasty fan painting is based on a poem by Su Shi (1037–1101) that urged, 'You must remember/that the best scenery of the year/is exactly now/when oranges turn yellow and tangerines green.'

heavy armature found in young plants is repellent to grazing animals and that the 'aura' of petitgrain oil, although antagonistic to insects such as locusts, is attractive to a particular kind of butterfly, which not only lays its eggs on citrus leaves but also pollinates the flowers.

However, the fauna of the forests where citrus species grow has greatly changed: the distribution of much of the megafauna from China to Australia has in human times been greatly restricted or, in the case of Australia, become extinct.[32] Humans have hunted these animals for some forty thousand years, but only in the last two to three thousand years has the pressure increased such that elephants and rhinoceros, for example, have been eliminated from much of their ranges. They have disappeared from central and southern China: elephants were once distributed from Syria to north-central China, Sri Lanka, Sumatra and Borneo. It is known that elephants not only relish citrus fruits, but also void apparently viable citrus seeds in their faeces.

Phillipps speculates, 'Rhinos disperse fruit up to 6 cm [2 in.] in diameter (*Citrus halimii*) and elephants disperse fruits up to 15 cm [6 in.] (*Citrus macroptera* [= *Citrus hystrix*, makrut lime]). Citrus are good floaters and therefore both species also fit my hypothesis for a circular dispersal path for many large Sundaland [south-east Asia to Borneo when united by lowered sea-levels] fruits which floated down the very large rivers and established on alluvial floodplains during colder periods and were carried back into the hills by megafauna when sea levels rose again.'[33] Phillipps more recently wrote, 'I carried out some flotation experiments on pomelos and found they could float unblemished for at least eight weeks in a bucket of water indicating that the thick pith and tough rind are probably an adaptation for water dispersal.'[34]

On the other hand, in a similarly large fruit, of a similarly bee-pollinated species, the orchard apple (*Malus domestica*), the original dispersal agents seem to have been bears, which also feed on the honey made by bees from nectar in the fruit-forests of the Tian Shan in western China.[35] Could it be that bears are also efficacious dispersal agents for some of the larger-fruited bee-pollinated *Citrus* species? There seems to be little documentation of any such interaction, but a number of videos on the internet show bears very dextrously eating citrus fruits, ripping open the hesperidia with their claws, discarding the bitter pericarp and swallowing the endocarp, complete with the seeds. Dr D. K. Hore wrote in 2010 that fallen fruits of *Citrus indica*, the closest relation of the citron and found in north-eastern India, were taken from the ground by both bears and elephants and that the seeds are dispersed in their dung, but that monkeys and bats also disperse them.[36] With different bear species across the range of the genus, perhaps they could be dispersal agents west of Wallace's Line (the biogeographical line running north–south through the Malay Archipelago), but what about east of it, for example in New Caledonia where there are no non-flying mammals? Perhaps bats and birds are important there. In short, there appears to be an array of dispersal strategies in the genus, but as with so much of the basic biology of citrus fruits, as commonplace and important as they are, there is still much to be done to understand them in their surviving wild populations of Asia and Australasia.

FROM EAST TO WEST

The Western story starts with the citron, now little known except for its use in the Jewish Feast of the Tabernacles (see p. 50). Although assumed to have been native in India, this plant has been cultivated and spread so far for so long that it is now difficult to identify any wild populations. The Greek philosopher Theophrastus (see p. 39) and many of the early botanists used the name *Malus Medica*, explaining that this was the citrus of the Medes and Persians. When men of Alexander the Great's army swept over the Persian empire they found the citron widely cultivated there; it was the first citrus known in the West.

The appearance of sour oranges and lemons reputedly coincided with the opening of the Roman–Indian trade route from the Red Sea in the last years of the first century BCE. The Indian name for the first became the Byzantine Greek *nerantzion*. Oranges and lemons were taken across the Mediterranean with the eighth-century spread of Islam. North African migrant farmers brought them to Sicily for Moorish gardens and to Spain (Al-Andalus), where citrus trees were planted in the Alhambra. The Crusades, beginning towards the end of the eleventh century, saw the sour orange and the lemon, but also the lime, appear in European literature. By the middle of the thirteenth century the sour orange, the lemon and the citron were well established in Liguria.

By the end of the fifteenth century, citrus-growing had diffused throughout the courts of Europe. The sweet orange followed later, the fruits being so valuable that, by 1700, the Chinese were exporting them individually wrapped in paper all over the Malay Archipelago. Citrus cultivation in cooler climates led to the development of orangeries, or '*green*houses' for these *ever*greens. In the Dutch Golden Age, patriotism led nobles of the House of Orange to collect as many different kinds of citrus as they could.

CITRUS TODAY

In botany to epidemiology, mythology to perfumery, politics and economics, there have been almost endless confusions in names, identities and origins. Now, however, we are faced with citrus greening disease, also known as huanglongbing, one of the most severe plant diseases in the world. It can affect any kind of commercial citrus. Once a tree is infected with the disease, there is no known cure. Although the disease is not harmful to humans, fruits from infected trees are not suitable for consumption because of their bitter taste.[37]

Despite the intricate origins of commercial citrus in Asia, the future may lie with lesser-known species from the Pacific.

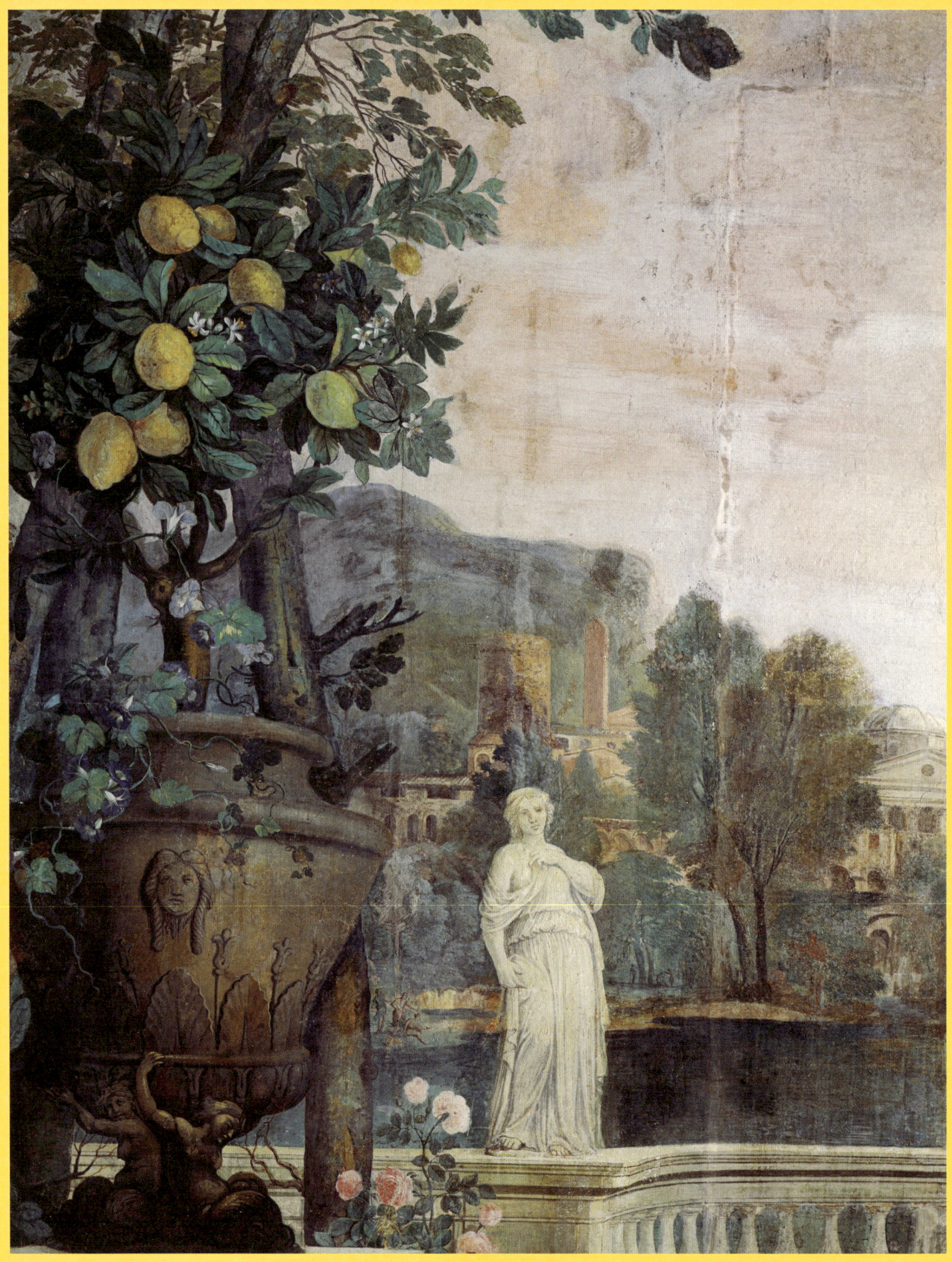

ONE

The Ancient World

CHINA AND THE WESTERN PACIFIC (including Australia) are the two main centres of *Citrus* species diversity in the wild. Although species diversity in *Citrus* is highest in the western Pacific, it is the mainland Asian species, long associated with the sophisticated societies there, which have had the most significant and most intimate relations with human beings – leading to modern citrus crops. According to Han-dynasty (206 BCE–220 CE) tradition, no citrus was cultivated in the days of the legendary Emperor Yao (claimed as 2356–2255 BCE) and his successor Shun,[1] so that, according to some authorities, the oldest citrus reference is actually from India, in the sacred Sanskrit text *Vajasaneyi Samhita* (part of the *Yajurveda*, one of four canonical writings, or Vedas, of Hinduism) elaborated before 800 BCE, in which the citron (and possibly the lemon) is called *jambhila*.[2]

The earliest records of citrus cultivated in China – and perhaps anywhere, because of doubts over this Indian record – are in the *Yu Gong* (Tribute of Yu [the Great, a semi-mythical ruler *c.* 2200–2100 BCE]), being chapter six of the *Shujing* (Book of Documents), one of the five classics of Chinese literature and perhaps compiled by Confucius (551–479 BCE). It is certainly the oldest Chinese geographical document known.[3] It was a kind of Domesday Book of the first half of the fifth century BCE, though the text is considered to be archaic and could well embody an older oral tradition and information from wooden or bamboo documents long lost.

The *Yu Gong* deals with the mountains and waterways within the nine provinces of the Shang dynasty (1600–1046 BCE), complete with their revenues, soil and vegetation types, commodities and produce. Referring to Yangzhou Province of the time (an area of eastern China including the modern Shanghai) and Qingzhou (now in Shangdong), it is recorded there that oranges and pomelos were sent up to the capital, Shangjing (modern Xi'an), as tax or tribute.[4]

By the Han dynasty, citrus was much grown and in the *Shiji* (Records of the Grand Historian) of Sima Qian (145–91 BCE), a historian of the period, it is recorded that 'people who own a thousand orange trees in Shu [present-day Sichuan], Han or Jiangling [Hubei].... may live just as well as a marquis enfeoffed with a thousand households'.[5] Indeed, around 350 BCE, it had already been said by a statesman that 'the State of Chu must necessarily gain wealth from its groves of orange and pomelo trees',[6] showing that citrus groves were already of great economic significance at this early date. It seems that citrus fruits were a speciality of the Jiangsu region, lying between the Huai and Qiantang rivers – in other words the lower Yangtze Valley and Delta[7] – in that small oranges (mandarins?) were sent from there. Certainly, there were many later references in Chinese literature to oranges and pomelos and it has been concluded that they were being grown on a commercial scale in Hubei, Anhui and Hunan Provinces in the third century BCE.[8]

That fruits from afar were still greatly valued very much later is shown by the fact that Sui Emperor Yang Guang (569–618 CE), in whose time the 270-km (168-mile) long Grand Canal was constructed between Beijing and Hangzhou, had 'tangerines' sent from Sichuan with their stems sealed in wax.[9] For transport such 'tangerines' were wrapped in paper from at least the beginning of the ninth century.

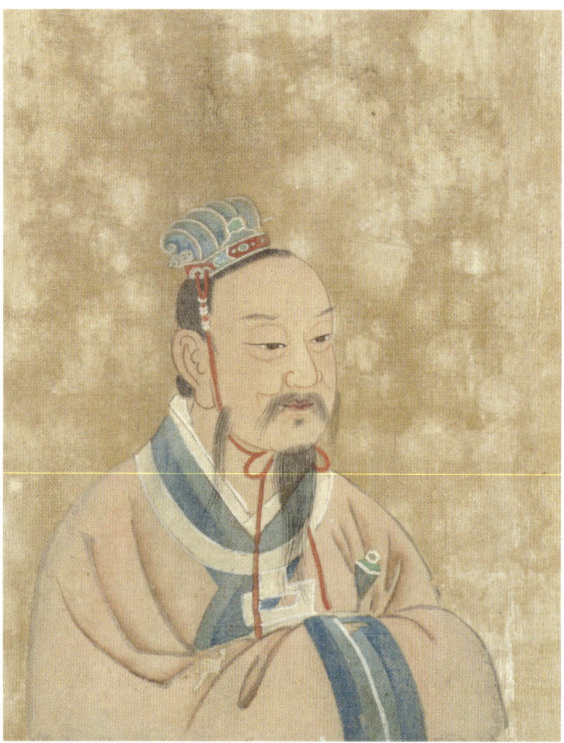

Portrait of Da Yu, or Yu the Great, unknown artist, late 19th to early 20th century. The *Yu Gong* (Tribute of Yu) was published in the 4th century BCE. It contains one of the oldest references to citrus cultivation anywhere in the world.

THE BITTER ORANGE (*CITRUS TRIFOLIATA*)

The thorny, bitter orange *Citrus trifoliata* is a truly wild Chinese species. It is the only citrus with trifoliolate (with three leaflets) leaves. It is also deciduous, with good autumn colour, and is the hardiest of all – it can be grown widely in the UK, for instance. In the south-eastern USA, where it is grown as an impenetrable hedge (as it is in Japan), it was promoted by Barrier Concepts Inc. of Tennessee in the 1980s,[10] such that 'security-minded customers' including the CIA, the Secret Service and the military planted it around their premises. It has since become an aggressive colonizer of natural as well as disturbed vegetation.[11]

As it contains high concentrations of auraptene, bitter orange is resistant to a scourge of citrus: the tristeza virus (see p. 216). It has therefore been widely used as a grafting rootstock in commercial citrus-growing, though today such stocks are often hybrids between it and oranges (citranges, *Citrus × insitorum*). The fertile shoots develop from the previous season's dormant buds and they bear singly, or in pairs, flowers up to 8 cm (3 in.) in diameter, the largest in the genus, in early summer. So strikingly different is it that it was for some time referred to its own genus, *Poncirus*,[12] even though it readily hybridizes with many *Citrus* species. Molecular work has confirmed that it belongs in *Citrus*, where Linnaeus placed it in 1763. He based his description on an illustration in Engelbert Kaempfer's (1651–1716) *Amoenitatum exoticarum* (1712), a classic European work about Japan and Japanese plants.

The twelfth-century *Ju Lu* (Register of Citrus, see p. 33) deals at length with grafting techniques known since at least the Tang dynasty (618–907 CE) and, by 1502, there is a clear record of oranges being grafted on bitter orange, as they still are today.

Grafting, today widely used to control infections – such as the tristeza virus – by breeding resistant rootstocks (see p. 217), inducing dwarfing – as with both apples and citrus – or reducing long non-fruiting juvenile phases in seed-grown citrus, may have been inspired by the observation of natural grafts between shoots.[13] These are known to occur not only within a particular tree or different trees of the same species, but, as root-grafts, between individuals of different species.[14] Such 'double-organisms' are, of course, not true hybrids.

THE FUNDAMENTAL IMPORTANCE OF TRUE HYBRIDS

By 1500 CE, several Chinese and other centres were far more sophisticated than anywhere in the West:

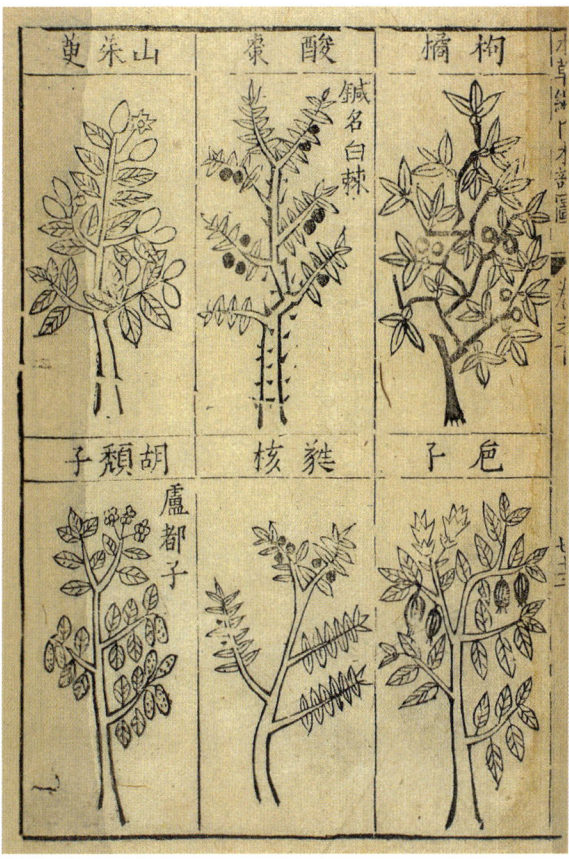

Woodcut from *Bencao Gangmu* by Li Shizhen, Ming dynasty (1368–1644). This pharmaceutical encyclopaedia describes *Citrus trifoliata* (top right), noting its medicinal benefits.

the silks of southern China were far superior to European textiles; porcelain far above the pottery of Europe, while ocean-going ships in China could outstrip European vessels. The rich lived in luxury likely exceeding anything in the West.[15] By this time most of the oriental crop plants had been tried out in every other region. Because of their tough skins and resistance to decay, citrus fruits could be transported over long distances for such 'trials'.

The unexplained evolutionary significance of citrus fruits hanging fresh on trees for months and months, with their seeds undispersed and pristine due to antifungal and insecticidal compounds in the rind, has been the key to the development of modern citrus crops. The longevity of the fruits meant not only that the plants could be successfully introduced to countries far from their native lands, but also that they could be brought into close contact with other *Citrus* species that were previously geographically isolated.

It is clear that humans have been associated with citrus plants for a very long time and that, in bringing wild species and selected forms of them together from outside their natural ranges, hybrids have arisen.[16] This occurred spontaneously in Asia, and subsequently (and sometimes intentionally) in the orangeries and botanic gardens of Europe (see Chapters 2 and 3), besides in modern large-scale hybridization programmes in all warm parts of the world, particularly the Americas (see p. 210).

All *Citrus* species have eighteen chromosomes and interspecific hybrids are readily formed. In most other kinds of plants such hybrids have very reduced or no fertility, but hybrid citrus seeds can be set through a phenomenon known as apomixis. This is seed formation without fertilization of an egg nucleus by a pollen nucleus (both with nine chromosomes). Instead, embryos arise directly from asexual tissue (with eighteen chromosomes), so that desirable forms can readily be propagated from seeds as predictably stable 'pure lines'. Moreover, chance mutants in other asexual (somatic) tissues can also be fixed in this way.

In addition, apomictic plants can occasionally undergo normal sexual reproduction and form hybrids with other such apomicts, yielding new apomictic clones. Apomixis is usually the cause of polyembryony, where seeds can yield two or more seedlings.[17] At the beginning of the eighteenth century, the Dutch tradesman and microscopist Antonie van Leeuwenhoek (1632–1723), 'the father of microbiology', showed that certain citrus seeds often have several embryos, the first time the phenomenon of polyembryony was described.

In many citrus, especially in cultivation, one or more asexually produced embryos, besides sexually generated ones, can develop from asexual tissue (nucellus) in the ovule. These embryos are in effect clones of the mother plant. The genes responsible for the switch to this asexual reproduction have recently been identified.[18] Polyembryony is particularly common in the makrut (the lime leaves of Thai cooking) or leech lime (*Citrus hystrix*) and the calamondin (*Citrus × microcarpa*, see p. 36). Three to twelve embryos develop from the nucellus alongside the normal one, as was first described in detail by the German cytologist Eduard Strasburger (1844–1912) in 1878.[19]

It was in China, then, that the combination of the then undescribed phenomena of apomixis, hybridization and somatic mutation led industrious growers, unwittingly, to create the basis for today's global citrus industry.

Although the phenomenon of apomixis in citrus has long been known, only with the pioneering work of the Polish-American botanist Reinhard Scora (1928–2016) was there revelation. Born at Mokre, Silesia, Scora migrated to America and his research there shone light on centuries of confusion on the origins and naming of citrus crops. From studies in the 1970s of certain enzymes and their substrates, besides essential oils in the peel, Scora and his colleagues showed that there is no such thing as a wild orange, a wild lemon or a wild lime, but that these are all derived from hybridizations between truly wild species, the hybrid lines then fixed by apomixis. Subsequent DNA analysis has confirmed and expanded their findings and shown that the truly wild species underpinning the modern industry are: the mandarin (*Citrus reticulata*) and the pomelo (*Citrus maxima*), from China; the citron (*Citrus medica*), likely from north-eastern India; and the makrut lime (*Citrus hystrix*), from the Malay Archipelago.

The fact that the different kinds of citrus known to ancient China were known by names of single written characters suggests that they had indeed been recognized as distinct long before the characters were developed.[20] The Chinese had single written characters for the truly wild species, like citron, mandarin and pomelo, but the characters also include *cheng* for the sweet orange and *luan* for the sour, neither of which is known outside cultivation or is a wild species.

In terms of classification, the Cambridge biochemist Joseph Needham (1900–1995) pointed out[21] that in 'quite early times' the Chinese had already understood very well that all the citrus in China belonged to one group (*lei*), which is reflected in the most recent classification proposed,[22] even though the fruit ranges in size from 'that of a marble to that of a football'. Moreover, other citrus crops introduced later into China were treated as part of the same group. As early as 280 CE Chang Hua was pointing this out and Kuo I-Kung wrote in *c.* 450 CE in his *Kuang Chih* (Extensive Records of Remarkable Things), 'Of *kan* there are twenty kinds. The [fruits of] *huang* [yellow] *kan* have only one seed [each]',[23] which Needham suggests means that before 450 there were 'seedless' citrus cultivars being grown in China. These could only have been propagated by vegetative means.

THE MANDARIN (*CITRUS RETICULATA*)

It would appear, then, that *Citrus* hybridizations are of very ancient origin. Until humans started moving these crop-plants around, the species were largely geographically isolated, though they are very often interfertile. Bringing them together, and the pollinating activities of bees and other insects, led to the origin of many interspecific hybrids, the earliest perhaps being mandarin-pomelo crosses, which have given rise to the *Citrus × aurantium* complex comprising all the oranges, both sour and sweet, tangerines and the commercial mandarins (and much later grapefruit, tangelos, ortaniques etc.; see pp. 161 and 201–2).

The wild mandarin[24] has small orange-yellow fruits, sometimes sweet-flavoured, with a readily removed thin peel, the pomelo having enormous, pale-yellow fruits with a rather sweet-sour taste and very thick peel not readily removed. The true wild mandarin, unadulterated with pomelo genes, is today found only in certain mountain ranges in southern China, while the pomelo may have been wild in tropical south-east Asia extending to south-west China. Hybridizations between these two must have occurred time after time in China, some of the resultant hybrids more like the original mandarin, some more like the pomelo, and then crosses between such lines, besides somatic mutants, giving rise to a whole range of sweet and sour oranges. The well-known, thornless, *chinotto* (i.e. from China; *Citrus × aurantium* 'Chinotto') fruits, an essential ingredient in Italian *amari*, notably Campari, is apparently such a somatic (bud) mutation,[25] like the very similar 'Myrtifolia', found several times on sour-orange trees.[26]

The name mandarin comes from the 1580s word for a Chinese official, via Portuguese *mandarim* or older Dutch *mandorijn*, these names in turn derived from Asian words encountered by Europeans in the East. According to one review, 'mandarins' have been cultivated for about four thousand years.[27] The major production areas were Sichuan, Hunan and Hubei, extending to Jiangxi and Anhui in eastern China by the third century, then to Jiangsu and Zhejiang in the east and Yunnan and Guangxi before the ninth century and Guangdong, Fujian and Taiwan since then. Before the Tang dynasty, there were three main cultivar groups – '*Huangpi ju*', '*Huang gan*' and '*Zhu ju*' – with three more selected during that dynasty – '*Hong ju*', '*Ru ju*' and '*Sha gan*', all of which are said to exist today. It is unlikely that any of these are the original pure mandarins.

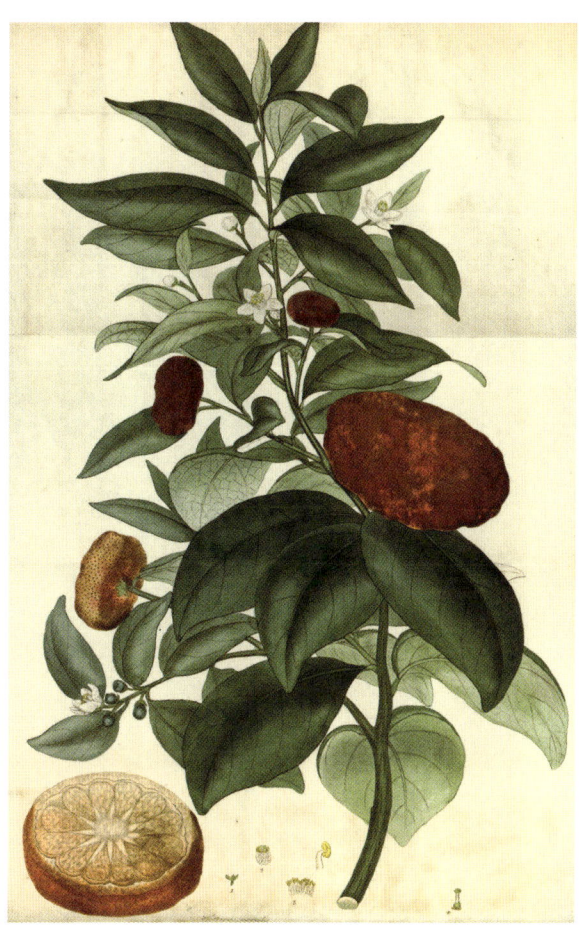

Mandarin, engraving from *The Botanist's Repository* by Henry Andrews, 1810. Mandarins were not successfully introduced to the West until about 1805, when two cultivars from Guangzhou were growing at Wormleybury, Hertfordshire, England.

The original mandarins are now very restricted in distribution, comprising populations with different degrees of wild, semi-wild and domesticates known from near Mangshan and Daoxian (Hunan Province), Hezhou (Guangxi Province), Congyi (Jiangxi Province) and areas in Guangdong Province.[28] They seem never to have been introduced to the West. Whether or not the species had a greater distribution before the widespread clearance of forests in southern China cannot now be ascertained, but suffice it to say that in terms of crop wild relatives (wild plants that are closely related to cultivated ones), the populations of pure mandarins are of the highest importance in maintaining genetic diversity for future breeding work in the citrus industry.

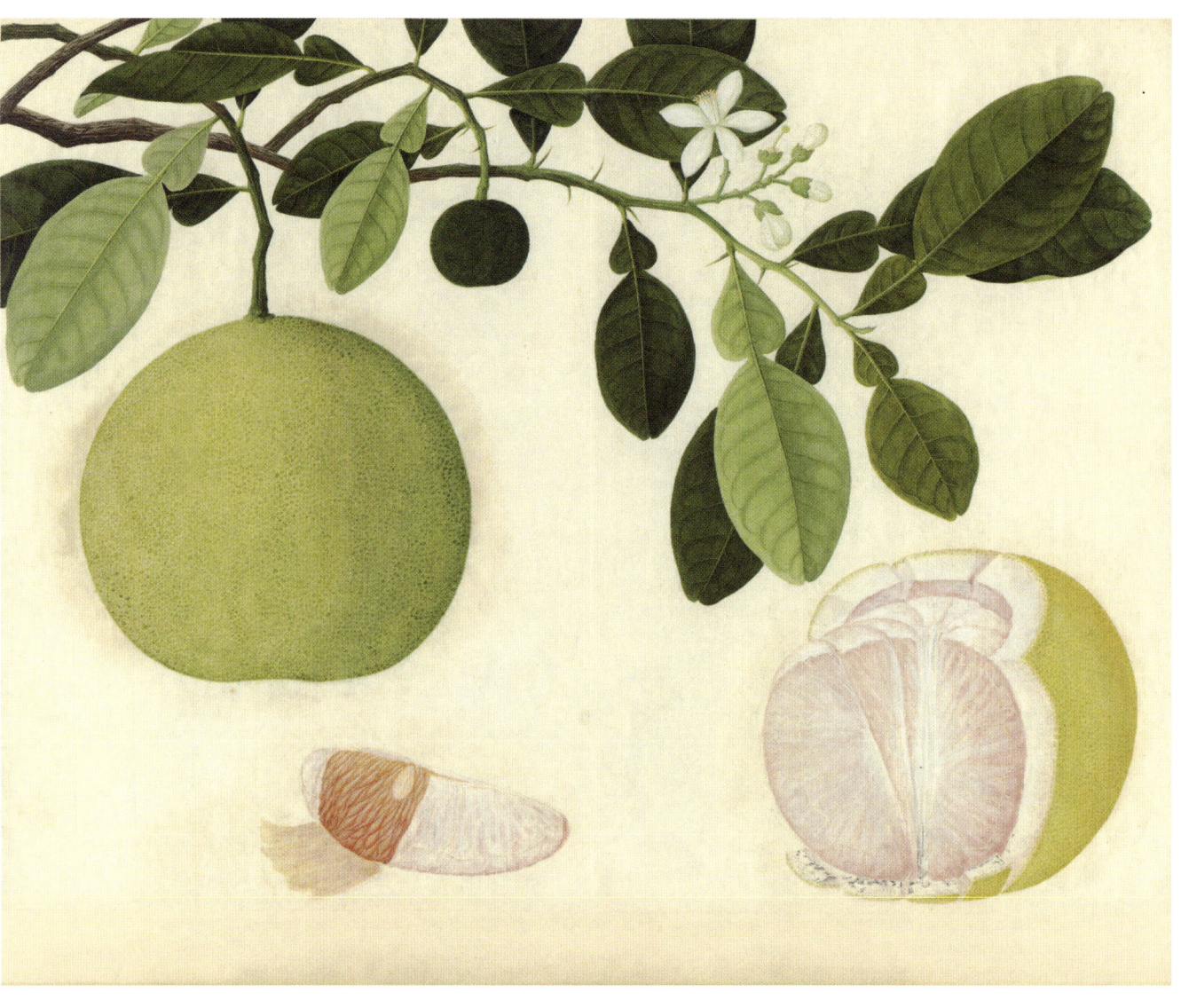

ORANGES (*CITRUS × AURANTIUM*)

In 1178 CE appeared the *Ju Lu*, the first monograph, in the modern sense, of citrus by Han Yanzhi, Song-dynasty Governor of Wenzhou in Zhejiang Province on China's east coast.[29] The *Ju Lu* is divided into three parts, the first two giving descriptions of eight *kan* oranges, fourteen *chu* oranges and four *chheng* oranges, with references also to the pomelo (*yu*) and the bitter orange (*chih*), in all twenty-eight taxa, detailing for each the tree's habit, the form of the branches and the leaves, the size and shape of the fruits, colour, taste, oil content, peeling properties, the numbers of segments and seeds – in effect a remarkably modern monograph.[30] The third part deals with husbandry, including harvesting and storage as well as medicinal uses. A section is devoted to citrus diseases and protection against them.

THE POMELO (*CITRUS MAXIMA*)

True *Citrus maxima* is purportedly native somewhere in south-east Asia, but convincing examples of wild populations have yet to be identified. The name pomelo (or pummelo) derives from the Dutch *pompelmoes*,[31] which also gives us the English 'pompelmoose' or 'pampelmouse' as well as French and Tamil names. It is said to be a portmanteau combining 'limoes', derived from the Portuguese *limões*, plural of *limão* (lemon), under which the Portuguese then included all citrus they encountered in the East (and from which the original Malay word *limau* is still used in a general sense), with the Dutch 'pompoen', pumpkin, being an allusion to the large size of the fruit. 'Pompelmoes' is therefore a contraction of 'pompoenlimoes', or 'pumpkin-like citrus'.[32]

The pomelo has been widely introduced and innumerable cultivars selected. There are several still grown today, both acidic and sweet and varying greatly in shape.[33] Pomelos have become naturalized in Fiji, particularly along riverbanks, as the fruits are buoyant,[34] giving some hints as to their ancestor's original ecology. In view of its significance in the ancestry of both oranges and lemons, the identification of wild strands of this progenitor is urgent – if indeed any such have survived the

Pomelo, by a Cantonese artist of the 'Straits School', *c.* 1800. In Chinese, the word for pomelo is homophonous with the word for blessing; today the pomelo is is a conspicuous element of the mid-autumn Moon Festival in China, eaten with mooncakes.

destruction of riparian habitats in south-east Asia, not least due to the American napalm of the Vietnam War.

The pomelo reached Europe through the Arab trading links with China (see p. 58). In modern nomenclature, Linnaeus made it a variety (in fact it is a parent) of the orange (*Citrus × aurantium*), naming it var. *grandis* and referring to Hans Sloane's *Voyage to ... Jamaica* (1707). A successful doctor and naturalist, Sloane (1660–1753), whose collections were the original basis of what is now the British Museum, had seen pomelos in Barbados, where they were known as shaddocks. They were named after Captain Philip Chaddock (or 'Shaddock'), governor of Bermuda from 1637 to 1640, who introduced them, probably directly, from Asia.[35]

In seventeenth-century Ambon in the Moluccas, Georg Eberhard Rumpf,[36] compiling his monumental *Herbarium Amboinense* (see p. 106), gave full descriptions of various pomelo cultivars, describing the fruit and its uses:

It is also a marvellous fruit for sea Journeys, for one can keep them unspoiled for a long time, as long as they were treated gently when they were taken off the tree, and did not crash onto the ground. They are strung up, and eaten at Sea, where they refresh the stomach and quench one's thirst.[37]

Unknowingly, the importance of what we now know as vitamin C had been grasped (see p. 152).

The pomelo came to be considered the 'tree of knowledge'[38] in the Garden of Eden.[39] In the early seventeenth century, the Swiss botanist brothers Johann and Gaspard Bauhin called it *pomum Adami* (Adam's apple), though this was likely a confusion as it is probable that the true *pomum Adami* (part of the pomelo-mandarin hybrid complex), with its characteristic fruit shape, rather than pure pomelo, was that cultivated in Europe and thence spread to its colonies. As so often in citrus nomenclature, though, '*pomum Adami*' has been applied to a number of different kinds of citrus, so that interpreting the literature on it is fraught with difficulty.

Yet again, confusingly, the name pomelo, or pompelmoes, is now often used in the USA and elsewhere for the grapefruit, which is an eighteenth-century hybrid, also in the pomelo-mandarin hybrid complex (see p. 161). The French had pomelo as *pampelimousse* (now *pamplemousse*) from 1696 and they founded the Jardin des Pamplemousses (today's Sir Seewoosagur Ramgoolam Botanic Garden) at Pamplemousses near Port Louis, Mauritius, a settlement apparently named after 'pomelo'

The chin kan *[kumquat] fruit has a golden colour, a very fine-grained skin, and a spherical form; with its ruddiness it is agreeable and enjoyable to handle. One eats it without peeling off its golden coat.... Formerly the people of the capital did not value them very highly, but after the empress Wencheng became very fond of the fruit the price there went up considerably.*

HAN YANZHI, *JU LU* (REGISTER OF CITRUS), 1178 CE

Flowering and fruiting stems of the kumquat (*Citrus japonica*), zincograph, after Walter Hood Fitch, drawn from a plant exhibited at the Horticultural Society in London in 1876.

Many of the earliest references to citrus anywhere in the world come from China. A particularly intriguing observation is recorded in a mid-3rd century BCE text, the *Han Feizi*: 'When the orange tree crosses the Huai River it turns into [the bitter orange, *Citrus trifoliata*]'. This hints at the possibility that grafting techniques were already employed, since the bitter orange is often used as a rootstock on which oranges are grafted, and in areas in which only the first is hardy, the scion will die and the stock grow out. In other words, the orange will disappear and the bitter orange will manifest itself. Early Chinese classification systems also indicate that it was well understood that citrus plants belonged to one group despite huge variation in size and shape of the fruits. A 12th-century CE monograph on citrus, the *Ju Lu*, recognized twenty-eight kinds of citrus, of which *kan* included the kumquat.

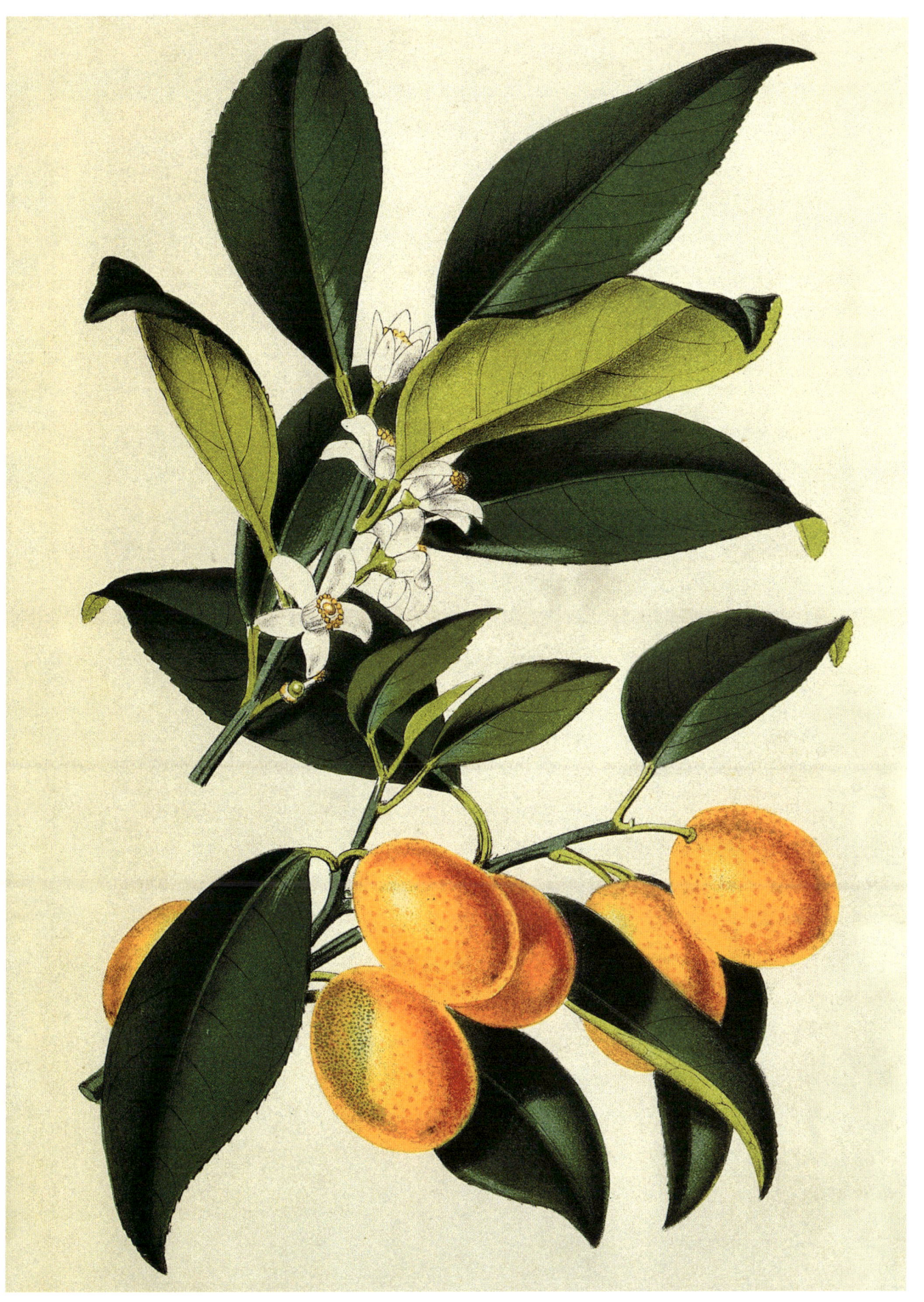

plantations there. Established in 1770, the Jardin is the oldest extant botanic garden in the Southern Hemisphere. It would appear that the Mauritian 'pomelos' came with Dutch gardeners from the Cape, to which the ultimately Chinese stock would have been introduced by the Cape Colony's founder Jan van Riebeeck (1619–1677).

THE KUMQUAT (*CITRUS JAPONICA*) AND THE CALAMONDIN (*CITRUS × MICROCARPA*)

The name kumquat or cumquat is, according to the *Oxford English Dictionary*, a Cantonese dialect form of the Chinese *kin ku*, 'gold orange'. Wild populations of the kumquat, a variable species in cultivation, are found in southern China, and forms with different shapes have been selected and given species names.[40] The kumquat was given its Latin name by a Swedish pupil of Linnaeus, Carl Pehr Thunberg (1743–1828), who between 1775 and 1776 was employed by the Dutch East India Company as a surgeon at a trading base on the artificial island of Dejima, Japan. Despite severe restrictions on leaving the island and interacting with locals, Thunberg was able to record another three hundred species growing in Japan.[41] From nursery stock he established a collection of trees and shrubs on Dejima, which were shipped back to the Hortus Medicus (now Hortus Botanicus) in Amsterdam. Like many other Chinese plants taken into cultivation and then spread to Japan, the kumquat was first known to the West from Japanese gardens and was thus given the misleading epithet '*japonica*'. The kumquat has formed hybrids with the mandarin

in cultivation, giving rise to the calamondin (*Citrus × microcarpa*). At the recommendation of Alexander von Humboldt, the Russian Academy at St Petersburg sent botanist Alexander von Bunge (1803–1890) on a new Ecclesiastical Mission to Beijing, which he reached on 17 November 1830.[42] Among the 420 species of plants he collected in the next few months was the calamondin, which he found in 'Beijing hothouses' and named *Citrus microcarpa*, not realizing it was a hybrid. A few years later it was described anew, as *Citrus mitis*, from plants grown in the Philippines and, indeed, it has long been a major fruit crop there and known as calamansai from Tagalog *kalamansi*.

THE ICHANG PAPEDA (*CITRUS CAVALERIEI*) AND THE YUZU (*CITRUS × JUNOS*)

Other hybrids that arose in China include those between the wild mandarin and another native tree, the Ichang papeda (*Citrus cavaleriei*), one of which is the yuzu, though some forms of yuzu apparently also have some contributions from kumquat and pomelo according to recent analyses.[43]

Citrus cavaleriei is a south-west Chinese tree growing up to 10 m (33 ft) tall, with pale-red flower buds, thick-skinned, very sour fruits 10 cm (4 in.) or more in diameter, and leaves resembling those of the makrut lime (*Citrus hystrix*), with which it has been much confused in the literature. Its name commemorates a missionary, Pierre Julien Cavalerie (1869–1927),[44] who collected a large number of plant specimens in China such that more than two hundred allegedly new plant species were named after him.

Yuzu, the Japanese name derived from Chinese,[45] is according to Tanaka[46] identical with the so-called Kansu orange of China and has long been grown in Japan for medicinal uses, for flavouring food and drinks and for dressing salads. In Japan it also figures in Toji, a winter solstice tradition begun in the eighteenth century: hot baths, *yuzu-yu*, are taken with whole yuzu fruits added to the water.[47]

The first European to note yuzu being grown in Japan was the German doctor-botanist Philipp Franz von Siebold (1796–1866),[48] who introduced many Japanese plants to European gardens, sadly including the pestilential Japanese knotweed (*Reynoutria japonica*), perhaps retribution for what would now be considered biopiracy. In 1830, Siebold named the yuzu *Citrus medica* var. *junos*, but he correctly considered all mainland Japanese citrus to be

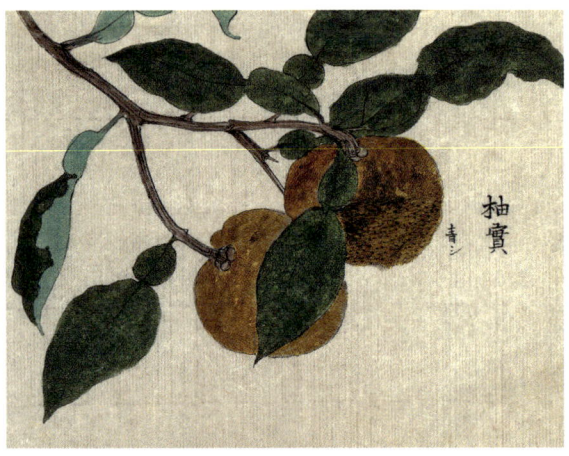

Yuzu (*Citrus × junos*), from an album of Japanese watercolours of flowering plants and fungi, n.d. The yuzu is a cross between the wild mandarin and the Ichang papeda.

introduced from China, claiming it was 33 BCE 'in which they were first offered by the Korean prince to the Japanese emperor'.[49] In fact, yuzu has been known in Japan for over 1,100 years, as it was during the Tang dynasty that this very hardy plant was introduced via Korea to Japan, where most is cultivated today.[50]

Several other Japanese citrus cultivars analysed[51] are also Ichang papeda/mandarin crosses (ichandarins), some also with contributions from other taxa. Such include *Citrus × tamurana* (hyuganatsu, konatsu) and *Citrus × sudachi* (sudachi, a speciality of Tokushima Prefecture), among a whole array of hybrid 'mandarins' so favoured in Japan.

JAPAN AND THE TACHIBANA (*CITRUS × TACHIBANA*)

The first records of any citrus from Japan are in the *Wajinden* (Treatise on the Wa People; 280–297 CE), the thirtieth volume of the Chinese history *Sanguo Zhi* (Records of the Three Kingdoms), and note that the thin-skinned citrus fruits are considered inedible. In the eighth century CE, Japanese nobleman Ono Yasumaro (d. 723) recorded details of an expedition in 61 CE by one Tajimamori, sent by Emperor Suinin to collect a magic fruit that produced a beautiful scent forever.[52] It took ten years to find, and on Tajimamori's return, he found that the Emperor had died at the age of 104 – he was so mortified he promptly died too, perhaps by killing himself. Legend has it that he thus introduced the 'mandarin' known as *tachibana*, a word said to be a corruption of his own name, but it is not clear that this fruit was what is now *Citrus × tachibana*, known to be a cross between the wild mandarin and the not yet formally described '*Citrus ryukyuensis*' native to the Ryukyu Islands.[53]

The Shishinden (ceremonial hall) of the imperial palace in Kyoto has a tachibana and a Japanese flowering cherry planted in front, in deliberate contradistinction to the mandarin and Chinese flowering cherry, thereby asserting Japanese nationalism. Today, sprigs in fruit are figured on the five-hundred-yen coin of Japan, while the flower is the principal motif in the medal of the Order of Cultural Merit awarded by the emperor on Culture Day (3 November) for contributions to Japanese culture. The Shinto festival of Hinamatsuri (Girls' Day, 3 March) is marked by displays of models of the Japanese royal family and attendants of the Heian period (794–1185 CE) arranged in tiers, the fourth tier having models of tachibana trees.[54]

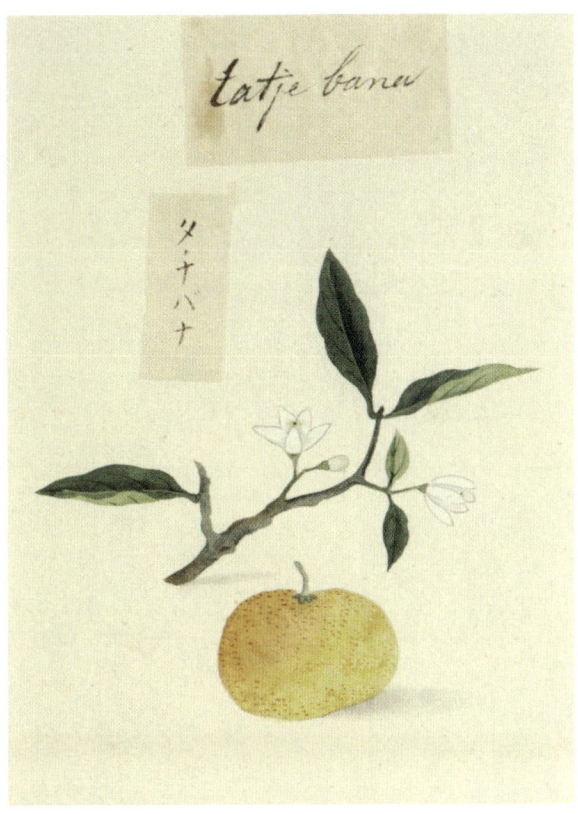

Tachibana, watercolour, Kawahara Keiga, 1823–29. A hybrid involving the wild mandarin, the tachibana was celebrated in poems and figured in coats of arms of aristocratic families in ancient Japan.

THE CITRON (*CITRUS MEDICA*)

One species introduced from south Asia, traditionally held to be India, was to have a far more profound effect than any of the Chinese species so far discussed, save the mandarin and the pomelo, with which it formed hybrids to give the lemon. This species is the citron, the first citrus fruit known to Europeans. By comparison with other *Citrus* species, the citron is a rather small tree or shrub and has large, simple serrate leaves without the winged petiole (leafstalk) typical of oranges, and with fruits up to 30 cm (12 in.) long weighing more than 2 kg (4 lb).

The citron seems to have been grown in Sumeria (now part of Iraq) in the third millennium BCE, when the Sumerians already had sophisticated irrigation systems suited to such crops. Assyrian sculptures found in the Sumerian city of Nippur are said to

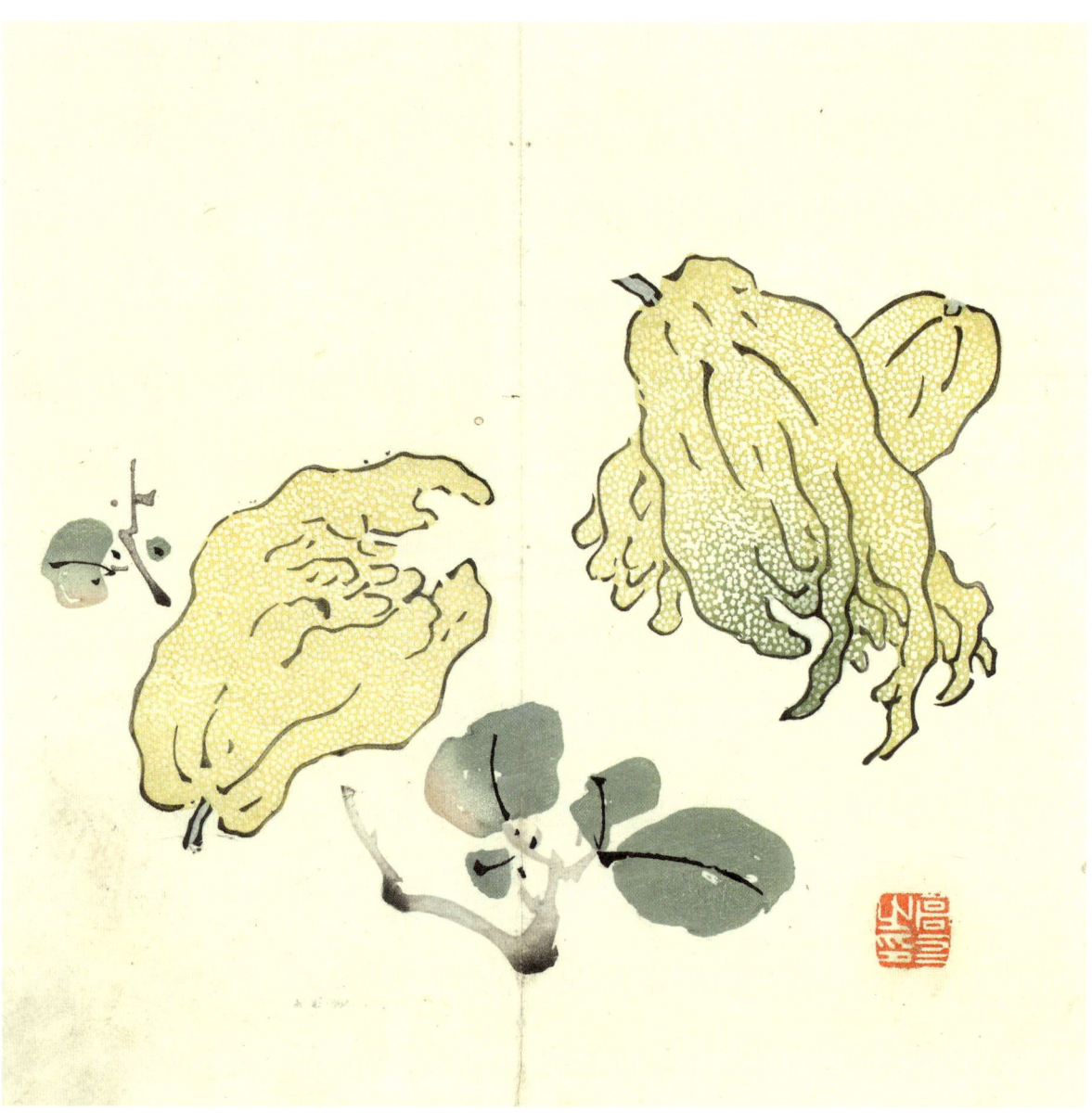

Fingered citron, woodblock print, Gao You, *c.* 1633. The shape of the 'Buddha's hand' citron is the result of spontaneous genetic mutation, unlike some other *bizzarria* (see p. 81).

incorporate representations of the citron[55] and it has been argued that the Sanskrit name *matulunga[ka]* is the root of the Assyrian *iltakku* (or *ildaqqu*), also called *adaru*, suggesting the tree was well known in Mesopotamia. More recently it has been concluded that the Sanskrit word is also the root of the Pahlavi *wa drang* or *w'tling*, and thence Persian *tarandj*, perhaps linked to the Aramaic *'atrunga* and *terunga*, likely the basis of the Arabic *utrundj* and *turundj*.[56]

Certainly, there are ancient links between Iran and citrus, likely citrons. The classical Persian fairy tale *The Girl of Narenj and Toranj* includes the narrative of a girl's death and her resurrection from inside a fruit, a citron (or maybe a sour orange), perhaps harking back to ancient reverence for some kind of botanical deity. Such fruits also figured in customs like marriage ceremonies, being symbolic of happiness and fertility (an association continued in Europe into at least the nineteenth century).[57] The story is held to be a parallel to, or even the origin of, one of the fairy tales collected by the Neapolitan poet Giambattista Basile (1583–1632), namely *The Three Citrons* (also known as *The Love for Three Oranges*). This was the basis for Carlo Gozzi's commedia dell'arte *L'amore delle tre melarance* (1761), and that, in turn, for Sergei Prokofiev's opera *The Love for Three Oranges* (1921), in which fairy princesses emerge from oranges.

The citron, due to its pleasant and refreshing perfume, the longevity of the fruit, as well as its culinary and medicinal properties, also became implanted in religious rituals in parts of Asia, where it must have been grown long before. In India, it became associated with worship of the Hindu god Ganesha, while the god of riches Kubera, assimilated into Buddhism, is figured with citrons held in his hands in both Tibetan and Javanese sculpture, as he is in Jainism, one of the oldest religions still being practised.

Also still offered in Buddhist temples and presented as a New Year gift in China today is the fruit of an ancient group of cultivars (Fingered Group, *fo shou*) in which the hesperidium comprises partially separate, finger-like carpels and resembles a fruit attacked by bud mites. In China, fingered citrons are said to promote good fortune (*fu shou*), happiness and longevity; in the West they are sold by florists as 'Buddha's hand'. It has even been suggested that some fruits in the Assyrian sculptures in Nippur also represent such cultivars, though this is perhaps wishful thinking. In any case, by the eighth century CE, they were recorded from China,[58] where they are used today to scent rooms and clothes as well as to flavour sweets and tea.[59] There are a number of distinct cultivars, including the dwarf ones seen in pots in Europe, such as 'Aihua' (meaning dwarf in Chinese), which is particularly favoured. They were already important in the Min kingdom (909–945 CE) and have been figured in many artefacts in bamboo, jade, ivory and bronze, as well as paintings and drawings. Such mutants are otherwise only known in lemons.[60]

In India, unripe citron fruits are used in pickles and preserves; in China they are prominent in traditional medicine. Commercially, the most significant part is the peel, so the fruits are halved, pulp removed and peel immersed in seawater or salted water for about forty days.[61] After de-salting and boiling, the peel is candied in strong glucose or sucrose solutions, usually in France, England and USA, where it is then used in jam or fruit bars. Citron juice in Malaysia is used instead of expensive, imported lemon juice, while citron water made in Barbados is sent to France to flavour wine and vermouth.

THEOPHRASTUS AND ALEXANDER THE GREAT

From the third millennium BCE onwards, there was a regular trade between the Persian Gulf and the Indus, with camel caravans between northern India and Babylon and on to Syria and Egypt. In 284 CE discussion of a consignment of citrons from the Roman Empire to China is held to indicate that no kind of citron was native in China.[62, 63] However, the citron exhibits enormous variation in Yunnan[64] and, according to some authors, there are indeed 'wild' stands in China, with some cultivated selections bearing fruits weighing 10 kg (22 lb) or more (e.g. the 'Ning'er Giant').[65] However, citrons have been in cultivation for so long that it is now difficult to ascertain where they may have originated. Indeed, although it is still unclear where the citron grows truly wild, if anywhere, today, India is usually cited, but the citron has been transported widely – as far south-east as Indonesia in pre-European times, for example. Emanuel Bonavia, a Maltese doctor in the Indian Medical Service, wrote:

I am still in doubt whether it is indigenous in India. It does not appear to have any ancient Sanskrit name and the number of varieties, if they are variations, on the western sea-coast is suggestive. It is curious that they should be found in the area which came into most contact with foreigners.[66]

Joseph Needham held that the citron reached the Mediterranean from 'Babylonia [largely present-day Iraq] and Palestine', while later authors take it back to Persia (modern Iran).[67] Tradition has it that the citron was introduced by the armies of Alexander the Great (356–323 BCE). Although some twenty contemporaries are believed to have written books about Alexander,[68] not one of these accounts is known to have survived, but, from reports and quotations from these lost works, it is known that within five years of the murder of his father, Philip II of Macedon, Alexander overthrew the two-hundred-year-old Achaemenid empire of Persia, to rule over a territory of five million square kilometres (two million square miles).[69] Alexander would die at thirty two – his life story to be romanticized and embellished with achievements even beyond that massive heritage. One such is the introduction of the citron from India to the West.

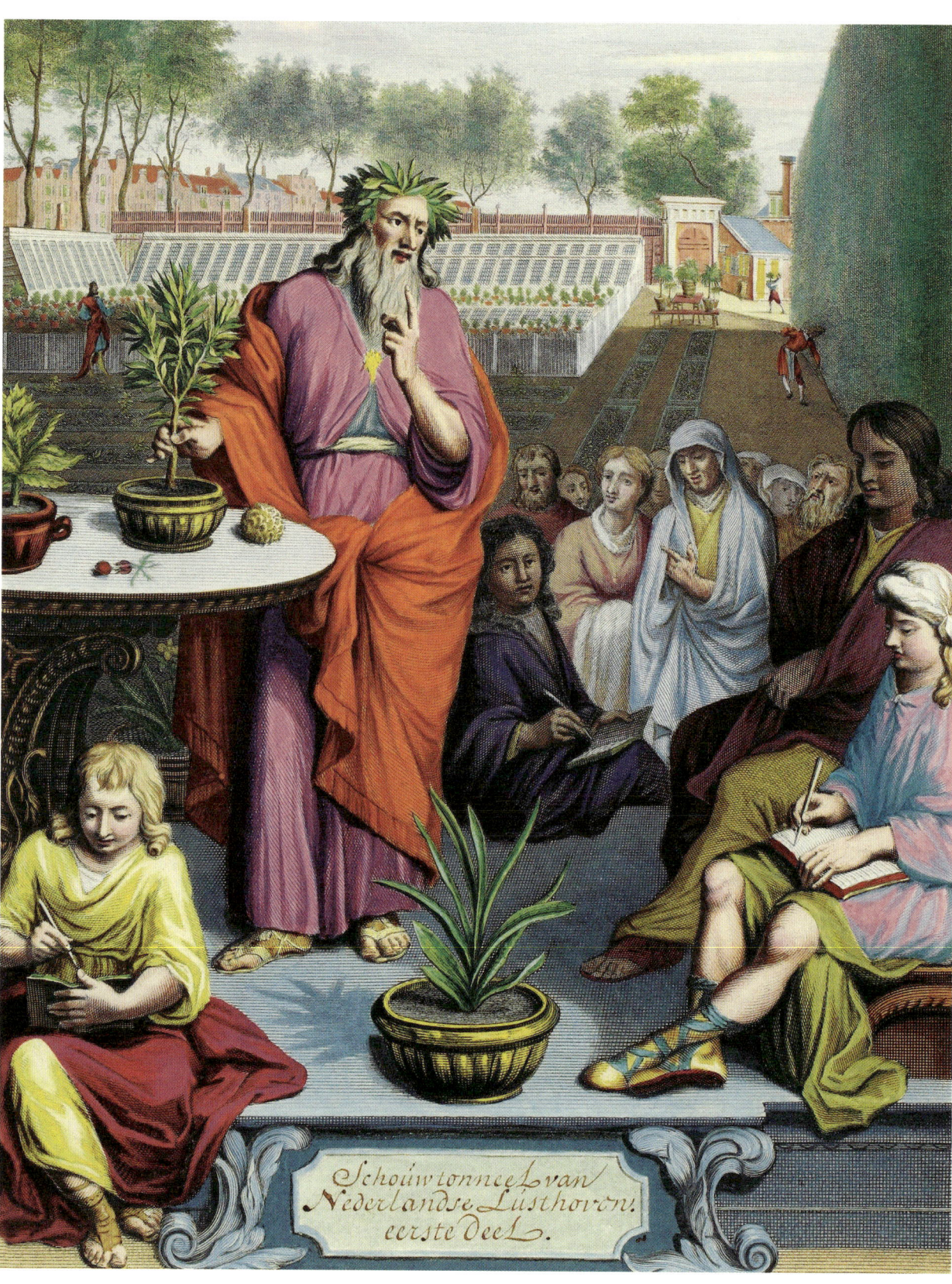

Schouwtonneel van
Nederlandse Lusthoven.
eerste deel.

The first of the Greeks to write on plants was Hippocrates, the 'Father of Medicine', but his work concentrated on their medicinal properties. The germane works of Aristotle (384–322 BCE) have been lost, so it is Aristotle's pupil Theophrastus of Eresus (370–286/262 BCE) who is the 'Father of Botany'. The first that the Western world recorded of any citrus fruits is to be found in his account of the citron in his *Enquiry into Plants*, begun some time after 350 BCE. Theophrastus's works, like those of Pedianus Dioscorides (see p. 47), were the basis of plant descriptions to be found in the early European herbals (see pp. 80–85).[70]

Theophrastus[71] was born Tyrtamus on the Aegean island of Lesbos and was sent to the Academy in Athens where he likely became a follower of Plato, who died when Tyrtamus was twenty-two years old. Tyrtamus then became an associate and close friend of Aristotle who, around fifteen years older, had been Plato's pupil for some twenty years and now set up the Peripatetic School at the Lyceum; he gave Tyrtamus the name Theophrastus (for the 'godlike' style of his speech). Among Aristotle's earlier pupils was Alexander of Macedon (later 'the Great').

With Aristotle's death in middle age, Theophrastus inherited his library, reputedly the richest then known and including the manuscripts of Aristotle's own works, and the botanic garden that Aristotle had established in Athens – to be enhanced by Theophrastus, who took his work forward for the next fifty years.

Cultivated by slaves, the garden was where Theophrastus, while teaching in a school of over two thousand pupils and with no opportunity for travel abroad, was exposed to the wide range of living specimens of exotic plants that figure so markedly in his works. His detailed accounts of the biology of plants, from the germination of seeds to the appearance of mature plants at different times of the year, were the result of intense and continuous observation, likely his own.

Theophrastus prepared some 227 treatises, the best known being *Historia plantarum*, as *Enquiry into Plants* later became known in Latin, in nine

books with the fragment of a tenth, and his *De causis plantarum*, six of the original eight parts still surviving. He incorporated not only his own empirical observations but also material from Aristotle's manuscripts, documenting not merely the utilitarian but also the philosophical. For the first time in the West, plant structure and the relationships between different plant species, the precursors of modern plant morphology and taxonomy respectively, were clearly set out. Embedded in the plagiarizing Pliny the Elder's *Natural History*, it came down to the Middle Ages, but only with the translation into Latin in Rome by Theodorus Gaza (*c.* 1398–1475 CE) at the invitation of the Pope was Theophrastus's (including Aristotle's) body of empirically gathered information brought to light. Theophrastus wrote:

And in general the lands of the East and South appear to have peculiar plants, as they have peculiar animals; for instance, Media and Persia have, among many others, that which is called the 'Median' or 'Persian apple' [citron] … The 'apple' is not eaten, but is very fragrant, as also is the leaf of the tree; and if the 'apple' is placed among clothes, it keeps them from being moth-eaten. It is also useful when one has drunk deadly poison; for being given in wine it upsets the stomach and brings up the poison.[72]

Theophrastus emphatically called the citron the apple of the Medes, an ancient Iranian people who had an empire extending from modern-day eastern Turkey almost to the Indus. This is the origin of the 'medica' in the modern scientific binomial *Citrus medica*, coined by the Swedish naturalist Carl Linnaeus in his defining book *Species plantarum* of 1753.[73] He took the species name from 'Malus Medica' of Gaspard Bauhin's *Pinax theatri botanici* (1623), which refers to Theophrastus and many of the early botanists' use of 'Median'. Pliny the Elder (23/4–79 CE) took this forward in his *Natural History* of 77–79 CE by using the Latin word 'malus', a name originally used for the domestic apple (*Malus domestica*), but the concept expanded in a kind of carpological classification that now seems faintly absurd,[74] though perfectly logical then, to take in other fruits as they were introduced from the east, with 'Malus Armenaica' for the apricot (*Prunus armenaica*) and 'Malus Persica' for the peach (*Prunus persica*), for example. 'Malus arantia' [sic] was used for the orange, 'Malus limonia' for the lemon.

Although Theophrastus's botanical work is still considered fundamental today, botany occupies perhaps only 5 per cent of his writings, which include treatises on astronomy, mathematics, ethics and politics among many other things besides plants.

Theophrastus teaching students, engraving from *Toneel van Nederlandse Lusthoven*, 1718. Theophrastus's account of the citron in his *Enquiry into Plants* is the first record of any citrus in the Western world.

Coloured engravings from *Hesperides* by Giovanni Battista Ferrari, 1646.

According to Greek mythology, a tree bearing 'golden apples' reputed to bestow immortality
was given to the goddess Hera upon her marriage to Zeus. She kept them jealously hidden
away in a blissful garden at the western end of the Earth, guarded by the Hesperides (daughters
of the evening) and the dragon Ladon. Heracles was later said to have stolen the apples, in
some versions slaying Ladon (below). Though believed to have been quinces (related to true
apples), the 'apples' came to be interpreted as citrus fruits as those were spread across the
Mediterranean. This rendering was perpetuated in the work of later writers such as Ferrari,
whose botanical *magnum opus* included an illustration of the Hesperides bringing
oranges down the Tiber to Rome (opposite).

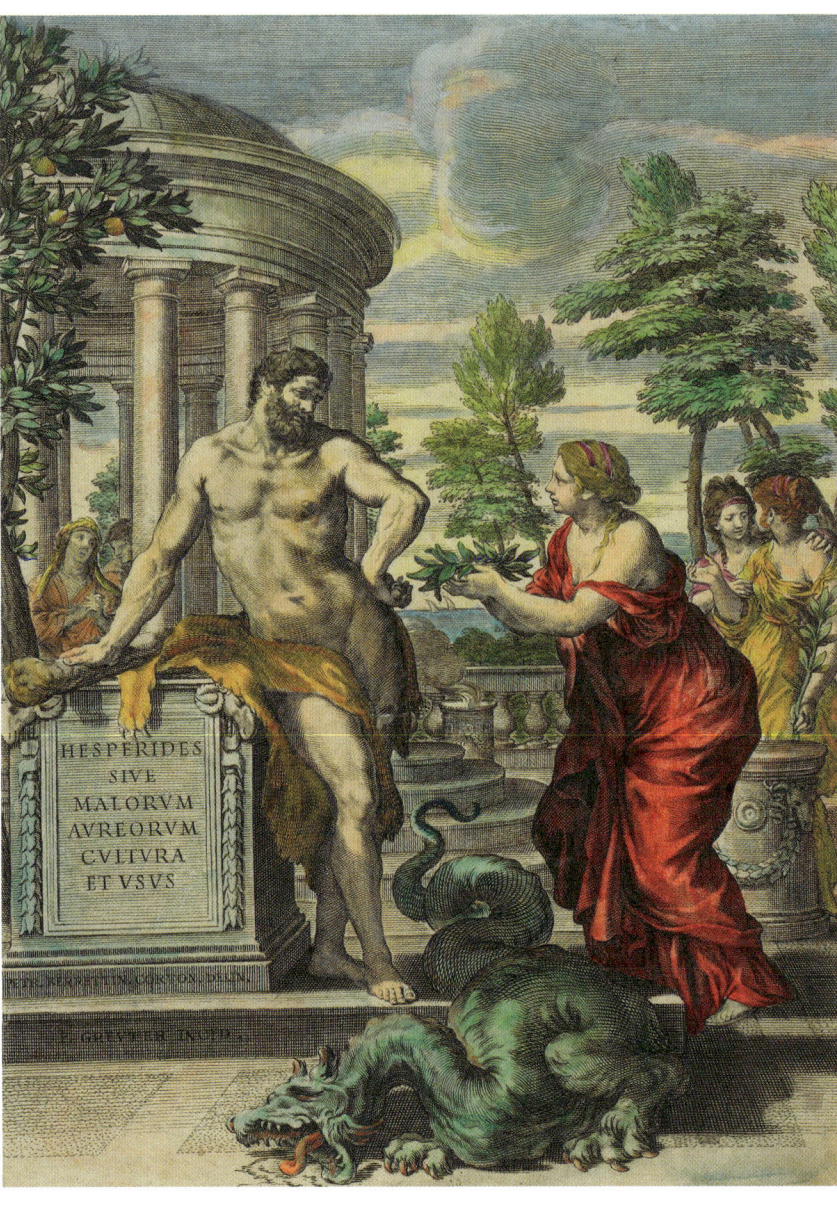

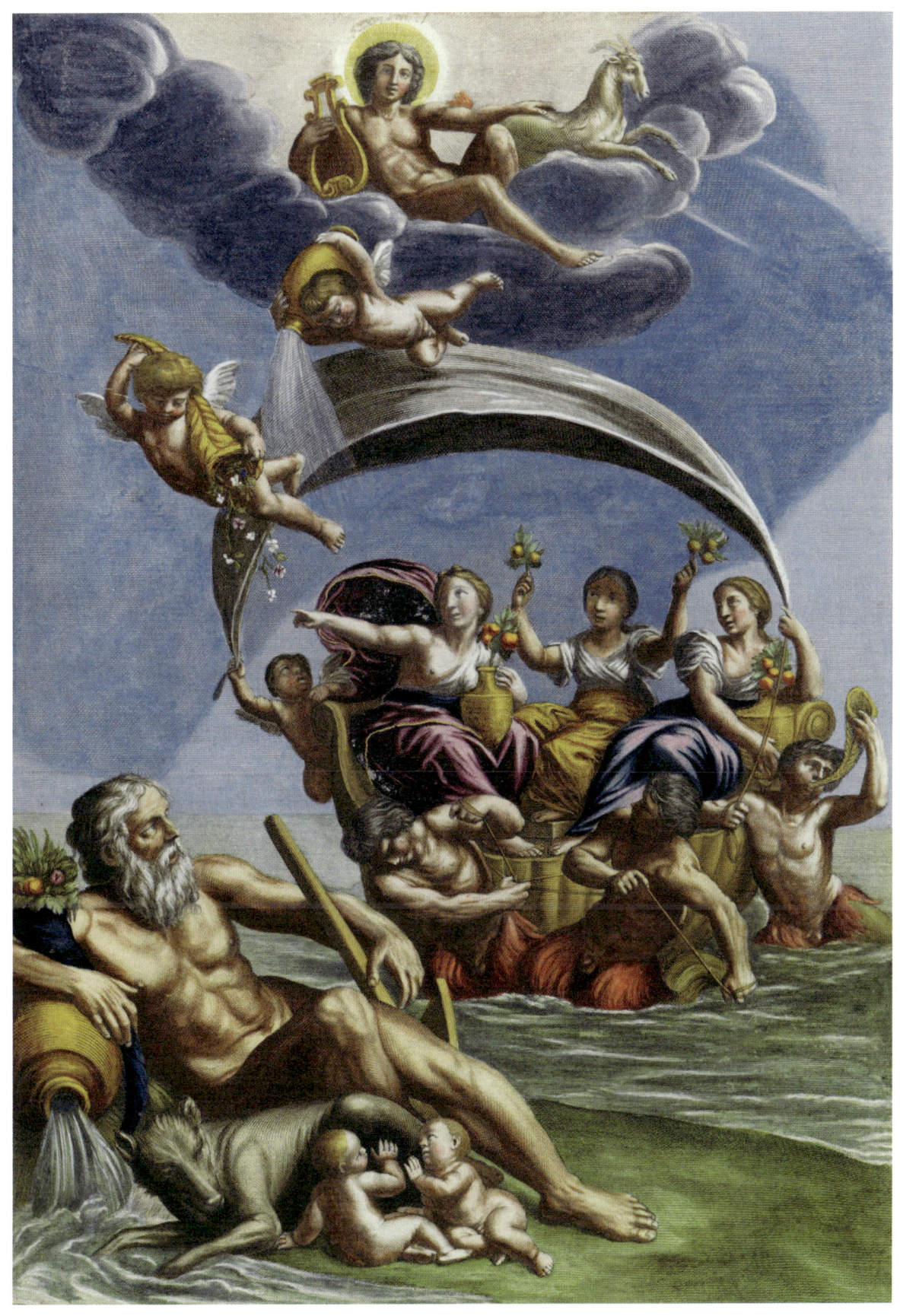

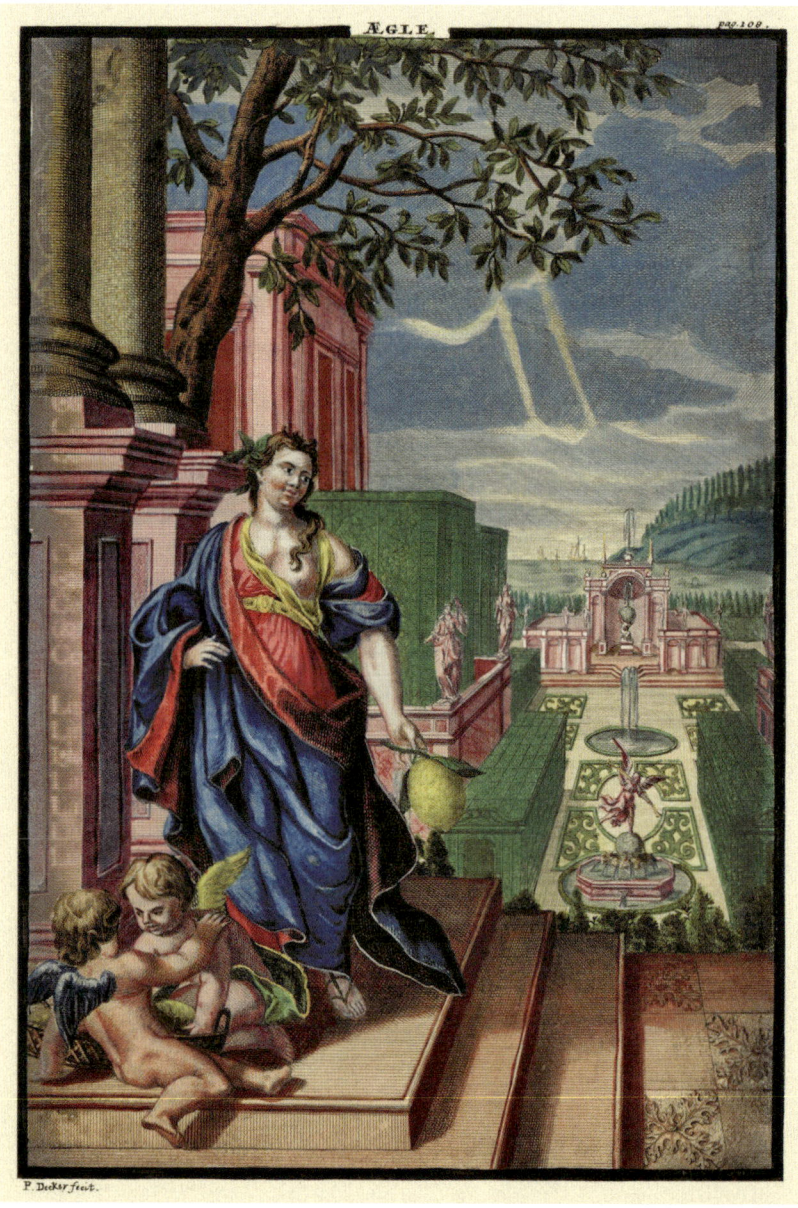

ABOVE: Colour engraving from *Nürnbergische Hesperides* by Johann Christoph Volkamer, 1708. **OPPOSITE:** *The Garden of the Hesperides*, Frederic Leighton, 1891.

The Hesperides were nymphs, the daughters of Nyx (Night) or in some versions Atlas (the Titan god who held up the sky for eternity). The three (sometimes seven) sisters are usually named as Aegle, Arethusa and Erytheia (sometimes Hespere or Hesperethusa). So entrenched had the citrus–Hesperides connection become in European thought by the Renaissance that both Ferrari (see previous spread) and Volkamer divided their works into sections named after the mythological sisters. On the frontispiece to Volkamer's second volume, Aegle can be seen holding a citron (above). The confusion was compounded by the early nineteenth-century naming of two other genera of the citrus family (Rutaceae): *Aegle*, which includes the bel tree or Bengal quince (*Aegle marmelos*), a tree of India and Myanmar sacred to Hindus; and *Hesperethusa* (now *Naringi*, a rendering of the Hindi *narangi*).

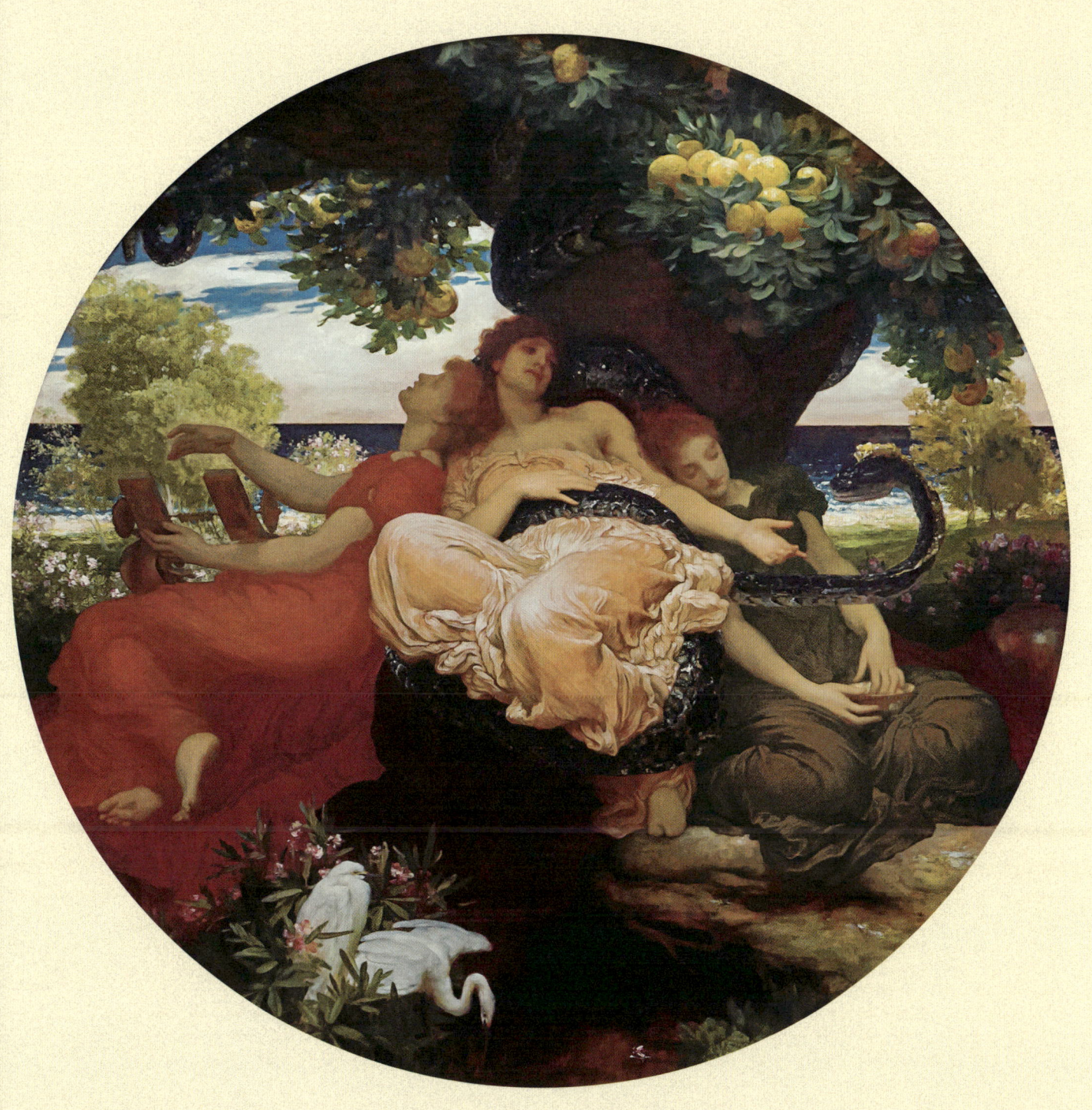

At a time when plants and everything else were thought of as existing because of their relationship to human beings, the empirical scientific approach of Aristotle and Theophrastus, so familiar and seemingly obvious today, was not unanimously acclaimed at the time. Theophrastus's (and Aristotle's) pupils were reportedly ridiculed for their field excursions, on which they studied objects of natural history without known uses, an attitude still not lost today.

It is likely that citron fruits, brought back from Persia, had been for sale in Greek markets long before Theophrastus's work. The way he wrote suggests that his readers would have been familiar with the fruit, though he seems emphatic that the trees were then only known to him as coming originally from Persia. Indeed, recent excavations in Israel suggest strongly that the Persians introduced citron there,[75] probably in the fifth century BCE when the Jewish 'temple state' was instituted.

On a hill above modern Jerusalem, the Ramat Rahel site comprises the remains of a palace with an elaborate irrigated garden. From the linings of the watercourses, replastered over time and so capturing the pollen shed by trees and other plants of the period, archaeobotanists have been able to identify the plants that were being cultivated there. Besides olives, grapes and figs, widely grown species but also not indigenous to the area, there were remains of local fruit trees, ornamentals and imported trees, including exotic species such as walnuts and citrons in a deposit dated to the Persian period (the 400s BCE). Perhaps these exotics were imported from remote parts of the Persian empire by the ruling elite and grown as status symbols. Indeed, it seems plausible that efforts would have been made to grow the citron, then greatly valued for its uses in medicine, scent, and as an insect deterrent.

It is feasible that the cultural significance of the citron led to its becoming a crucial feature of the Jewish Tabernacles festival (see pp. 50–53). According to Jewish tradition, the Jews brought the citron to Israel after their exile from Egypt. The French archaeologist Victor Loret (1859–1946) argued that it is figured on the walls making up the so-called 'botanical garden', likely the site of an initiation ritual,[76] in the temple of Amun-Re at the Karnak Temple north of Luxor, which is dated to the reign of Thutmosis III, approximately three thousand years ago. Most modern scholars dispute this identification.

Whatever the details of the spread of the citron from India in ancient times, Theophrastus's observations remain important today. He noted, for instance, that the citron bears its fruits 'at all seasons; for when some have been gathered, the flower of others is on the tree and it is ripening others'.[77] Indeed, unlike most citrus grown in the Mediterranean, which flower in spring, the citron is notable for its continuous flowering and fruiting. He continued, 'Of the flowers ... those which have, as it were, a distaff [i.e. pistil] projecting in the middle are fertile, while those that have it not are infertile'. This was the first description of the typical reproductive system in citrus, where bisexual flowers, with both stamens and pistil, produce fruits, while those without a pistil produce merely pollen, a clear allusion to sex in plants nearly two thousand years before northern Europeans recorded it.

Theophrastus's work and writing have had a profound effect on scientific thinking: even as late as the Renaissance the descriptions of plants in the first printed herbals were 'almost word for word translations of the ancient paragraphs of Theophrastus'.[78] As he in effect wrote up the botanical findings of Alexander the Great's invasion of India when incorporating them in his *Enquiry into Plants*, the introduction of the citron from India has sometimes been attributed to Alexander's invading Macedonian army.

Alexander's expedition began in 327 BCE, overrunning the Achaemenid empire (the 'First Persian Empire') and reaching its easternmost domains in what is now Pakistan and Afghanistan. During Alexander's invasion of India in 326 BCE, his army encountered tropical plants including the banyan tree (*Ficus bengalensis*) and mangroves, the astonishing trees living at the edge of the sea.[79] This expedition, the first European encounter with tropical vegetation, revealed the jackfruit, the mango and the banana, while the troops found people clad in cotton rather than the linen, hemp and wool they were familiar with.

It is more than likely, then, that Alexander's army, spending so much time in what is now Iran on the way to and from India, became familiar there with the citron, which must have been brought from across India by traders or other travellers long before. Indeed, in his *Enquiry into Plants*, Theophrastus wrote of the fruit as the 'Median' or 'Persian' apple – not 'Indian'.

The oldest archaeological evidence of citrus seeds is from the early second millennium BCE of northern India, while citrus-wood charcoal of 1400–1300 BCE is claimed for Karnataka in south-west India.[80] Nothing in the Mediterranean is known before c. 1200 BCE, where some mineralized seeds found in Hala Sultan Tekke in Cyprus have been claimed to be citrus. More certain is evidence of citrus peel, as indicated by citrus-specific polyphenols found in organic residues in a wine-jug used as an offering for the dead and preserved in a cremation grave of the sixth-century BCE necropolis of Monte Siraï in southern Sardinia. This suggests that Phoenician settlers were responsible for the spread of plants from the eastern to western Mediterranean, as they were in southern Italy in the ninth century BCE, so closely linked with the Cypriot influence with regard to trading iron and associated technology.

Other first-millennium evidence comes from Cuma near Naples, the site of the city of Cumae (Kyme), one of the earliest Greek colonies in Italy. The amount of *Citrus* pollen there is remarkable, bearing in mind that *Citrus* species are insect-pollinated, suggesting that the trees must have been intensively cultivated locally. Pollen is recorded from many other younger sites. Moreover, investigations in 2013 have revealed mineralized and carbonized citron seeds in the pre-Roman Samnite levels under the Temple of Venus in Pompeii, as well as seeds and peel, perhaps of lemon, in central Rome from *c.* 100 CE.[81] Both finds suggest that citrus fruit was precious and used in sacred ceremonies.

The scholarly Juba II, King of Numidia and Mauretania (*c.* 48 BCE–23 CE) in north Africa, reportedly identified the Classical golden apples of the Hesperides as citrus, presumably citron, fruits. This interpretation was repeated by the Roman poet Martial (*c.* 38–41 CE) in the second half of the first century CE.[82] Around the same time, the Greek grammarian Pamphilus of Alexandria wrote that the Spartans offered the scented yet inedible 'apples of the Hesperides' to their gods and the citrus interpretation stuck, even though in the mid fourth century BCE they were considered, far more plausibly, to be quinces (the south-west Asian *Cydonia oblonga*, which has aromatic fruits with a superficially similar shape).

ROMANS

Virgil (70–19 BCE) gave a poetic description of the citron and its qualities: 'with it the Mede treats his noisome breath, and cures the asthma of the old';[83] in the reign (54–68 CE) of Nero, the fruit figured in the infamous banquet in the *Satyricon* of Petronius (d. 66 CE).[84] The fruit was commonplace by the time Pedianus Dioscorides (*c.* 40–90 CE), a Greek physician and botanist living in the Roman colony of Anazarbus

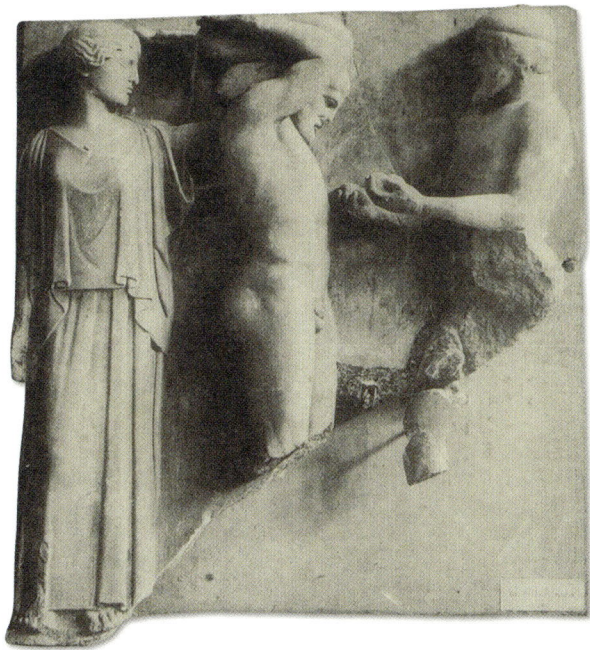

Atlas brings Heracles the golden apples, relief, *c.* 460 BCE. In some versions of the Hesperides myth, Heracles tricks Atlas into gathering three 'golden apples' for him. By the 1st century CE the 'apples' had become identified with citrus.

(today's Anavarza in southern Turkey), wrote in his pharmacopoeia, now known by its Latin name *De materia medica*, 'Those which are called Median, Persian, or *kedromela* and in the Latin *citria*, are known to all for it is a tree that bears fruit throughout the whole year one under another.'[85]

Moreover, in the ruins of Pompeii there are frescoes and mosaics depicting citrus trees,[86] suggesting they were being cultivated by the beginning of the first century CE, and possibly much earlier. Excavations there in 1973 revealed terracotta pots containing roots referred to citron.[87] However, neither Columella, in his *De re rustica* (first century CE) nor the earlier Varro in *Res rustica* or Cato in *De agricultura* mention citrus-growing, suggesting it was not widely practised in Italy. Pliny wrote of the exotic *malus Assyria*. However, by the second century CE, the price of a citron in the eastern Mediterranean was no more than that of the fig, so that citrons could hardly still have been a rarity.[88] By the time of Palladius (in his late fourth- or early fifth-century CE *Opus agriculturae* [*De re rustica*]), it seems *citreum* was being grown in Sardinia and Naples. Palladius also discussed

The citron or Assyrian apple, called by others the Median apple, is an antidote against poison ... Because of its great medicinal value various nations have tried to acclimatize it in their own countries ... but it has refused to grow except in Media and Persia.

PLINY THE ELDER, *NATURAL HISTORY* XII, 15–16, 77–79 CE

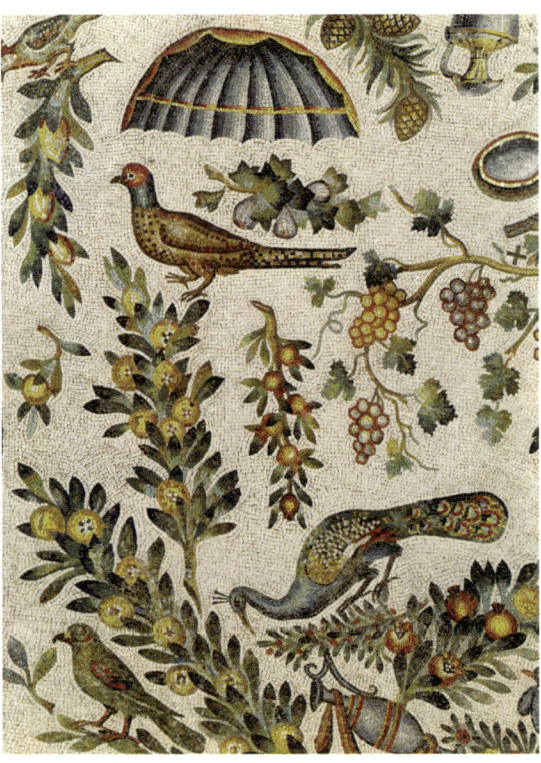

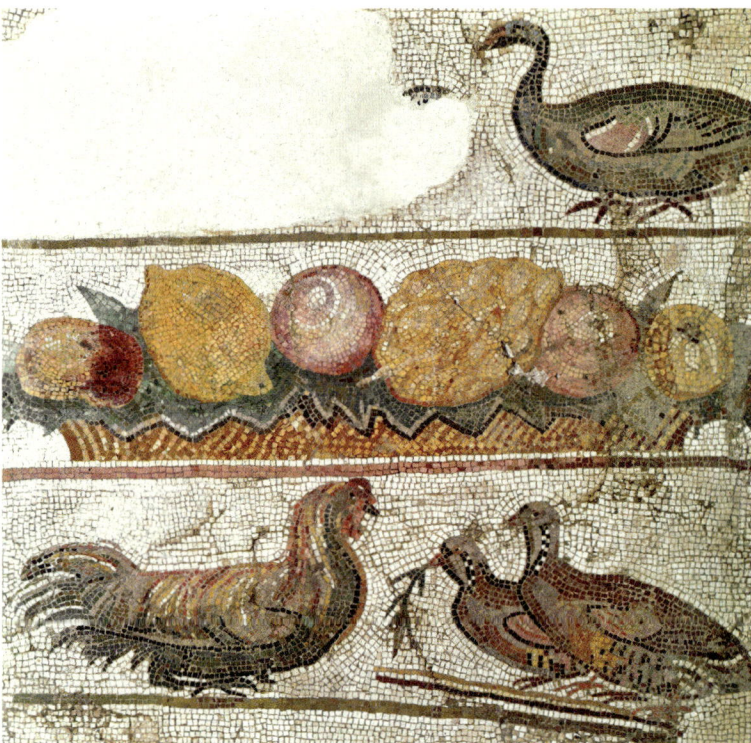

ABOVE, LEFT: Citrus branches on the ceiling of the Church of Santa Costanza, Rome, mosaic, mid-4th century CE. **ABOVE, RIGHT:** Lemon and citron, Rome, mosaic, late 1st century CE. **OPPOSITE:** Lemon tree in the House of the Orchard, Pompeii, fresco, *c.* 25–20 BCE to 40–50 CE.

While the citron was well known to the Romans, particularly for its medicinal qualities, its cultivation in Italy seems not to have been widespread; other forms of citrus were very rare. However, depictions in art provide tantalizing glimpses: a fresco from Pompeii (destroyed by the eruption of Vesuvius in 79 CE) appears to depict a lemon tree in fruit. This suggests the lemon was then in cultivation rather than merely imported. By the 4th century there appear citrons, lemons and perhaps oranges, all attached to leafy branches, in a mosaic on a vaulted ceiling in the Church of Santa Costanza in Rome – though the foliage could easily have been added to imported fruits drawn from life.

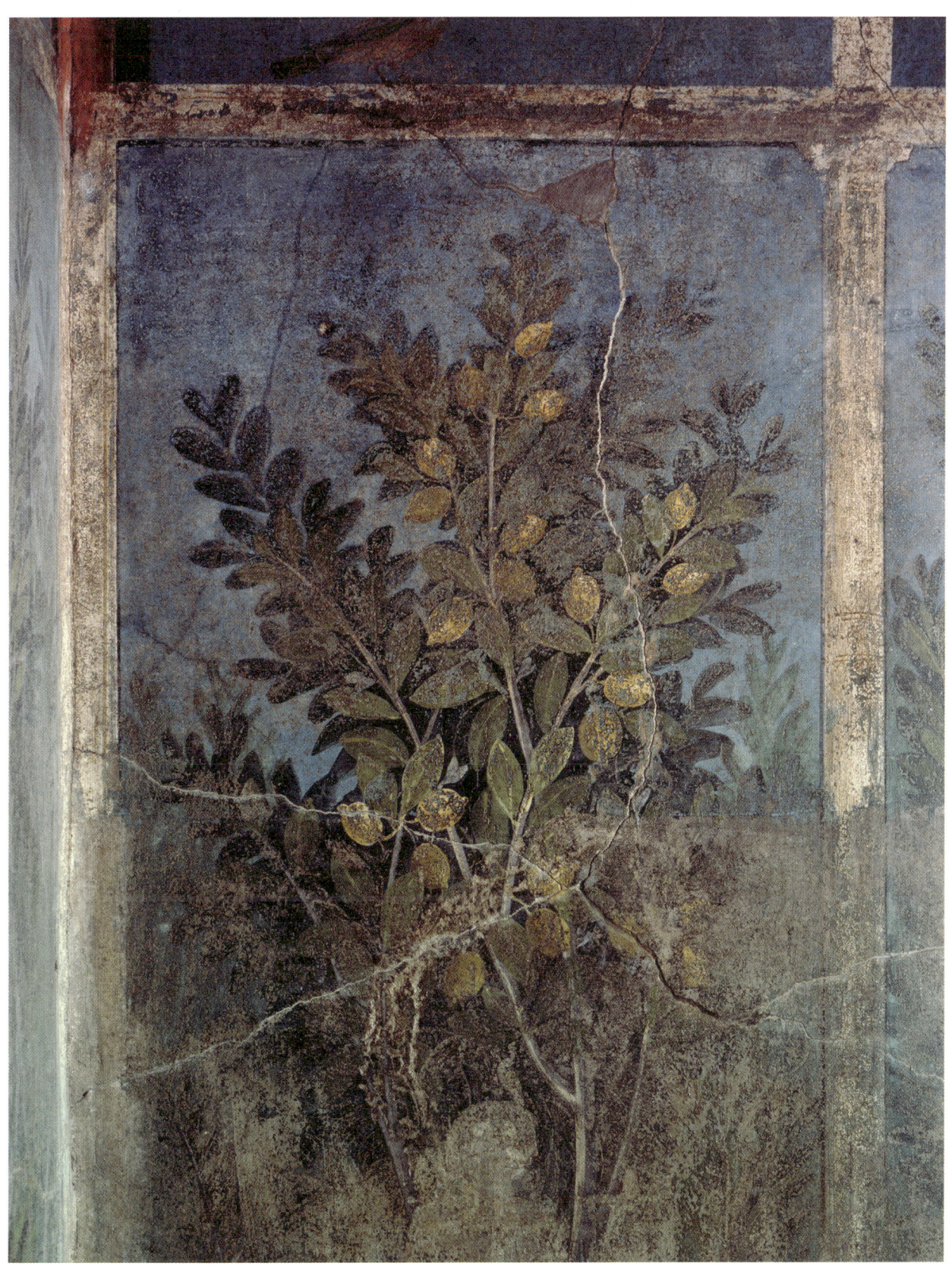

a *citretum* in northern Italy where the *citreum* was grown under a roof to protect it from cold.[89] Perhaps this was a proto-orangery (see pp. 90–96).

THE FEAST OF THE TABERNACLES

The Jewish Feast of the Tabernacles, or Sukkot, is a kind of harvest festival and is likely adapted from a rite practised by the Canaanites, from whom the Jewish people are thought to be descended.[90]

The Hebrew word *sukkot* is the plural of *sukkah*, a tabernacle or booth, which is a flimsy, walled structure covered with thatch, often of date-palm fronds. The *sukkah* is reminiscent of temporary field accommodation used by farm workers during harvesting, and it also recalls the temporary housing

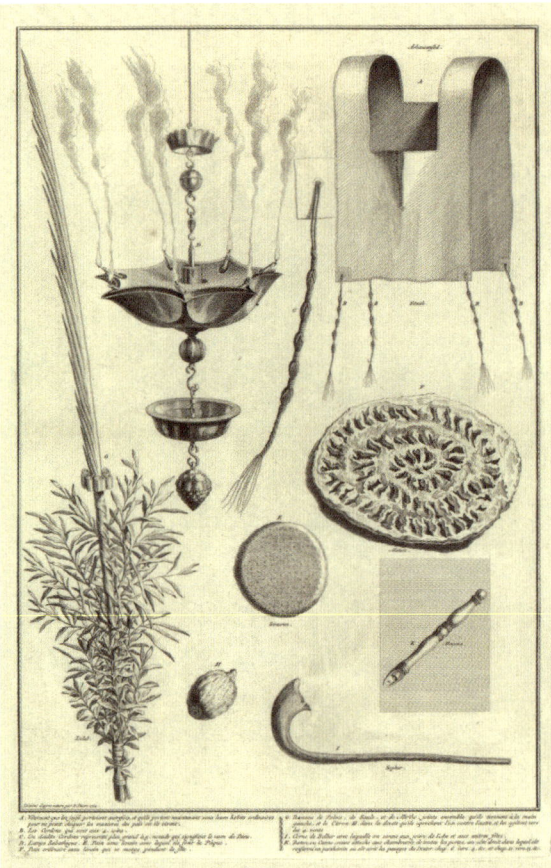

Jewish religious items from *Ceremonies et coutumes religieuses de tous les peuples du monde* by Bernard Picart, 1724. The *lulav* (date palm) and *etrog* (citron) seen here, bottom left, are two of the 'four species' used in the celebration of Sukkot.

in which the Israelites are said to have lived during their forty years in the desert after their Exodus from slavery in Egypt (reportedly based on thirteenth-century BCE events). The holiday now has wider importance, as the construction of the ephemeral *sukkah* signifies the temporary putting aside of materialism to focus on spirituality and hospitality.

The festival lasts seven days (eight in the diaspora) and, throughout, meals are taken inside the *sukkah*, where some also sleep. It is one of the three holidays ordained by the Torah, when the Israelites were expected to travel to Jerusalem. On each day of the festival, a blessing is recited over the *etrog* (certain cultivars of citron[91]) held in one hand, and the *lulav* (an immature leaf of the date palm), bundled together with *hadass* (myrtle) and *aravah* (willow shoots), in the other. The use of these four plants (*arba'at haminim*, literally 'four species'), waved in the air during the ceremony, follows Leviticus (23: 40, after 538 BCE): 'On the first day [of Sukkot] you shall take a *peri etz hadar*, palm fronds, branches of leafy trees and river willows, and you shall be happy before the Lord your God for seven days.' The four plants represent the sins to be avoided: palm = pride, willow = slander, myrtle = nosiness, citron = lust. However, the citron is not specifically mentioned there or anywhere else in the Bible, though the New English Bible translates *peri etz hadar* as 'the fruit of citrus-trees', perhaps because it is mentioned in the noncanonical *Book of Jubilees* (160–150 BCE).[92]

The word *hadar* is said to be of mixed etymology, literally 'the' (Hebrew, *ha*) and 'tree' (Iranian, *dar*); the *dar* is argued to be cedar (as in Sanskrit *devadaru*, 'tree of the gods'), *Cedrus deodara*, a tree holy for Hindus.[93] Shemuel Tolkowsky (1886–1965), an agronomist and diplomat,[94] held that *hadar* originally meant the cone, that is the fruit, of the *dar* tree. Assyrian bas-reliefs depict such cones used in water libation rites, while cuneiform texts confirm its being venerated in what is largely now modern Iraq. The cone was possibly what was originally used in the Jewish festival, perhaps through Babylonian influence. According to Tolkowsky, the citron was a replacement for the cone, which had apparently become associated with 'bacchanalia'.[95] So, the *Mishnah*, the second-century CE code of Jewish oral tradition, interpreted the *peri etz hadar*, the fruit of the tree, as a citron. In modern Hebrew, *hadar* certainly refers to citrus. However, it has since been argued that the word came from *adaru*, as used in Assyria for the citron;[96] the modern word *etrog* was introduced later, apparently from Aramaic via Persian *turunj*. It has also been suggested that, rather than being a cone substitute, the long-lasting and exotic citron replaced other fruits like *Mimusops laurifolia*, originally from Ethiopia, but long cultivated in Egypt and called *persea* by historian-warrior Josephus (*c.* 37–100 CE).[97]

Bearing in mind the harvest-festival origins of Sukkot, it has more recently been argued that 'during the seventh Hebrew month, which correlates to our September and October, the time of Sukkot, grapes, figs, dates, and pomegranates were either ripe for harvest or already harvested, and the olive harvest was just beginning. These ripe tree-fruits were most likely then the *peri etz hadar*'.[98] Perhaps, then, incorporating elements from far away, the ceremony evolved from one with the cedar cone to be replaced by tree fruits in general, then, possibly by entrepreneurial growers and merchants, the more spectacular, long-lived fruit, the citron.[99]

Forms of the citron, known in Hebrew today as *etrogim*, became increasingly important – and increasingly associated with nationalism. A coin depicting the *etrog* survives from the first century CE[100] and, by the time of the Bar Kokhba revolt (132–135 CE) against Emperor Hadrian, Judean tetradrachm coins, made from Roman ones stamped with Hebrew symbols, had both the *lulav* and the *etrog*.[101] From 40 BCE onwards the citron was one of the most common motifs in Jewish mosaics, tombs and inscriptions generally.[102]

In Judaism citrons have been considered symbolic of the heart while the citron tree has been a contender for the tree of knowledge of good and evil. By 1003 CE[103] the citron had long been established in Italy and was much cultivated in Salerno, south-east of Naples. In that year, fruits (*poma cedrina*) were sent to the Norman nobles who had saved a Salerno prince from the Saracens. By then, Liguria to the north had for many centuries provided the Jews of Italy and northern Europe with citrons for Sukkot, and still does. By the seventeenth century, Jewish merchants from Poland, Germany and Austria were travelling to San Remo in August every year to buy citrons and palm branches for the festival.[104]

For citrons to be considered kosher they must be neither grafted nor hybridized, as well as fulfilling certain cosmetic requirements. Certain cultivars[105] do not lose the remains of the style and stigma of the flower – these remnants known as a *pitam* in Hebrew – and were considered by some the most kosher of all, but this is debated, the more important feature being a persistent stalk (peduncle). An important cultivar today is 'Diamante',[106] grown in

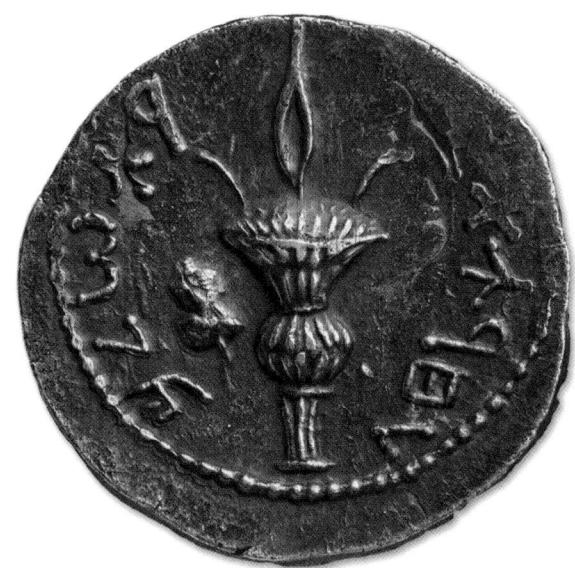

Silver tetradrachm with *lulav* and *etrog*, 132–135 CE. During the Bar Kokhba revolt, the *etrog* (citron) became a symbol linked to Jewish identity – and rebellion.

a small part of the Tyrrhenian coast, the 'Riviera dei Cedri' of southern Italy, where Diamante is a town in Cosenza province.[107] It yields the best-quality peel for the food industry and is favoured for the Sukkot because of its shape and retained stalk, and because it is not grafted. Grafting in general was being practised in the Mediterranean at least as early as the fifth century BCE, but some kinds of grafting were forbidden in the *Mishnah* as it appears to have been considered, mistakenly, a form of interbreeding.[108] After the introduction of the lemon to Europe, lemons were used as a rootstock for citrons, to increasing concern so that in the sixteenth century it was not permitted by Jewish Law. Nonetheless, by the end of the eighteenth century, *etrogim* from allegedly grafted trees in Corfu dominated the market. In 1845, their use, and the use of grafted citrons in general, was challenged (by those with vested interests, their orchards being elsewhere), but, grafted or not, they were eventually considered acceptable.[109] Today elaborate *etrog* containers made of silver or wood are used for holding the fruits for the festival.

Whatever its true origin and distribution, the citron in the Middle East did not lead to any major citrus crop, but, much further east, this apparently Indian plant was brought into contact with Chinese citrus with important consequences.

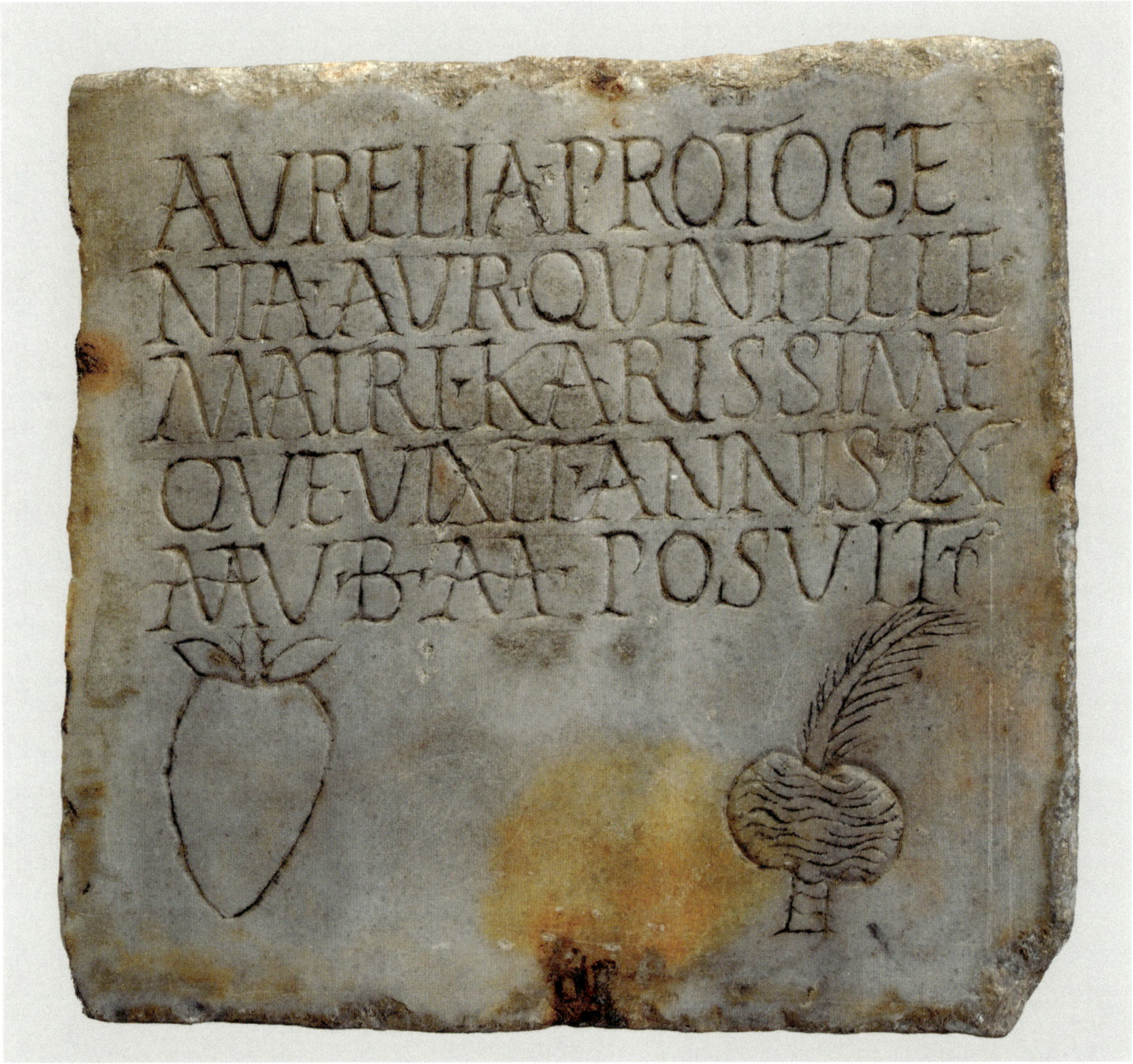

AVRELIA PROTOGE
NIA AVR QVINTILLE
MATRI KARISSIME
QVE VIXIT ANNIS IX
MVB AA POSVIT

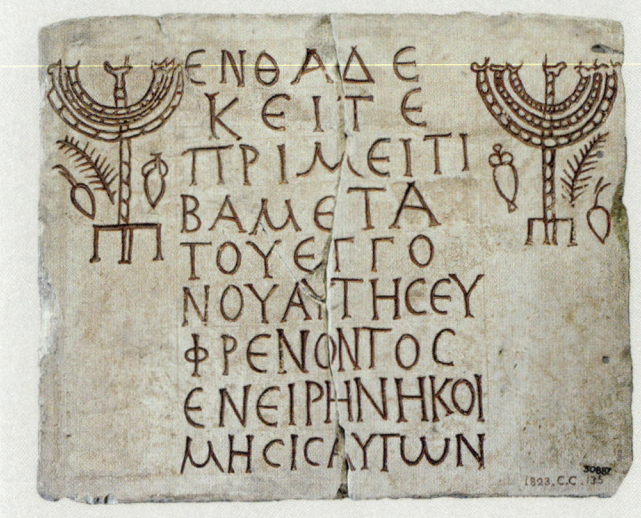

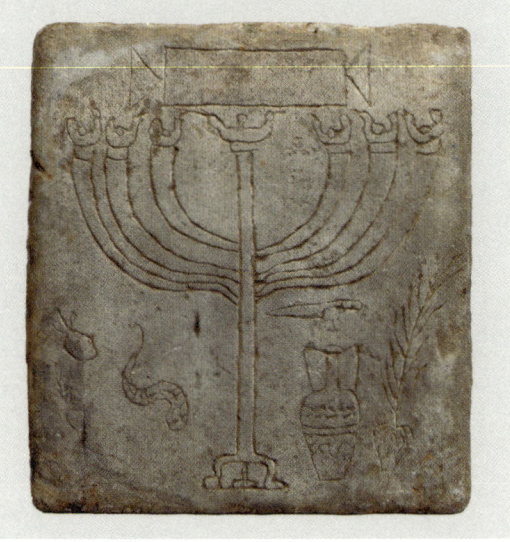

On the first day you shall take a peri etz hadar, *palm fronds, branches of leafy trees and river willows, and you shall be happy before the Lord your God for seven days.*

LEVITICUS 23:40

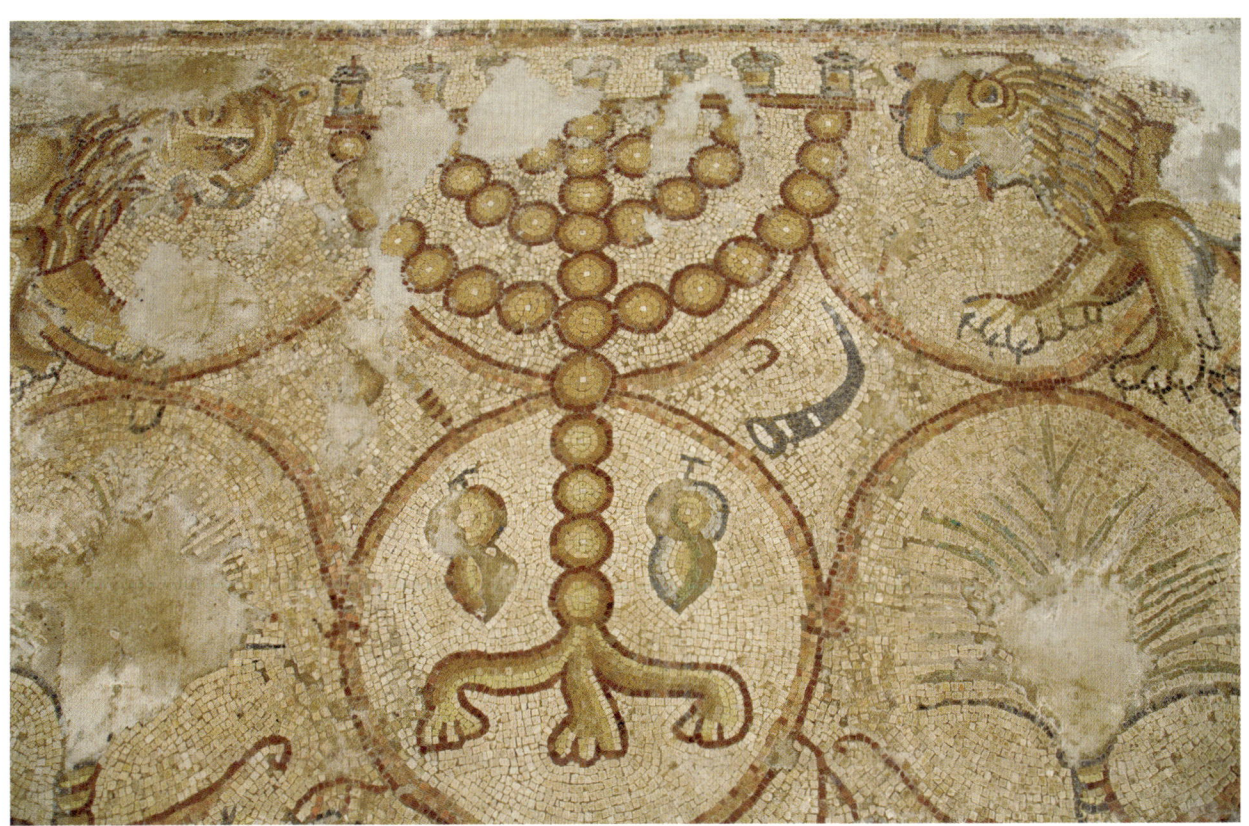

OPPOSITE: Jewish burial plaques, Rome, 3rd–4th centuries CE (?). **ABOVE:** Mosaic floor in the Maon Synagogue, Negev Desert, 6th century CE.

To be suitable for use in the celebration of Sukkot, the *etrog* must meet several criteria. It must be 'beautiful', with a nice shape and no marks; those with a *pitom* (remnants of the style and stigma of the flower) or a *gartel* (indentation around the centre, visible in the *etrogim* depicted in the mosaic above) are favoured. But at different times through history, fruits could not be from a plant that had been grafted or hybridized. The importance of *etrogim* to Jewish identity is evidenced by their depiction on the tombs of Jews buried in Rome (opposite) during the 3rd and 4th centuries. As a key element in the celebration of Sukkot, a major Temple festival, the *etrog* symbolizes the hope of national redemption and the rebuilding of the Temple.

CITRON HYBRIDS

Along with Buddhism, the citron may have been taken along the Silk Roads to China in the first centuries of the first millennium CE (while the apple was going in the opposite direction), though citron could of course have been native in south-west China and brought north directly much earlier. Almost inevitably the citron became involved in the gigantic, and probably completely unintentional, hybridization programme, which is the origin of the citrus industry as we know it. Citron seems always to have been the pollen (male) parent, but it is not clear which of the crosses with Chinese species arose first or where. The simple hybrids with the pomelo have been called lumias, though several other plants have been included under that vernacular name, so interpreting the literature on these, as so much on citrus, is problematic.

Much clearer, perhaps, are the hybrids between citron and pure mandarin (*Citrus reticulata*), which no doubt appeared in different places and at different times. Well-known sorts, including back-crosses with the mandarin, are the rough lemon, Tahiti orange, Rangpur lime (Canton lemon) and Volkamer lemon, many of them used as rootstocks resistant to the largely aphid-spread citrus tristeza virus. Pehr Osbeck (1723–1805) noted in the 1750s that the Volkamer lemon was the first such hybrid recorded growing in Europe. Named after Johann Christoph Volkamer (see p. 135), it was featured by Giovanni Ferrari in his *Hesperides* of 1646 (see p. 98) and reportedly arose in Italy, not as an import from China.[110] The fruits of the Rangpur lime, named after a city in what is now north-west Bangladesh, are very acidic. The tree was apparently not introduced to the USA until 1902, though it has long been naturalized in Australia. In 2006 production began in Britain of a Rangpur lime-flavoured gin, which has now become fashionable. 'Otaheite' (Tahiti orange) is a sweet-flavoured cultivar often seen today in the USA as a pot plant for winter decoration, plants producing bright orange-yellow fruits when only 30 cm (12 in.) tall. It was introduced to France c.1812, but was reported as being grown in England by 1805 under the name 'Citrus limetta sinensis';[111] the epithet 'otaitensis' was first used for such a hybrid and is unfortunate, as the well-known Tahiti lime (see p. 157) has nothing to do with it.[112]

Far more important was the origin of *Citrus × limon*, the lemon. It has now been proved that lemons arose, no doubt many times independently and in different places, as crosses incorporating three *Citrus* species: citron (*C. medica*), pomelo (*C. maxima*) and mandarin (*C. reticulata*). It is not clear whether these were crosses between pomelo and rough lemon (*C. × otaitensis* i.e. *C. reticulata × C. medica*) or between citron and oranges (*C. × aurantium* i.e. *C. reticulata × C. maxima*) or between a 'lumia' (*C. medica × C. maxima*) and a mandarin – or two of these, or all three. Different crosses and back-crosses in this complex gave rise to sweet lemons (including the limetta), the bergamot and many other hybrids best treated in cultivar groups in order to accommodate cultivars with different fruit shapes and flavours, e.g. Sweet Lemon Group and Bergamot Group.[113]

Lemons, or at least their juice, were royal tribute in China by 971 CE,[114] but they are not mentioned in the *Ju Lu*. The earliest Chinese record is said to be that by Fan Chengda (1126–1193) in his *Gui Hai Yu Heng Zhi* of 1175, the *li-mung* being 'the size of a large plum; again it resembles a small orange, and is exceedingly sour to the taste'.[115] Three years later, Zhou Qufei's *Lingwai Daida* noted that lemons were being used in Guangzhou and that they seemed to have come from the south, as indeed the name *li-mung*, compared with the Malay *limao*, suggests. The English word lemon, first used in English as *lymon* c. 1400, apparently comes thence via *limun* (Persian) and *laimun* (Arabic), the Arabic *lim* being the general word for fruits of the citron kind.[116]

By 1299, lemon sherbet (lemonade comprising sugar, lemon juice and water) was being made in China under the Mongols[117] and by the early seventeenth century was commonly drunk in Turkey as well as Persia. Before this, the Cairene Jew Hibat-Allah Zayd Ibn Jami (*fl.* 1117–1193), personal physician to Saladin, wrote a *Tractatus on Lemons*, later translated into Latin by the Italian physician and Arabist Andrea Alpago (c. 1450–1521) and published posthumously early in the sixteenth century – by 1758 it had been printed three times.[118] The strong Persian narrative suggests that it may have been the Persians, in touch with south-east Asia, who brought the culinary lemon to China. It is not inconceivable that some of the hybrids in the lemon complex arose in orchards in Persia itself, where the citron had been long established.

Lemon (*Citrus × limon*) from *Histoire naturelle des orangers* by Antoine Risso and Pierre-Antoine Poiteau, 1818–22. Lemons are crosses between citrons and cultivars in the orange complex. In the eighteenth century, lemons were used to curdle milk, perhaps first by Quakers, to prepare what is now called lemon curd. Citric acid in the lemon juice denatures proteins in milk, which creates the typical texture of the spread.

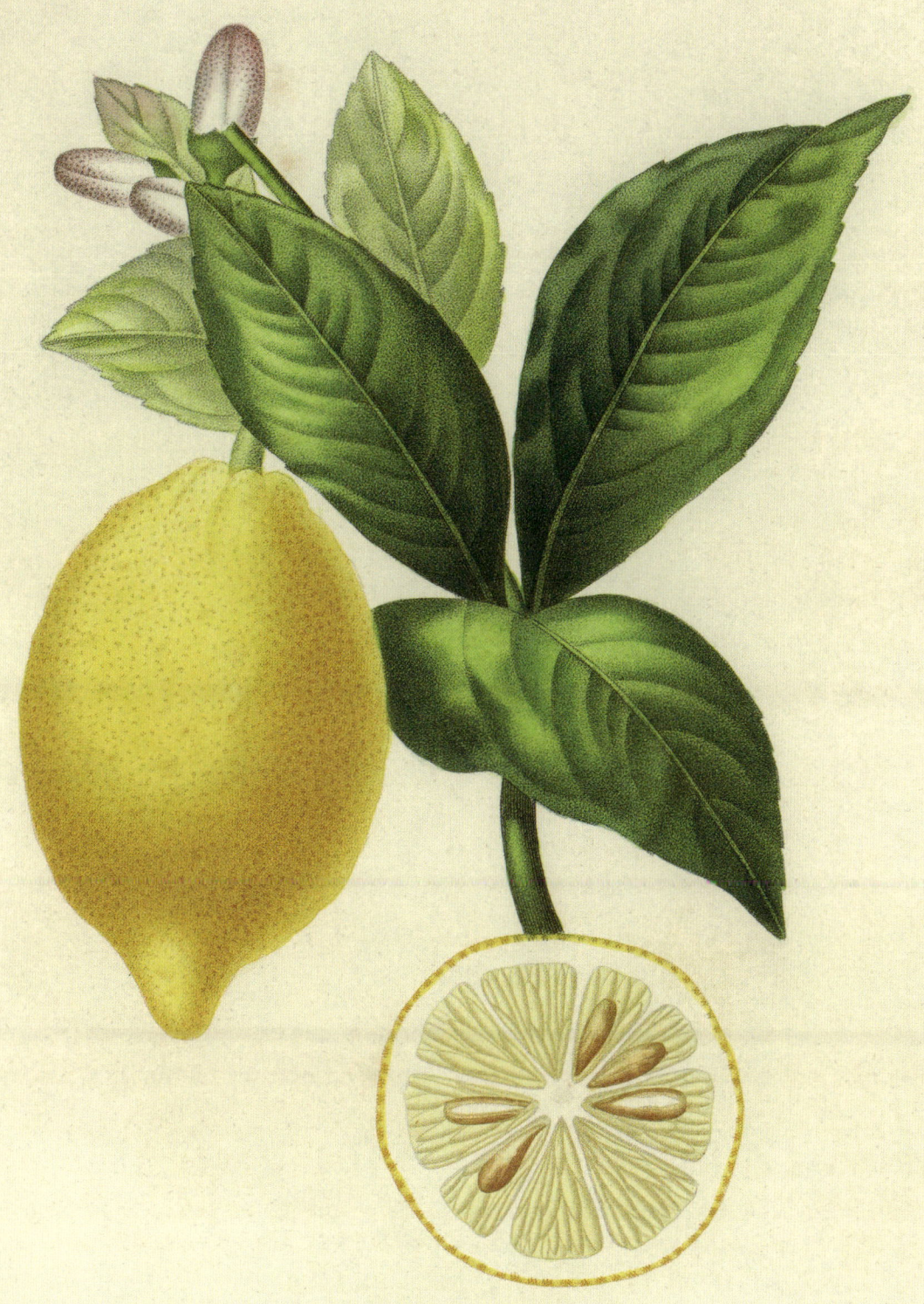

LIMONIER ORDINAIRE.

Limone ordinario.

Poiteau pinx.t　　　Tab. 84.　　　*Gabriel sculp.t*

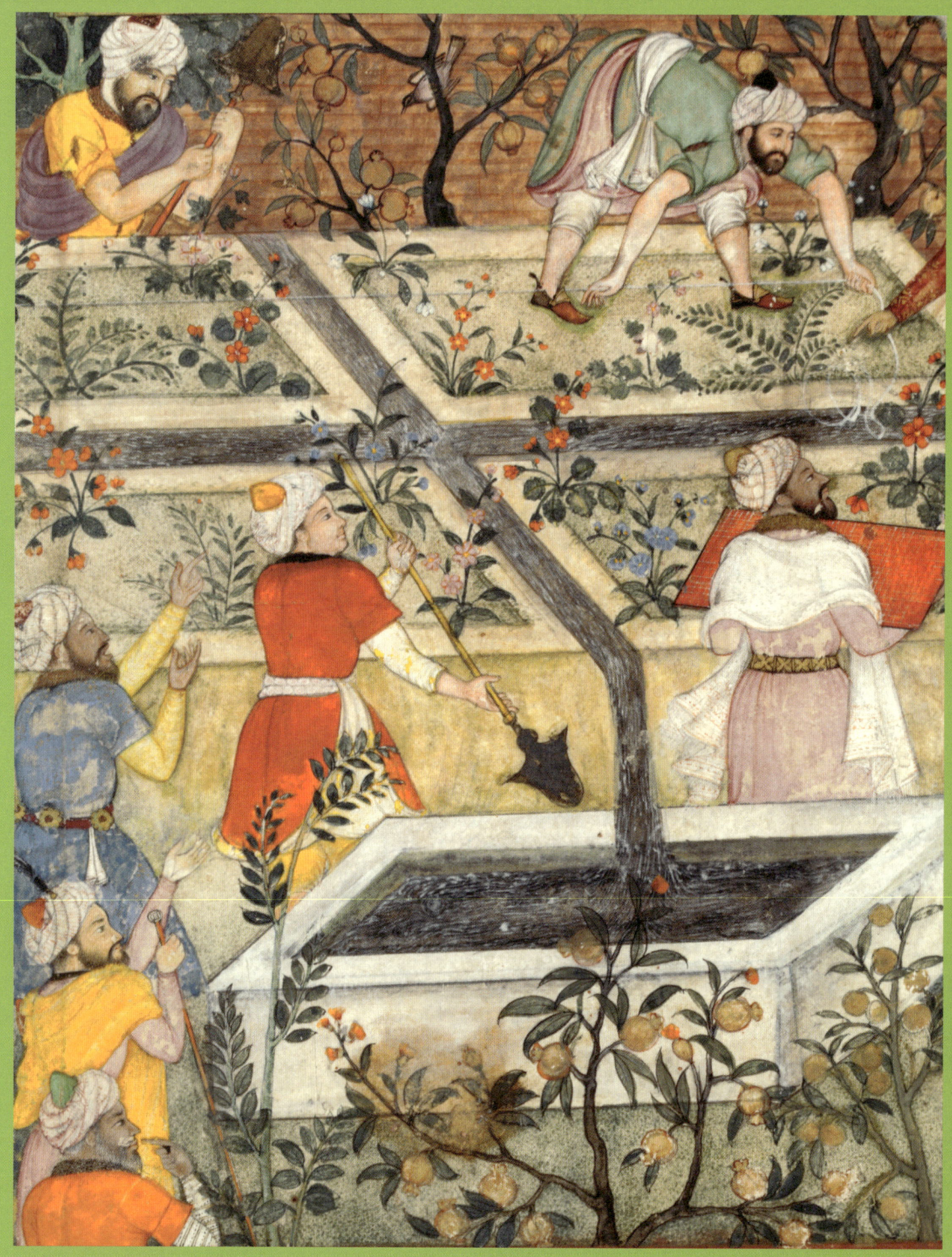

TWO

Herbals to Hesperides

THROUGH STRATEGIC MARRIAGES and urban networks the Medici stealthily eclipsed the other powerful families in Florence and rose to dominance.[1] Four Medici became popes, while two were queens of France, and the family eventually assumed the Duchy of Florence and, in 1569, the Grand Duchy of Tuscany. With wealth initially from textiles, they ran the Medici Bank, the largest in Europe; they were prominent not only as statesmen and bankers, but also as patrons of science and the arts – including horticulture.

When a coat of arms was devised for the founder of the bank, Giovanni de Bicci de' Medici (1360–1429), the principal feature was eight orange-red spheres, whose origin and meaning have long intrigued historians (see pp. 78–79). One interpretation is that they represent citrus fruits.[2] These fruits, later reduced to five, were to become the permanent principal motif of the arms. The origin of the name Medici is unknown, though *medici* is the plural of *medicus*, a doctor; *medici* was also the name for citrus fruits in medieval and Renaissance Italy.[3] Whether or not there is a link to oranges, with the arms some kind of visual pun, it is unquestionable that the family was later to embrace citrus-growing as an art form in the design of their magnificent gardens. But where had these citrus come from?

ISLAMIC SCHOLARSHIP

Although citrons were grown and lemons (crosses between oranges and citrons) as well as oranges (crosses between pomelos and mandarins) apparently known in the Classical world, all were spread across the Mediterranean with Islam (as were cotton and sugar cane). Though the significance of the citron to Jewish people may have been important in the spread of citrus in general even before that (see p. 50).[4]

The Middle East had long been in contact with China through trade along the Silk Roads, but, by the emergence of Islam in the seventh century CE, the maritime route became critical in the exchange of goods and ideas, with Muslims setting up trading posts and other settlements in Chinese ports. By 878 they were among some 100,000 foreign merchants in China.[5]

In *c.* 851, Sulaiman al-Tajir, a trader from Siraf (today's Bandar-e Siraf, Iran) in the Persian Gulf, returned from what is now Guangzhou in China and recorded the importance of citrus fruits in the East –

as well as describing tea-drinking, porcelain-making and the Chinese use of fingerprint 'signatures' for immigrants.[6] Later, Chinese ships regularly visited Siraf. They 'brought thither all the commodities the East could furnish'[7] and could well have introduced different kinds of citrus to the Middle East – if the Arabs had not already done so.

The Arab world was soon experienced at growing, including grafting, citrus trees. In 904 CE, the writer on agronomy, poisons and alchemy Ibn Wahshiyya (born in what is now Iraq, d. 930) finished his most influential work, the *Kitab al-Filaha al-Nabatiyya* (Book of Nabataean Agriculture).[8] Written down and edited by one of his students around the time of his death, it was the first book on agriculture in Arabic, though it was unknown to European scholars until 1835.

Ibn Wahshiyya claimed his work was based on 1,500 leaves of Chaldean parchment, some 20,000 years old, but this fanciful attribution was likely merely in support of a movement to extol pre-Arabic Mesopotamian culture, which the Iraqi Nabataeans wanted to have acknowledged in the face of its usurpers.[9] Recent scholarship has concluded that it was actually derived from Syriac sources of the fifth or sixth centuries, those in turn ultimately from Greek and Latin texts, but elaborated in a local context. They included details of grafting citrons (sometimes fancifully with olives and pears), but also the belief that oranges arose from them. Copies of the book were made and distributed across the Arab world, so that such beliefs were then repeated in other works, including agricultural texts prepared in Andalusia.

During the rule of 'Abd al-Rahman III (891–961), caliph of Cordoba, Dioscorides' *De materia medica* (see p. 47) was translated into Arabic; in 1518 in Toledo it was translated into Latin. The next year, far to the east in the Muslim domains, Babur (1483–1530), great-great-great-grandson of Timur (Tamerlane), wrote *Baburnama*, his memoirs, covering not only his political success in Samarkand and the Fergana Valley of what is now Uzbekistan, then his move to Kabul and finally northern India, but also much cultural, geographical and biological information, including

---

Illustration from *Farah's Encyclopaedia of Nature* by al-Muttahhar ibn Muhammad al-Yazdi, 17th century. Arabic and Persian works such as this one furthered the dissemination of citrus knowledge.

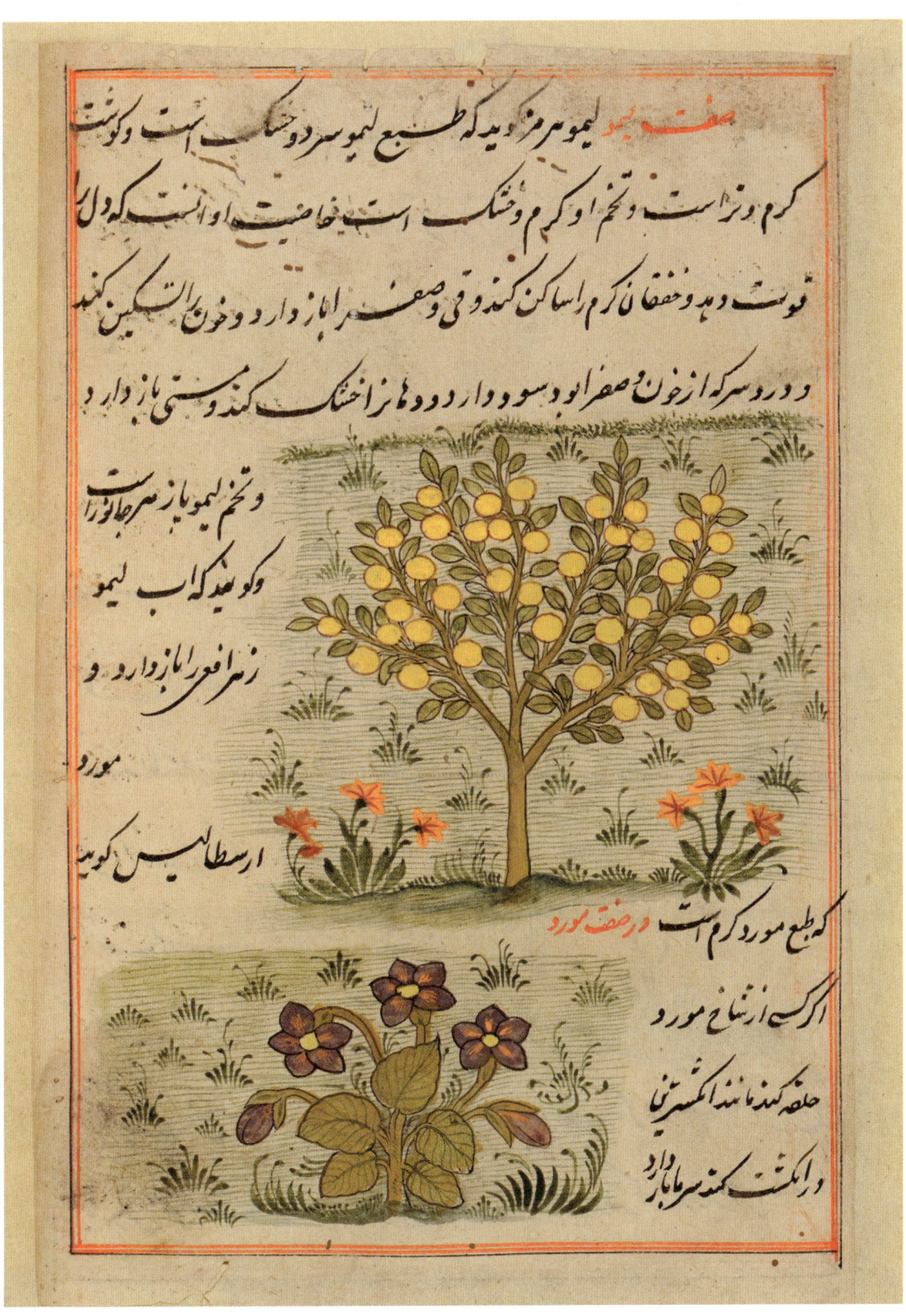

فاذا زاد العصير نصفه فهذا الشراب موافق لوجع الحلق والجنب والرئتين
والاسر والنفث ولزته لغم غليظ في حلقه يصفي اللون وكثر النوم

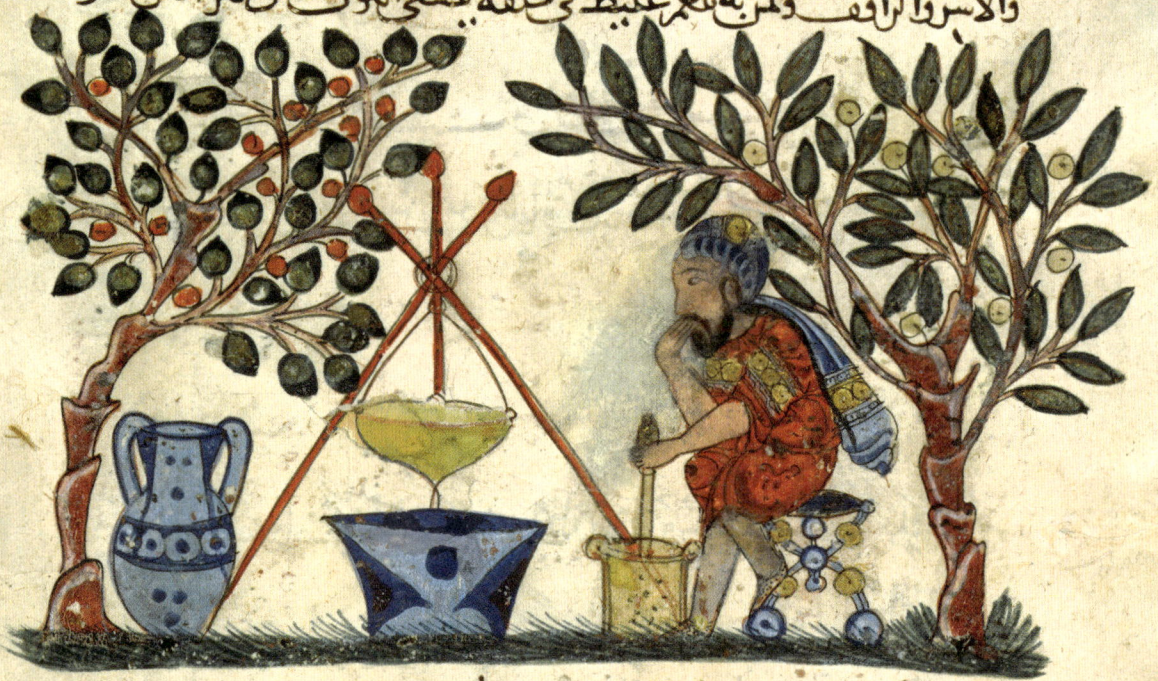

وليس له غائله موافق للمثانه والكلا ٤ ٤ ٤

صنعه شراب للزكام والسعال ::

وزم البطن واسترخا المعده خذ من ربع اوقيه واصول سوس ثن اوقيه
وفلفل ابيض بع ثن اوقيه دقه جميعا واربطه خرقه واجعله في مثله فناط شراب
طيب واتركه ثلثه ايام ثم صفه وارفعه في اناء نطيف اشربه منه بعد العشا

Those which are called Median, Persian, or cedromela and in the Latin citria, are known to all … The fruit itself is somewhat long, wrinkled, resembling gold in colour, smelling sweet with heaviness, with seed similar to a pear.

DIOSCORIDES, *DE MATERIA MEDICA*, 1ST CENTURY CE

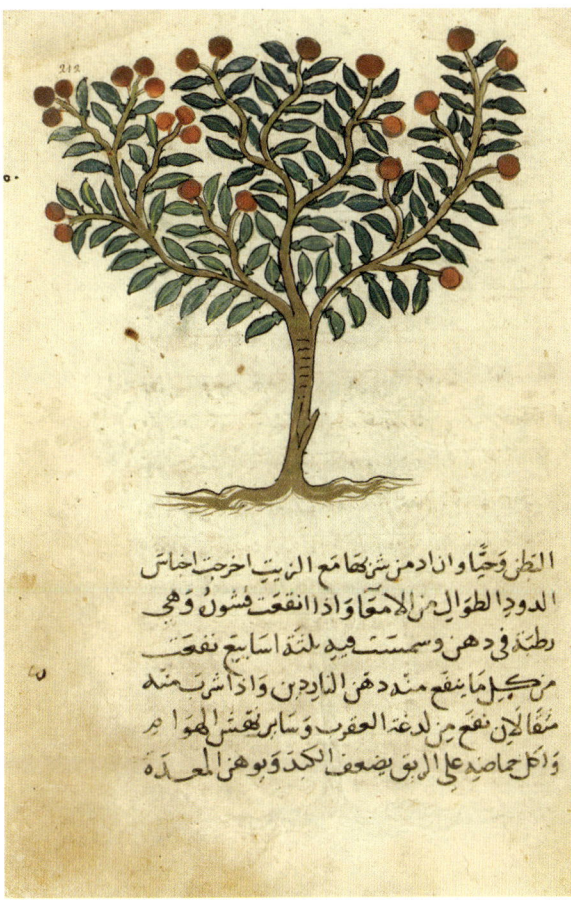

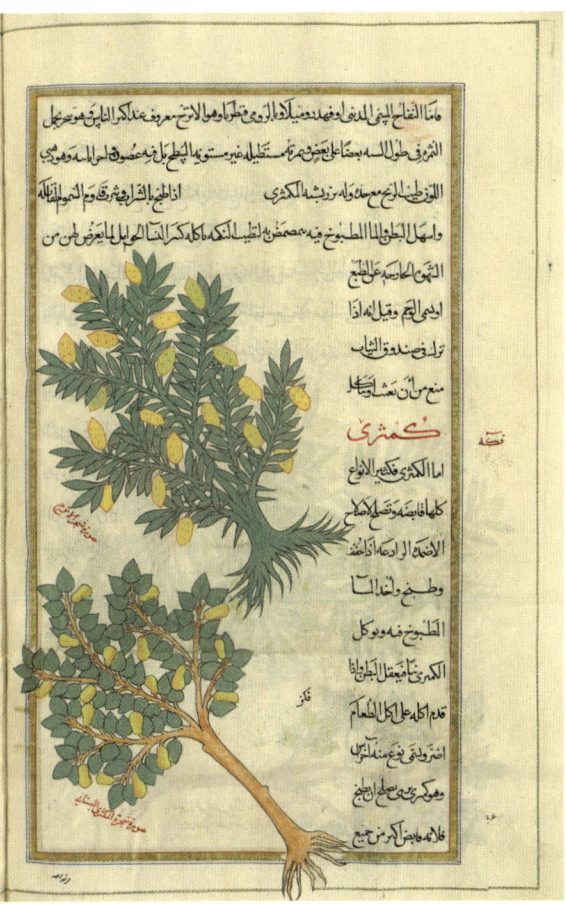

OPPOSITE: Physician preparing an elixir, from an Arabic edition of *De materia medica* by Dioscorides, 1224 CE. **ABOVE, LEFT:** Sour orange tree, from *Masalik al-absar* by Ibn Fadlallah al-Umari, 14th century. **ABOVE, RIGHT:** Citron (top), from an Arabic edition of *De materia medica* by Dioscorides, 9th century CE (translation), 1889–90 (illustrations).

Dioscorides was a Greek physician and botanist, whose encyclopaedia on herbal medicine was written around 50–70 CE and drew on trees and other plants found in the eastern Mediterranean. His text was among the most influential European pharmacological sources of knowledge. He also had a profound influence on medicine in the Islamic world: his *De materia medica* was disseminated in lavishly illustrated Arabic editions as well as Latin and Greek throughout the medieval period, its popularity continuing in later French, Italian, German and Spanish translations. Scholars such as Ibn Fadlallah al-Umari drew on Dioscorides as sources for their own works (above, left). The widening of global trade led to the spread of citrus plants, which was accompanied and perpetuated by the exchange of knowledge.

Nature: the pulp is cold and humid in the third degree, the peel is dry and warm in the second. Best kind: those that are perfectly ripe. Benefit: the candied peel soothes the stomach. Harm: difficult to digest. To remedy the harm [take] with the very best wine.

ENTRY ON SOUR ORANGES IN IBN BUTLAN, *TACUINUM SANITATIS*, 11TH CENTURY

Illustrations from European editions of *Tacuinum sanitatis* by Ibn Butlan, 11th century.
Top row: possibly from Verona, 1380–99; middle row: 1445–51; bottom row: Venice, *c.* 1480.

Now better known by its Latinized name, the *Taqwim al-Sihha* (Maintenance of Health) was compiled by the 11th-century doctor Ibn Butlan of Baghdad. It was translated into Latin and became very popular all over Europe throughout the Middle Ages, as much for its lavish illuminations as its text. The treatise describes the elements of good health and provides remedies. An array of fruits, vegetables and other plants are described and listed alongside their positive and negative effects. Three kinds of citrus are mentioned: citron (*nabach i cedrum*, left); sour orange (*cetrona*, centre); and lemon (*citra*, right).

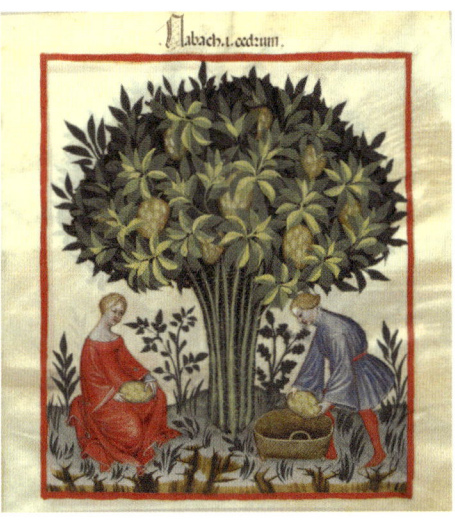

Nabach i cedrum.

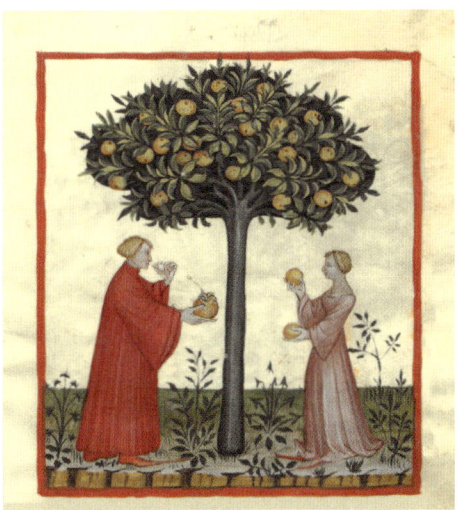

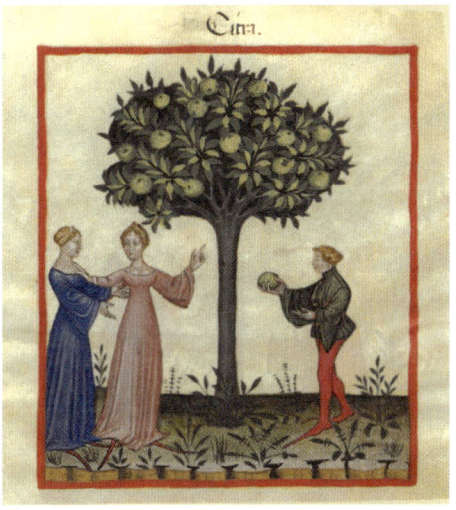

Cira.

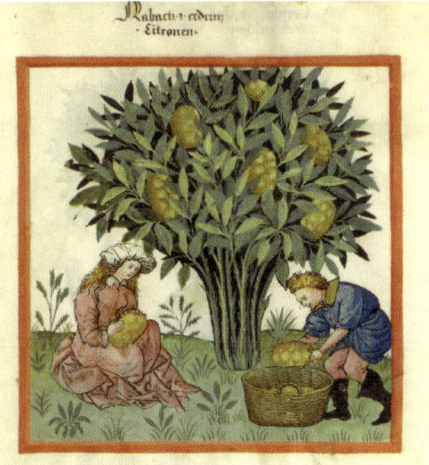

Nabach i cedrum
Citronen.

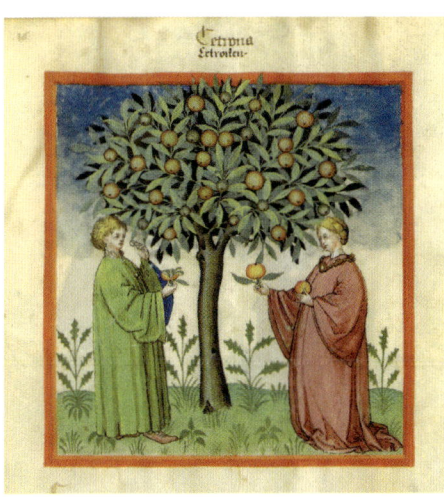

Cetrona
Citronen.

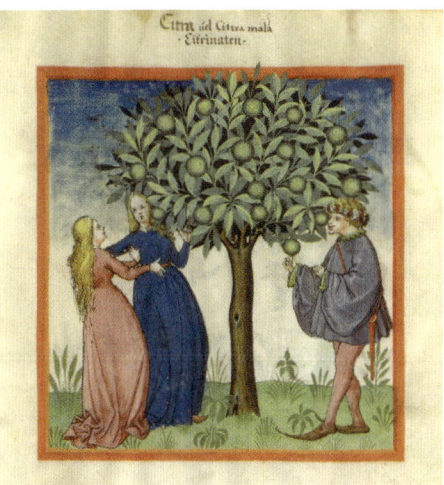

Citri siel Citrea malá
Citrinaten.

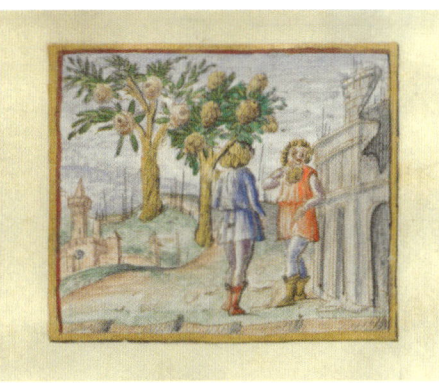

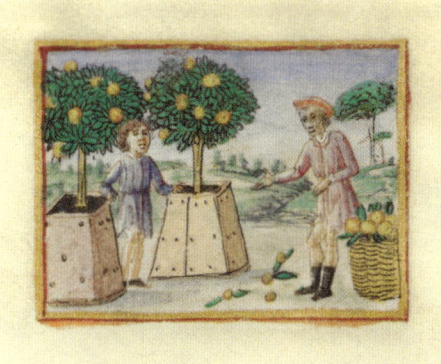

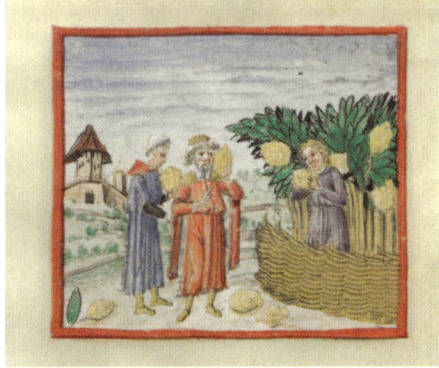

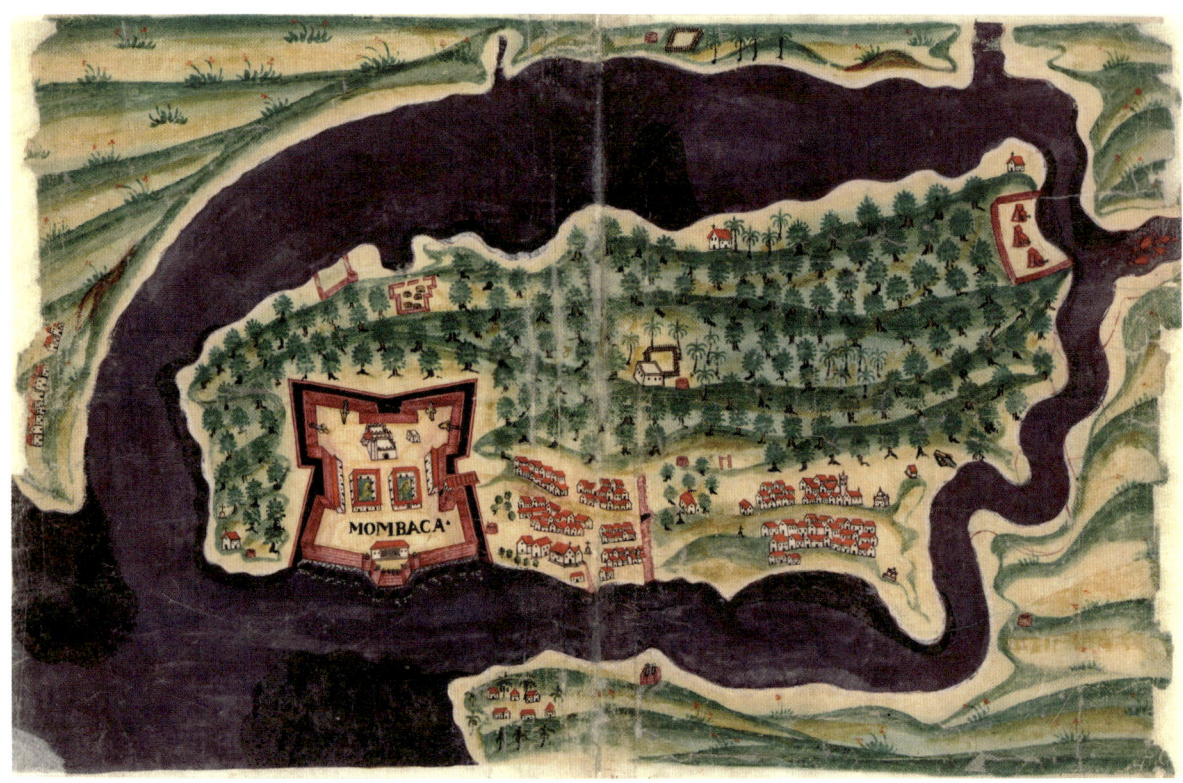

on sweet and sour oranges and limes and citrons.[10] Babur's descendants were the Mughal dynasty, which ruled in Delhi until the beginning of the British colonial era in 1858.

Ibn al-Awwam (d. 1158), an Arab landowner and writer, described citrus-growing in his book on agriculture, a major compilation written in Seville, Spain, in the second half of the twelfth century.[11] His work cited 112 earlier authors and drew especially on that of Ibn Wahshiyya. It dealt not only with the citron, the sour orange and the lemon, but also quoted the work (*c.* 1070–75)[12] of another Iberian scholar, Abu I-Khayr al-Ishbili, 'the tree-planter', in discussing the pomelo, then named *zamboa*.[13]

ORANGES (*CITRUS × AURANTIUM*) AND LEMONS (*CITRUS × LIMON*)

According to al-Mas'udi's *Golden Meadows* of *c.* 943, the sour orange reached Oman (via India) after 912,[14] and it was spread from there to Iraq and Syria, and then to Palestine and Egypt and across the Maghreb. Orange and lemon trees, probably from North Africa, reached Sicily with the Arab invasion of 831 and were established in Moorish gardens there. In

Map of Mombasa, Kenya, *Livro das Plantas de todas as fortalezas* by António Bocarro, 1635. Although the Portuguese were not the first to introduce sweet oranges to Europe, they saw them in Mombasa in 1498. The fruits later became known as 'Portugal' oranges in Europe; they are still known as *portocale* in Romania and *portokali* in Greece.

Al-Andalus (what is now Spain), they were planted in the Alhambra, and thus the forebears of the familiar sour Seville oranges became established.

The Arabic name *naranj*, from which 'orange' derives, is cognate with the Persian *narang* and Sanksrit *na ranga*. *Na ranga* is probably derived from a Dravidic or Tamil root meaning fragrant, no doubt originally referring to the citron (see p. 37).

The sweet orange (the group of *Citrus × aurantium* cultivars closer to their parental, sweet mandarin than to their other parent, the sour pomelo) seems to have reached the Arab world from China and India via Oman *c.* 912.[15] By 1470, at the latest, such oranges were probably in Europe, due to Genoese contacts in the Levant.[16] A bill of sale for 15,000 *citranguli* issued in 1472 has been held to show that sweet oranges were by then well known in Italy, though that term was used for oranges in general.

By 1566, oranges were widely grown at Hyères on the French Riviera, though plants are reported as having been available in Nice as early as 1336.[17] Although possibly introduced there by Crusaders, a likely source was nearby Liguria where, by the middle of the thirteenth century, the sour orange, the lemon and the 'lima', besides the citron, were well established. Liguria became the centre for supplying citrons for the Jewish festival of Sukkot, as well as for the production of essences, citric acid, lemons and sweet oranges. Sweet oranges were mentioned by many early-sixteenth-century authors,[18] including historian Leandro Alberti (1479–1552), who was in Italy in 1528, writing of Salerno, 'We see there citrons, lemons and orange trees of all the species. Some have sweet, some have sour fruit, and, finally, others, producing fruits of a medium taste.'[19]

Traditionally, the introduction of sweet cultivars to Europe has been associated with Portuguese trade following the travels of Vasco da Gama (c. 1460–1524). When the Portuguese reached the east African coast via the Cape of Good Hope in 1498, da Gama described the Kenyan port of Mombasa as 'a city of great trade with many ships ... a large boat laden with fowls, sheep, sugar canes, citrons, lemons and large sweet oranges, the best that had ever been seen'.[20] They likely also saw the rough or Mazoe lemon (named after a central African river, now Mazowe), *Citrus × otaitensis*, the apparently Chinese cross between the citron and the mandarin.

The sweet orange thus became known as the 'Portugal orange' (in Arabic and Persian languages *portugal* means oranges generally), though, by the middle of the seventeenth century, better cultivars of 'China orange' were being eaten in Europe. Those from China were so popular that by 1700 they were not only being exported but also grown all over the Malay Archipelago.[21] In England, these oranges were sold to theatre-goers by Eleanor 'Nell' Gwyn (c. 1650–1687), who later became an actress, and then the mistress of Charles II.

From the beginning of the tenth century, Arabic sources were also referring to the lemon,[22] calling it the *limuwa* or *laymun*, names said to have been derived from the Arabic word for lime, *lim*, itself derived from a Malay name for citrus. No doubt via Arab trade with China and then African peoples, lemons reached deep into central Africa long before Europeans arrived there.

The first Arabic mention of true lime (*Citrus × aurantiifolia*) is some time in the eleventh to twelfth centuries. The lime (see p. 155) was not known or planted in the West until the strengthening of seventeenth-century trade with the Malay Archipelago. Today *limon* is the Spanish name and *limão* the Portuguese one for the lemon, *lima* in both being the word for the lime.

Early in the nineteenth century, the Italian botanist Giorgio Gallesio (1772–1839)[23] wrote that the lemon was originally considered a medicinal plant, also used as a seasoning, later becoming the basis of *limonade*, which reached Italy in the fourteenth century and then France in the seventeenth century. The word lemonade has not been recorded in English before 1663, though *lemonado*, the Spanish word, was in English usage as early as 1640.[24] In Paris, shops for the sale of *limonade* were established, whose proprietors were known as *limonadiers*, who united as a body of tradesmen in 1678.

The title *limonadier* was later applied to owners of Paris cafés in general, perhaps the most famous of them being *La Muse limonadière*, Charlotte-Jacquéline Reynier Bourette (1714–1784),[25] daughter and wife (twice) of *limonadiers*. A poet and playwright, she ran her Paris café L'Allemand (later Café des Muses) in rue Croix-des-Petits-Champs. It became a noted meeting place for literary figures, including Voltaire, who exchanged letters in verse with her.

CRUSADES AND CONQUEST

The Crusades, beginning towards the end of the eleventh century, brought Arab civilization into collision with European, so that only then did the sour orange and the lemon, but also the lime, appear in European literature as the Crusaders brought them back to Italy and France. The first record of Englishmen encountering citrus seems to be in the camp of Richard the Lionheart in the winter of 1191–92, when his crusaders 'refreshed themselves with the luscious fruits' in the orange groves of Jaffa in Palestine.[26]

According to Gallesio,[27] the French theologian, crusader and chronicler Jacques de Vitry (c. 1170–1240) recorded 'the Adam's apple [likely a kind of lemon], the lemon, the citron, and the bigarade [sour orange] found in Palestine by the Crusaders and regarded as new trees foreign to Europe.' The Flemish friar Thomas de Cantimpré (1201–1270) likely encountered Vitry in Liège during his education,

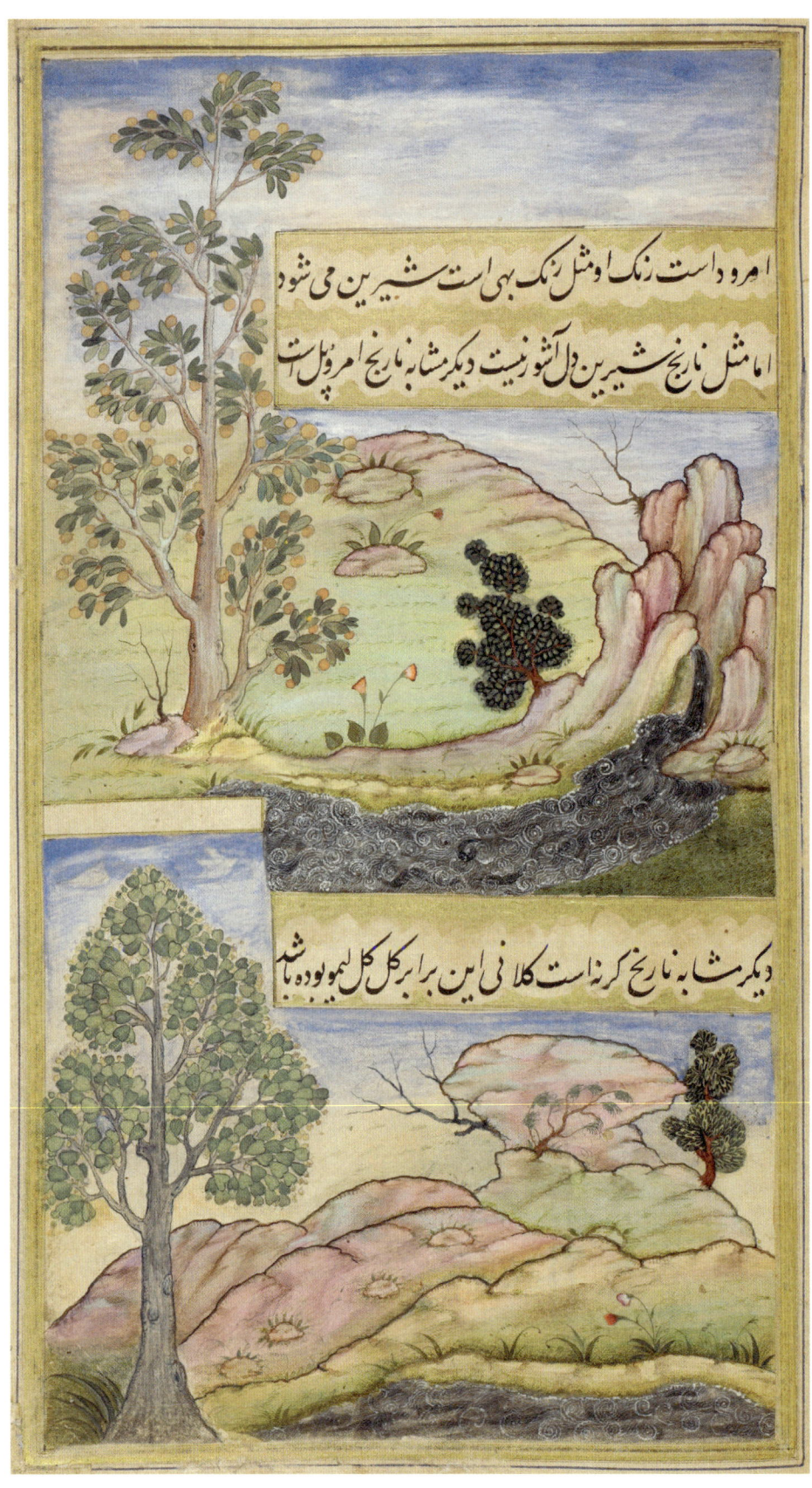

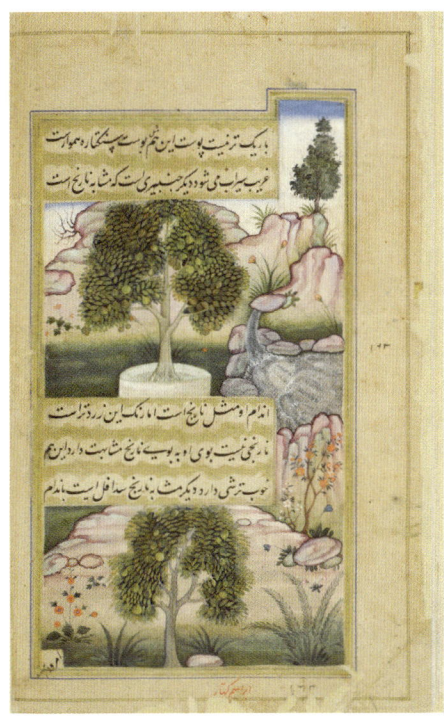
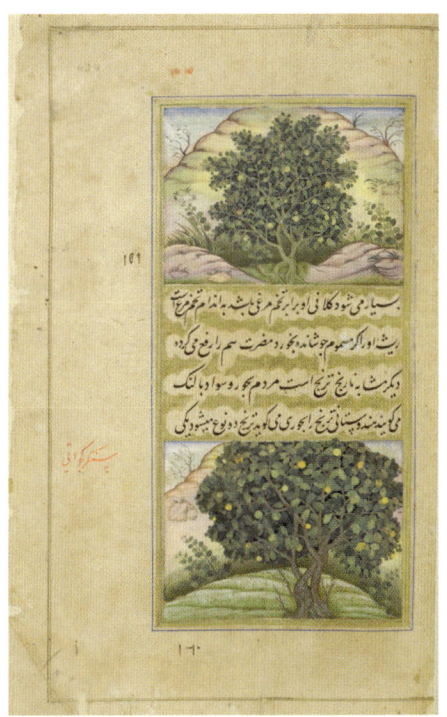
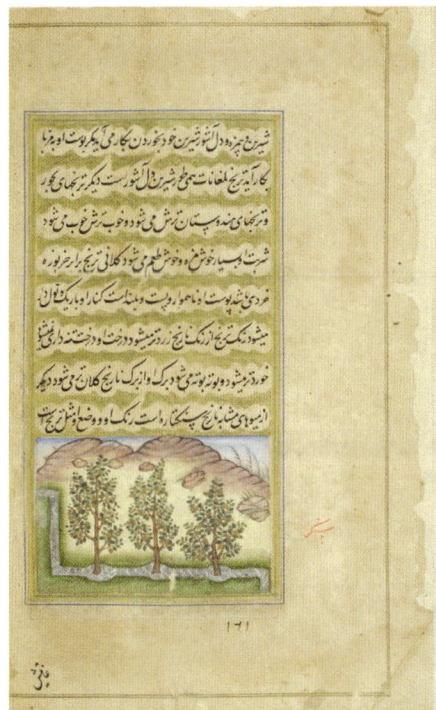
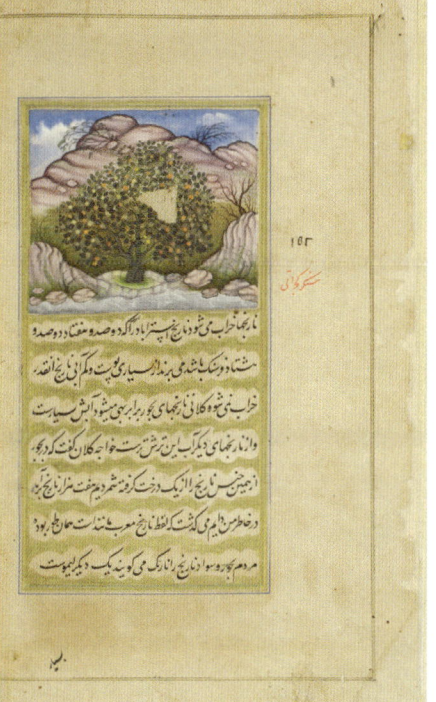

Citrus trees in a Persian translation of the *Baburnama*, *c.* 1590.

Originally written in Chagatai Turkic, the Mughal emperor Babur's memoirs were translated into Persian and survive in several illustrated copies. He recorded his military victories but also vivid observations on history, geography, astronomy – and fauna and flora. Citrus illustrated here include both sweet and sour oranges, limes and lemons. In the bottom left folio above sweet orange trees are depicted in an irrigation channel.

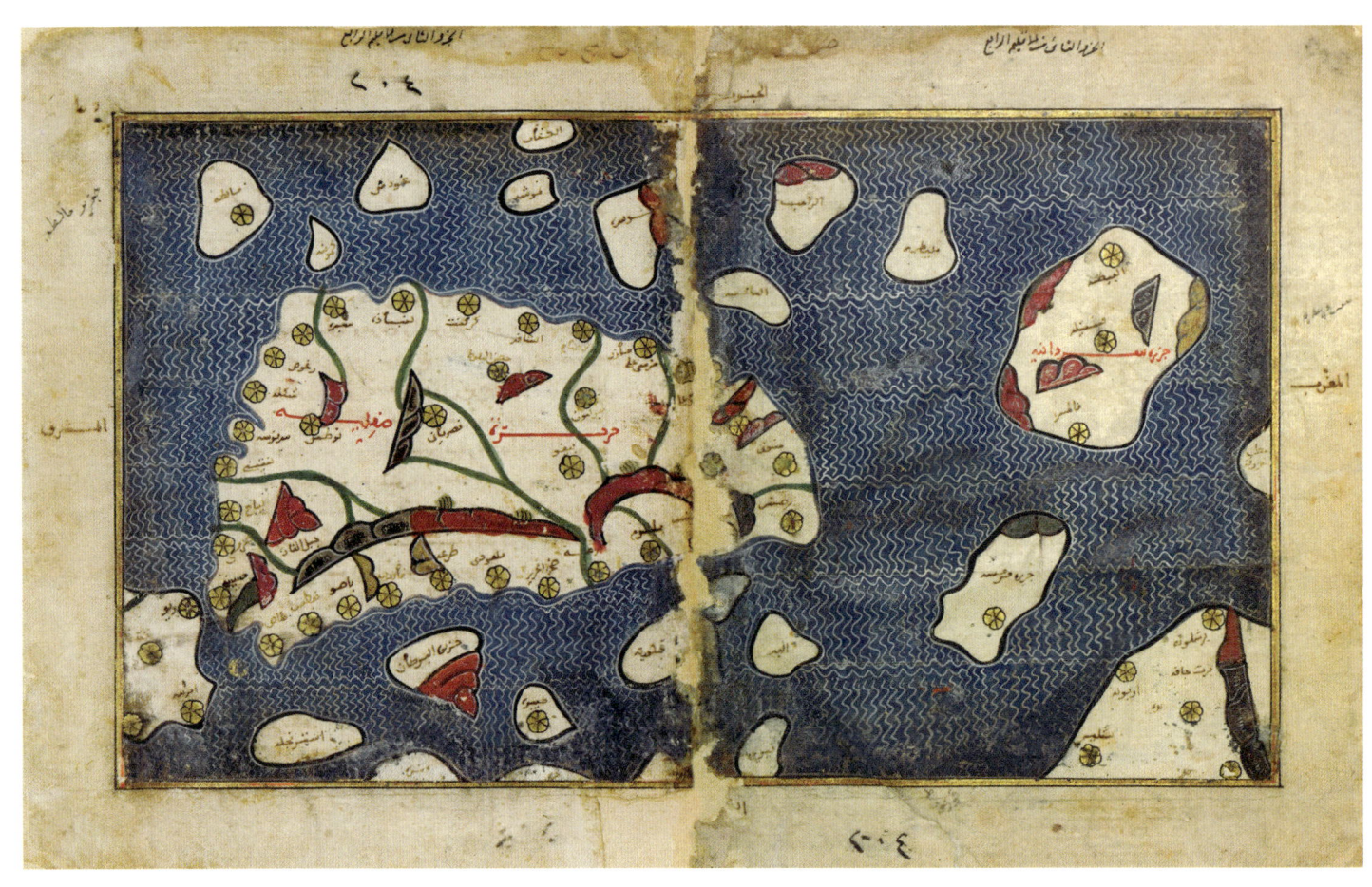

Map of Sicily (above) and world map (opposite) from *c.* 1325 and 1553 editions of the *Tabula Rogeriana* by Muhammad ibn Muhammad al-Idrisi, 1154.

Fifteen years in the making, the full title of this impressive geographical treatise is *Nuzhat al-mushtaq fi ikhtiraq al-afaq* (literally, 'the excursion of one eager to penetrate the distant horizons'). However, it is usually referred to as the 'Book of Roger' after Roger II of Sicily, in whose court it was produced. To accompany his maps, al-Idrisi included descriptions of the flora, fauna, culture and politics of each region. Though Sicily had been conquered by the Normans, the practices and scholarship established under Arab rule were continued – including the cultivation of citrus. As a meeting point between cultures, facilitating trade and the exchange of ideas, Sicily occupies an important position in the history of citrus.

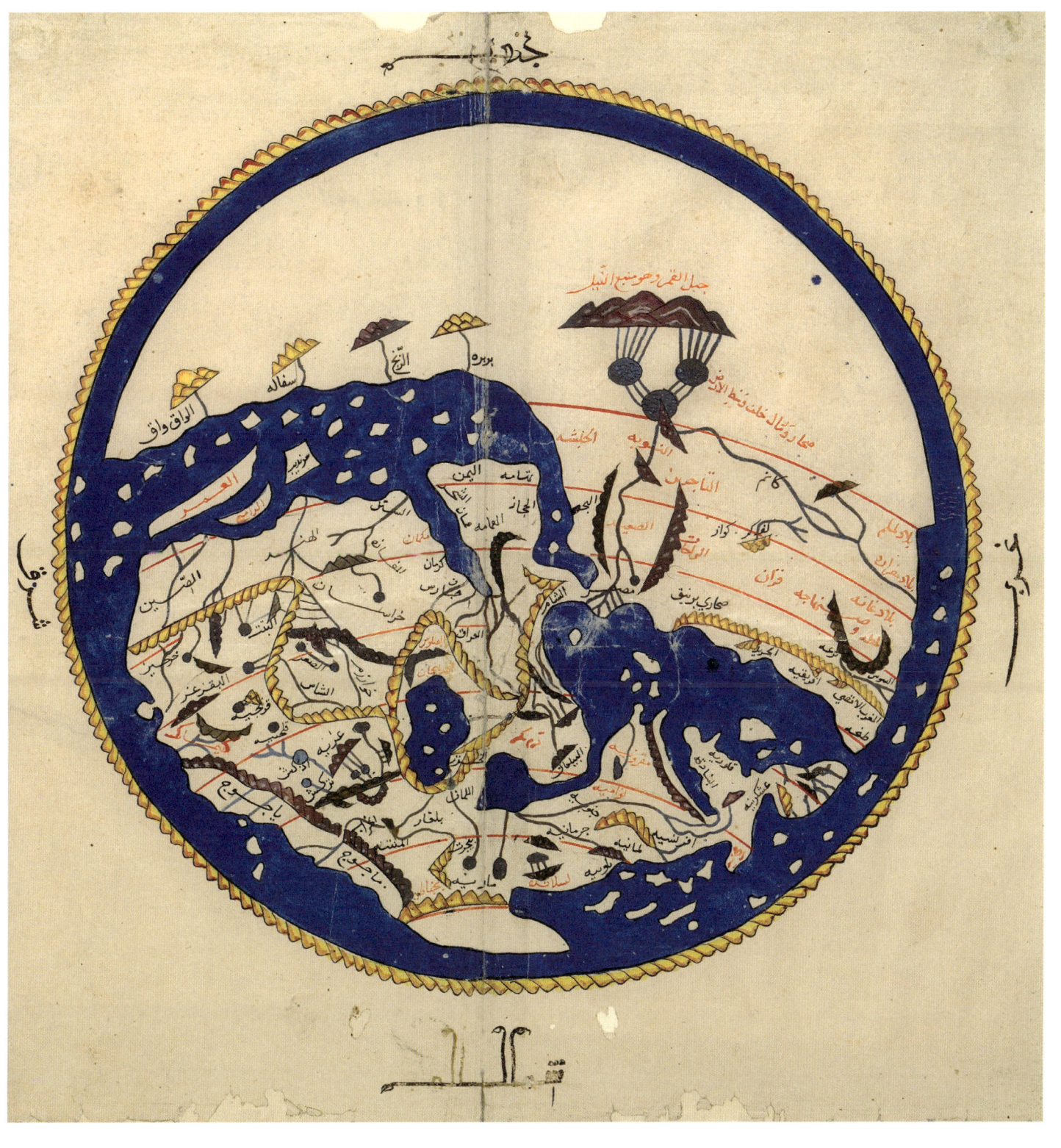

and in his *De natura rerum*, an encyclopaedia of natural history finished at Leeuven in 1244, Thomas wrote of both the 'Adam's apple' and the citron.

In 1060, the Norman Count Roger de Hauteville (1031–1101) and his army of knights invaded Sicily; by 1091 the island was completely conquered. The invaders continued the practices of the usurped Arabs, not only in agronomy but also in science generally. Among the court of Roger II of Sicily (1095–1154) in Palermo was the Muslim geographer and mapmaker Muhammad al-Idrisi (1099–1165/6), who was descended from Hasan ibn Ali, the grandson of the prophet Muhammad.

In 1154, after some fifteen years' work, al-Idrisi completed a comprehensive geographical treatise written in Arabic and Latin for the king, who died just a few weeks later. Its full title is *Nuzhat al-mushtaq fi ikhtiraq al-afaq* (literally, 'the excursion of one eager to penetrate the distant horizons'), but the book is usually known as the *Tabula Rogeriana* (Book of Roger). It is essentially a biogeographical account of the known world, divided into seven climatic zones covering plants (including citrus), animals, agriculture and much anthropological information, compiled from both personal experience and interviews with other travellers. Associated with the text are seventy longitudinal section maps, the whole work based on the premise that the world is a sphere. Astonishingly, al-Idrisi calculated the circumference of the Earth to be 37,000 km (23,000 miles) – it is actually just over 40,000 km (25,000 miles).

By the thirteenth century, citrus was being exported to Egypt from the Crusader states in the Middle East.[28] The wife of Edward I of England, Eleanor of Castile (1246–1290), who went with her husband on a Crusade between 1270 and 1273, imported oranges to England, as well as olive oil and Spanish customs like furnishing walls with tapestries and using floor-rushes. The first full Latin description of citrus was that given by the German philosopher, scientist and bishop Albertus Magnus (1193–1280) in his *Libre de vegetabilibus* (*c.* 1260),[29] published under the Latinized name of Arangus.

As the Christians moved into the southern parts of Muslim-occupied Iberia, the Arabic *naranj* became *aranja*, today's *naranja* in Spanish and *laranga* in Portuguese. However, the original meaning, referring to the citron, is retained as *naroinja* in Catalonia and *aranja* in the Balearic Islands, the name for the sour orange there being *taronja*.

RENAISSANCE PAINTING

During the rise in mainland Italy of a naturalistic painting style, one great exponent, the highly influential Duccio (*c.* 1260–1319) of Siena, often set his rather Byzantine-style biblical figures in architectural scenes resembling the Tuscan hills. However, he and his followers added elements that were then thought of as archetypically Palestinian, notably orange trees despite the fact that these did not actually grow in the land where Jesus Christ lived and taught. The 'lollipop' trees painted in the *Triumph of Death*, a fresco of *c.* 1335–40 by Francesco Traini (*fl.* 1321–1363) in Camposanto, Pisa, could perhaps be identified as such, and they seemed to become a staple of biblical scenes. Andrea Mantegna's (*c.* 1431–1506) *Agony in the Garden* (1457–59), now in the Musée des Beaux-arts in Tours, France, depicts citrus, and the 1460 version in the National Gallery, London, features an orange tree complete with its characteristic winged petioles.

Citrus featured increasingly prominently in pictures of the Virgin Mary. The Virgin's throne in *Madonna and Child in a Garden*, painted *c.* 1460–70 by Cosmè Tura (1433–1490), has orange trees around it; while an orange tree dominates the background of *Madonna of the Orange Tree*, painted *c.* 1496–98 by the Venetian artist Giovanni Battista Cima da Conegliano (*c.* 1459–*c.* 1517). Under a canopy of fruiting orange twigs, an angel offers the Christ child an orange in *Madonna and Child with an Angel* (*c.* 1510–20) by Giovanni Agostino da Lodi (*fl.* 1467–*c.* 1525). In the late Renaissance, the architectural centre of Paolo Veronese's (1528–1588) *Annunciation* (1578) is flanked by orange trees in tubs. The orange tree seems to have become associated with purity and virginity.

After the Renaissance, the chaplet of roses worn up until then by girls getting married was replaced with orange blossom.[30] At the wedding of Queen Victoria of England (1819–1901) to Prince Albert in 1840, the bride wore a satin dress trimmed with orange blossoms, as well as a wreath of them, signifying chastity.[31] By the end of the century, in books on the 'language of flowers', orange blossom symbolized 'the virginity of those who have pledged themselves in love'.[32] On the Aegean island of Chios, when marriages take place at Carnival time, orange-blossom crowns are worn by both bride and groom.

Nonetheless, the Virgin Mary did not have a monopoly. For example, *Saint Sebastian* by Dosso

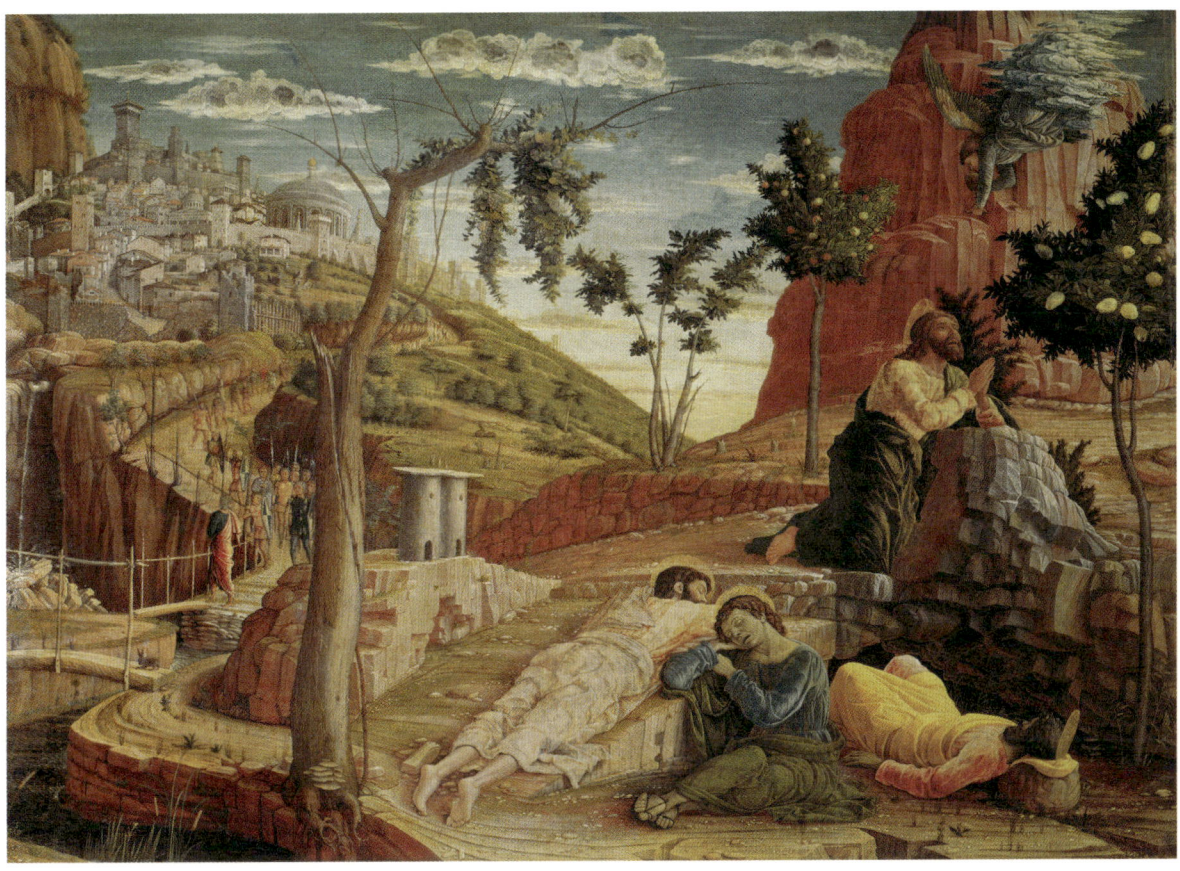

Agony in the Garden, Andrea Mantegna, 1457–59. Citrus trees became a recurring feature in biblical paintings during the Renaissance as they were then thought typical of the Holy Lands. Here the characteristic leaf-shape of citrus is clearly visible on the orange tree in the background.

Dossi (*c.* 1489–1542) has the martyr tied up, in an apparent state of ecstasy, beneath an orange tree, while *The Holy Family* of the Flemish Joos van Cleve (1485–1540/1541) has the Christ child grasping an orange fruit, interpreted by some commentators as reflecting the bittersweet nature of love. Citrus appears in the magnificent, multi-panelled Ghent Altarpiece (*Adoration of the Mystic Lamb*) painted by Jan van Eyck (*c.* 1380/1390–1441) and designed by his brother Hubert *c.* 1422. In the lower tier of panels, Van Eyck painted citrus trees and date palms in his rendition of New Jerusalem, but, most remarkably, the flanking panel to the right shows Eve holding the forbidden fruit: clearly not an apple but a Florentine citron (in fact a kind of lemon).

Citrus fruits became almost indispensable props in depictions of the Last Supper in the Italian Renaissance.[33] The sketches Juan de Juanes (1523–1579) made for his version, now in the Prado in Madrid, included an orange (no. 612) and two half-oranges (no. 99), but the finished picture just one half. The Last Supper painting that was recognized

in 2021 as coming from Titian's workshop has an excellent depiction of a citron on the table in front of the sleeping John the Apostle, while Jacopo Bassano's (*c.* 1510–1592) *Last Supper* of 1542 places an orange on the table at the centre of the painting.

The symbolism, if any, is unclear, unless alluding to the notion of bittersweet love. However, the Last Supper is now believed to have been the ritual feast of Passover Seder, which features particular foods symbolic of Jewish history and to which, coincidentally, oranges were added in recognition of women and gay people and other marginalized groups in the 1980s, the orange symbolizing the fruitfulness of inclusion.[34]

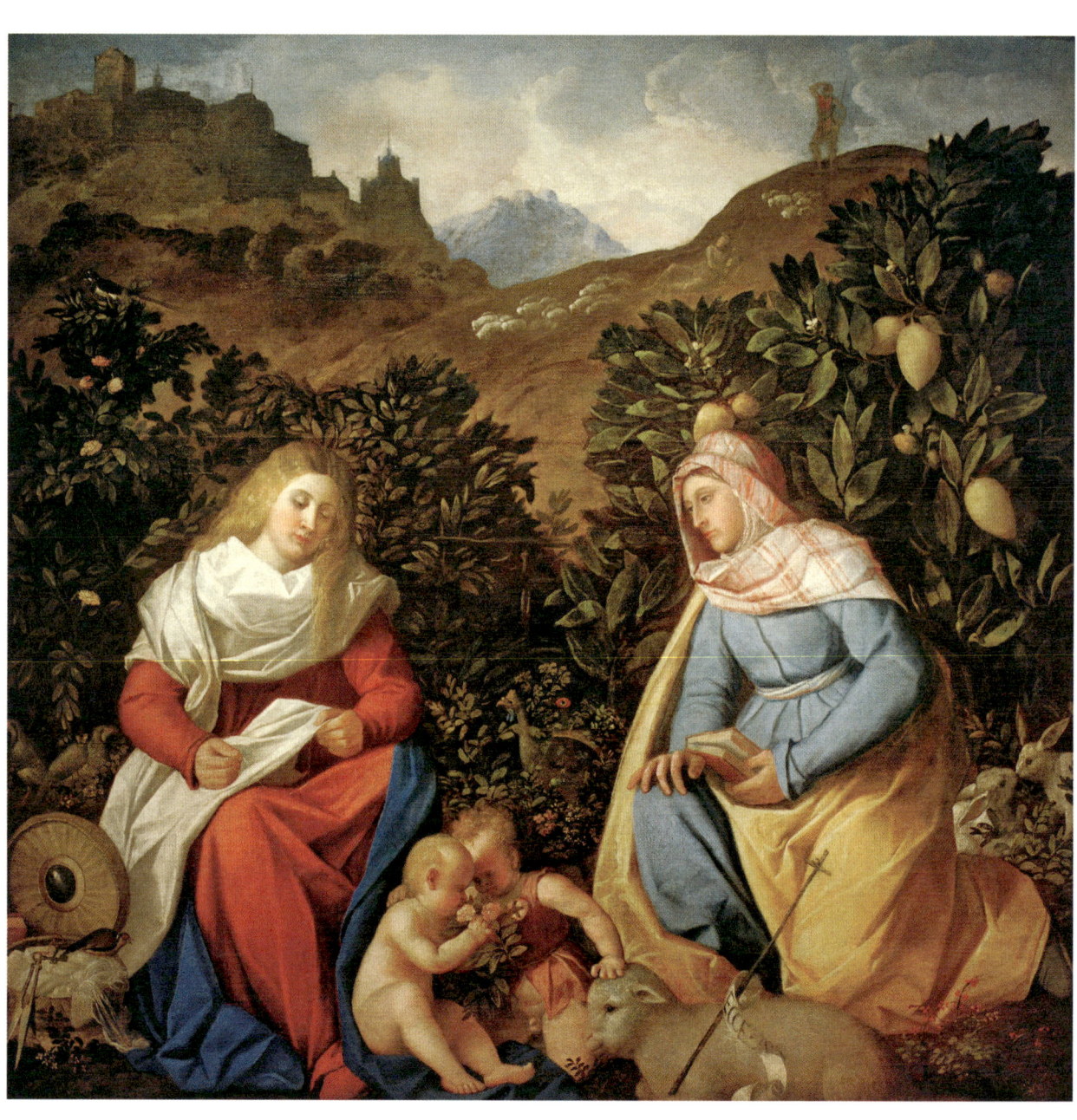

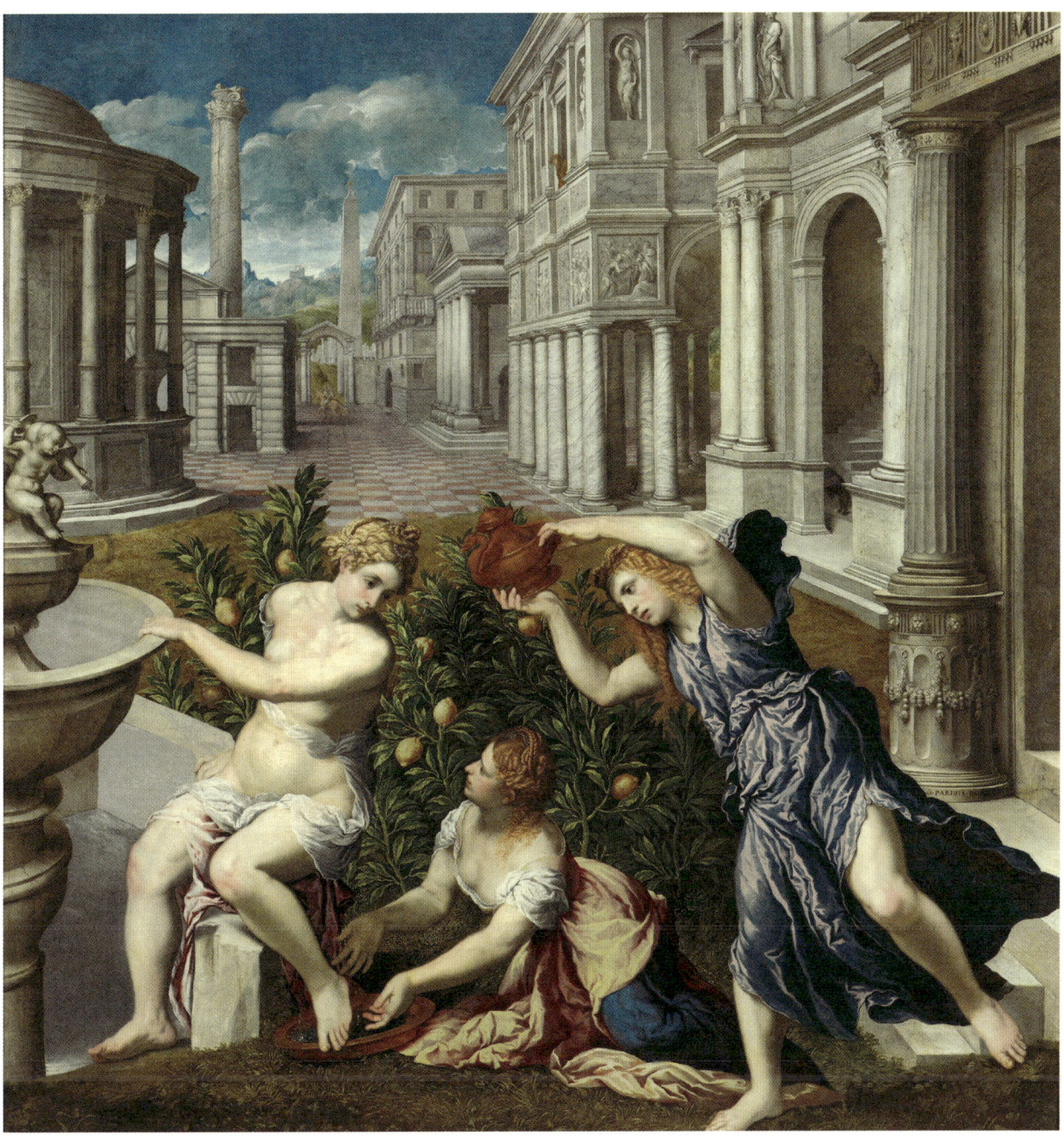

OPPOSITE: *Sacred Conversation*, Giovanni Cariani, 1525–30. **ABOVE:** *Bathsheba Bathing*, Paris Bordone, 1549.

Citrus has a long history in both Judaism (see pp. 52–53) and Christianity. The 'forbidden fruit' eaten by Adam and Eve in the Garden of Eden has often been interpreted as a citron, which was then dubbed the 'Adam's apple' (though this term has also been applied to pomelos, lumias and lemons). Citrus, in particular oranges, also became associated with the Virgin Mary and therefore purity or innocence. In Cariani's painting of the first meeting of Jesus and John the Baptist, citrus evokes a double theme: both Mary's virginity and Elizabeth's miraculous fertility. In Bordone's depiction of Bathsheba being watched by King David (just visible peering out of a top-floor window), the allusion takes on an erotic charge. As citrus was associated with the 'forbidden fruit' it hints at temptation and seduction.

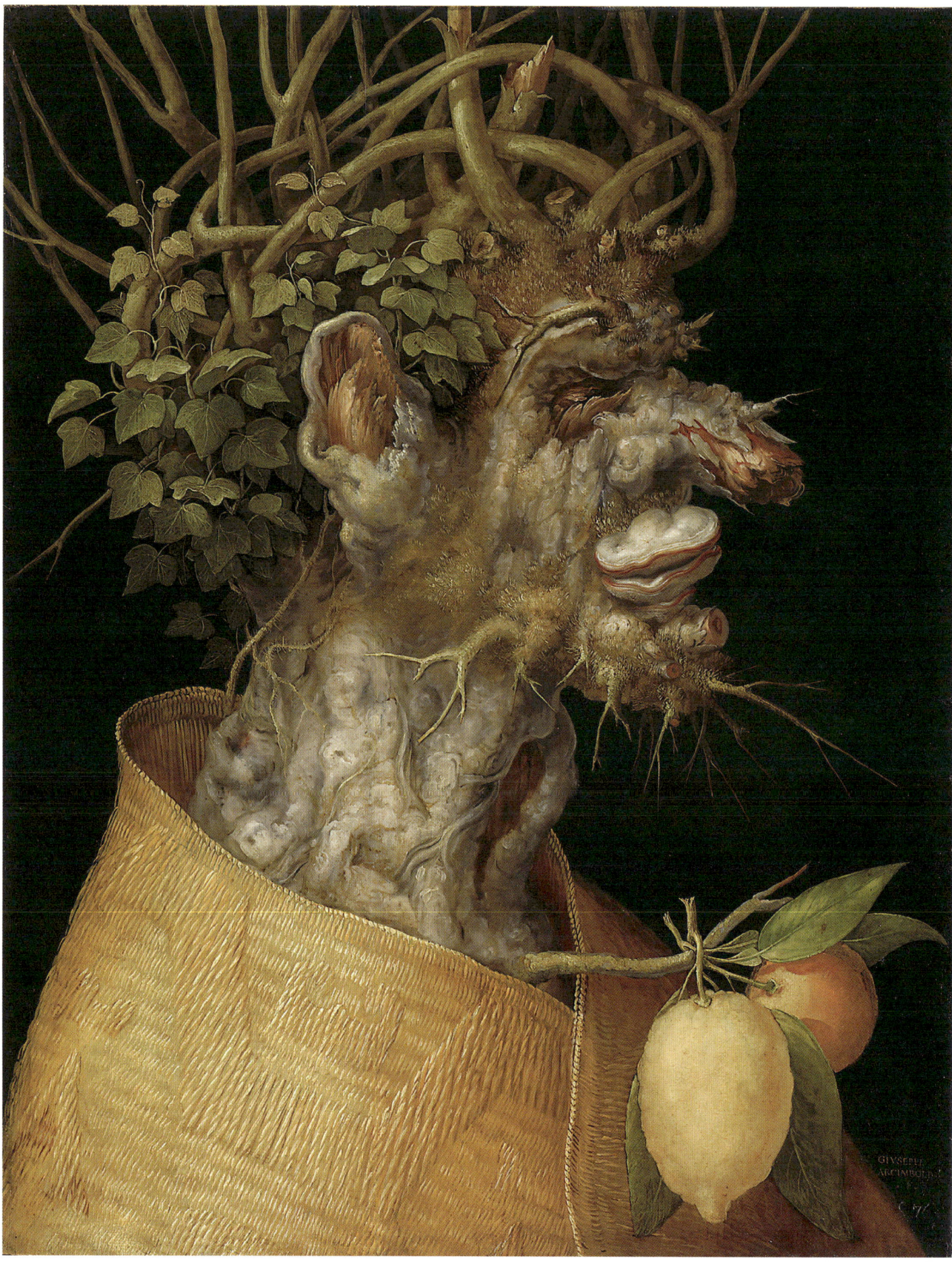

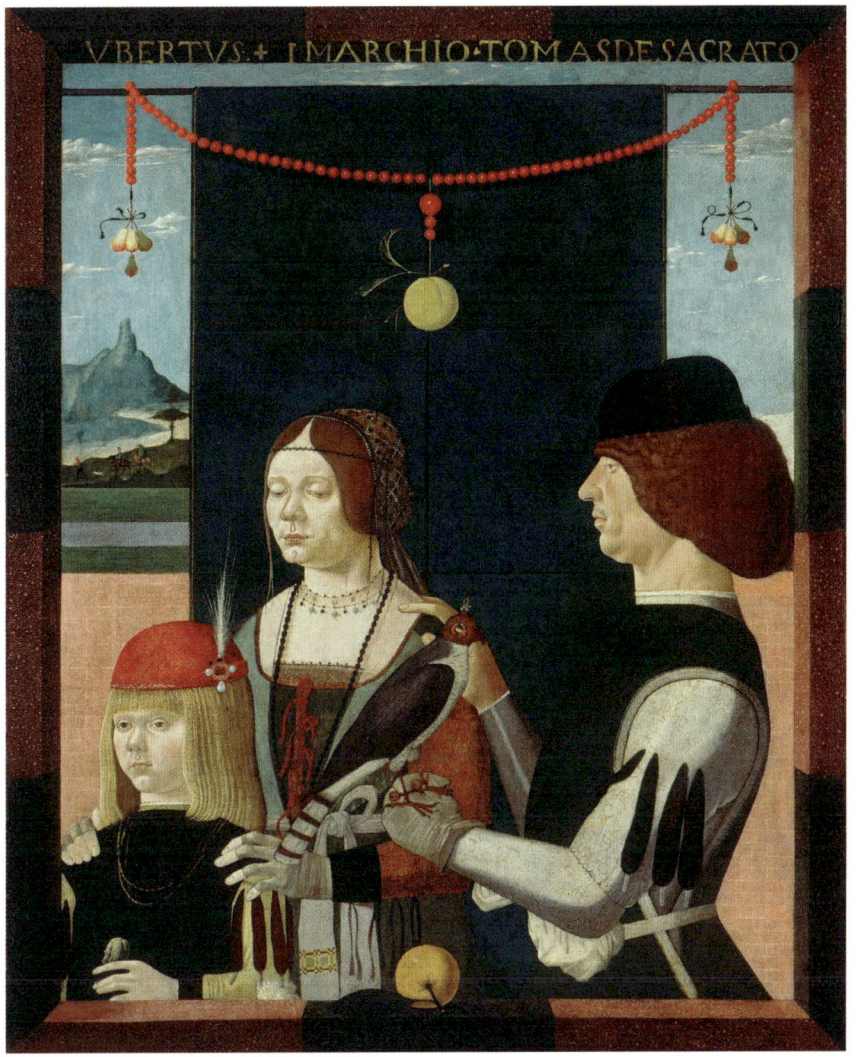

OPPOSITE: *Winter*, Giuseppe Arcimboldo, 1563. **ABOVE:** Family portrait, unknown artist (possibly Antonio da Crevalcore), *c.* 1480.

Arcimboldo's (1526–1593) playful portrait is part of a series in which each of the four seasons is represented by the plants associated with them. In contrast to the other seasons there is, appropriately, a conspicuous lack of fruit and vegetables in *Winter*, the orange and lemon being the only fresh fruits available during the Italian winter. This connection between humanity and nature hints at a harmony between people and the fruits and vegetables they have cultivated around them.

Citrus is used to a rather different effect in this enigmatic scene of the family of Uberto de' Sacrati from Ferrara, Italy. Along with their rich clothing and the presence of a hunting hawk with the family crest on its hood, the fruits placed and suspended around them indicate wealth – citrus requiring significant investment to acquire or grow in northern Italy's colder climes.

OVERLEAF: *Last Supper*, Jacopo da Ponte Bassano, 1542.

Among the familiar bread and wine, an orange occupies a central position in this depiction of the Last Supper. Citrus fruits appear again and again in paintings of the same scene, including in the versions by Titian and, most famous of all, da Vinci. However, despite the orange's seemingly symbolic prominence, its meaning remains elusive.

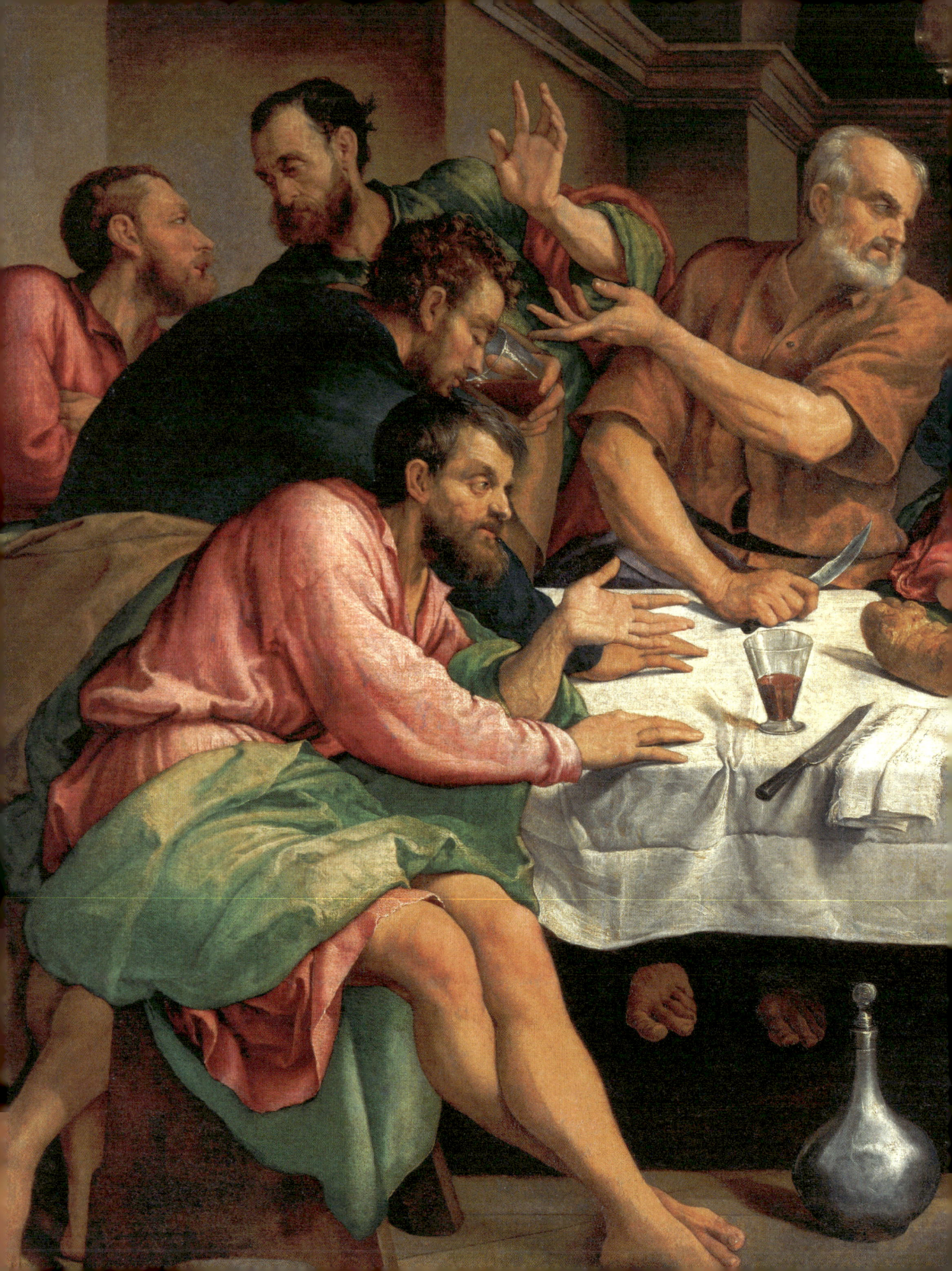

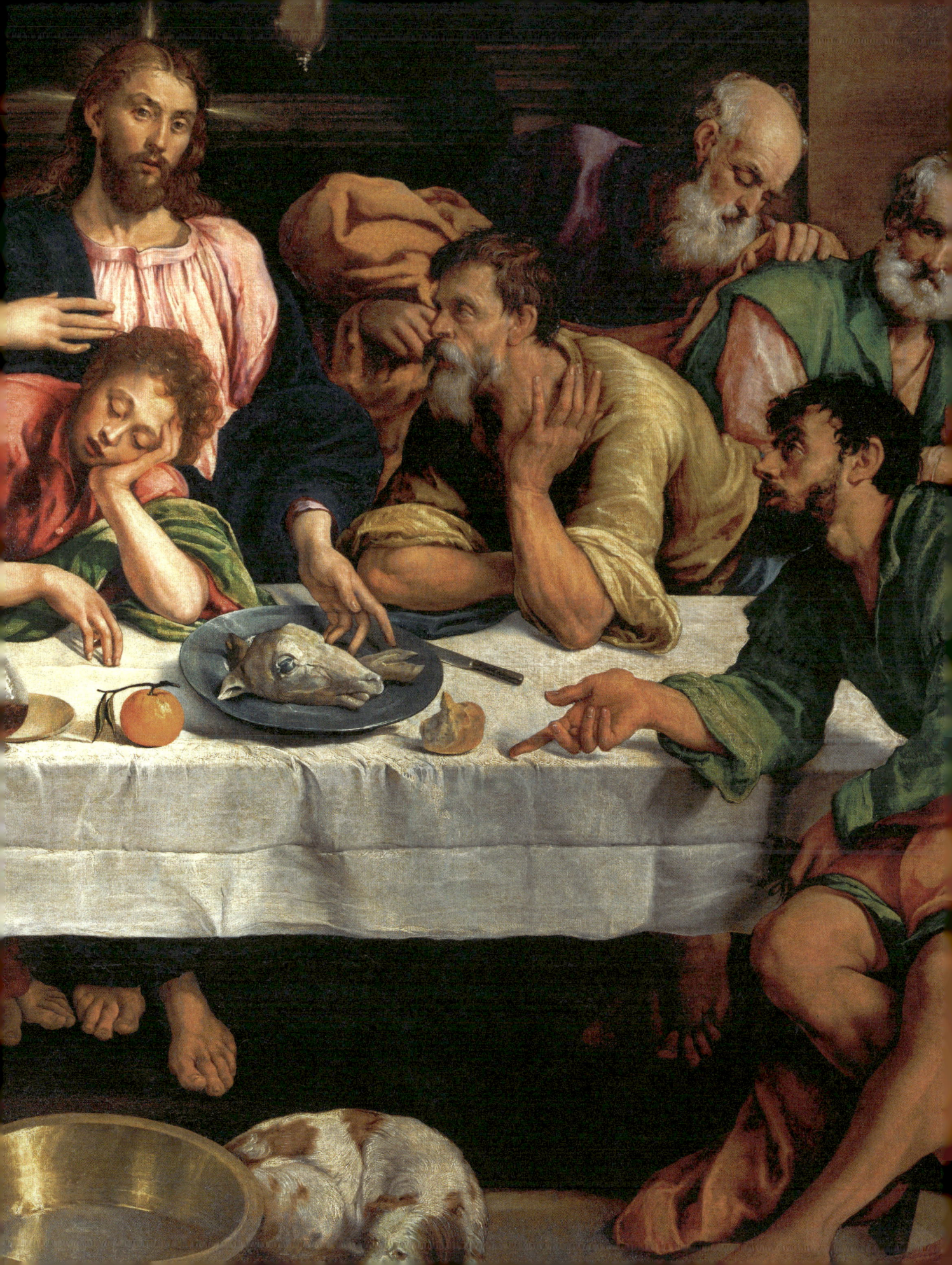

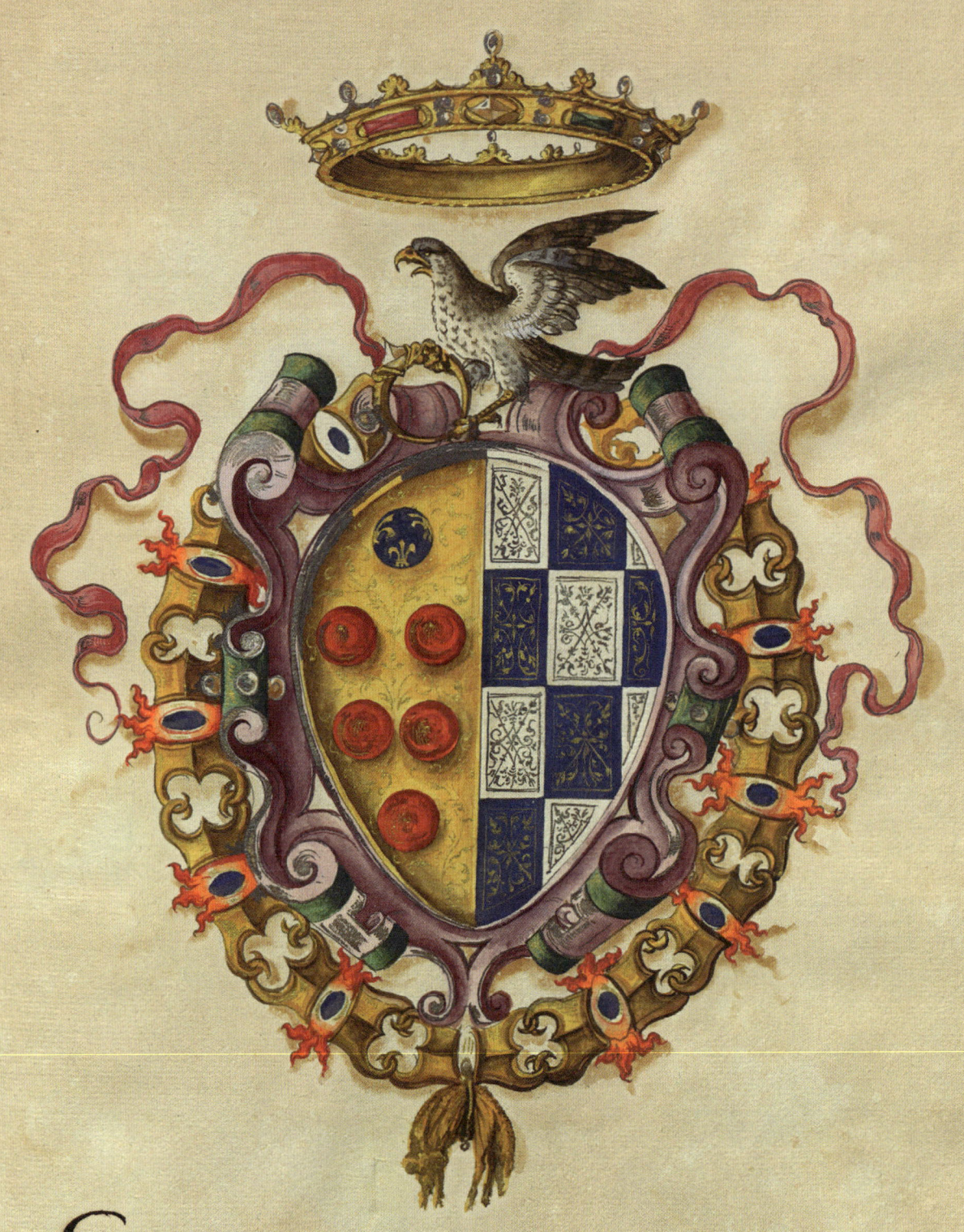

COSMVS MEDICES
DVX.II.FLORENTIE.

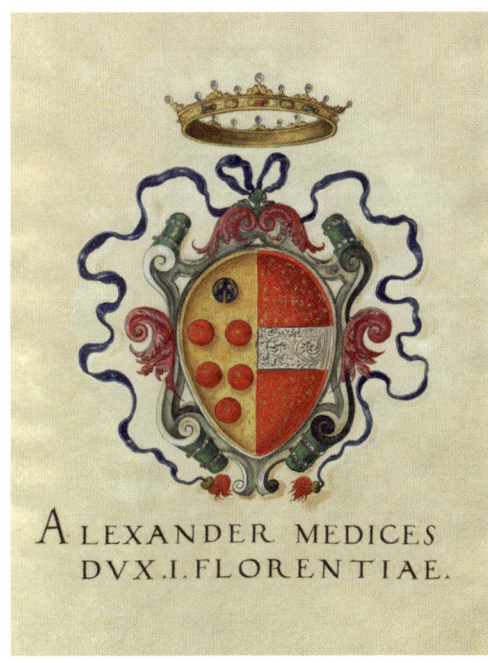

A LEXANDER MEDICES
DVX.I.FLORENTIAE.

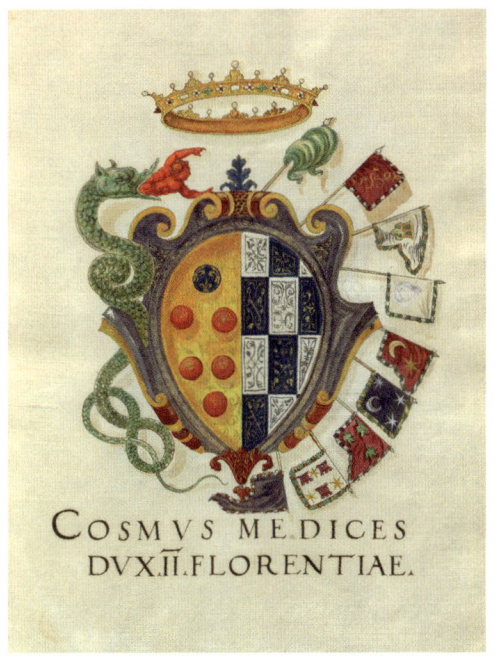

COSMVS MEDICES
DVX.II.FLORENTIAE.

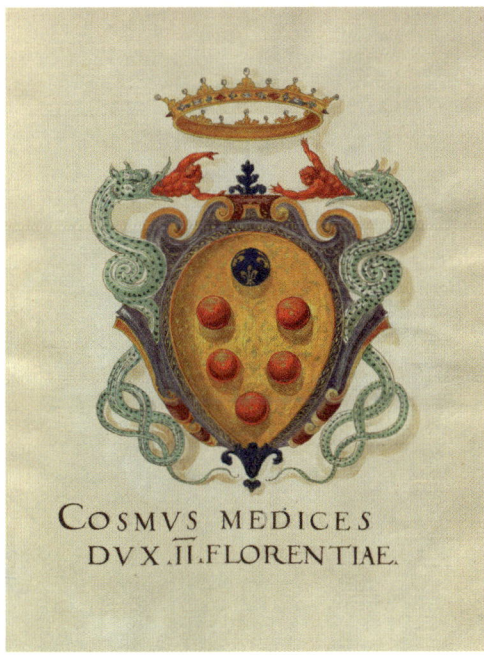

COSMVS MEDICES
DVX.II.FLORENTIAE.

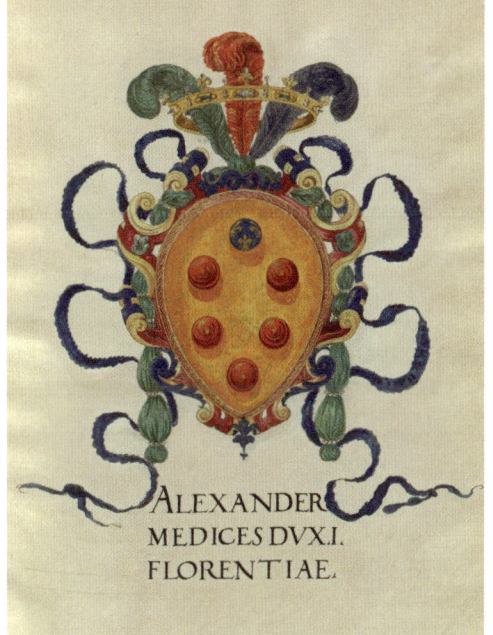

ALEXANDER
MEDICES DVX.I.
FLORENTIAE.

Medici family arms, *Insignia Florentinorum*, 1550–55.

The influence of the Medici family spread far and wide in Renaissance Italy; through trade, banking, politics and religion they consolidated their wealth and power. This extended to the world of citrus. Some interpret the orange orbs on their coats of arms as citrus fruits; however, this may not be so, and the true Medici–citrus connection lies in their magnificent gardens. Expensive to buy and grow, citrus collections became highly prized and were a fashionable way to show off. The Medici perfected this art form in their extensive collection at the Villa di Castello, which is said to have held over 600 varieties and species, including the fabled *bizzarria* (see p. 81).

HERBALS

Images of citrus in the emerging scientific literature of the time were less distinguished. With the 'invention' of printing in the West, by the mid-sixteenth century woodblock illustrations of the kinds of citrus being cultivated in Europe were included in the major botanical compendia of the time, though information was often copied book to book. Citrus appeared in the great compilation of real and imaginary animals and plants *Ortus sanitatis*, published in Mainz, Germany, *c.* 1491 by Jacob Meydenbach. Another more-or-less contemporary plate, also from Mainz, seems to show the citron, or perhaps a lemon.

It was the early southern European botanists who made the first clear descriptions and woodcuts of citrus, because they were able to work with living trees. Pietro Andrea Mattioli (1501–*c.* 1577), a doctor from Siena, wrote a commentary on the work of the Greek physician-botanist Dioscorides, his *Commentarii* first appearing in 1554, including recognizable images of citrus as grown in Italy at

the time. These plates, as so often then, were copied for the works of later compilers. By comparison, the richly illustrated book of hours created by Jean Bourdichon (1457/9–1521) for Anne of Brittany (1477–1514) contains far more realistic representations of exotic plants, including citrus (see p. 82). Clearly these were drawn from life, from trees cultivated in France.

THE MEDICI

Outdoor collections of citrus were in one sense an extension of the concept of the cabinet of curiosities, with as many kinds as possible – including teratologous forms – all displayed together.[35] By the end of the fifteenth century, citrus-growing had already diffused throughout the courts of Europe. The finest collection was that of the Medici.

In 1477, Lorenzo (1463–1503) and Giovanni (1467–1498) di Pierfrancesco de' Medici bought the Villa di Castello, built fifty years before near Sesto Fiorentino in the hills north-west of Florence. Eventually one of the most famous of all paintings, Sandro Botticelli's (*c.* 1445–1510) *Primavera* of *c.* 1480, now in the Uffizi in Florence, would hang there. The gigantic picture[36] depicts mythological figures in an orange grove with some 190 different plant species, reminiscent of the Flemish *millefleur* tapestries of the time. In 1537, when the teenaged Cosimo I de' Medici (1519–1574) had established his authority as the Duke of Florence, he retreated to Castello, which was restored and enlarged by architect Giorgio Vasari (1511–1574). The upper terrace of the walled garden was devoted to citrus trees from southern Italy, but Cosimo's eldest son, Francesco I de' Medici (1541–1587), collected and added rarer ones.

By the late sixteenth century the collection expanded into the Boboli Gardens of the Medici's Florence residence, the Palazzo Pitti. In the time of Grand Duke Pietro Leopoldo (1747–1792), wax models of citrus fruits were made and displayed in La Specola, the oldest public museum in Europe.[37] Also exhibited were four citrus paintings in oil on canvas over 2 m (6 ft 6 in.) wide, made by Bartolomeo Bimbi (1648–1729) in 1715 for Villa della Topaia, Cosimo III de' Medici's retreat. They portray some 116 different kinds of life-size citrus. Plaster casts, wax models and pictures of fruit were exhibited at La Specola until the mid-nineteenth century and in 2015 some 223 plaster models were rediscovered in Florence, in an office of the Boboli Gardens.[38]

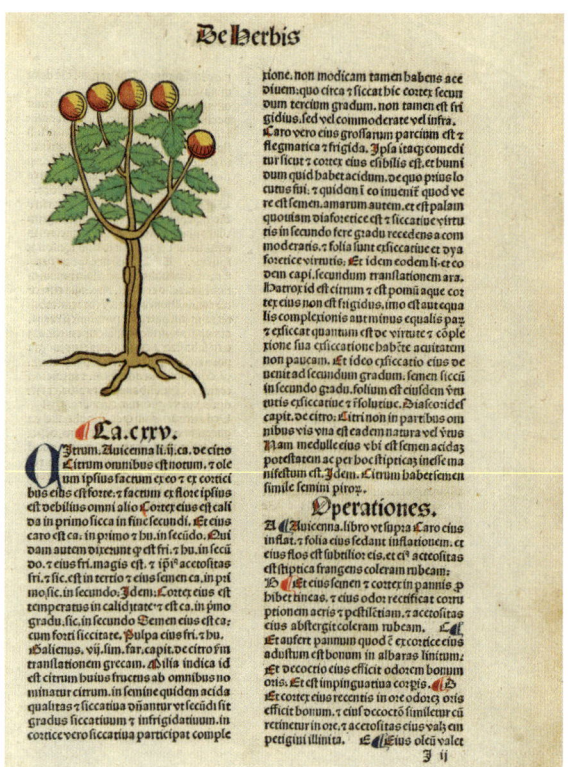

A citrus tree, *Ortus Sanitatis* published by Jacob Meydenbach, *c.* 1491. An early example of a herbal containing descriptions of plants and how to use them medicinally.

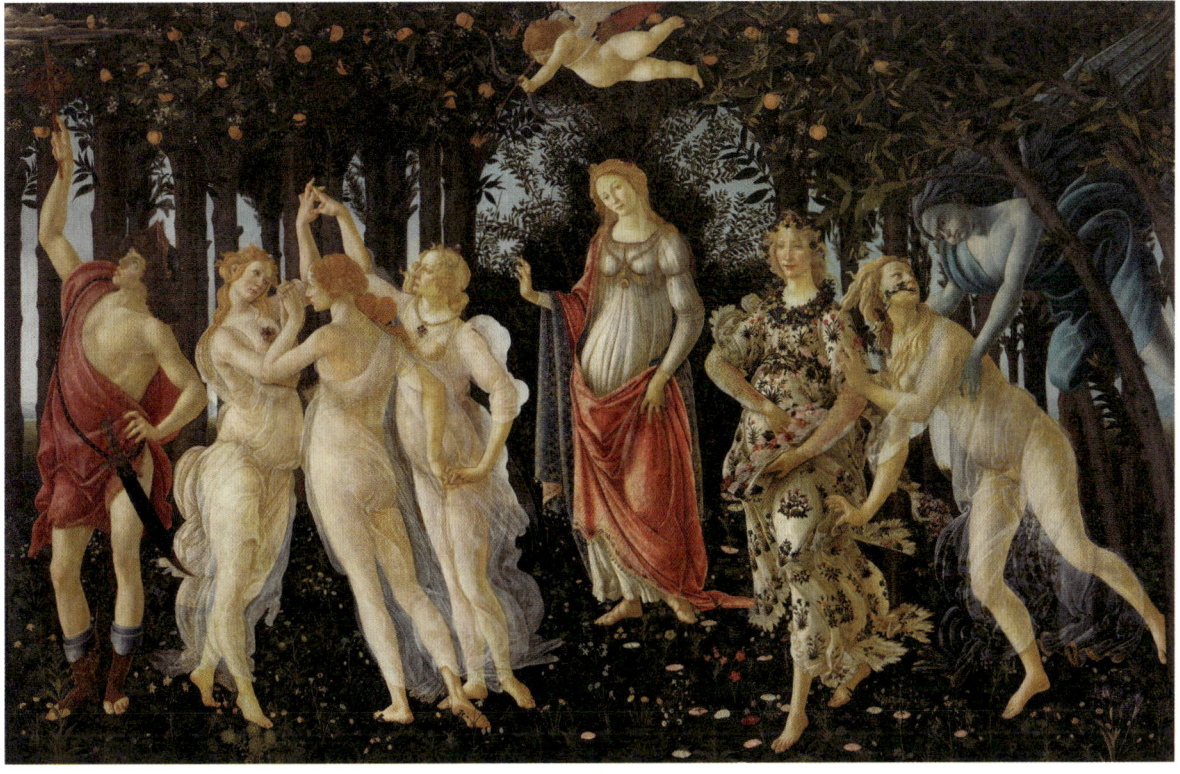

BIZZARRIA

In 1672, the Grand Duke of Tuscany Cosimo III (1642–1723) appointed to the directorship of the botanic garden in Pisa a locally trained doctor, Pietro Nati (1625–1715). Nati published a dissertation, *Phytologica observatio de malo limonia citrata-aurantia florentiae vulgo: La Bizzarria* (1674). The work caused a sensation. It described a citrus tree that bore sour oranges, citrons and 'mixed fruits'. As Gallesio wrote in 1811:

The tree looks like a bigarade [sour orange]. Its leaves are shaped sometimes like those of the orange, and often like those of the citron, sometimes uniting the two ... The fruit follows the caprice of the rest of the tree. One sees sometimes a bigarade in the form of a lemon; others are mingled lemon and orange, at times round, sometimes having a nipple at the summit. Others have the skin of an orange and pulp of a [Florentine] citron. These trees bear also citrons of many forms, of which some unite the citron and the orange, and, finally, there are fruits of which the outside and inside show four parts crossed, of which two are citron and two are orange, while by the side of these are oranges perfectly formed, without the least mixture ... There is a further caprice of this tree even more singular - that of a citron coming from a bud

Primavera, Sandro Botticelli, *c.* 1480. Likely commissioned for the wedding of Lorenzo di Pierfrancesco de' Medici, he being the model for Mercury (far left), his wife that for Flora (third from right). Citrus was often linked to weddings, fertility and love.

---

which grew at the axil of an orange leaf, and conversely the orange from a bud of which the leaf is a citron.[39]

It is traditionally held that the *bizzarria* was first noticed in about 1644, by a gardener, in the collection of the rich Panciatichi banking family at the Villa Torre degli Agli, now in the western suburbs of Florence. Though now lost, that garden is commemorated in the name of the street, Via della Torre degli Agli, while the citrus itself is remembered in the name, Via del Giardino della Bizzarria. The gardener 'multiplied the new variety by the graft, and made it quite profitable to himself. He making a mystery of its origin, everybody thought the wonder was owing to the industry of the gardener, who had mingled by the graft the buds.'[40]

Such a monster made people question the nature of species and the whole concept of hybridity. It was perhaps one of the first recorded

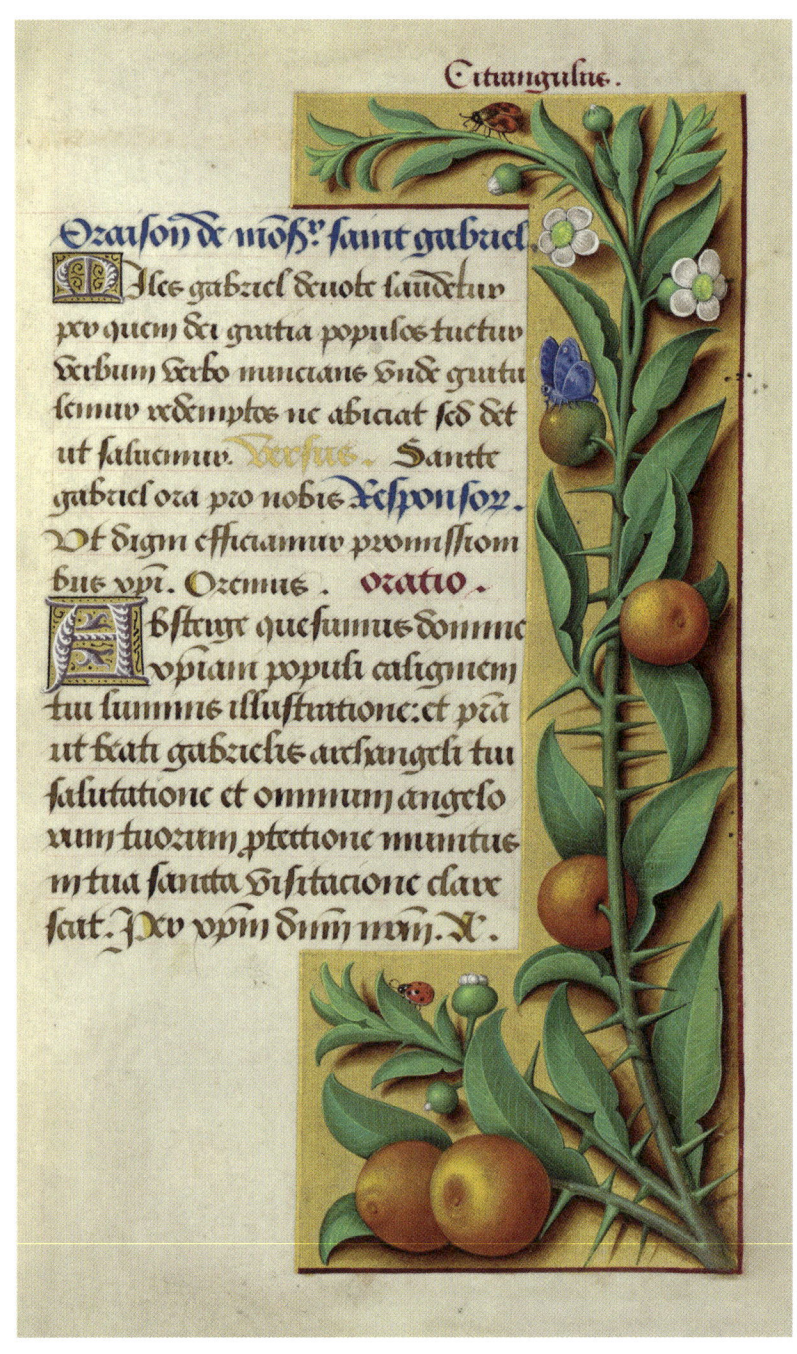

ABOVE: 'Citrangulus', *Grandes Heures of Anne of Brittany*, Jean Bourdichon, 1503–8. **OPPOSITE:** A cucurbit and citron (right) in *Tractatus de Herbis*, an Italian herbal, *c.* 1440.

Herbals, providing detailed descriptions of plants and their medicinal, culinary and even magical properties, became increasingly important in the Renaissance. The accompanying woodcuts (see p. 80) were often crude and further debased by being copied from one text to the next. Illustrations by southern European botanists (opposite) were often clearer, as the authors were able to observe living plants. However, the book of hours created for Anne of Brittany (above) eclipsed any herbal. Clearly the 'citrangulus' (sour orange) depicted here was drawn from life.

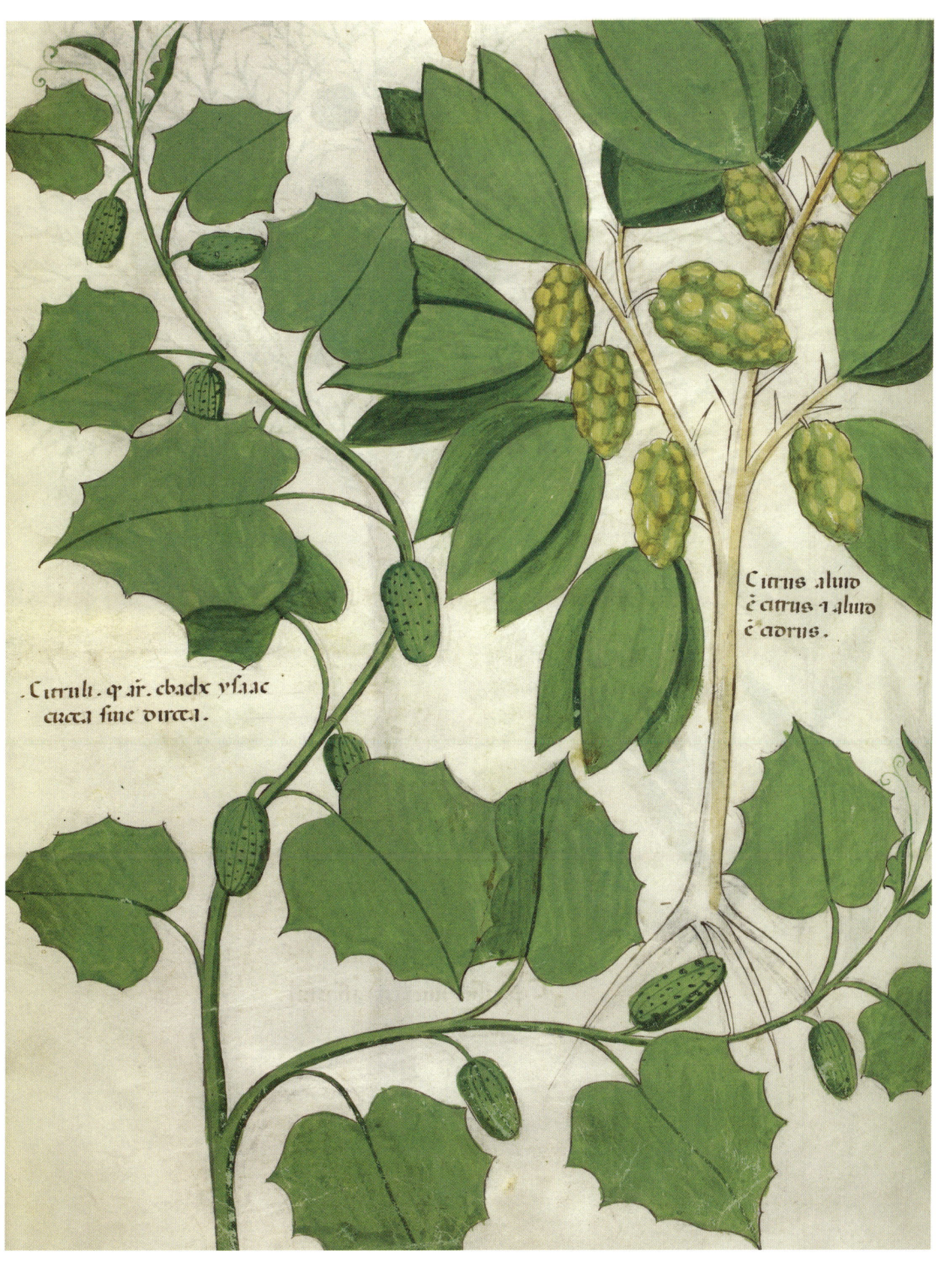

Citrus aluid
é citrus ꞁ aluid
é adrus.

Citruli. qͥ aͥr. chache yſaac
cucta ſiue oírcca.

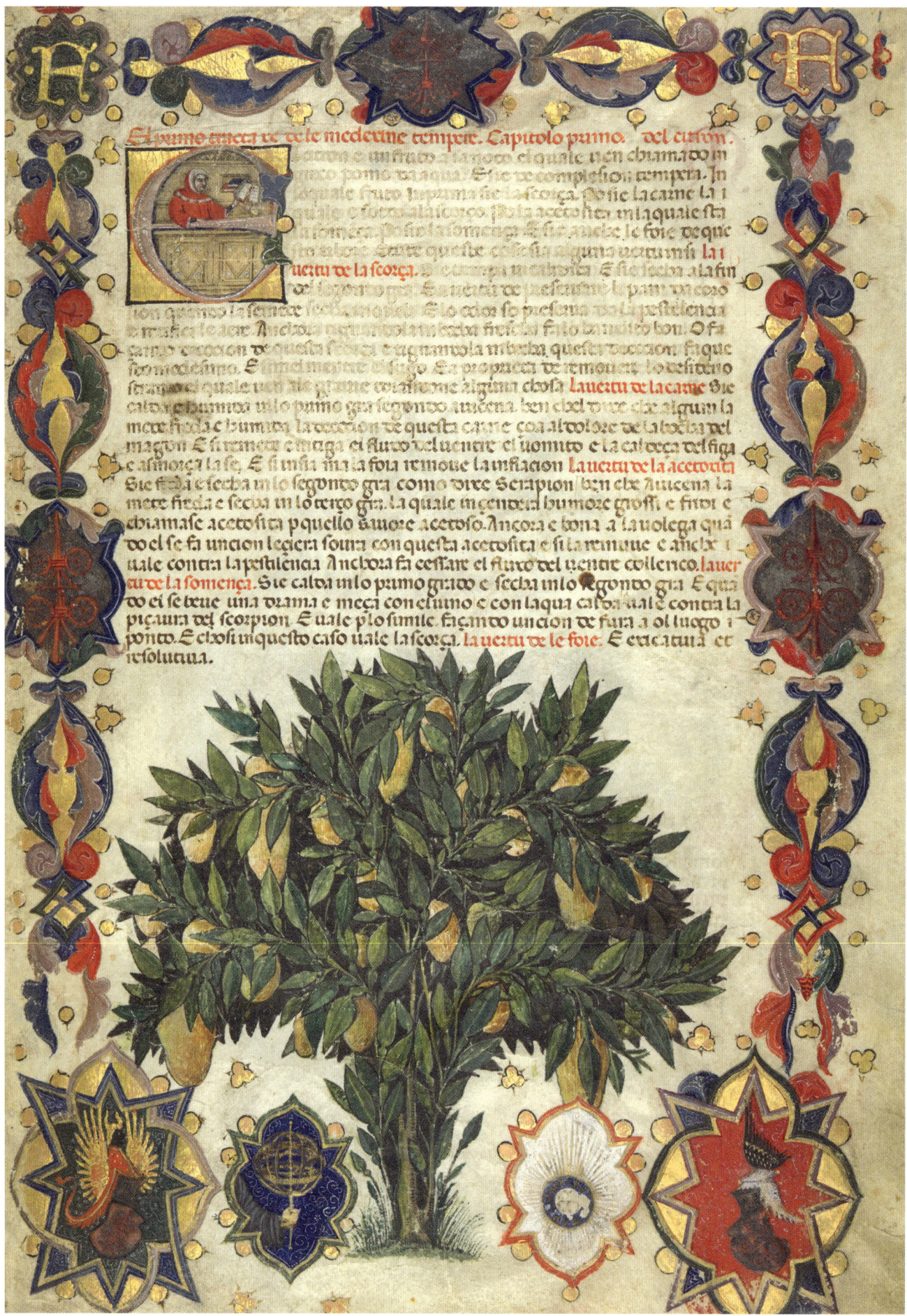

Citron tree on the frontispiece to the Carrara Herbal, Padua, *c.* 1400.

This magnificently illustrated herbal is another example of how medico-botanical knowledge was shared and transmitted across cultures throughout the Middle Ages and into the Renaissance. The book was created as a translation from Latin into the Paduan dialect for Francesco II da Carrara (1359–1406), Lord of Padua. The Latin was itself a translation of an Arabic pharmacopoeia of the 12th or 13th centuries by Serapion the Younger (Ibn Sarabi) of Al-Andalus (Spain). The original Arabic work is now lost and survives only in translation. Essentially a compilation of existing knowledge, it provides information on the medicinal properties of plants, animals and minerals. Interestingly, while Serapion's original sources were mostly Arabic scholars, he also quoted information from ancient Greek physicians such as Dioscorides and Galen (129–216 CE).

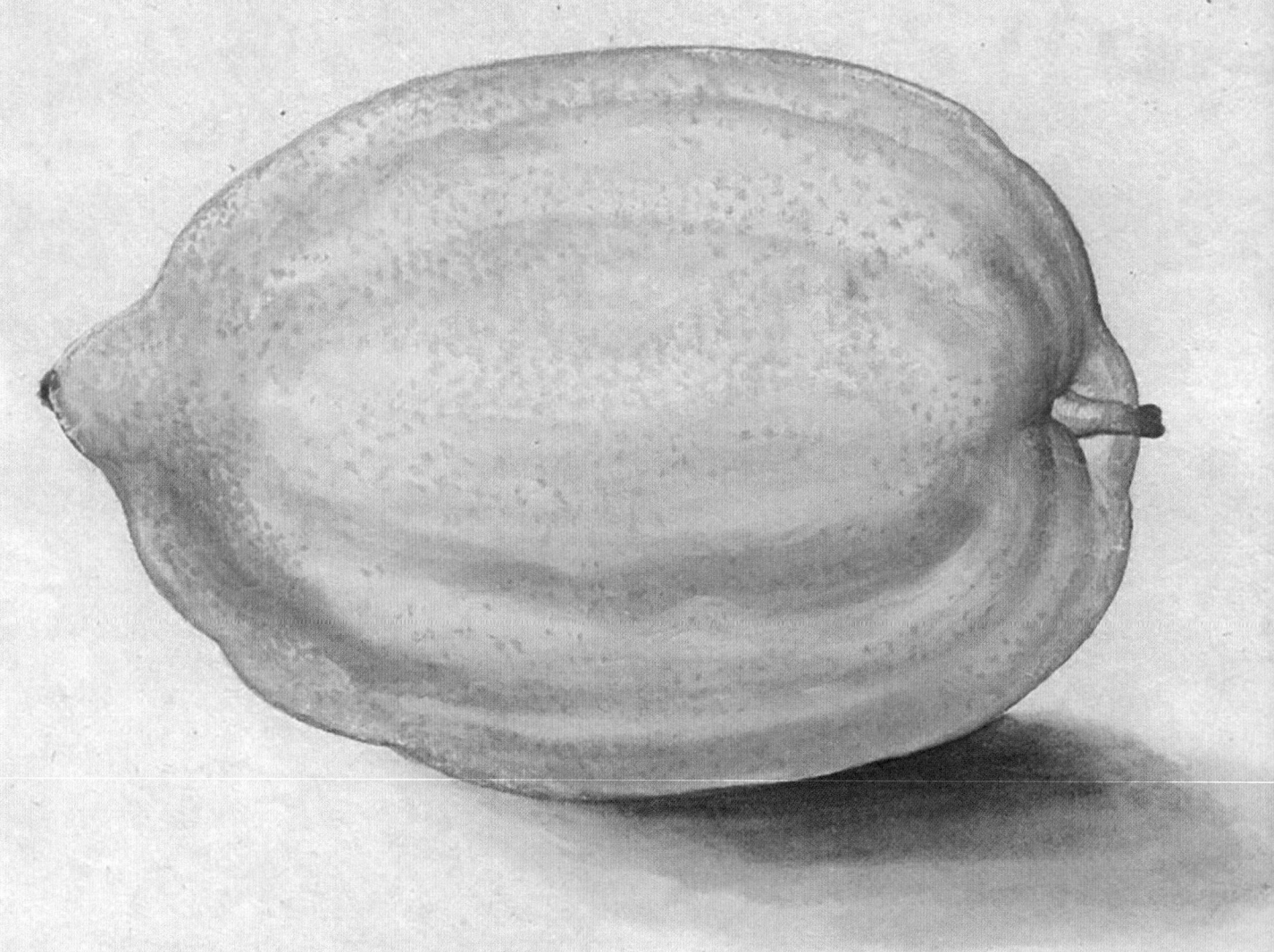

Citron (*Citrus medica*), watercolour from *Libri Picturati*, *c*. 1550–95.

Though celebrated for its artistic skill and scientific value, the origins of this
collection of watercolours remain somewhat obscure. They were likely produced
under the patronage of Flemish nobleman Charles de Saint Omer (1533–1569) with
contributions by his friend Charles de l'Écluse (see p. 119). The absence of foliage
and flowers strongly suggests that the artist (who may have been the painter Jacques
vanden Corenhuyse) knew only the fruit, likely then as an exotic import to Bruges.
He recorded its weight as 30 oz (850 g).

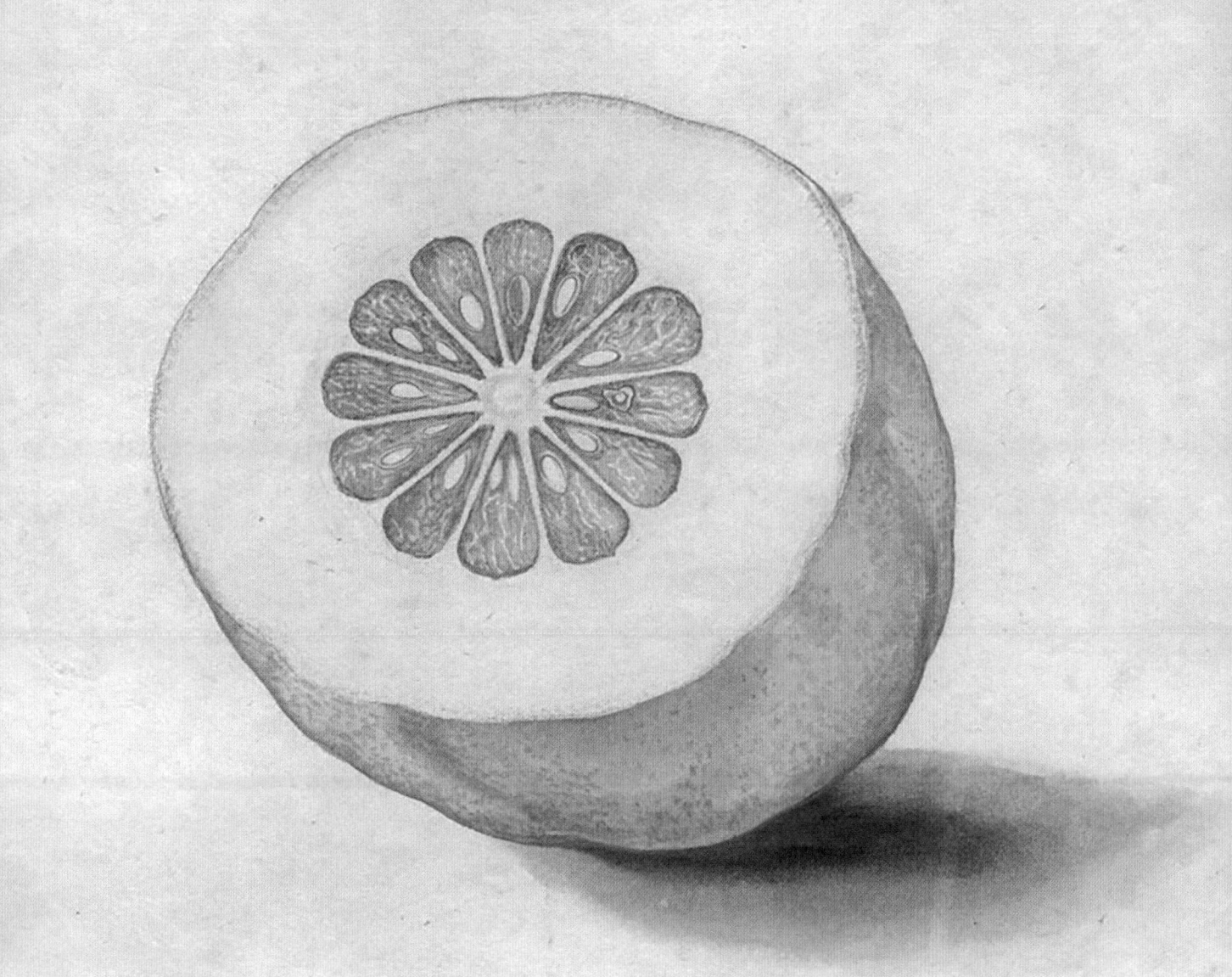

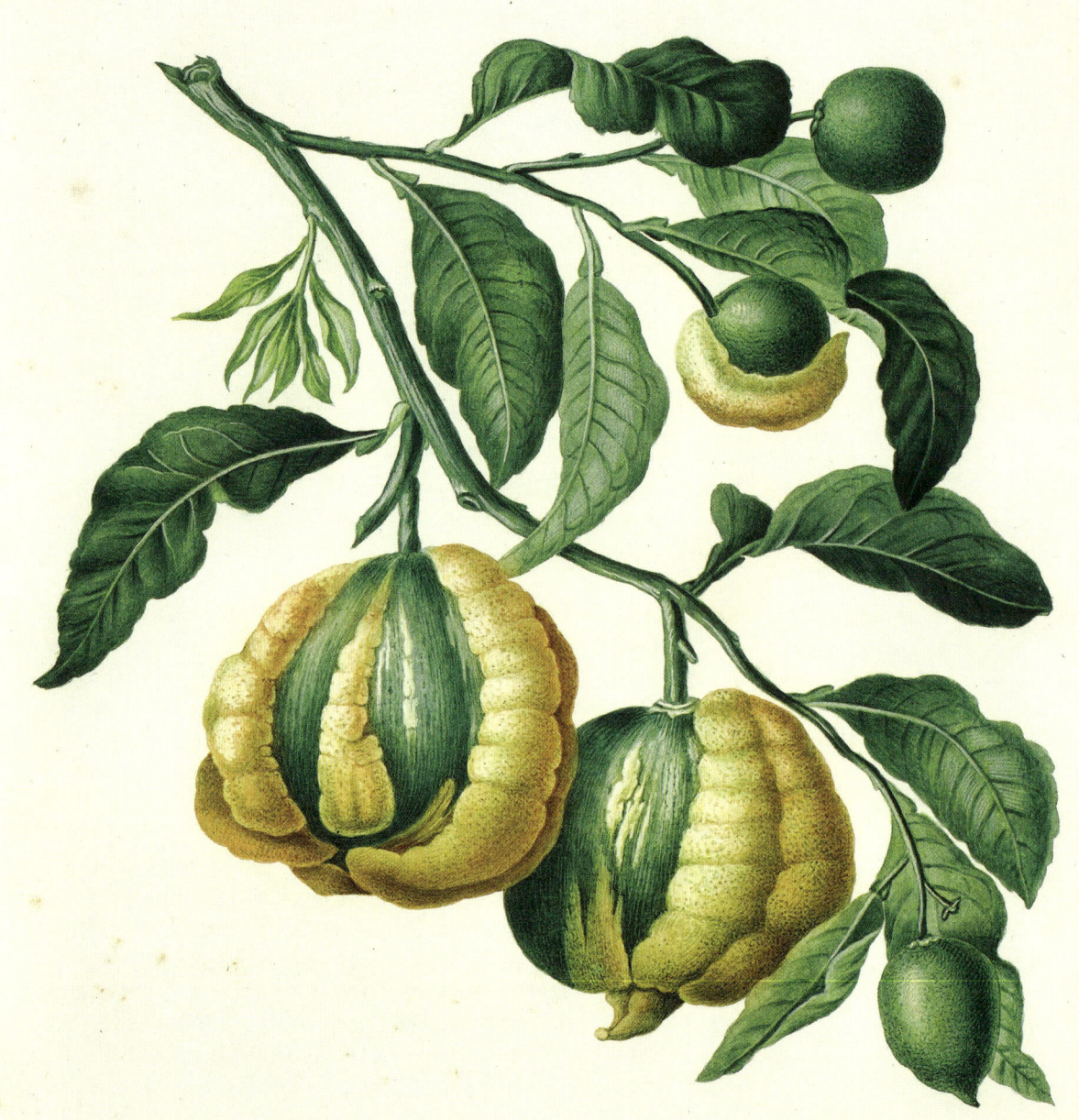

CITRUS AURANTIUM *Limo-citratum*
Bizzaria

Giovanni Gori dis.   *Giuseppe Seva inc.*

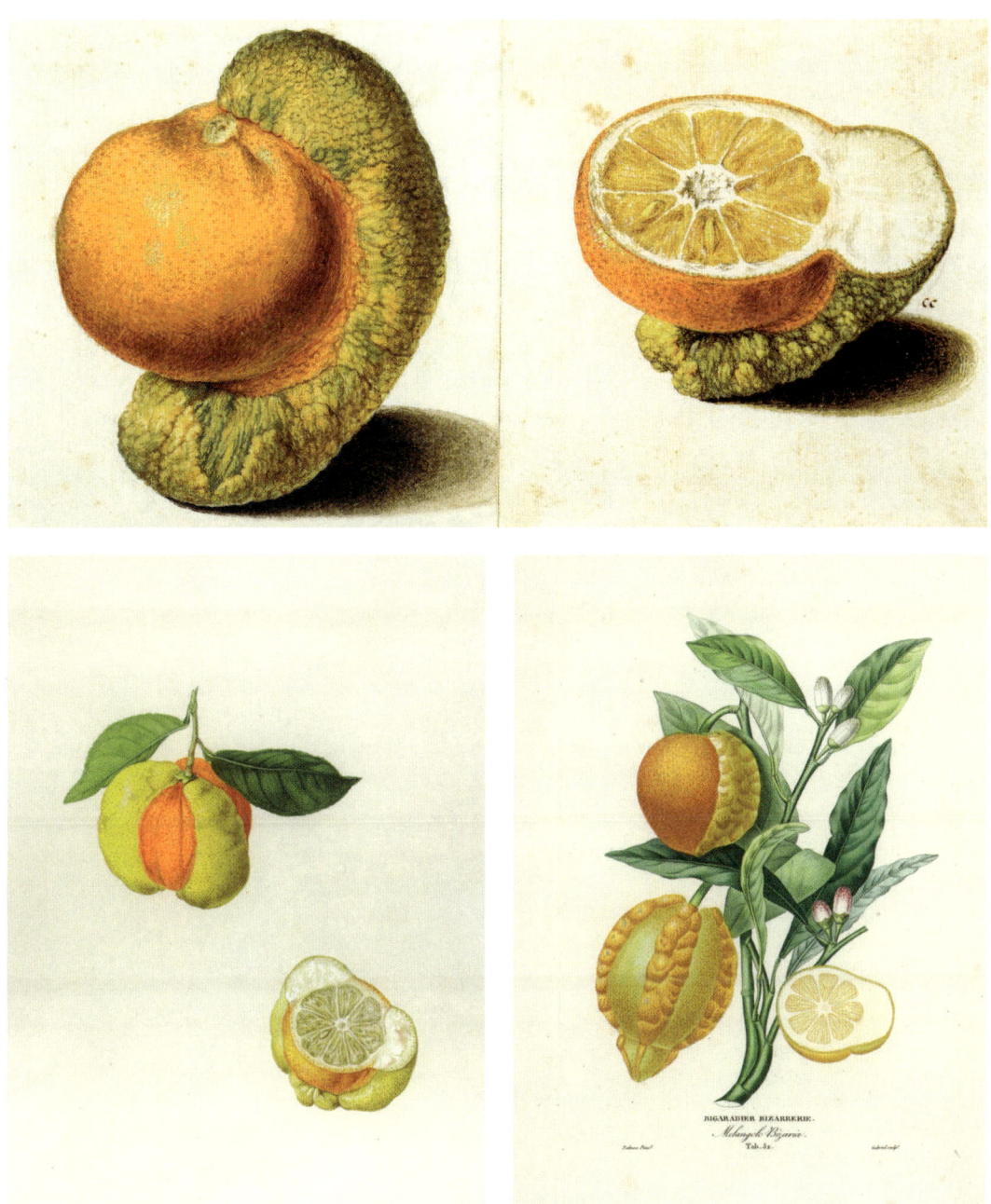

OPPOSITE: 'Citrus aurantium bizzarria', from *Raccolta di fiori, frutti e agrumi* by Antonio Targioni-Tozzetti, 1822–29. **TOP:** 'Aurantium callosum', watercolour, Vincenzo Leonardi, before 1646. **ABOVE, LEFT:** Citrus chimaera, watercolour, Joseph Knapp, 1843–57. **ABOVE, RIGHT:** 'Bigaradier bizarrerie' from *Histoire naturelle des orangers* by Antoine Risso and Pierre-Antoine Poiteau, 1818–22.

The most famous and envied citrus of all those in the Medici gardens were the *bizzarria*. The original *bizzarria* was a chimaera arising from a failed graft of a Florentine citron on a sour orange, both in fact sexual hybrids.

of what are now called plant chimaeras (or so-called graft hybrids), which have shoots comprising tissues from two different kinds of plant growing together. An anonymous Englishman was able to buy such a fruit, reportedly from Genoa, in Paris in 1660,[41] and chimaeras were being grown in England by 1672, Germany 1684, and the Netherlands by 1687.

The Italian *bizzarria* arose by accident, from a Florentine citron being grafted on a sour orange.[42] The graft seems to have failed, but some tissue of the citron scion became enveloped by that of the orange stock. This *bizzarria* was once commonly grown in Tuscany and propagated by grafting, but then thought to be lost until it was rediscovered in the 1970s by Paolo Galeotti, the head gardener of the Villa di Castello and the Boboli Gardens. It has subsequently been propagated and is reported as growing in both the Boboli Gardens and elsewhere in Florence.

Other chimaeras (now collectively referred to by the general term *bizzarria*) are known, including even earlier records, such as that of a sweet orange grafted on to a sour one bearing both sweet and sour oranges described by the polymath Giambattista della Porta (1535–1615) in his *Phytognomica* of 1588. In 1831, a sweet orange bearing fruit with a longitudinal band of citron was growing in an espaliered orangery belonging to Claude Fion (1778–1844) in Paris.[43] In the twentieth century, a frost-damaged sweet orange grafted on a sour one exhibited the same effect,[44] and in 1986 appeared *Citrus* 'Hormish', this time involving the bitter orange (*Citrus trifoliata*) as the rootstock and a Ponkan mandarin the scion, resulting from a frost-damaged graft in China. It bore trifoliolate leaves, like the stock, but also unifoliolate, like the scion, as well as bifoliolate ones. Other such Chinese chimaeras involving different combinations of grafted citrus have been recorded since.

Japanese botanists working on a graft chimaera between a sweet orange and a mandarin found that, although some characteristics, like the colour of the rind and juice sacs of fruits, were like those of the parental forms, developing independently, some others, like leaf and fruit size, sugar content and acidity, were due to interactions between the two genetically different tissues.[45] They suggested that such interactions between genetically different cells caused variation at the DNA level and could be a source of genetic variation. In 1868, no less than Charles Darwin (1809–1882) suggested that heritable changes could be induced by grafting, calling it 'graft hybridization', with genes moving from the stock to the scion.[46] This was part of his provisional hypothesis

of 'pangenesis' in which 'gemmules' from various parts of the plant (or animal) body, rather than merely from 'the male and female organs' could be transmitted to the next generation. His idea has often been glossed over in a somewhat condescending way.[47]

Subsequently, many experiments on various genera of plants in many parts of the world seem to have demonstrated that this phenomenon is real, although most botanists have been more than sceptical that it could apply in crops such as citrus, despite the Japanese findings. Nonetheless, such 'horizontal gene transfer' is now respectable in plant science, while it is amusing to recall that the very word 'gene' is derived from the same root as Darwin's pangenesis.

BERGAMOT AND EAU DE COLOGNE

Unlike the original Medici *bizzarria*, at least some other of their three-hundred-year-old citrus trees in pots have survived at the Villa di Castello. This is despite the lemon houses being converted into a military hospital – the pots left outside – in World War I. During World War II the inventory and lead labels were lost, and some of the trees were even turned into firewood.[48]

Growing different kinds of citrus with incomplete sterility barriers in such rich collections side by side led, as it had in China long before, to the formation of sexual hybrids through cross-pollination by nectar-seeking bees, and their fixation as true-breeding clones through their peculiar form of asexual production of seeds (see p. 30). This rash of crosses led to a chaotic outpouring of names for the resulting, sometimes evanescent, hybrids, which heralded the taxonomic confusion for which the genus *Citrus* is notorious – to be dispelled only by molecular research in the twentieth and twenty-first centuries.

Many of these were mere back-crosses between hybrid crops and one of their parents, perhaps the most important in Italy, besides the Florentine citrons, being the bergamot, appearing in mid-seventeenth-century Calabria. Why it is called bergamot – a name sometimes associated with the

'Limon bergamotto' (bergamot), engraving from *Nürnbergische Hesperides* by Johann Christoph Volkamer, c. 1708. The bergamot, a back-cross between the lemon and the orange, is prized for its strong fragrance and is used in perfume and to flavour teas.

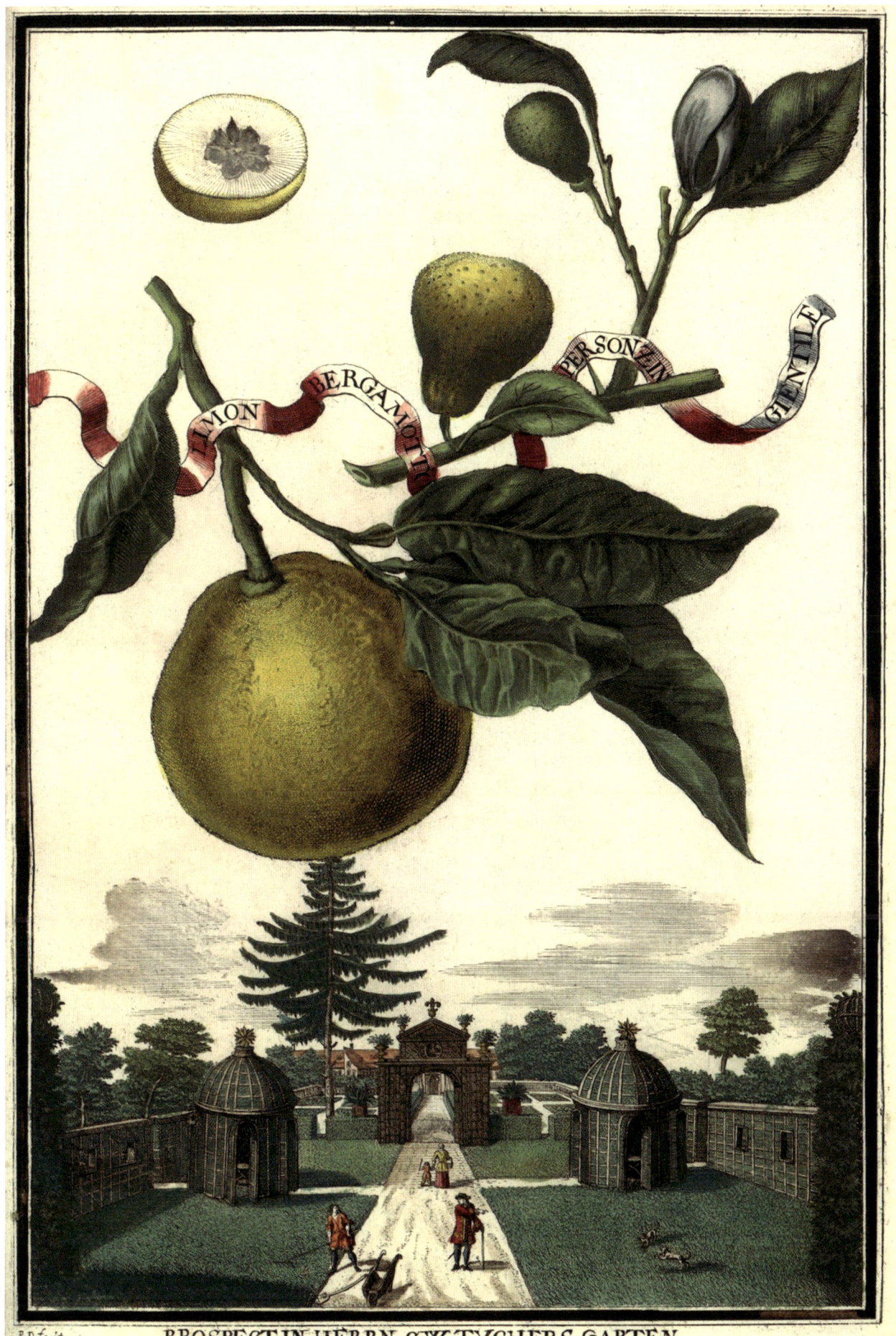

PROSPECT IN HERRN C.W. TVCHERS GARTEN.

Italian town of Bergamo, where it has not been grown – is unclear, though the name of bergamot pears (Italian *bergamotto*) is derived from Turkish *beg armudu*, suggesting the citrus name may refer to its somewhat pear-shaped fruit. The bergamot citrus is a back-cross between the lemon and the orange group. The extraordinarily powerful scent of bergamot led Giovanni Maria Farina (1685–1766), a perfumier who had migrated to Cologne (then a free city within the Holy Roman Empire), to formulate in 1708 eau de Cologne, based largely on bergamot oil. Initially extremely expensive, eau de Cologne was remarkable in blending, in a reliably uniform way, many individual essences. The precise combination is still a commercial secret, though reportedly involves oils from lemons, oranges and limes besides, most importantly, neroli oil from the petals of sour orange flowers. After attempts by the Mülhens family of Cologne to capitalize on Farina's name led to litigation and closure of their businesses, they developed their own 'Eau de Cologne 4711' in 1881. It was named after Glockengasse No. 4711, a number allocated to their building by Napoleonic occupying forces in 1796. At the same time, 'cologne' has become a general term for scent.

Bergamot oil is also the flavouring used in the production of Earl Grey tea, while orange pekoe tea has nothing at all to do with citrus (see p. 118). Bergamot was used to improve the taste of poor-quality teas, so the reason for the later association of bergamot tea with the earls Grey is much disputed. According to the Grey family, the 'tea was specially blended by a Chinese mandarin' for Charles, 2nd Earl Grey (1764–1845), to suit the water supply at the family seat in Northumberland.[49] It may have had more humble beginnings, the earliest advertisements being posted by a Northumberland tea merchant, William Grey, in 1852, the 'earl' later added to 'make the tea seem more posh'.[50]

Neroli oil is yielded by Seville orange flowers and is named after Marie Anne de La Trémoille (1642–1722), who, as princess of Nerola (a town now absorbed into Rome), is credited with making the essence fashionable by using it to scent her bath and her gloves. In the Nice markets in 1879, Seville orange flowers fetched three francs per kilogram, and those of sweet oranges two francs; during the season (over a month), 15–18 tonnes were sold *every day*, each tonne yielding a kilogram of oil.[51] 'Neroli bigarade oil' from the oranges is now one of the most widely used floral oils in modern perfumery and is largely produced in Morocco and Tunisia. It is also said to be used in flavouring Coca-Cola.

LEMON HOUSES AND EARLY ORANGERIES

Cultivation of citrus in Italy soon spread as far north as Lake Garda, east of Milan, where the Franciscan friars of Gargnano were growing citrus as early as the thirteenth century. By the next century, the columns of their cloister were decorated with carvings of lemons, complete with their flowers and leaves.[52] This is the furthest north commercial citrus-growing has ever reached. By the nineteenth century, the area had a strategic advantage over more southerly regions in that it was closer to northern European markets. By then, lemons, wrapped in tissue paper, as the Chinese had long done for citrus, were good for at least six months and were exported as far as Hungary, Poland and Russia. By the twentieth century, with improved transport systems, southern Italy eclipsed the local industry, while the synthesis of citric acid reduced the demand in any case. Lemon houses, or *limonaie*, were constructed at Lake Garda to protect the trees in winter: they still survive, though largely as roofless rows of poles along the terraces of cultivated land.

Citrus cultivation in the cooler climates to the north was the major force leading to the development of orangeries, or greenhouses, in Europe. And they became high fashion, especially during the Dutch Golden Age, where patriotism to the House of Orange encouraged the nobility to collect as many different kinds of citrus as they could (see Chapter 3). Such protective houses had an ancient pedigree, as the first seem to have been built for citrons when these fruit trees were being taken north beginning in the fourth century.[53] In the 'bad season', the trees were sheltered under wooden roofs, though the technology seems to have been forgotten until some time before 1336. It has been argued that as Humbert II de la Tour-du-Pin (1312–1355), the extravagant (and last) Dauphin of the Viennois, bought twenty sour orange plants in Nice, he must have had some facility for citrus-growing at his estate in frosty Beauvoir-en-Royans.[54]

When the Spanish King Alfonso V of Aragon (1396–1458) became King Alfonso I of Naples in 1442, he is said to have been instrumental in the establishment of orangeries in Italy, whence the fashion spread north. Alfonso sent to Valencia for not only plants but

Engraving after Guido Reni from *Hesperides* by Giovanni Battista Ferrari, 1646. The Hesperides nymphs plant a Renaissance garden. Ferrari's depictions of Classical mythology included contemporary architecture.

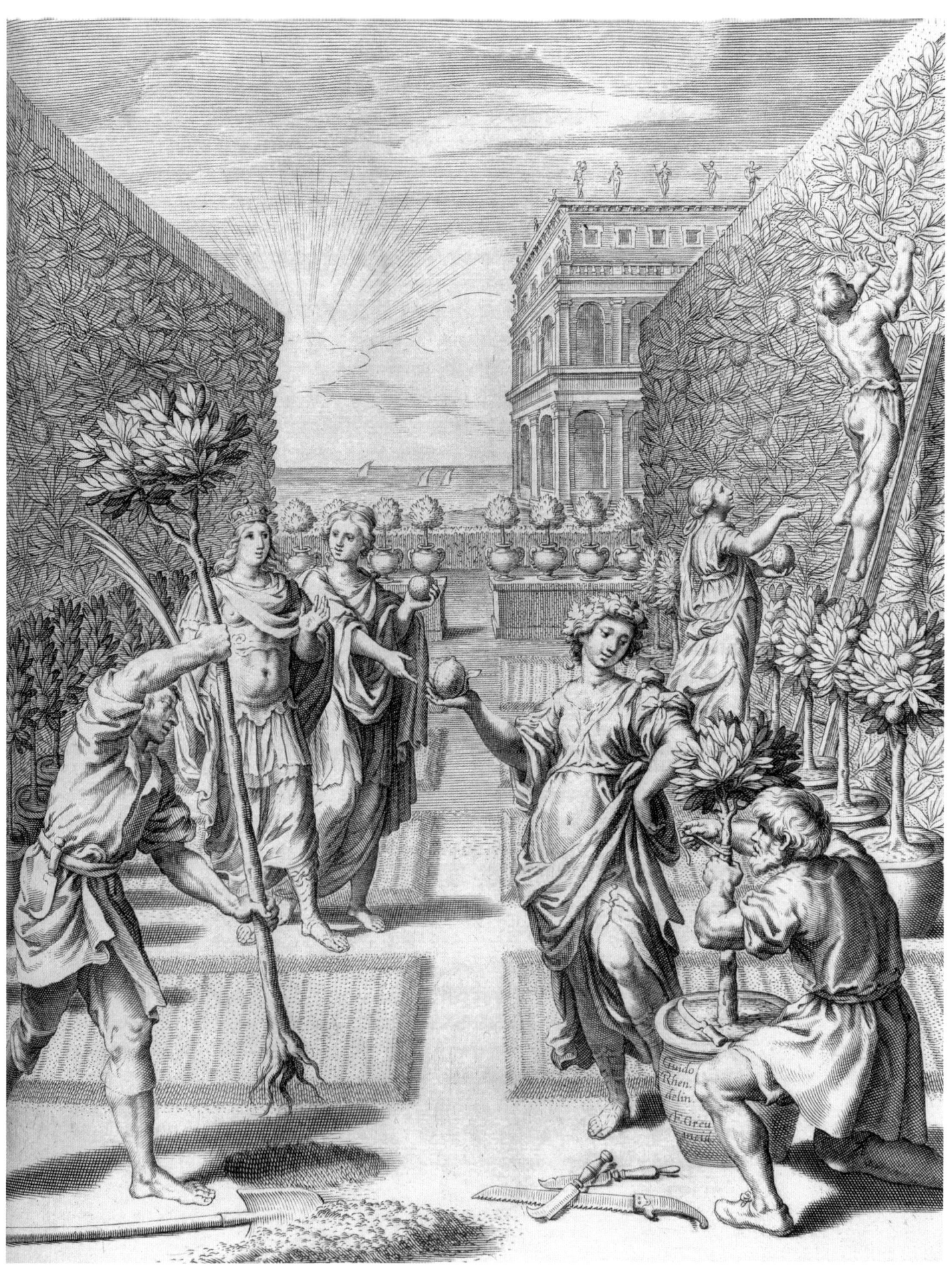

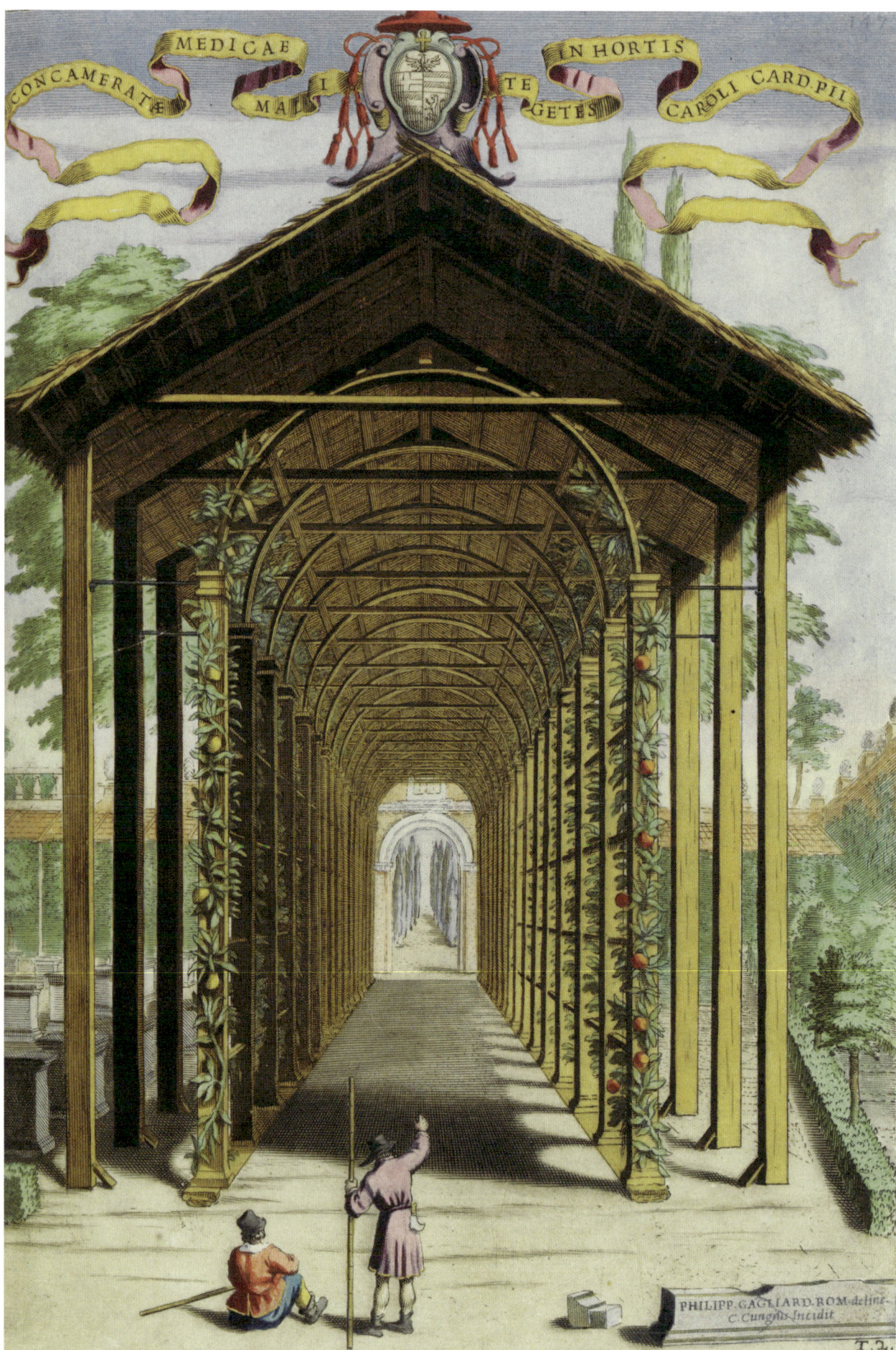

MEDICAE IN HORTIS CONCAMERATAE MAII TEGETES CAROLI CARD. PII

PHILIPP. GAGLIARD. ROM. deline.
C. Cungio Incidit

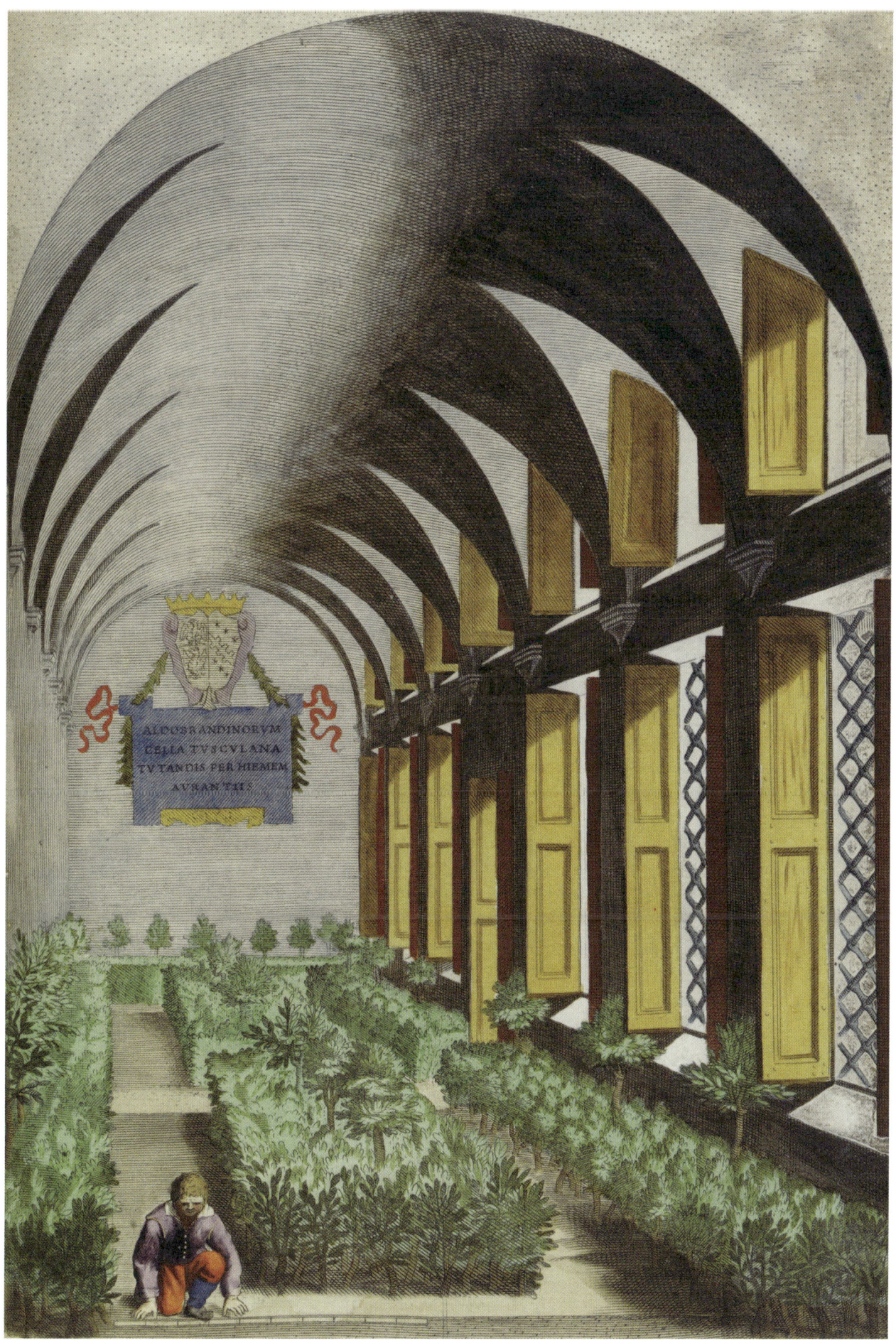

ALDOBRANDINORVM
CELLA TVSCVLANA
TVTANDIS PER HIEMEM
AVRANTIIS

All summer long, upon the mountain slopes steep by the lake, stand the rows of naked pillars rising out of the green foliage like ruins of temples ... as if they remained from some great race that had once worshipped here.

D. H. LAWRENCE ON THE *LIMONAIA* OF LAKE GARDA, *TWILIGHT IN ITALY*, 1916

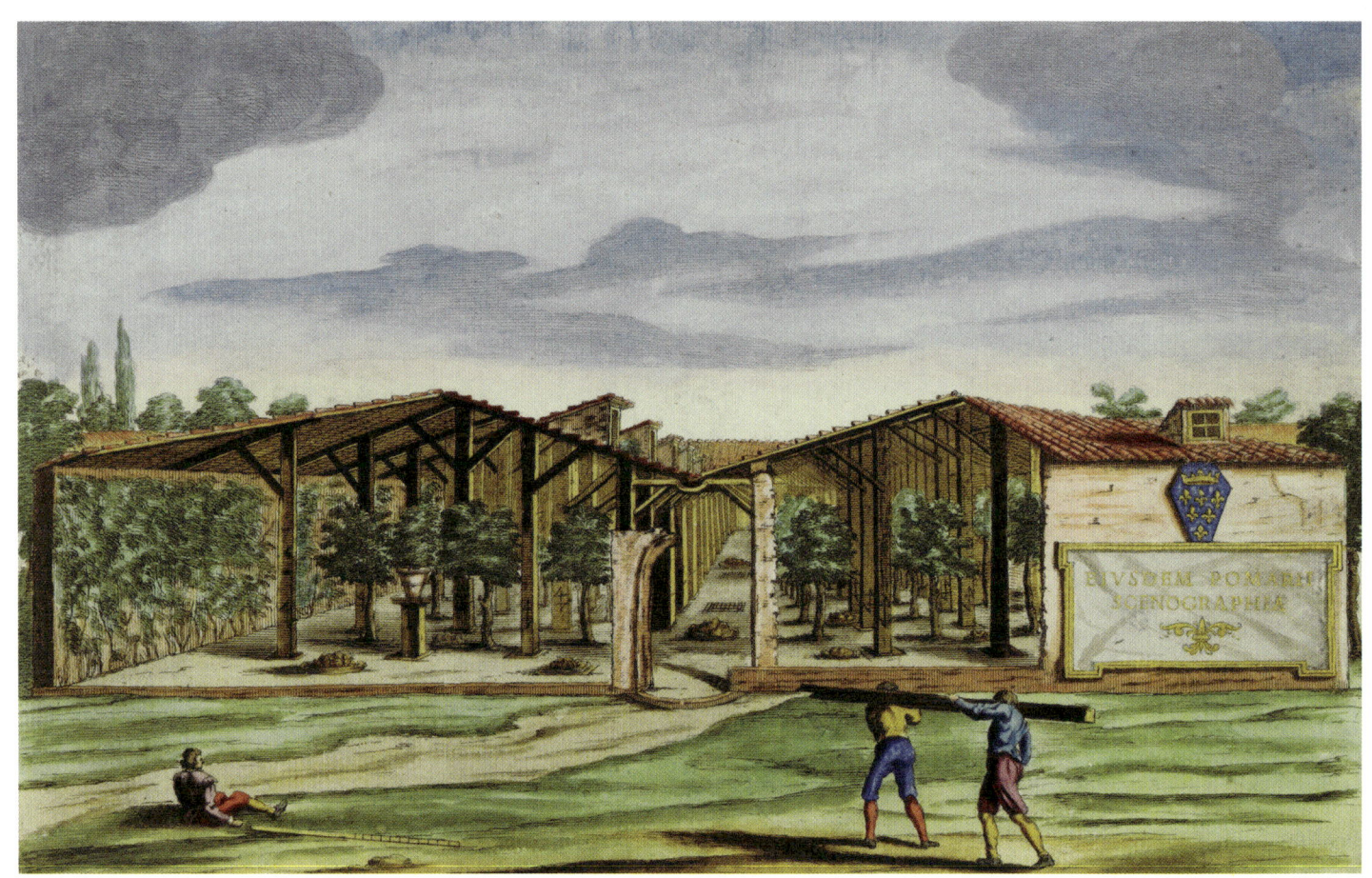

PREVIOUS PAGES, ABOVE AND OPPOSITE: *Limonaia* and orangeries, engravings after Filippo Gagliardi from *Hesperides* by Giovanni Battista Ferrari, 1646.

As the fashion for citrus-growing and collecting gained momentum, protective structures were required to protect the trees both in winter and summer. These varied from the vaulted orangery of the Aldobrandini family in Frascati (p. 95) to kinds of shed with slatted wooden walls that prevented the trees from becoming shrivelled in the summer sun (p. 94). But orangeries, as these structures later became known, were more than horticultural necessities. As Renaissance humanism encouraged the revival of the culture of Classical antiquity, the association of citrus with the 'golden apples' of the Hesperides became more deeply entrenched. The cultivation of citrus became a way to embrace these ideals – but only for those who were well versed in Classical literature. Orangeries were a status symbol, not only of the wealth required to grow the exotic fruits, but also of intellectual cachet. Ferrari's monumental work included several plates of real orangeries owned by the enlightened.

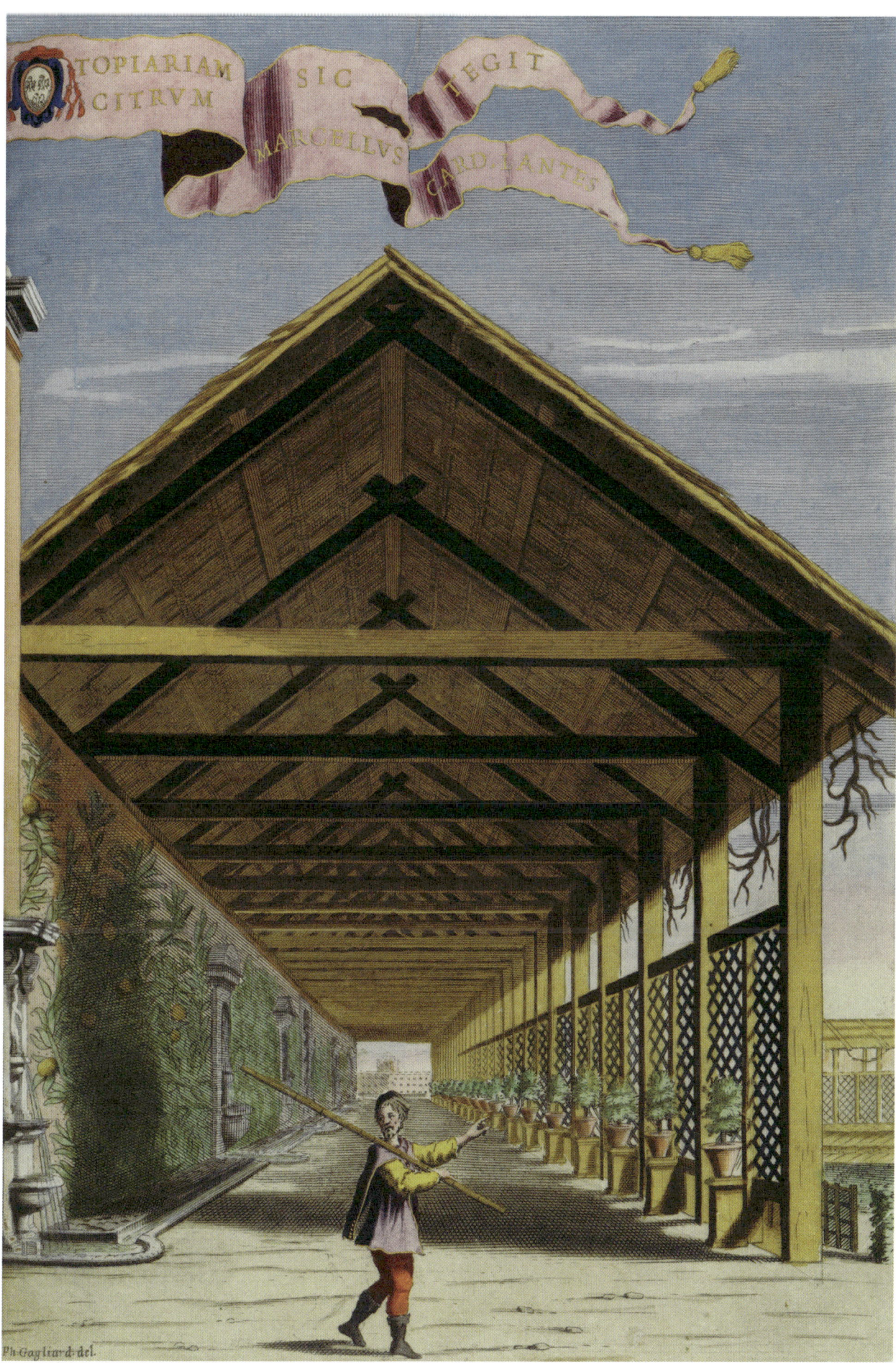

TOPIARIAM CITRVM SIC TEGIT MARCELLVS CARD. LANTES

Ph. Gagliardi del.

also gardeners to tend them. When the French King Charles VIII (1470–1498) conquered Italy in 1494–98, he saw the fruits of Alfonso I's initiative in Alfonso II's gardens of his summer residence, Villa Poggio Reale (now lost) just outside Naples, laid out along the lines of his ally Lorenzo de' Medici's Villa di Castello. Charles took back to France one of the gardeners and the gardens' designers to update the gardens at his residences, including his birthplace, Chateau d'Amboise, thereby bringing citriculture – and inspiration for what was to become the French formal garden – to the French court.

It has been forcibly argued that, ultimately, the fashion for citrus and orangeries is rooted in the intellectual milieu of Renaissance humanism of the fifteenth and sixteenth centuries, a movement focused on the revival of the culture of Western antiquity through studying and translating the works of Classical writers.[55] With regard to citriculture, this involved, on the one hand, focus on the completely confused 'golden apples' story (see p. 42), and, on the other, the importing of plants and expertise from Spain and Sicily, where citrus had long been cultivated. The Italian humanist architect, poet and priest Leon Battista Alberti (1404–1472) wrote on classical gardens, emphasizing the importance of citrus trees among other evergreens like myrtle (*Myrtus communis*) and bay (*Laurus nobilis*). For the humanists, orangeries were the winter home not only for citrus and other evergreens but also for humans, where they could interact with the plants. As the art historian Helmut-Eberhard Paulus (b. 1951) wrote, 'The task of the orangery was to promote the golden fruit and the evergreen flowering citrus tree as an attribute of man and his immediate social environment. Orangeries were allegorical spaces, which took on a deeper meaning only through the philosophical interactions of the people present in them.'[56]

FERRARI'S *HESPERIDES*

It was only when the growing of citrus in orangeries, for whatever motive, became fashionable all the way from southern to northern Europe that there was any European treatise on citrus to surpass the Chinese *Ju Lu* of 1178 (see p. 33). Now there were live specimens to draw, measure, weigh and dissect in magnificent collections, in the open and under glass, so that in 1646 there appeared in Rome the magnificent synthesis of citrus *Hesperides sive De malorum aureorum cultura et usu* (Hesperides, or, The Cultivation and Use of Golden Apples) by the Sienese Giovanni Battista Ferrari (c. 1584–1655).

He is remembered now for just two works: his *Hesperides* and the earlier *De florum cultura* (1633), a heavily illustrated treatise on flowers and their cultivation notable not only for its pioneering stress on the ornamental, rather than merely medicinal, qualities of plants, but also for including some of the earliest printed images based on microscope findings.[57] Ferrari also took great care over innovative nomenclature at a time when new plants from the 'Indies' were challenging for those dependent merely on the Classical sources.[58]

Ferrari's *Hesperides* was essentially a collaboration with Cassiano dal Pozzo (1588–1657), who was admired above all for his knowledge of natural history and antiquities, but also for his Paper Museum,[59] an ambitious project that included illustrations of living plants and animals, geological specimens and the known material remains of antiquity. Although none of it was published until the twentieth century, the Paper Museum was accessible to scientists, antiquaries and artists. A prominent collector and patron living in Rome, Cassiano had been born in Turin,[60] but he went to live with his father's cousin, the Archbishop of Pisa, an intimate of Ferdinando I de' Medici, Grand Duke of Tuscany (1549–1609). Cassiano was in the Pisa intellectual circle that included Galileo Galilei (1564–1642), centred on the university and its botanic garden. In 1622 he was elected to the first-established learned society in Europe, the Accademia dei Lincei. The revolutionary scientific work of the Accademia no doubt chimed with Cassiano's apparently unorthodox religious views and the principles of direct observation and experiment that informed his approach to natural history.

In 1631, Cassiano sent three manuscript accounts of flowers and citrus to the polymath Nicolas-Claude Fabri de Peiresc (1580–1637) at Boisgency, France. Peiresc replied in some desperation, 'I have no difficulty with the flowers and other plants, but in the case of the citrus fruits, I really cannot grasp either the differences between the fruits or between their names.' He pressed Cassiano to get his citrus drawings reproduced in Ferrari's *De florum cultura*, but they did not appear until after Peiresc's death – and then in *Hesperides*.[61] Both books had fanciful allegorical scenes (devised by Ferrari himself) as well as plant illustrations, almost all engraved by the Dutch printmaker Cornelis Bloemaert (1603–1692). Apparently intending a work, *Pomona*, to cover all fruits, Ferrari confined his second horticultural book to citrus fruits, a treatise compiled from 1635 onwards.[62] Rather than relying entirely on Classical texts as authority, his was, in part, an empirical approach, with fresh descriptions of the fruits – the colour and texture of the peel and the seeds – with notes on the vernacular names and the culinary and medicinal uses made of them, garnered by responses to a questionnaire he prepared to be distributed to growers by Cassiano.

Ferrari's *Hesperides* is celebrated not only for Cassiano's contributions but also for the work of other artists who provided the allegorical illustrations, including more than forty paintings by the studio of Nicolas Poussin (1594–1665). The original watercolours of the plants themselves were the work of the little-known Vincenzo Leonardi (*c.* 1590–*c.* 1646). Bloemaert's eighty fruit plates based on them depict eight kinds of citron, thirty-nine lemons, nine 'strange' fruits grouped with lemons, four limes and twenty oranges. Two of them were reproduced in the next century by Volkamer in his *Nürnbergische Hesperides* (1714; see p. 135). The *Hesperides* shows not only the fruits from the outside but also cross-sections, sometimes longitudinal, with the numbers of segments and seeds accurately portrayed in ways no one had done for citrus before. Although Leonardi's drawings were coloured, very few examples of the published book have original coloured engravings.

Much has been made by earlier commentators of Ferrari's bringing order to a scientific understanding of the genus, but it must be stressed that the fundamentals for his classification were far from scientific in the sense of the Accademia dei Lincei at the time. The book's very title perpetuated the confusion of citrus with quinces: the text opens with Heracles searching for the golden apples in the garden of the Hesperides at the west of the then known world (as the last of his twelve labours), a story Ferrari traced through the Classical literature. *Hesperides* documents around one thousand sorts of citrus, but, apparently mesmerized or constrained by his mythological sources, Ferrari divided the main part of the book into three sections, so that each of the Hesperides is celebrated. Aegle (now commemorated in the genus *Aegle*, which includes the bel tree or Bengal quince) presides over the citrons, Arethusa over lemons and Erythia oranges.

Ferrari's synthesis combined information drawn from not only botany and horticulture but also art, literature and ethnography,[63] and included recipes, herbal remedies and descriptions of the *limonaie* of the cognoscenti. One plate shows the vaulted orangery of the Aldobrandini family in Frascati (see p. 95), while others show sheds with

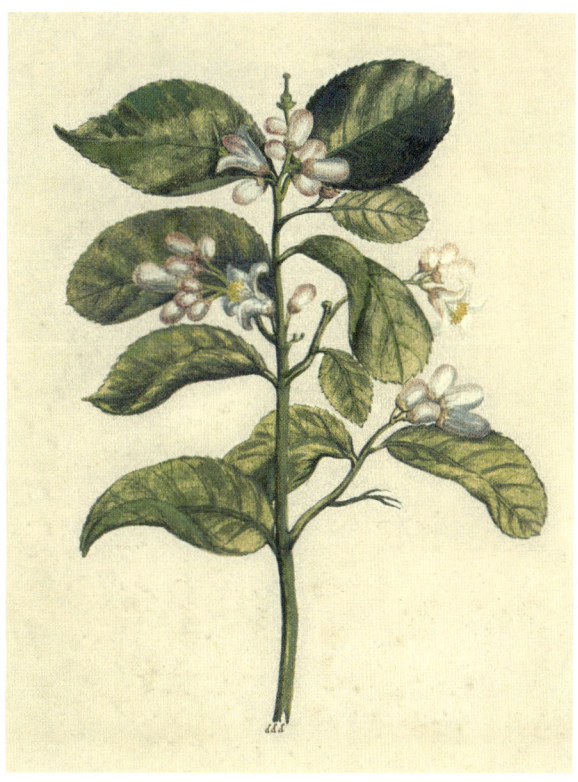

A flowering lemon twig, Vincenzo Leonardi, before 1646. Watercolours for Cassiano dal Pozzo's Paper Museum were reproduced in the *Hesperides* of Giovanni Battista Ferrari.

slatted wooden walls that prevented the trees from becoming shrivelled in the summer sun (see p. 97). Encyclopaedic in scope, the book drew together all that was known on the genus, even including the then current practice of 'citrus fights' in Reggio Calabria and Treviso. Since medieval times, citrus fruits have been used in various celebrations as at Reggio Calabria and Ivrea, Turin, with fruit from Sicily.[64] Such fights are still held, as in the Battle of the Oranges in Ivrea, where, according to the *New York Times*, 900 tonnes of oranges from southern Italy were thrown over three days in 2023. In Binche, Belgium, 'the richly costumed Gilles, mounted on stilts, bombard onlookers with oranges from baskets during Carnival.'[65] Rather more decorous are the annual citrus festivals held in many citrus-growing areas (see pp. 238–41).

Despite Cassiano's empirical approach to natural history, Ferrari was not immune to the fanciful ideas of pre-scientific thought.

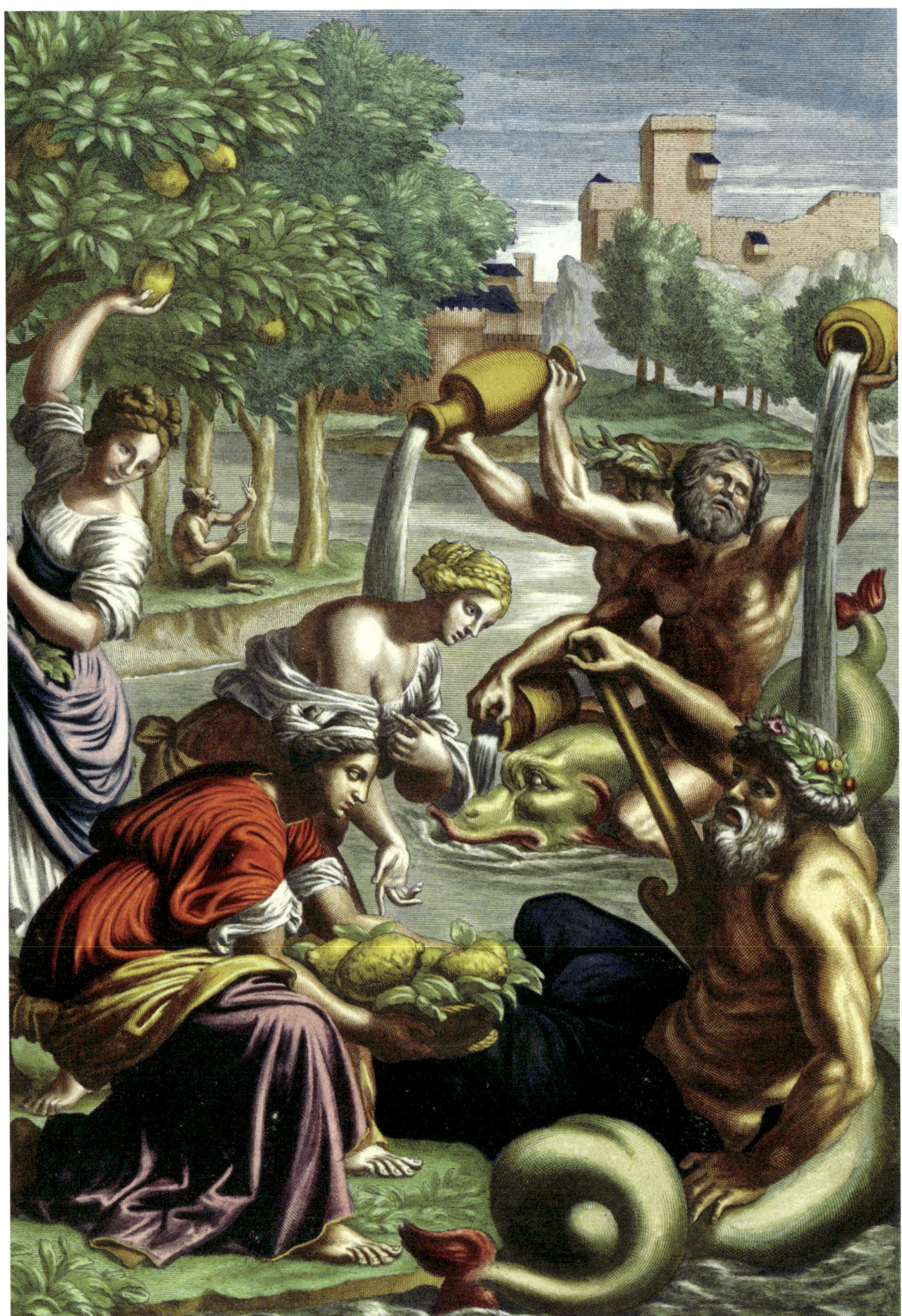

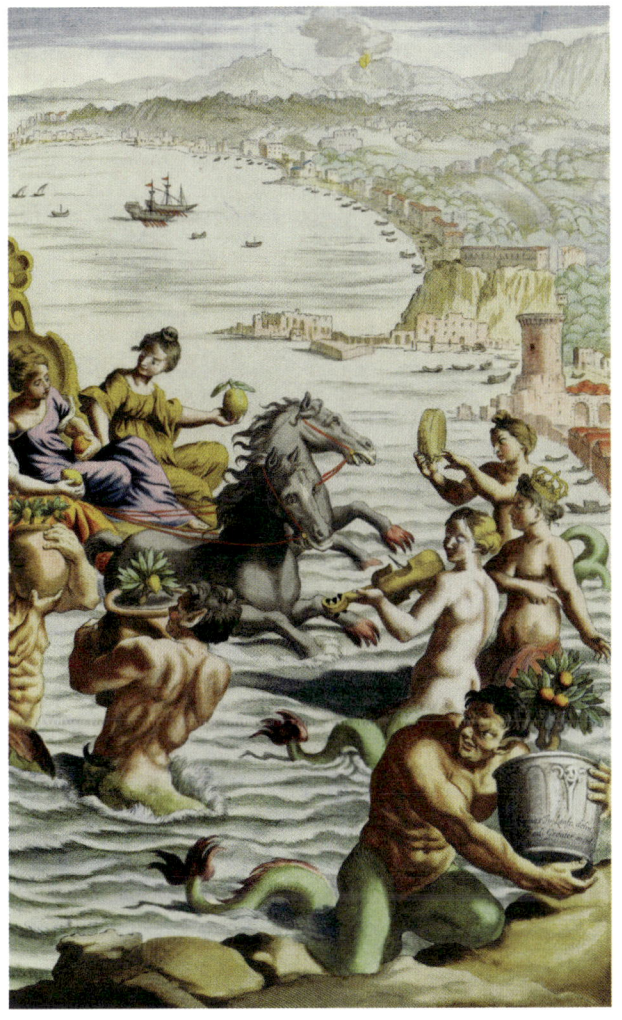

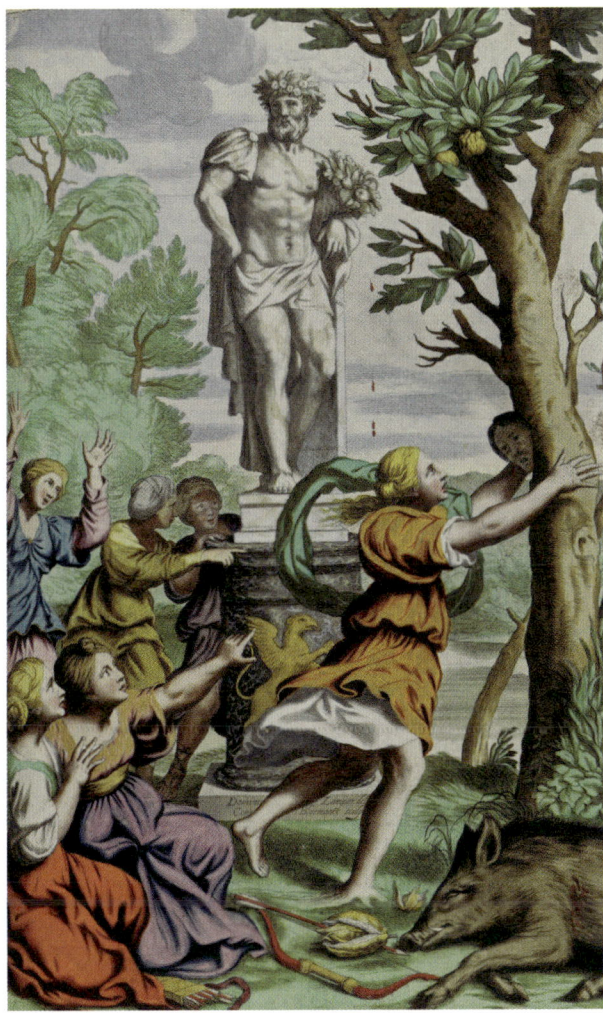

Allegorical engravings from *Hesperides* by Giovanni Battista Ferrari, 1646.

Ferrari was a Jesuit priest who taught Hebrew in a seminary in Rome before becoming horticultural advisor to the papacy. His book on citrus was the first in the West devoted entirely to the genus. Though Ferrari's approach often appears to be scientific – informed by direct observation – the work was punctuated by references to the Hesperides myth: the Hesperides nymphs presenting lemons to the god of Lake Garda (opposite) and the three sisters bringing lemons, oranges and citrons to Naples (above, left). Ferrari even included fantastical tales of his own invention to account for 'monstrous' forms of fruit, such as the fingered citron or the *aurantium distortum* (misshapen orange), with humans magically transformed into trees (above, right).

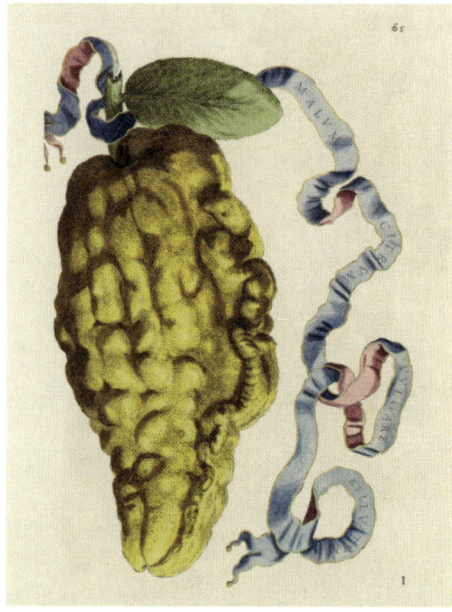

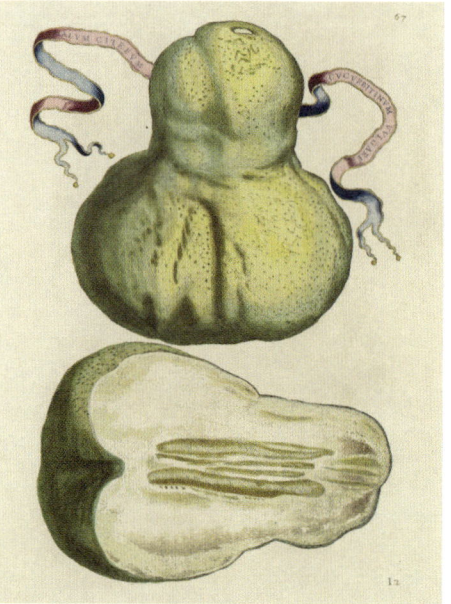

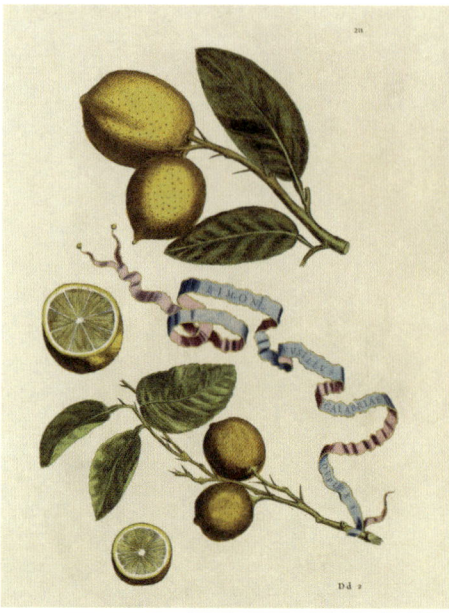

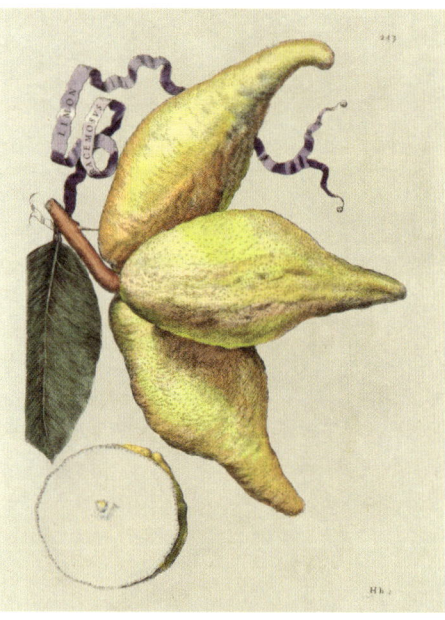

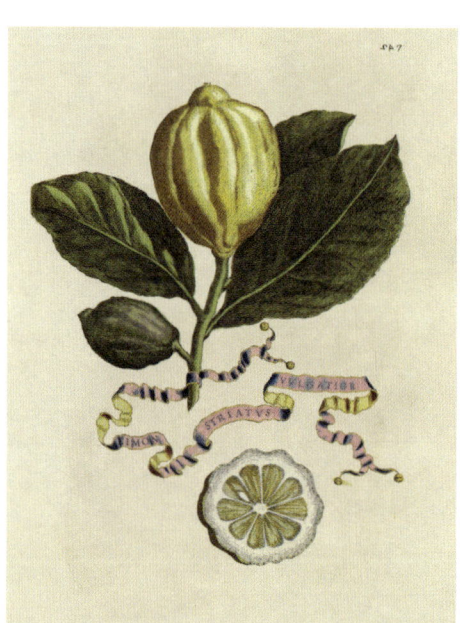

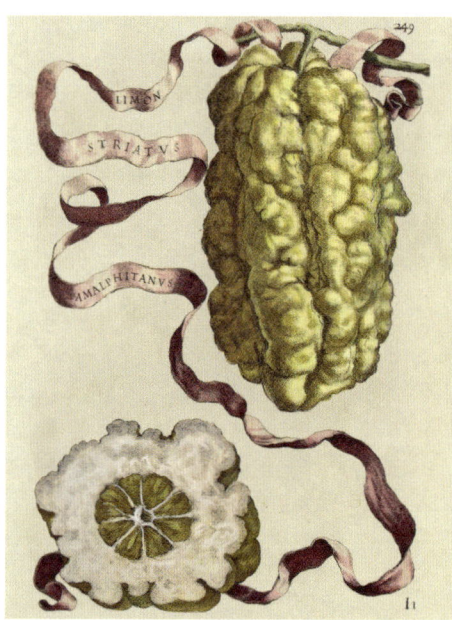

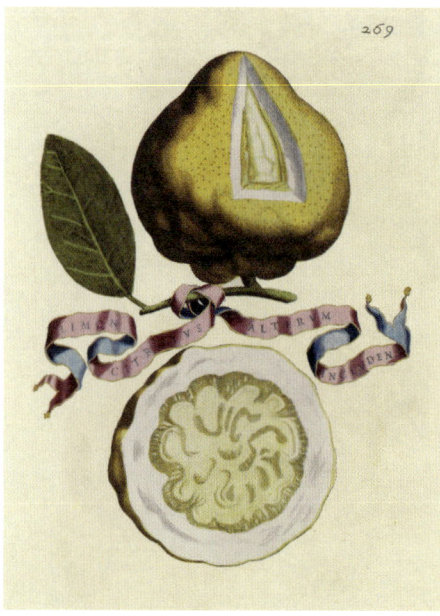

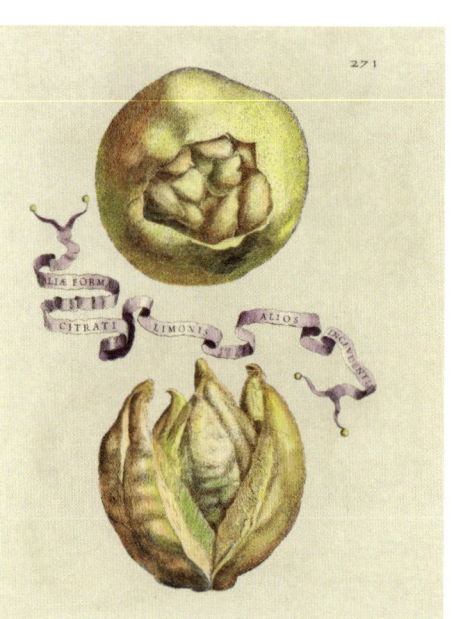

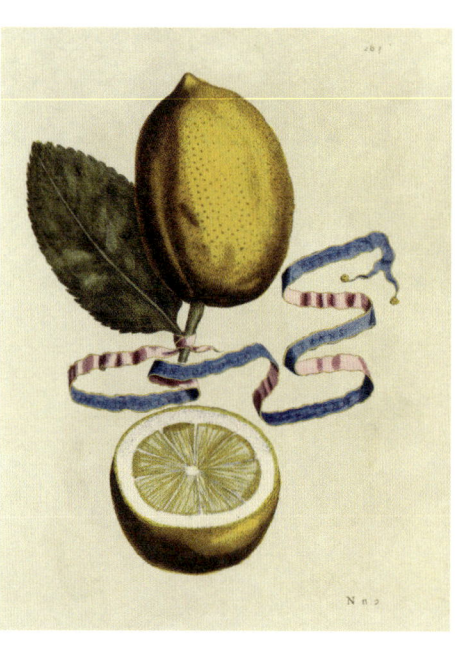

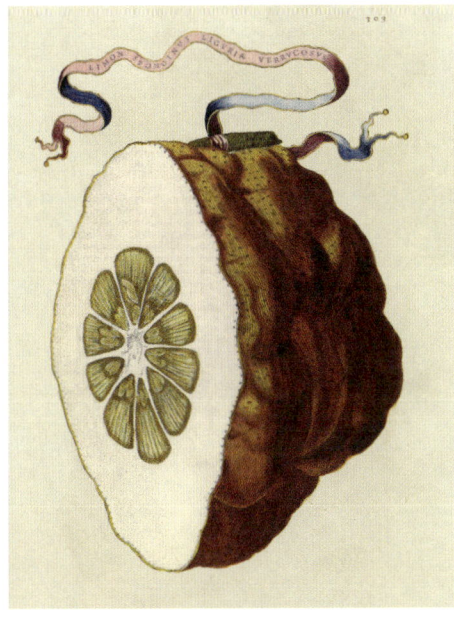

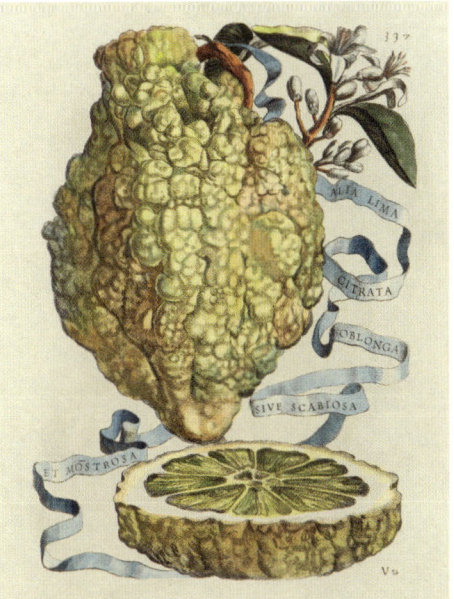

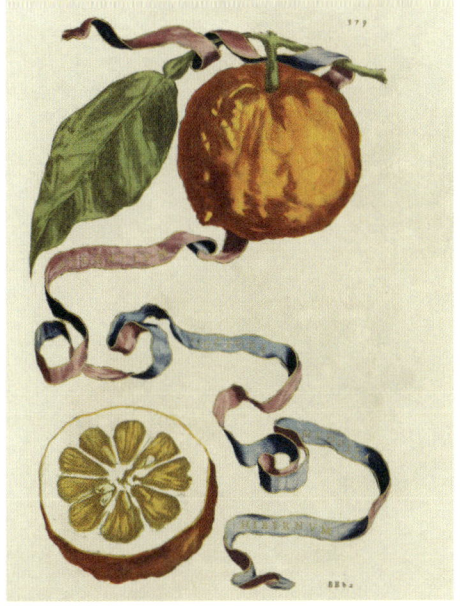

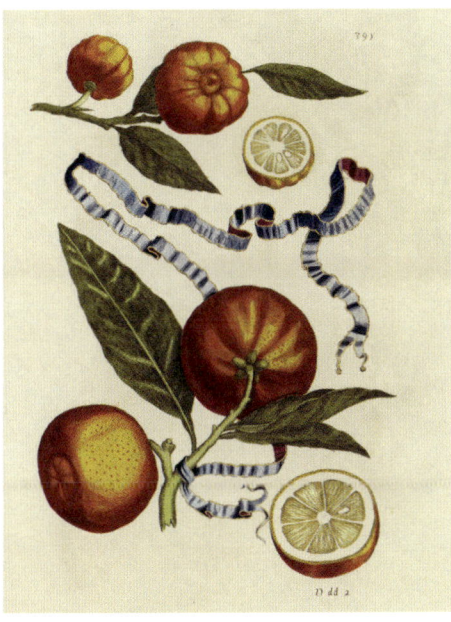

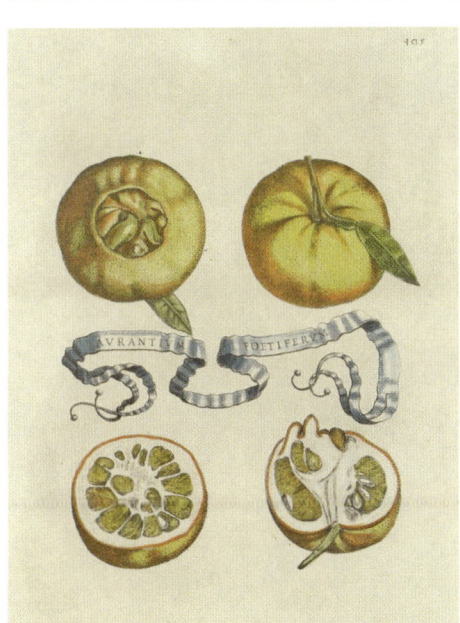

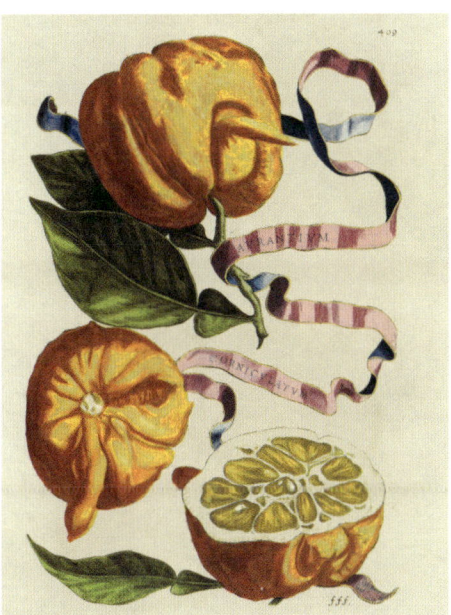

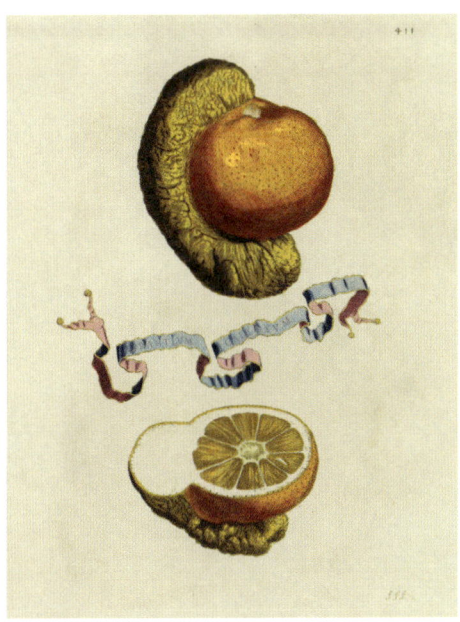

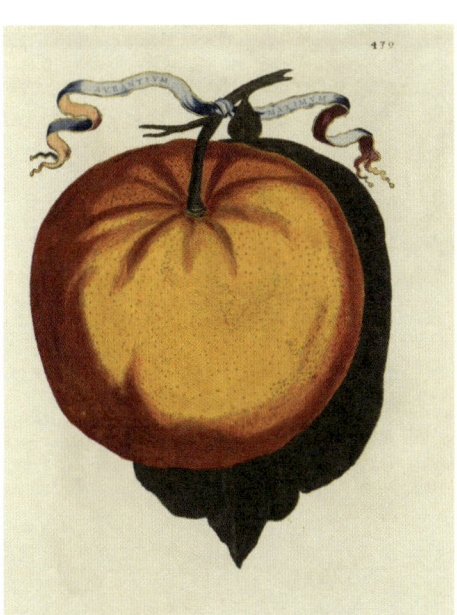

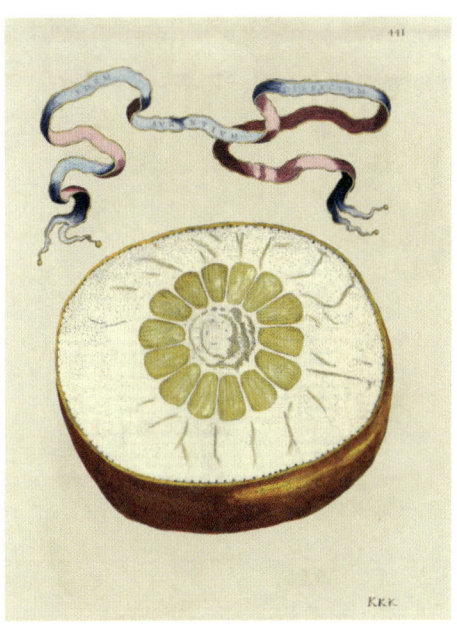

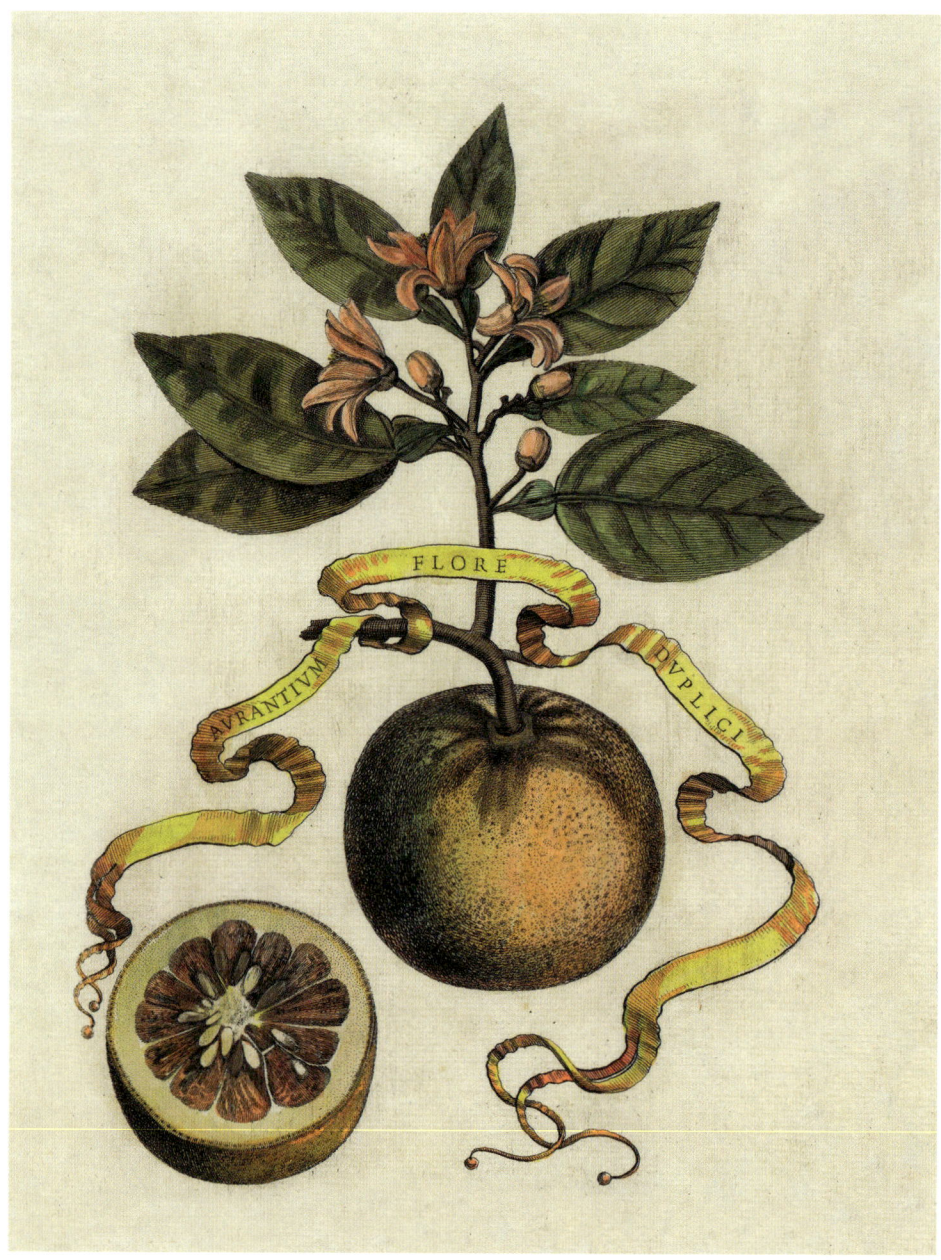

PREVIOUS PAGES, ABOVE AND OPPOSITE: Engravings after Vincenzo Leonardi from *Hesperides* by Giovanni Battista Ferrari, 1646.

The heart of Ferrari's master work are the eighty botanical plates, based on watercolours by Vincenzo Leonardi, depicting a wide range of cultivated citrus. The fruits are portrayed both whole and in section, along with separate studies of the flowers and leaves. Careful attention is paid to shape and size, and the colour and texture of the peel. The book is so detailed that is often possible to discern cultivars known today: Ferrari's *aurantium stellatum et roseum* is likely 'Selecta' ('Seleta'), a very common sweet orange in Spain and Portugal in the 17th century that is still grown, being an early introduction to Brazil.

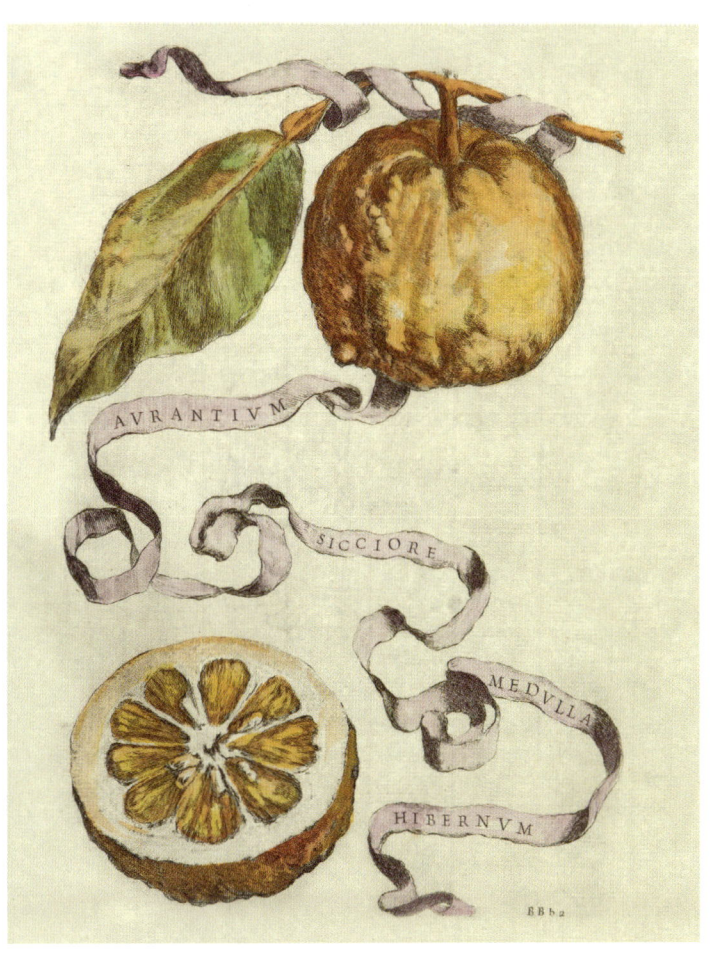

AVRANTIVM SICCIORE MEDVLLA HIBERNVM

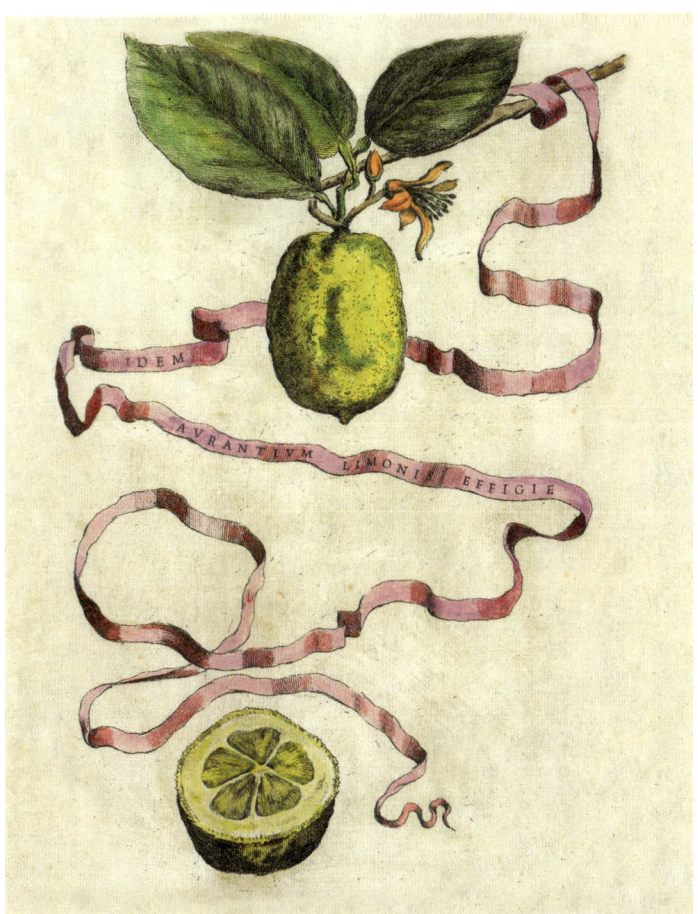

IDEM AVRANTIVM LIMONII EFFIGIE

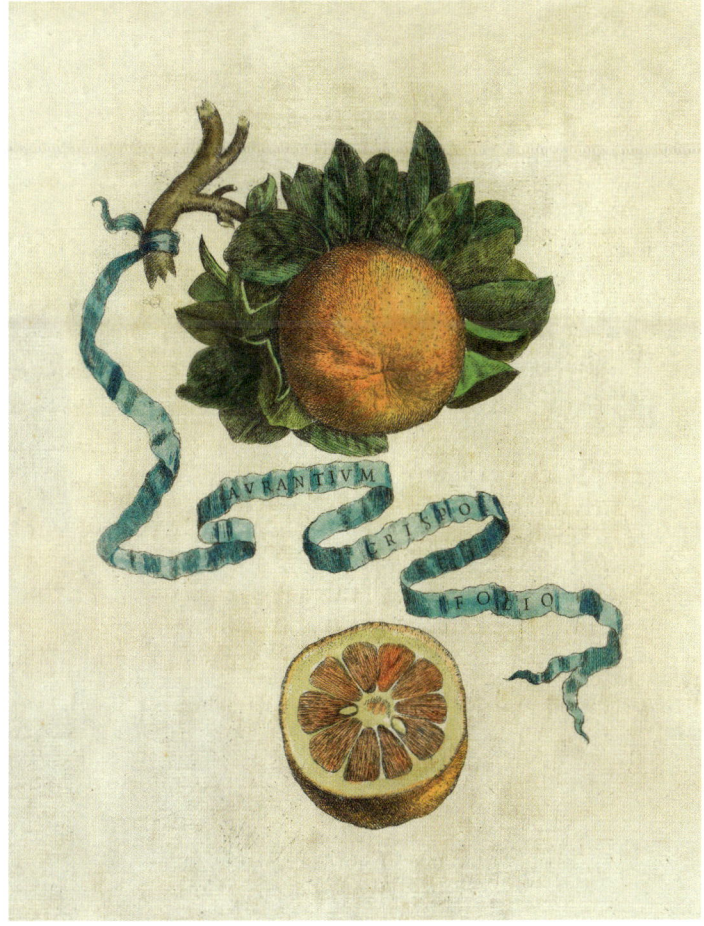

AVRANTIVM CRISPO FOLIO

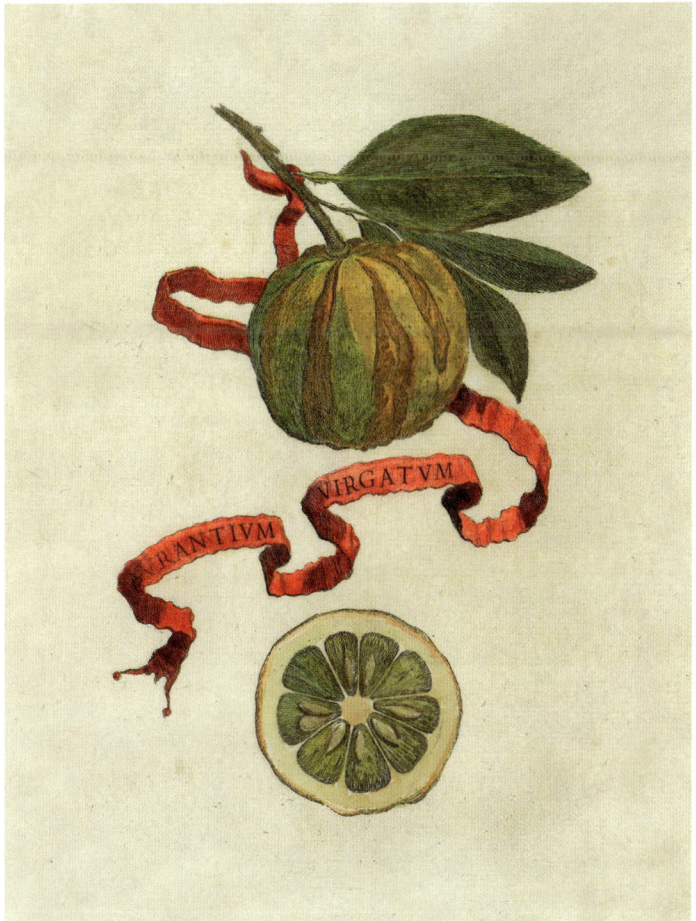

AVRANTIVM VIRGATVM

The fingered citron of Bloemaert's engraving after Andrea Sacchi shows the metamorphosis of a man into a tree, with legs taking root and hands transformed into fingered citrons, illustrating Ferrari's tale of Harmonillus, whose sweet singing voice irritated a witch to such an extent that she turned him into a citron tree. Ferrari noted that 'Nature plays no more licentiously in any other fruit' than in the fingered citron, but his explanation for its origin was equivocal, while he thought that navel oranges were the result of unusual soil conditions. For him, all such 'monstrous' fruits were instances of the *lusus naturae* (playing of nature), and he padded his text with poetic invention to explain them away.[66]

For such a lavish book, Ferrari needed patronage and endeavoured, through Cassiano and Poussin, to get that from Louis XIII of France, but no interest was shown and Louis died in 1643. Ultimately the book was dedicated to the city of Siena, Ferrari's home, and there is some evidence of Sienese support, though Cassiano himself likely made an important subvention, and it is possible that the publisher took on most of the financial risk.

RUMPF

The first European account of citrus as grown in tropical regions was written far away from Italy, in one of the Dutch colonial outposts in what is now Indonesia – and publication of it was far less felicitous than that of Ferrari's *Hesperides*. On the island of Ambon in the Moluccas, Georg Eberhard Rumpf (1627–1702) compiled his definitive *Herbarium Amboinense*, a work that established the basis for European study of south-east Asian botany and dealt with more than 1,300 kinds of plants from the region.[67] Rumpf was a German mercenary who had served with the Dutch West India Company and later signed up as a gentleman-soldier (*adelhorst*) with the Dutch East India Company. Leaving from the Dutch port of Texel in December 1652, he reached Batavia (today's Jakarta) in Java the following July and the company's settlement on Ambon in the Moluccas a few months later. The Dutch controlled the trade in cloves, nutmeg (and therefore mace, the fleshy aril around the nutmeg) and, by 1656, cinnamon.

Rumpf was fluent in Dutch, German, Portuguese, Malay and the Ambonese dialect, and had a 'working knowledge' of Hebrew, Ancient Greek, Latin and some facility with Chinese. By 1663 he was working on his book, helped by his local companion Susanna. In his comprehensive survey of native and introduced trees and other plants, Rumpf wrote up impressive accounts of citrus-growing not only on Ambon but also elsewhere in the region. He remarked on the interesting fact that oranges 'were not propagated by artful grafting onto [others] as some European Authors have falsely claimed: For these Apples grow throughout the entire Indies in a fixed shape, also among the Nations who do not know the art of grafting';[68] in other words, 'fixed genetic lines' or apomictic clones (see p. 30). He also described sweet oranges – as 'Chinese apples' – and what appear to have been mandarins, 'somewhat smaller and smoother, flat on top and bottom ... with a thin peel ... which can be easily removed', but noting that they were difficult to grow successfully in Ambon, though 'abundant in China ... and shipped everywhere'.[69]

By 1670, Rumpf was blind and was replaced in his posting. Early in 1674, during the celebrations for Lunar New Year, an earthquake and tsunami hit, killing 2,322 people, including Susanna and her youngest daughter. Rumpf had to have someone else read to him his original Latin manuscript (which had his own drawings), and then he translated it into Dutch to be taken down by an amanuensis, such that it was 'nearly finished' in 1687. Then on 11 January a major fire destroyed his house and all his possessions – the manuscript was saved, though the illustrations were lost. The first six of the twelve 'books' were finished in 1692 and sent to Amsterdam, but the Dutch ship carrying them was sunk by the French.

Fortunately, a copy had been ordered to be made by the Governor General, who also arranged for fresh illustrations (which Rumpf could not of course check for accuracy). The first nine 'books' were sent to Amsterdam in 1696 and the last three the following year, but, because Rumpf's manuscript was so comprehensive, it revealed a lot of commercial trade information, especially on cloves and nutmeg, and the company declined to print it in his lifetime.[70] The work was embargoed for forty years before Johannes Burman published his masterpiece with a 'commentary' in 1741–55. Before Rumpf died in the Moluccas, his collections were sold – to Cosimo III de' Medici, Grand Duke of Tuscany.

OPPOSITE: Frontispiece to the first volume of *Herbarium Amboinense* by Georg Eberhard Rumpf, 1741. This definitive work catalogued over 1,300 plants grown in Indonesia in some 1,661 folio pages. **OVERLEAF:** Villa di Castello, Giusto Utens, 1599. The Medici gardens at Castello had a major influence on the evolution of Renaissance gardens.

HERBARIUM
AMBOINENSE,

Plurimas conplectens Arbores, Frutices, Herbas, Plantas terreftres
& aquaticas,

QUAE IN AMBOINA,
ET ADJACENTIBUS REPERIUNTUR INSULIS,

Adcuràtiffime defcriptas juxta earum formas, cum diverfis denominationibus,
cultura, ufu, ac virtutibus.

Quod & infuper exhibet

VARIA INSECTORUM ANIMALIUMQUE GENERA,

Plurima cum naturalibus eorum figuris depicta.

**Omnia magno labore ac ftudio multos per annos conlegit,
& duodecim libris Belgice confcripfit**

GEORG. EVERHARD. RUMPHIUS,

Med. Doct. Hanavenfis, Mercator Senior, & in Amboina Conful, nomine
PLINII INDICI celeber, & Inluftris Societatis Academiae
Naturae Curioforum Germaniae Membrum.

Nunc primum in lucem edidit, & in Latinum fermonem vertit

JOANNES BURMANNUS,

MED. DOCT. ET IN HORTO MEDICO AMSTELAEDAMENSI PROFESSOR BOTA-
NICUS, ACADEMIAE CAESAREAE NATURAE CURIOSORUM SOCIUS;

Qui varia adjecit Synonyma, fuafque Obfervationes.

PARS PRIMA.

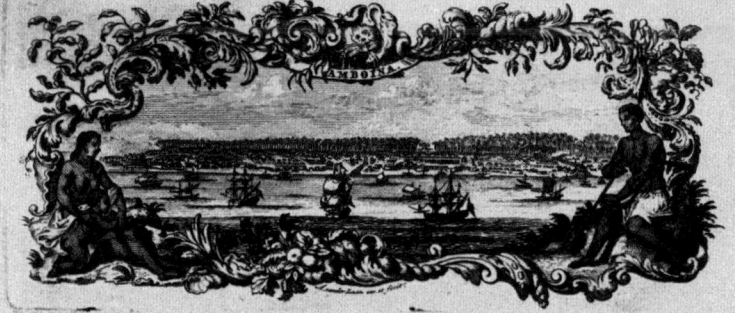

AMSTELÆDAMI, { Apud FRANCISCUM CHANGUION, JOANNEM CATUFFE, HERMANNUM UYTWERF.

HAGÆ COMITIS, { Apud PETRUM GOSSE, JOANNEM NEAULME, ADRIANUM MOETJENS, ANTONIUM VAN DOLE.

ULTRAJECTI, Apud STEPHANUM NEAULME.

M. DCC. XLI.

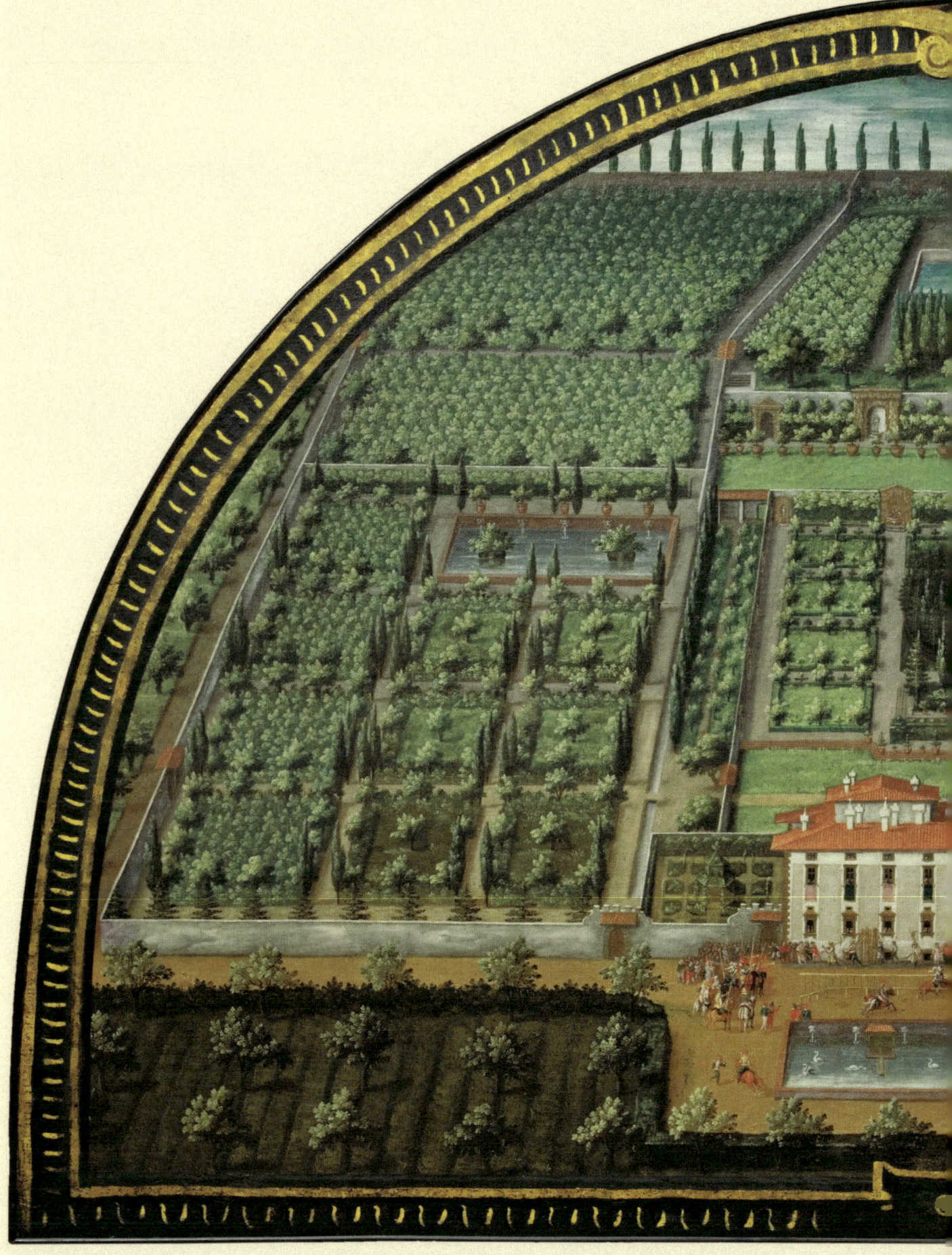

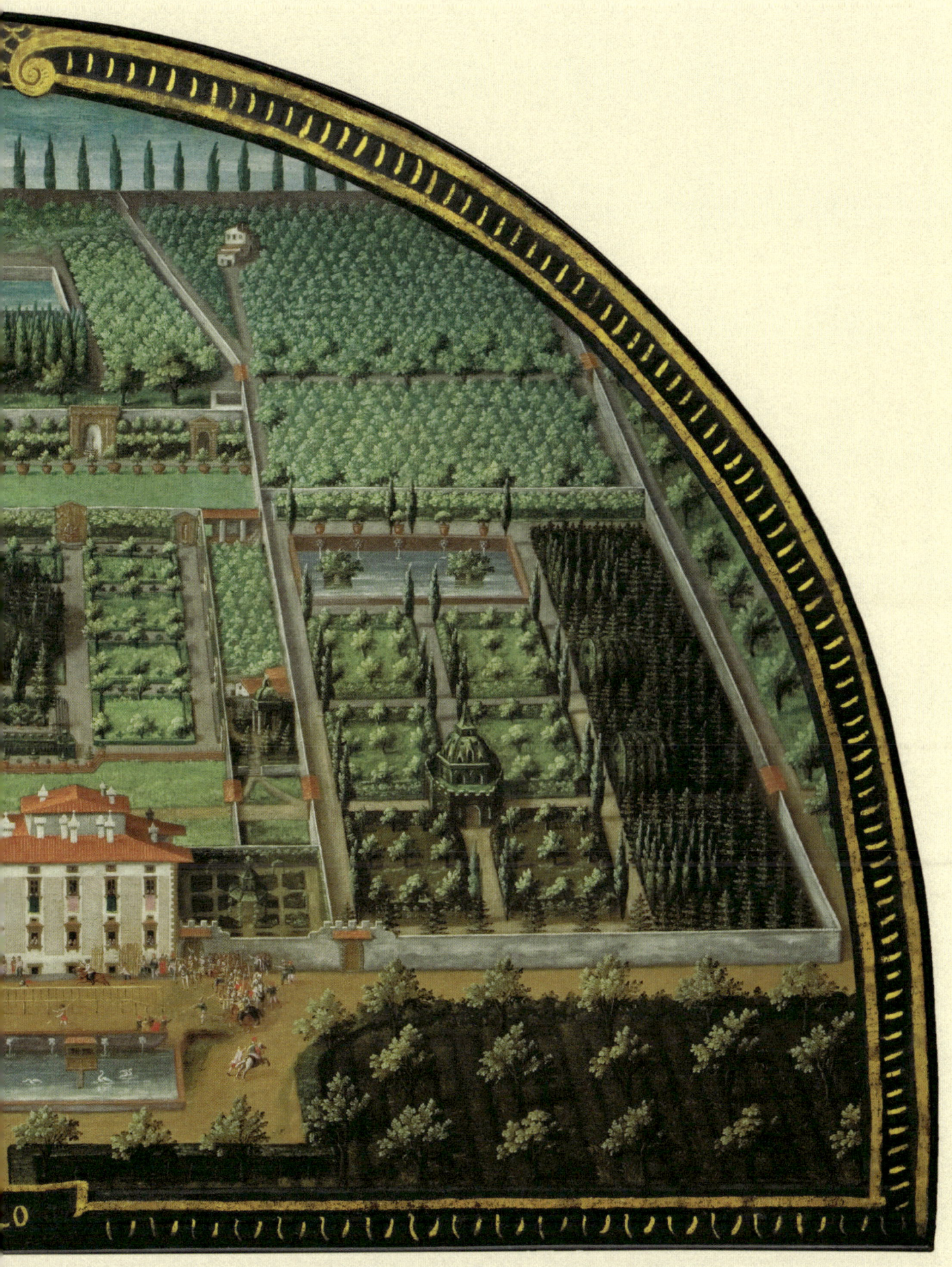

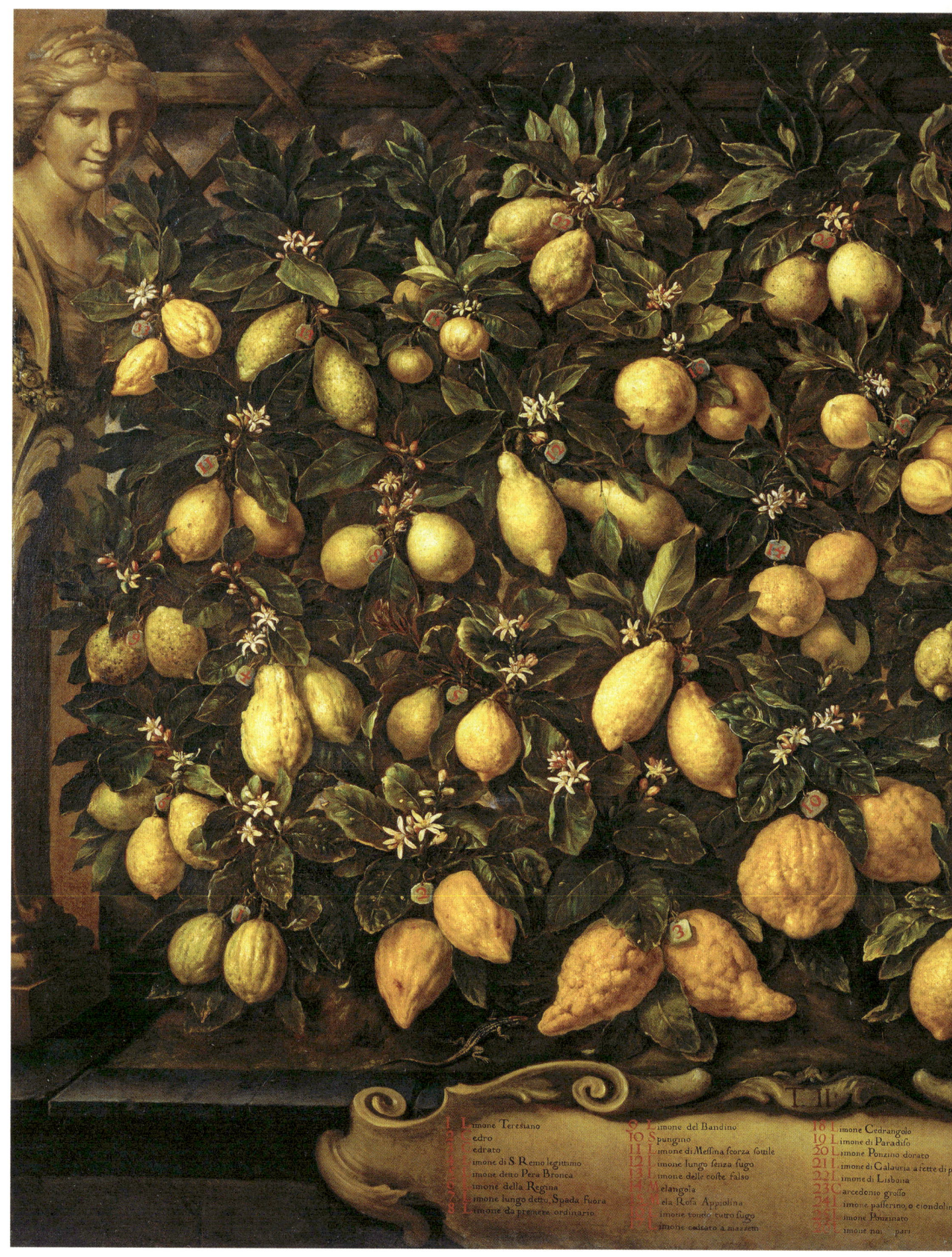

imone Teresiano
edro
edrato
imone di S. Remo legitimo
imone detto Pera Bronca
imone della Regina
imone lungo detto, Spada Fuora
imone da premere ordinario

L imone del Bandino
S pungino
L imone di Messina scorza sotile
L imone lungo senza sugo
L imone delle coste falso
M elangola
ela Rosa Appiolina
imone tondo tutto sugo
imoris castrato a mazzetti

16 L imone Cedrangolo
19 L imone di Paradiso
20 L imone Ponzino dorato
21 L imone di Calauria a fette di p
22 L imone di Lisbona
23 C arcedonio grosso
24 imone passerino, e ciondolin
imone Ponzinato
imone na   pari

Medici Citrus, Bartolomeo Bimbi, 1715.

1. Limone Teresiano
2. Cedro
3. Cedrato
4. Limone di S. Remo legittimo
5. Limone detto Pera Bronca
6. Limone della Regina
7. Limone lungo detto Spada fuora
8. Limone da premere ordinario

9. Limone del Bandino
10. Spungino
11. Limone di Messina scorza sottile
12. Limone lungo senza sugo
13. Limone delle coste falso
14. Melangola
15. Mela Rosa Appiolina
16. Limone tondo tutto sugo
17. Limone cedrato a mazzetti

18. Limone Cedrangolo
19. Limone di Paradiso
20. Limone Ponzino dorato
21. Limone di Calauria a fette di popone
22. Limone di Lisbona
23. Carcedonio grosso
24. Limone passerine, o ciondolino
25. Limone Ponzinato
26. Limone [...]

27. Limone S. Maura
28. Cedro di Paradiso senza sugo
29. Limone del Rio
30. Limone di Madonna Laura
31. Limone Barba d'Oro
32. Spada fuora mezzana
33. Spada fuora bastarda del fior doppio
34. Limone a scorza d'Arancio bizzarro

Under the patronage of Cosimo III de' Medici, Bimbi created four monumental canvases depicting 116 kinds of citrus then grown in Medici gardens. The paintings were originally hung in an antechamber next to Cosimo's own bedroom at the Villa della Topaia. Each painting also has a legend bearing the names of the different fruits, here predominantly lemons and citrons.

Medici Citrus, Bartolomeo Bimbi, 1715.

1. Arancio Appiccio
2. Arancio da fiori domestico
3. Arancio di Portogallo dolce
4. Arancio della China
5. Bizzarria
6. Arancio Turco
7. Pera Bergamotta razza d'Arancio
8. Arancio del fior doppio

9. Arancio da fiori Appiecie
10. Arancio del Gigante
11. Arancio scannellato
12. Mela Rosa
13. Pomo d'Adamo di Paradiso
14. Lumia [...]
15. Perettone di Gaeta
16. Limone Rollottino

17. Limone dolce di Spagna
18. Limone dolce di Portogallo
19. Limone dolce di Napoli
20. Lima falsa
21. Limoncello di Napoli Calautere
22. Limone scannellato Tondo
23. Peretta di S. Domenico
24. Limone spinoso

25. Lumia Cedrata
26. Arancino a Berretta di Prete
27. Portogallo di Spagna
28. Lumia Cedrata di quarta Classe
29. Lumia fatta a piatellina di terza
    Classe senza sugo
30. Lutnia fatta a Pera di seconda Classe
31. Arancino cedrato Perfetto

Bimbi's plant portraits are so accurate
that his paintings were used by Paolo
Galeotti in 1978 as an aid to identifying
the living, though much-neglected, citrus
trees surviving from the time of the Medici.
Most depicted here are kinds of oranges
and lemons, though the *bizzarria* is also
present (no. 5).

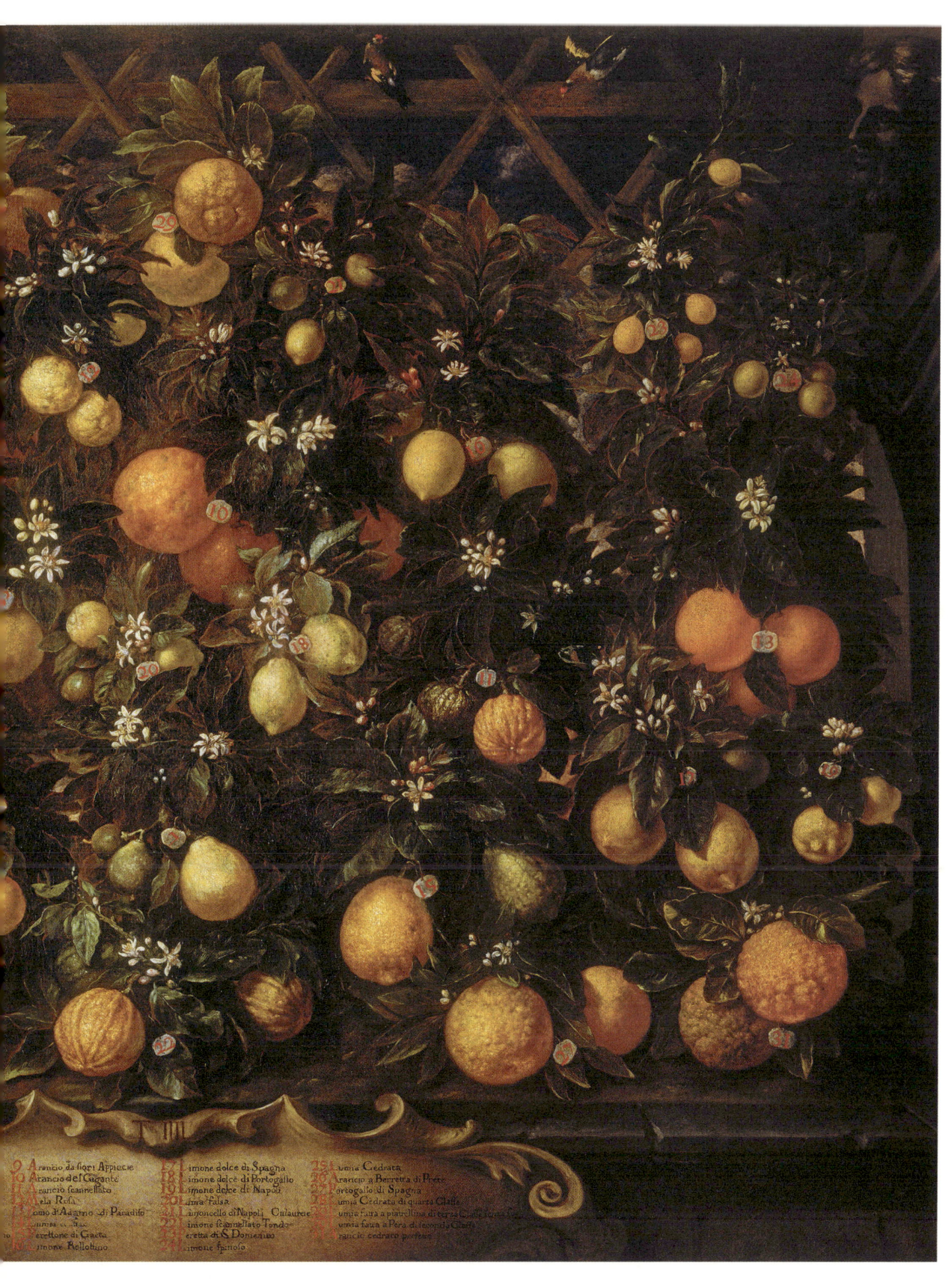

9 Arancio da fiori Appiecie
10 Arancio del Gigante
11 Arancio scanellato
12 e la Rosa
13 omo d'Adamo di Paradiso
14 mone di ...
15 erettone di Gaeta
16 imone Bollottino

17 Limone dolce di Spagna
18 imone dolce di Portogallo
19 Limone dolce di Napoli
20 imè Falsa
21 imoncello di Napoli Culaurae
22 imone scanellato Tondo
23 eretta di S Domenico
24 imone fanolo

25 uona Cedrata
26 Arancio a Berretta di Prete
27 ortogallo di Spagna
28 umia Cedrata di quarta Classe
29 umia falsa a piatrellina di terza Classe
30 umia falsa a Pera di seconda Classe
31 rancio cedrato perfetto

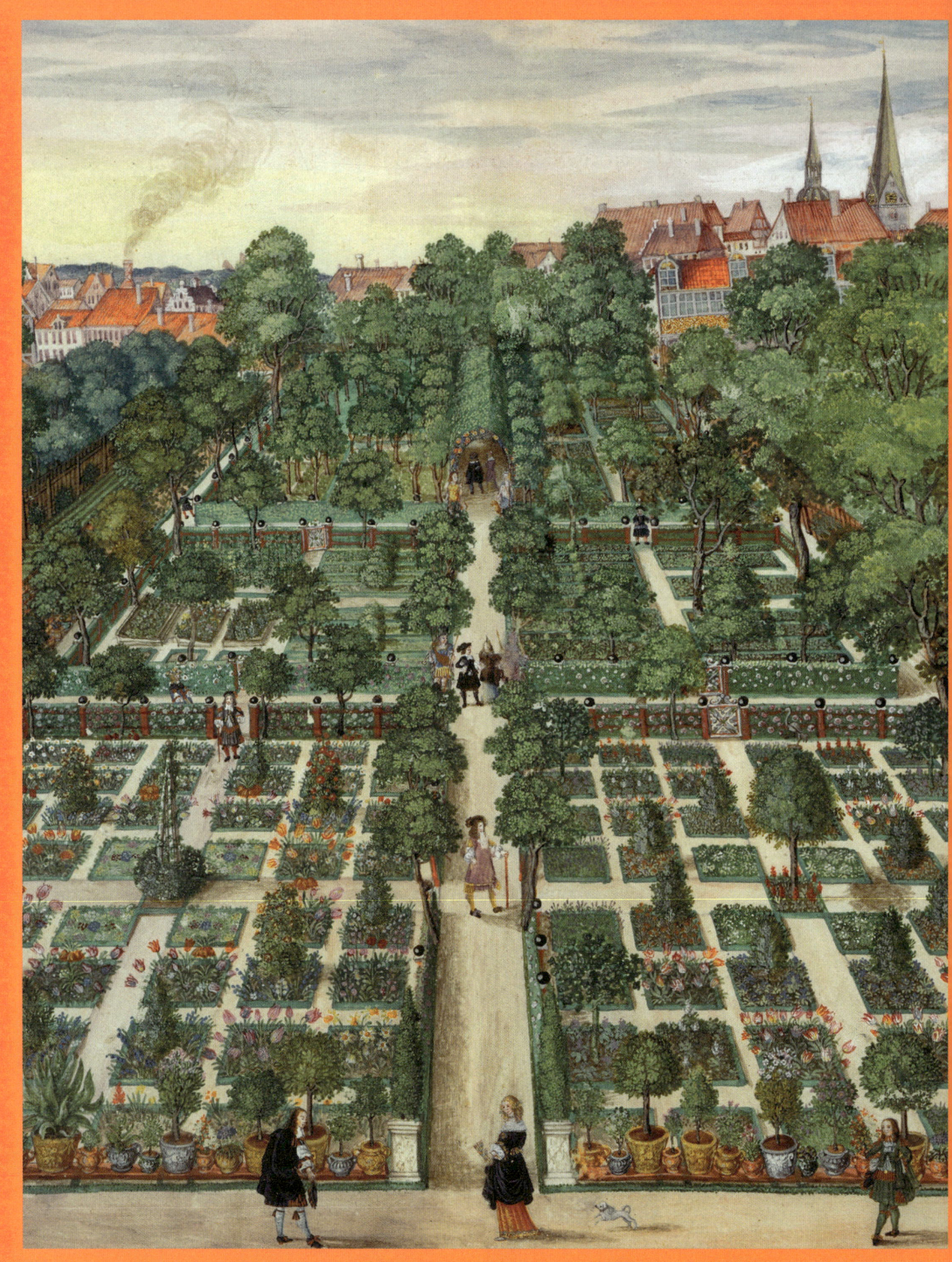

THREE

The House
of Orange

WILLIAM OF ORANGE (1533–1584), also known as William the Silent for his discretion and reticence, is the *Vader des Vaderlands* (Father of the Fatherland) for the Dutch.[1] Brought up as a Lutheran in what is now Germany, William was educated in Catholicism as a condition of his inheriting the title Prince of Orange in 1544. Going with the title was the principality of Orange in the present-day Vaucluse, France, besides land in what is now Germany and large estates in what is now the Netherlands and Belgium.

The colour orange had appeared in heraldry as early as 1526, but apparently was not given that name, after the fruit, until 1542. However, the colour officially adopted by William in his flag seems to have been a pun on the name of his town of Orange in south-eastern France, which has no connection with either the colour or the fruit. The town was anciently Arausio, founded in 35 BCE and named after the local Celtic water god.[2]

The Orange name then became attached to places in other countries when members of the Dutch royal family married other royals. In Brandenburg, the town of Bötzow and surrounding region was given to Louise Henriette of Orange-Nassau (Oranje-Nassau)

by Friedrich Wilhelm I of Brandenburg on their marriage in 1646. She promptly rebuilt the castle in the Dutch style and named it Oranienburg, and in 1653 the name was applied to the town too. The name Oranienburg has sinister connotations for many because here in 1933 the town housed one of the first Nazi concentration camps.

Nearby, the settlement of Nischwitz in Anhalt-Dessau became Oranienbaum in 1673, after Countess Henriette Catherine of the House of Orange-Nassau married Prince John George II.[3] In 1683, she had the Oranienbaum Palace built from the plans of Dutch architect Cornelis Ryckwaert.[4] The surviving orangery constructed between 1812 and 1818, one of the largest ever built, is 176 m (577 ft) long and 12 m (39 ft) wide.

When the Dutch set up a colony in North America in 1624, they called it New Netherland, its first settlement being Fort Orange (now Albany), the second New Amsterdam (now New York). Taken by the British in 1664, New York was recaptured by the Dutch in 1673 and renamed New Orange, but in the Treaty of Westminster of the next year it was returned to the British and became New York again – the Dutch having swapped Manhattan for the highly contested spice island of Run in the Moluccas (Indonesia) in 1667. Later, four municipalities in New Jersey, known as The Oranges (Orange, East Orange, South Orange and West Orange), were named to commemorate William III of Orange (1650–1702), who was from 1689 king of England, Ireland and Scotland.

The South African place names Orange River and Orange Free State (now Free State), associated with Dutch settlers, also commemorate the House of Orange, not the fruit, though the first stamps of the Oranje Vrij Staat (1868) had a design dominated by a fruiting orange tree, perpetuating the pun. There is also a town of Orange in New South Wales, Australia, named in 1828 by Major Thomas Mitchell (1792–1855) in honour of the Prince of Orange who was later to become Holland's King William II (1792–1849). Unlike, for example, Orange County (named in 1889) in California, this also has nothing to do with citrus: in fact, the town is renowned for its apples.

The Prince of Orange, William III of England came, jointly with his wife Mary, to that throne after the Glorious Revolution of 1688 and the removal of his Catholic father-in-law, King James II. James fled the country but his followers, the Jacobites, put up some resistance, particularly in Roman Catholic Ireland, where William himself led his armies to crush the Jacobites at the Battle of the Boyne on 1 July 1690.

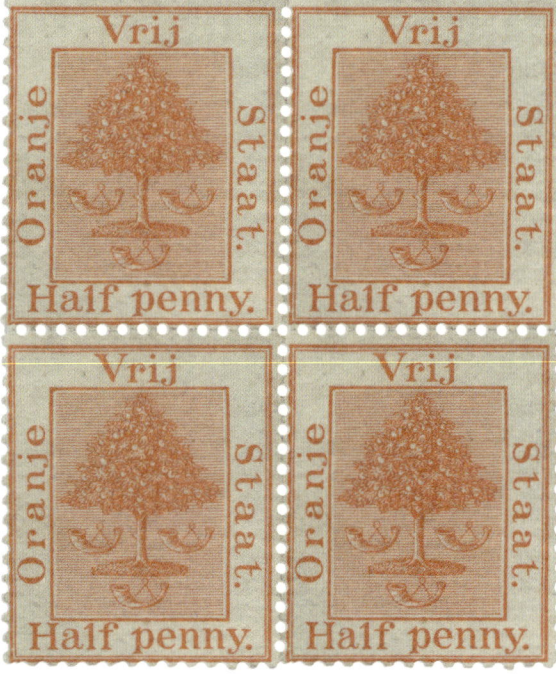

Halfpenny postage stamps of the Orange Free State, 1868. Many 'Orange' place names had nothing to do with the fruit. Nevertheless, the Orange Free State embraced the association.

The conflict is still celebrated by the Protestant Orange Order (Loyal Orange Institution), founded in 1795 as a measure to maintain the Protestant ascendancy in Ireland. Its flag, the Boyne Standard, said to have been used by William as his personal standard, is orange with the cross of St George and a purple star; it is carried in Order parades and marches. Because of his support for the Order and opposition to Catholic emancipation, the British politician Robert Peel (1788–1850), founder of the London Metropolitan Police, became known, perhaps inevitably, as 'Orange Peel'.[5] Later he recanted and, against the wishes of George IV, pushed a bill through parliament, resulting in the Roman Catholic Relief Act of 1829.

William the Silent's orange, white and blue flag was the basis for the national flag of South Africa from 1928 to 1994 and the current flags of both New York and Albany in the USA, though in the Netherlands the Prince's flag was banned in 1795, when Napoleonic forces took control of the country.[6] It was permanently suppressed in favour of the current red, blue and white flag in 1937, following a fresh revival in the 1930s by the National Socialist Movement. Nonetheless it continues to be used today by far-right groups in the Netherlands.

CITRUS TRADE

In the fourteenth century, Bruges and other Flemish towns were the hubs for overseas trade between the Low Countries and the Mediterranean;[7] tax on the trade was exacted at tollhouses. In Sluis, Flanders, the surviving toll tariff of 1252 (in a copy written in 1376) represents the earliest known Dutch document to name citrus fruits in trade: citrons, lemons and sour oranges. These were expensive and intended largely only as gifts for the rich, but were imported also for medicinal use; in 1410 four oranges cost more than a month's pay for a master carpenter. The link between luxury and citrus is exemplified by Jan van Eyck's *Arnolfini Portrait* of 1434 (see overleaf). Originally from Lucca, Italy, the Arnolfini family lived in Bruges and their riches came from trading luxury fabrics and tapestries. It is now widely held that the *Arnolfini Portrait* depicts Giovanni di Nicolao Arnolfini (*c.* 1400–1452) and his first wife Costanza Trenta. Oranges are shown on the windowsill and on a chest beneath it, perhaps alluding to Arnolfini's riches.

By the sixteenth century, though, oranges were cheap enough to be sold in fruit markets and alongside spices and drugs in the shops of apothecaries, because the cargoes of citrus had greatly increased. In January

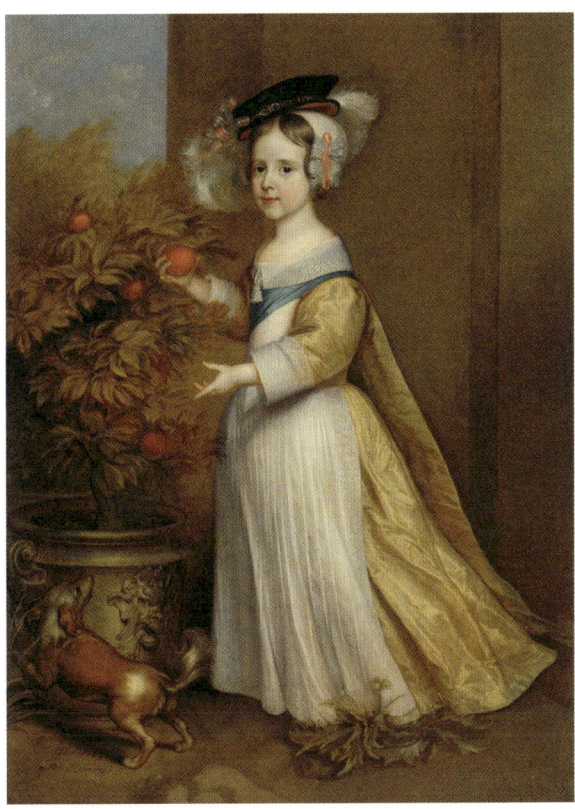

Portrait of William III, Prince of Orange, Adriaen Hanneman, 1654. The prince, aged four, is dressed in girls' clothing, as was then the Dutch custom; a potted tree alludes to the 'orange' pun.

1570 a single ship's cargo from Iberia included 264,000 oranges and 3,000 lemons; in April another arrived with 291,000, just from Laredo, Spain.[8] It was even claimed that the price of citrus fell below that of northern European apples. By the end of the century, citrus was prominent in recipes in cookery books and by the eighteenth century one such had citrus, in one form or another, in around a third of its recipes. Lemon juice was used not just as a replacement for verjuice, but also as a flavouring in its own right in puddings and cakes, sauces and garnishes. The prescribing of flowers in recipes meant that cooks must have had access to living plants.

Port of London records detail two cargoes from Spain unloaded in 1568, one with 40,000 oranges charged at £13 6s 8d, the second 40,000 oranges and lemons to the same value.[9] In cargoes unloaded from ships coming from Antwerp in the same year, the merchant Roger Warfyld had a consignment

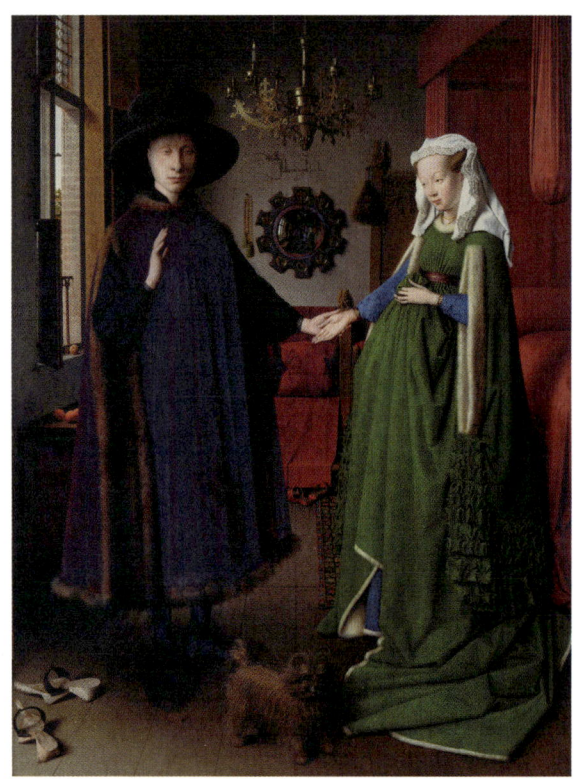

Arnolfini Portrait, Jan van Eyck, 1434. Perhaps a posthumous memorial to Arnolfini's first wife, Costanza Trenta, who died in childbirth in 1433; the oranges could be interpreted as medicine or symbolic of marriage and fertility, as in earlier paintings.

on the *Prym Rose*, including, among other things, '150 lbs marmalade', while on the *James*, William Luddington also had 150 lbs valued at £5.[10] What this marmalade was made from is unclear, because this sweet confection, originally rather like today's quince paste, was indeed first made from quinces (from the Iberian *marmelada*, *marmelo* being a quince) but, once again, as with the apples of the Hesperides, those were eventually replaced by sour oranges. In 1587 a 'conserve of oranges' appeared in a recipe book[11] and, by 1677, there were many such recipes for 'Marmlet of Oranges', which was essentially a paste sold in boxes. Sliced orange peel was added to apple jelly in glass jars and the name marmalade got transferred to that. In 1681 an apple-less 'marmalett of oringes' was being made in England, but it was the Scots who seem to have perfected marmalade-making in the form familiar today. The city of Dundee was a centre of marmalade mass production, started by James Keiller and his mother Janet, who added strips of peel to the

preserve and changed it to a spreadable consistency suitable for breakfast toast.[12] The company, James Keiller and Son, founded in 1797, existed in one form or another until 1992. On the less decorous side, by 1674 in London there were 'Marmalet Madams', who, the *Oxford English Dictionary* records, were 'strumpets', though it does not explain why they were so called.

Modern commercial marmalade is made from sour, or Seville, oranges, which have higher levels of pectin than do sweet oranges. The fruits, available during winter, are mostly exported to Britain. The city of Seville has tens of thousands of Seville orange trees in its streets, with the concomitant problem of the mess of fallen fruits. A proposed solution is to use the oranges to generate electricity for one of the city's water-purification plants, it being estimated that if all the fruit was used thus, the generated electricity could supply 73,000 houses.[13] Formerly, Seville oranges were *the* oranges in *duck à l'orange* (originally *canard à la bigarade*, *bigarade* being an old French name for them). They also provided the original flavours in certain bitters besides liqueurs like Cointreau and Grand Marnier (see p. 197). Belgian *Witbier* (white beer) is often spiced with the peel, though in the Netherlands that of the lemon is used. Seville oranges must have been very familiar to Elizabethan theatre audiences, because in Act 2, Scene 1 of William Shakespeare's comedy *Much Ado about Nothing* (1598) the sharp-tongued and quick-witted Beatrice declares of the treacherous Don John, 'The Count is neither sad, nor sick, nor merry, nor well; but civil count, civil as an orange, and something of that jealous complexion', the 'civil' being of course a pun on 'Seville'.

ORANGERIES

Dutch expansion into the tropical world in the seventeenth and eighteenth centuries led to a rising number of exotic plant products and living plants intolerant of cold temperatures being shipped back to the Netherlands. A high-quality black tea, 'pekoe', comprising leaves from new flushes (the earliest harvests), was introduced by the Dutch East India Company and so named from the Chinese word for white hairs due to the hairy covering (indumentum) of the young leaves used exclusively in its preparation. Although there is no hard documented evidence for it, it has generally been assumed that the name of the House of Orange was borrowed to promote 'orange' pekoe (the name still used today), since the tea contains no orange product at all.

Perhaps the greatest achievement in natural history by the colonial Dutch was the establishment of a network of botanic gardens, both private and public.[14] The first such in Holland was the Hortus Botanicus in Leiden, founded in 1590, with other universities following suit. With the arrival in Leiden of Charles de l'Écluse (1526–1609, introducer of tulips and potatoes), the first oranges were planted and an orangery constructed for them. Such special glass-fronted houses had the exotic 'greens' stationed outside in summer and taken back in during cold weather.

Buildings with large numbers of citrus trees had already become known as orangeries, and northern European designs were no doubt derived from structures built long before in Italy, but in the Dutch Golden Age (1588–1672) it is likely that patriotism to the House of Orange led to the nobility collecting as many different kinds of citrus as they could and showing them off in their country estates. It has also been argued that the vogue for orange carrots – Dutch carrots – as opposed to the original purple, red, white and yellow cultivars was a similar nationalistic phenomenon.[15]

The early orangeries in Holland and elsewhere in northern Europe were at first heated by open fires and were not ornate buildings. Such included those of Lord Burghley of Hatfield House, Hertfordshire, and Sir Francis Carew of Carew Manor (now Carew Academy), Surrey, who were probably the first to grow oranges in Britain, Carew having bought French trees in 1562.[16] He was the first in England to build an orangery, c. 1580, apparently a wooden structure with iron stoves providing the heating.

Lemon and orange trees had of course already been grown in gardens in France and Italy during the early sixteenth century, but the trees were protected from winter cold merely by large wooden shelters. The building erected in 1545 in the newly established Orto Botanico di Padova in Padua, Italy, is claimed as the earliest identifiable orangery.

Still Life with Pitcher, Jacob Foppens van Es, c. 1617–66. As the citrus trade expanded, the fruits – in particular lemons – became a staple of seventeenth-century still lifes.

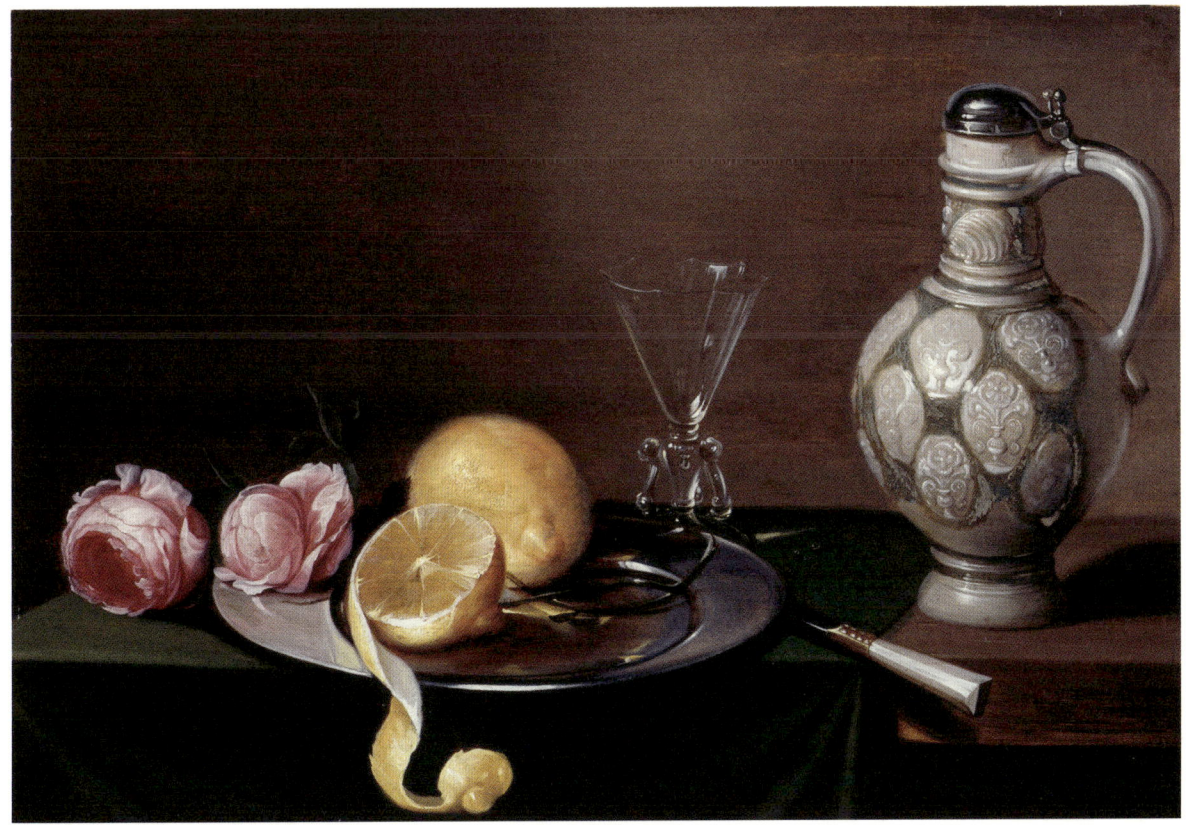

I drank a glass, of a pint, I believe, at one draught, of the juice of oranges, of whose peel they make comfits; and here they drink the juice as wine, with sugar, and it is very fine drink; but, it being new, I was doubtful whether it might not do me hurt.

SAMUEL PEPYS, DIARY ENTRY 9 MARCH 1669

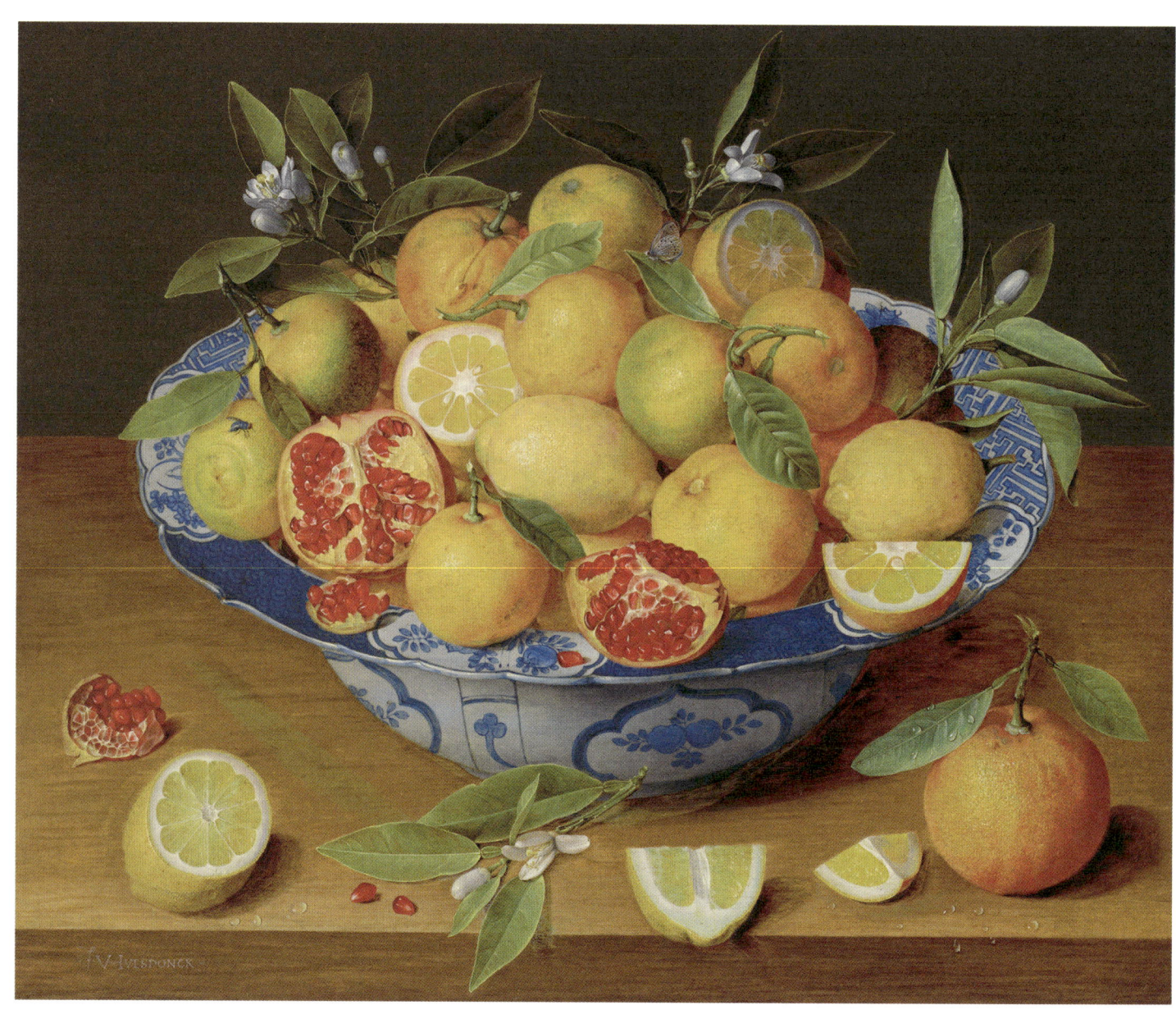

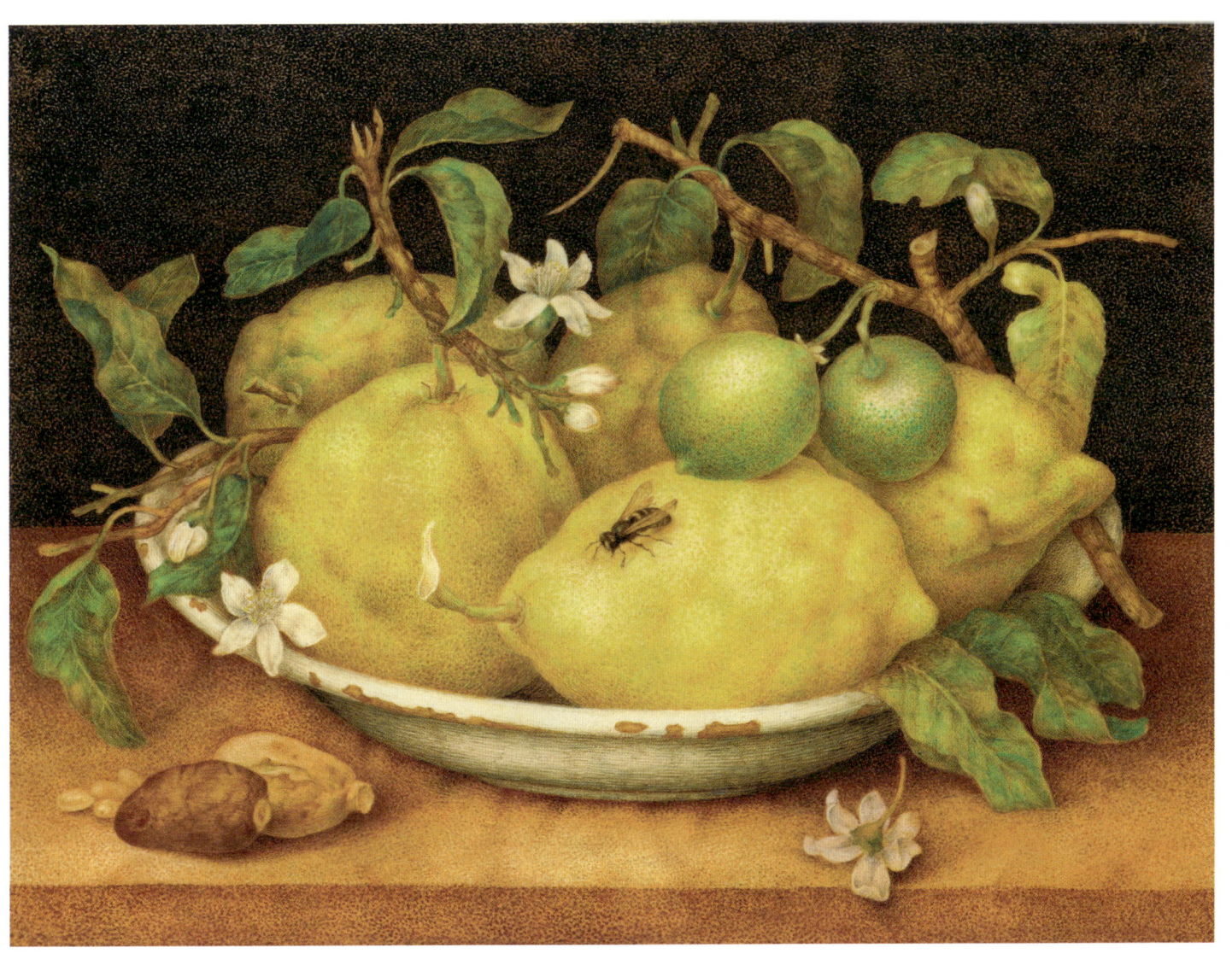

OPPOSITE: *Still Life with Lemons, Oranges and a Pomegranate*, Jacob van Hulsdonck, c. 1620–30. **ABOVE:** *Still Life with Bowl of Citrons*, Giovanna Garzoni, late 1640s.

As still-life painting became popular in Europe in the late 16th and 17th centuries, one subject in particular soon became ubiquitous: citrus. The challenge of depicting the colours and textures of the peel, flesh and leaves was no doubt relished by artists to show off their skill. But the craze also coincided with colonial expansion in the tropics and the import of exotic plants, as well as the growing fashion for orangeries. The omnipresence of citrus therefore reflected, and celebrated, the increasing prosperity of European powers. This was particularly true for the Netherlands, where the collecting and display of citrus also indicated loyalty to the ruling House of Orange.

49

Jardinier de parterre.　　Kunst u: Blumen Gärtner.

1. Roses rouges et blanches. 1. Roth u: weiße Rosen. 2. pie d'oeillets. 2. Nelcken stock. 3. narcisses. 3. Narzißen
4. lis blancs. 4. Weiße Lilien. 5. Jacinte. 5. gelber Stern. 6. perce neige. 6. Schnee ballen. 7. fleur Stomacal
7. Ohlmagen. 8 une tulipe fleurie. 8. eine offne Tulipan. 9. Monstreuse. 9. große Monstrosa d.o 10. une
Pionne. 10. Beonien Rosen. 11. Tulipe. 11. Tulipan. 12. Marguerites fleuries. 12. Margranten Blüth. 13. coque-
licoq. 13. Klaprosen. 14. Laurier. 14. Lorber Blätter. 15. unissoir. 15. eine Britsche. 16. rateau. 16. ein
Rechen. 17. Sardoir. 17. ein stoßeisen den weg zu butzen. 18. Ciseaux. 18. ein Gartenscher. 19. Serpettes. 19. ein Re-
ben meßer. 20. pivot à tirer au cordeau. 20. die eiserne Spitzen zu der Schnur. 21. Oranger. 21. Pomerantzen-
baum. 22. Citronnier. 22. Zitronen baum. 23. plan d'un parterre. 23. Aufriß von einem Garten. 24. Cuvier.
24. ein Gewächs Kübel.

Cum Priv. Maj.　　　　　　　　　　　　　　　　　　　　Mart. Engelbrecht excud. A. V.

No stupid or idle person is suited to being a gardener because he must have good scientific knowledge and a good eye ... A gardener must also be intelligent and thoughtful ... Further, a gardener should be industrious and untiring.

JOHANN CHRISTOPH VOLKAMER, *NÜRNBERGISCHE HESPERIDES*, 1708

'Flower Gardener' engraving, Martin Engelbrecht, from *L'Assemblage nouveau des manouvriers habilles*, *c*. 1730.

This intriguing illustration by Martin Engelbrecht (1684–1756), a contemporary of Volkamer, is part of a collection depicting different labourers, artists and craftsmen in their typical clothing and with the tools of their trades. The 'flower gardener' is flanked by an orange and a lemon tree and stands next to a diagram of a *parterre* or formal garden. The inclusion of these items in the typical accoutrements of a gardener is telling and indicates the extent to which the craze for citrus-growing and immaculately laid-out gardens had gripped the nobility of Europe.

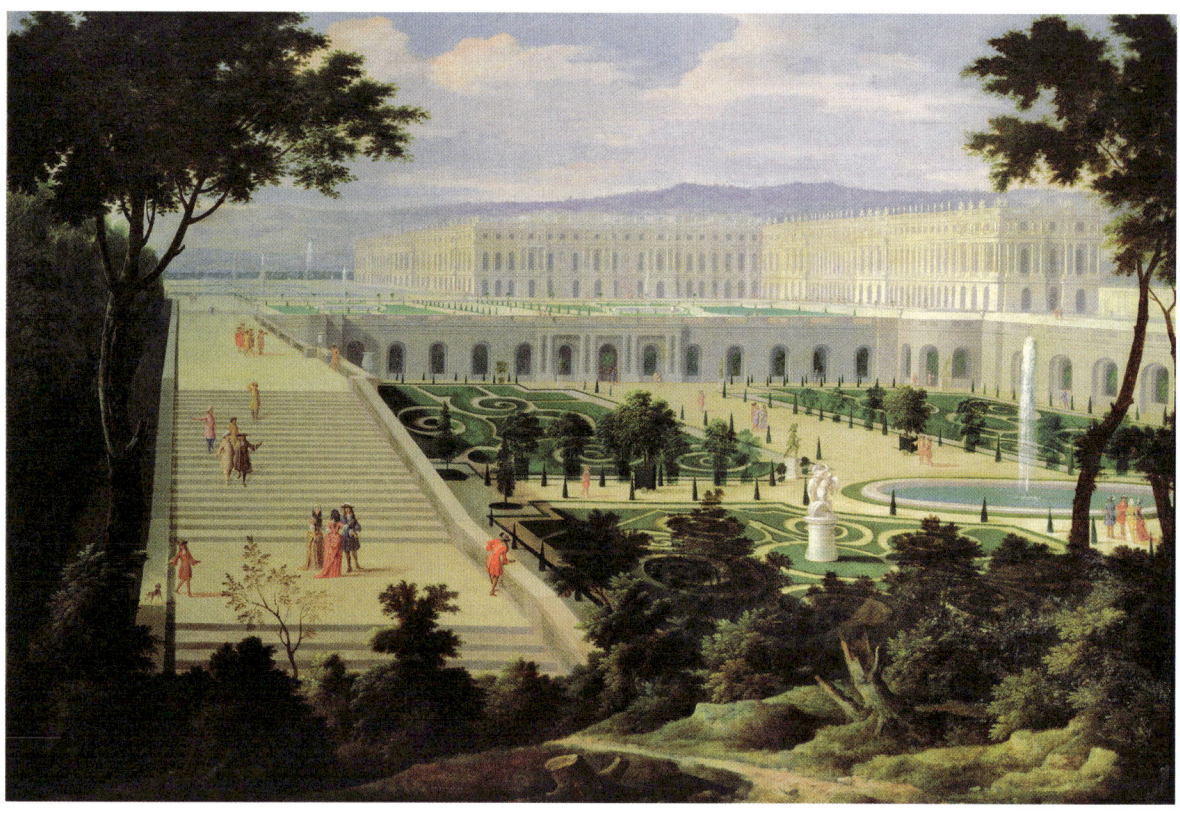

Simple, south-facing, glass-covered structures had been built before that, but those were not specifically for citrus. More elaborate structures were to be built later, because, after Charles VIII of France (1470–1498) conquered Naples in 1495, greenhouses began to be constructed further north in the Holy Roman Empire – as in Prague by 1535 and Vienna by 1542 – to accommodate the trees from the south.[17] By 1555 there was a stone and glass structure dedicated to citrus in northern France: the *orangerie* at the Château d'Anet, Eure-et-Loir.

Ferdinand I (1503–1564), Holy Roman Emperor, started the fashion for citrus-growing in central Europe, introducing plants to Vienna and Prague from Spain, where he had lived for most of his youth. As early as 1535, there is a record of a citrus gardener (*cetro*), Francesco, an Italian from Vienna, working in Prague, where, three years later, '*Lemoni, Pomeranzen, Citroni* [lemons, oranges and citrons]' were being grown.[18] At the castle in Prague, the citrus trees were at first overwintered under canvas in a basement; later, outside, in Vienna and elsewhere, temporary heated wooden structures (*Knock-Pomeranzenhäuser*) were built around the plants growing in the ground. These buildings sheltered other plants, largely true

Mediterranean ones like oleanders, bay trees and rosemary, and were used for recreation. The expense and trouble of constructing such buildings meant that a system of portable trees planted in boxes being moved into orangeries for the winter was preferred. Whichever system was used, citriculture was clearly a pleasure available only to the very rich.

From 1614 to 1619, the French Huguenot architect and engineer Salomon de Caus (1576–1626) built an orangery (*Pomeranzenhaus*) in Heidelberg, Germany, for the garden of the Palatine Count Friedrich V (1596–1632). The Hortus Palatinus was a magnificent Baroque garden – considered then to constitute the eighth wonder of the world – complete with grottoes, mazes and singing bird automata, the plants arranged by geographical origin.[19]

ABOVE: Palace of Versailles and the Orangery, Étienne Allegrain (attrib.), *c.* 1695. The citrus collection of Louis XIV was housed in the immense orangery. **OPPOSITE:** Illustrations from *Instruction pour les jardins fruitiers ...* by Jean-Baptiste de La Quintinie, 1690. Director of Louis XIV's fruit and vegetable gardens, La Quintinie recorded innovations in growing citrus in colder climates.

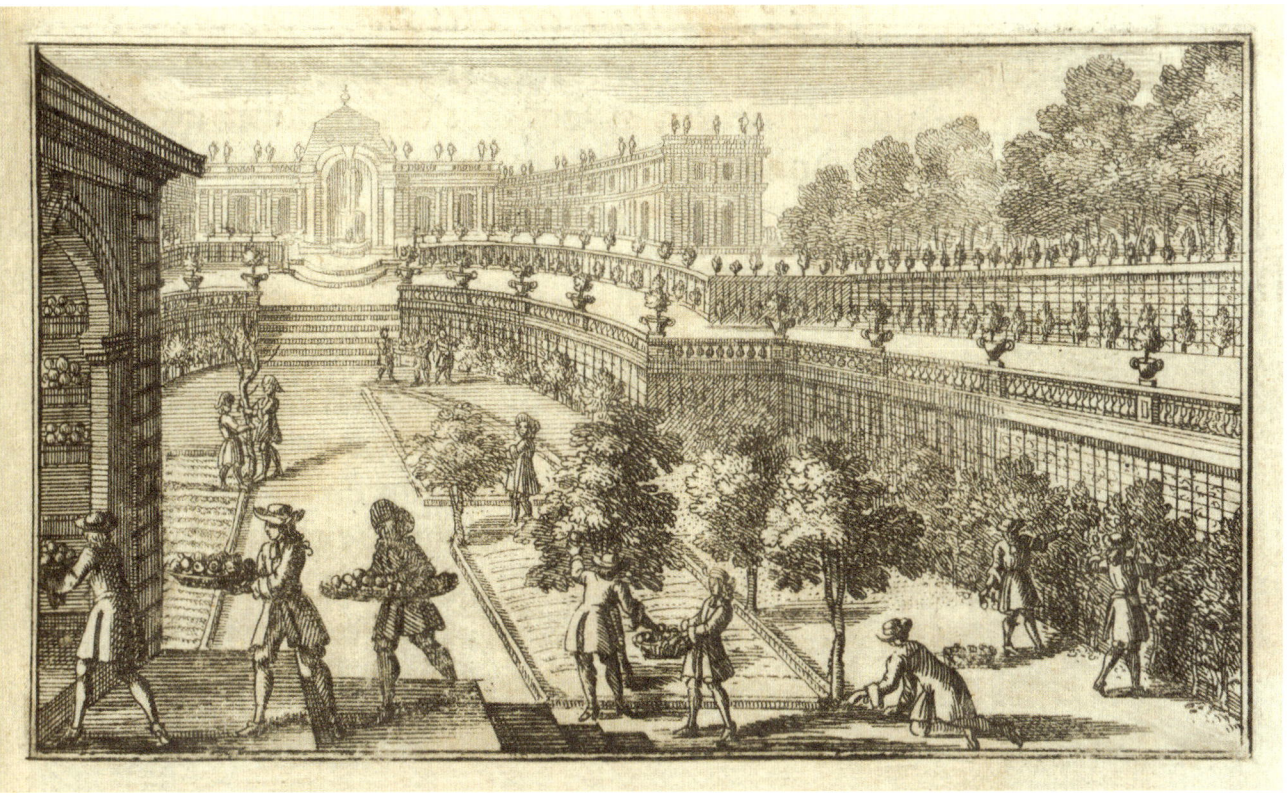

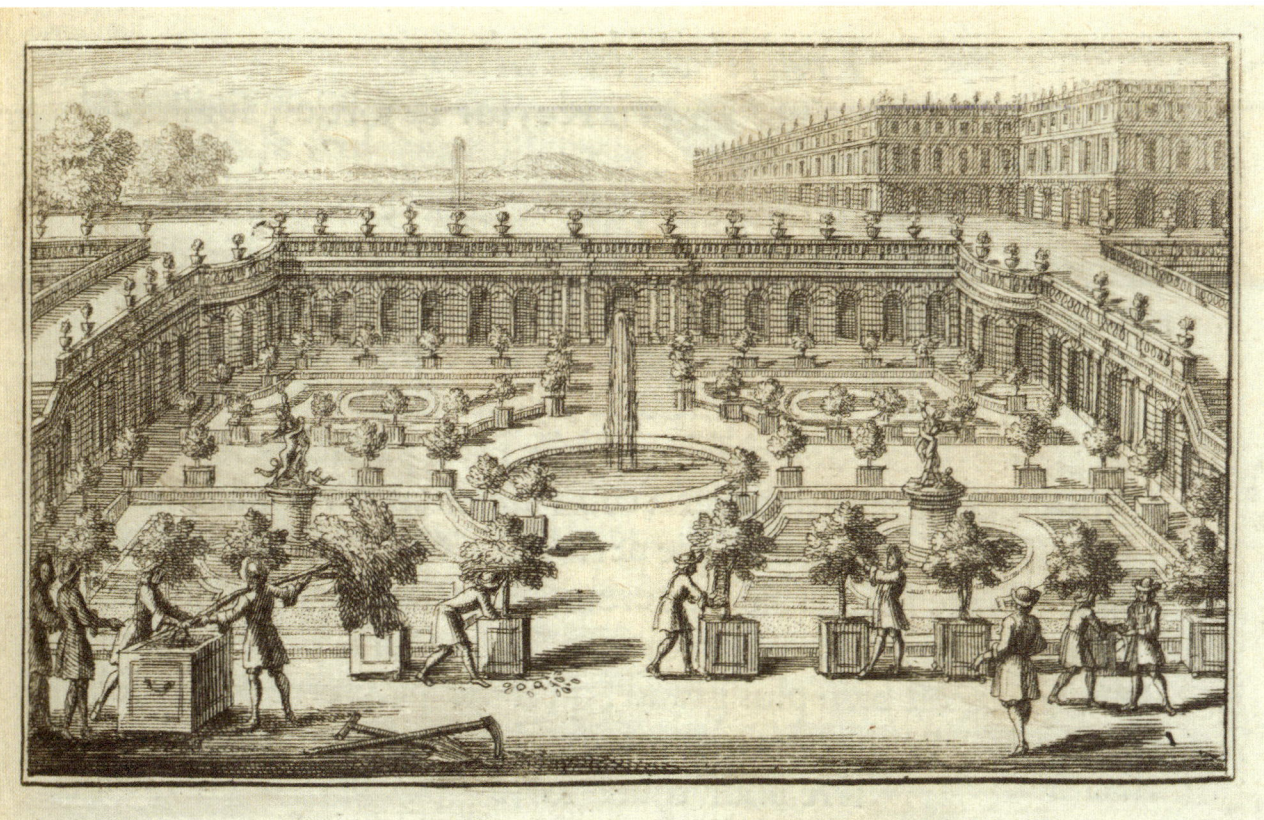

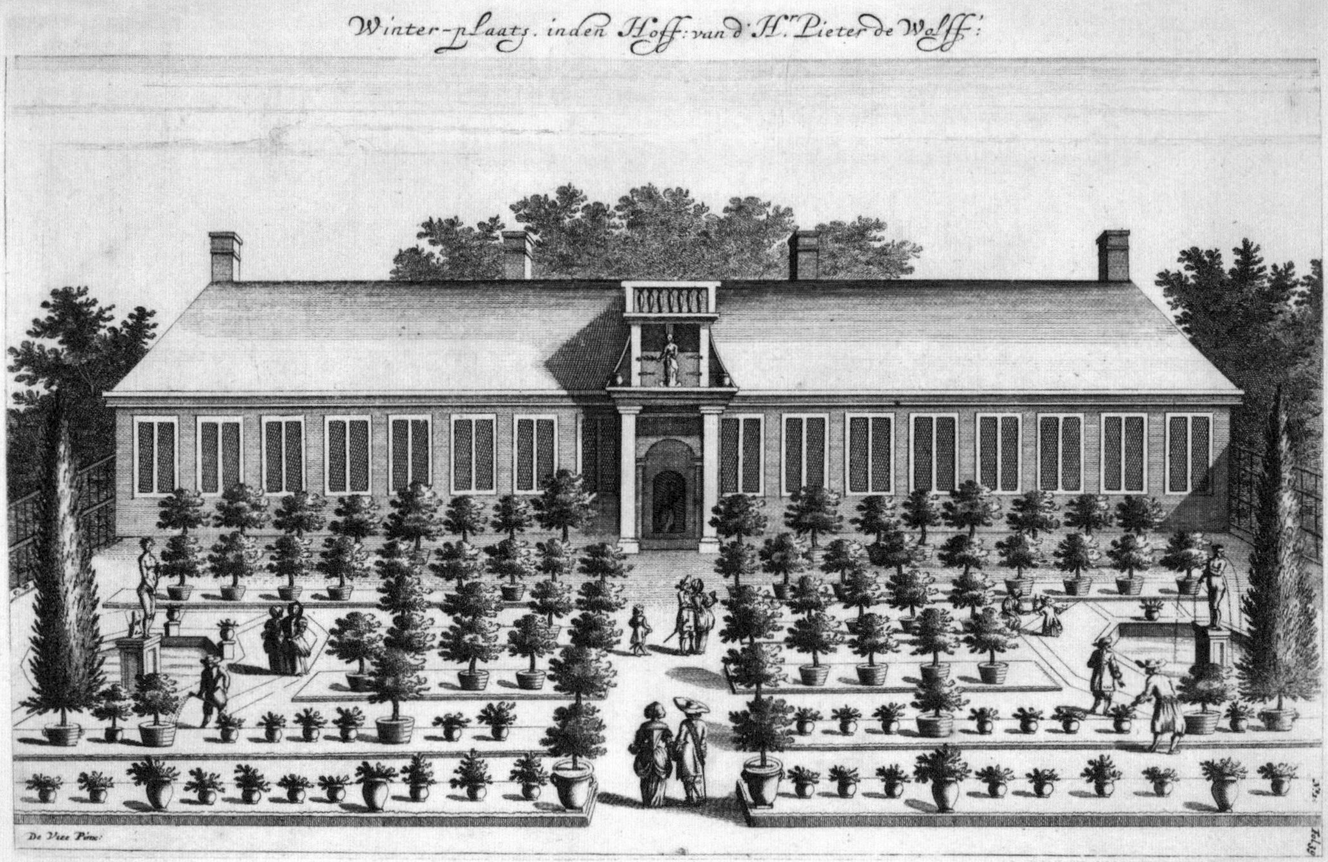

Winter-plaats. inden Hoff: van d: H.r Pieter de Wolff:

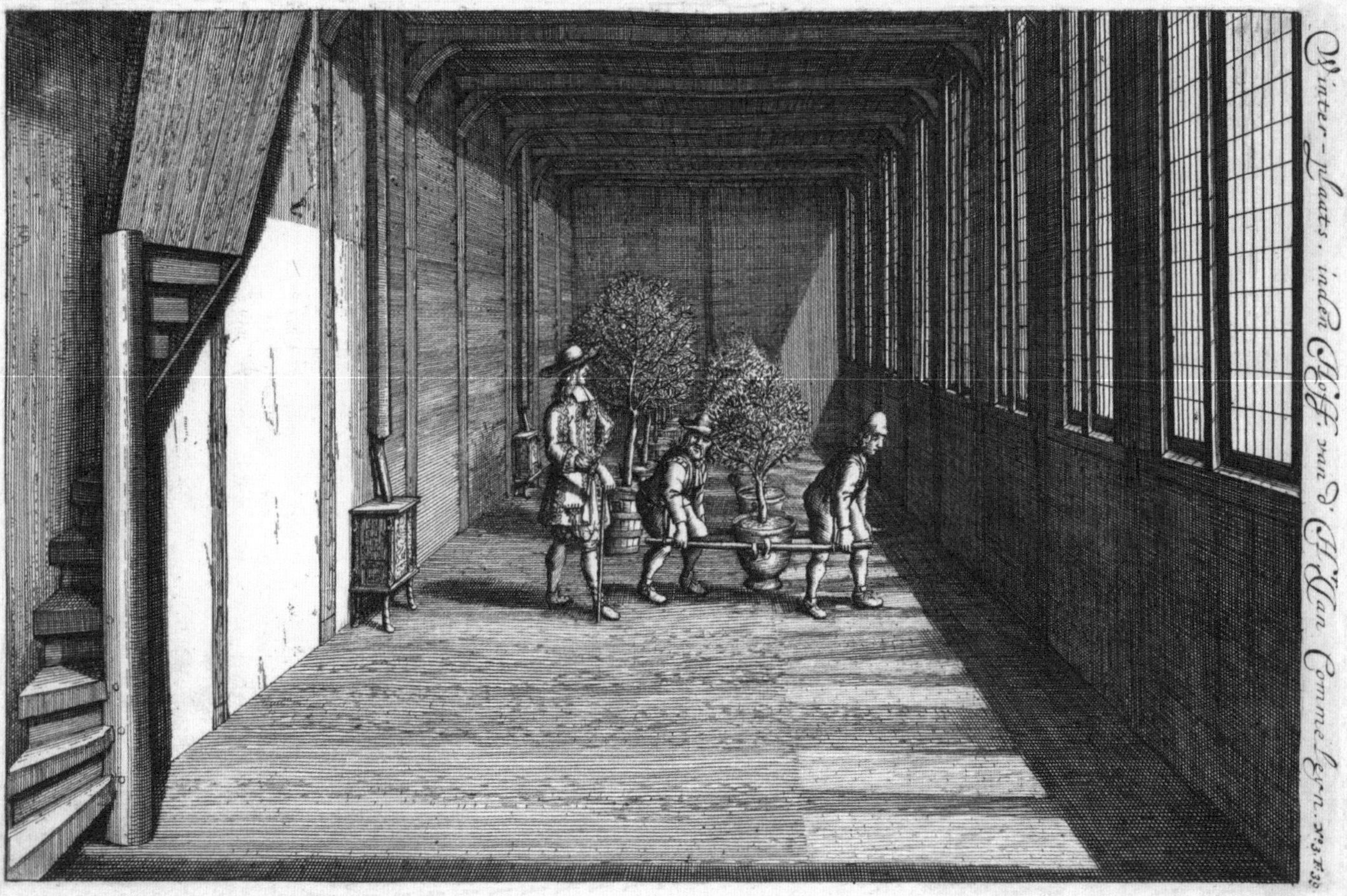

De Vee Pinx:

Winter-plaats. inden Hoff. van d: H.r Van Commelyn: N: 3 F. 128

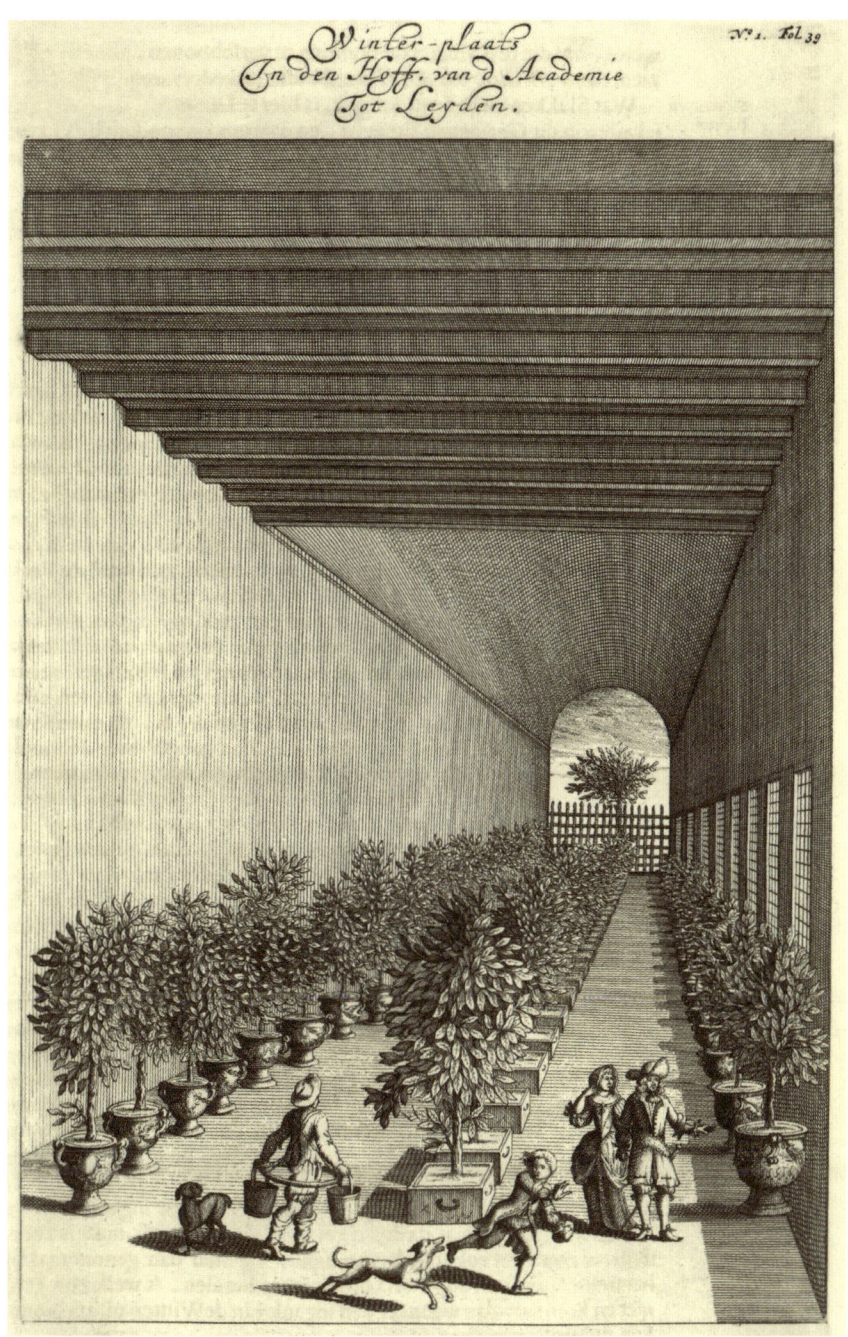

Dutch orangeries, engravings from *Nederlantze Hesperides* by Jan Commelin, 1676.

As 'citrusmania' raged through the Netherlands, Dutch gardeners began to devise new methods to support citrus cultivation during the winter months. Orangeries evolved from temporary protective structures into permanent buildings. Commelin depicted several such orangeries in his treatise on citrus-growing, including his own (opposite, below), in which a stove provided heat to the plants. In warmer weather they were moved outside using poles threaded through the handles on the pots. A summertime arrangement in front of the orangery of Pieter de Wolff (1647–1691), son of a prosperous Amsterdam silk-trader, is shown opposite, above. Though the botanic garden of the University of Leiden is also included (above), most of the orangeries that Commelin discussed belonged to individual rich people.

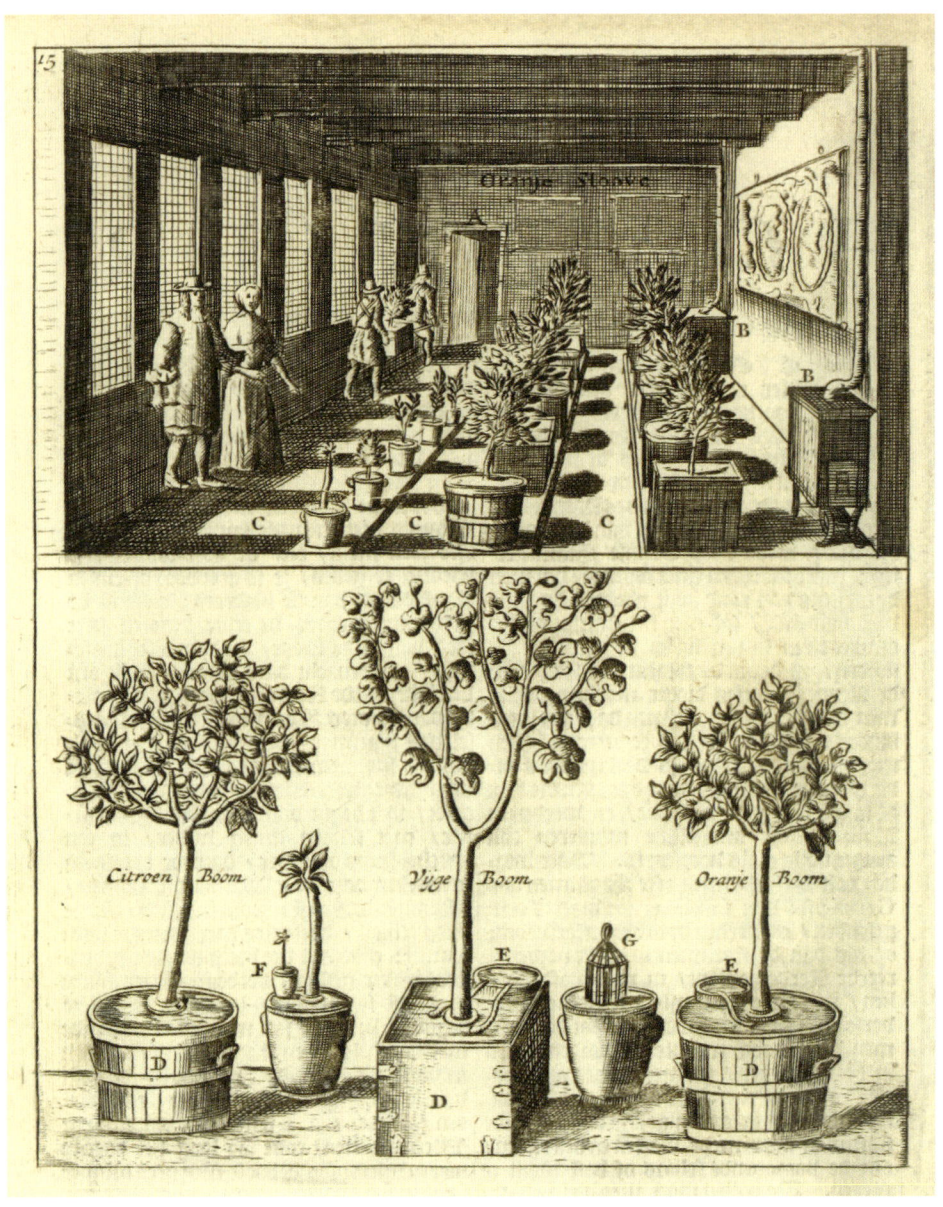

ABOVE: Orangery from *Den Nederlandtsen Hovenier* by Jan van der Groen, 1699. **OPPOSITE, ABOVE:** The queen's garden and the orangery at Het Loo Palace, Lorenz Scherm, 17th/18th century. **OPPOSITE, BELOW:** Entrance to the orangery at Gunterstein estate, Breukelen, Joseph Mulder, 1680–96.

Orangeries were soon *de rigueur* in Dutch society, from the princely palace of Het Loo (opposite, above) – built for Prince William III of Orange and his wife Mary – to those of private citizens (though only those rich enough to participate). One Magdalena Poulle had an orangery built at her country estate of Gunterstein and it was included in a series of engravings she commissioned to commemorate her home (opposite, below). There was also a surge in the demand for practical gardening manuals. Van der Groen (*c.* 1635–1672), gardener to William III, first published *Den Nederlandsten Hovenier* in 1669 and it remained in print for more than fifty years. Interestingly, the engraving shown here (above) demonstrates how the windows of early orangeries were often too small to meet the light requirements of citrus trees.

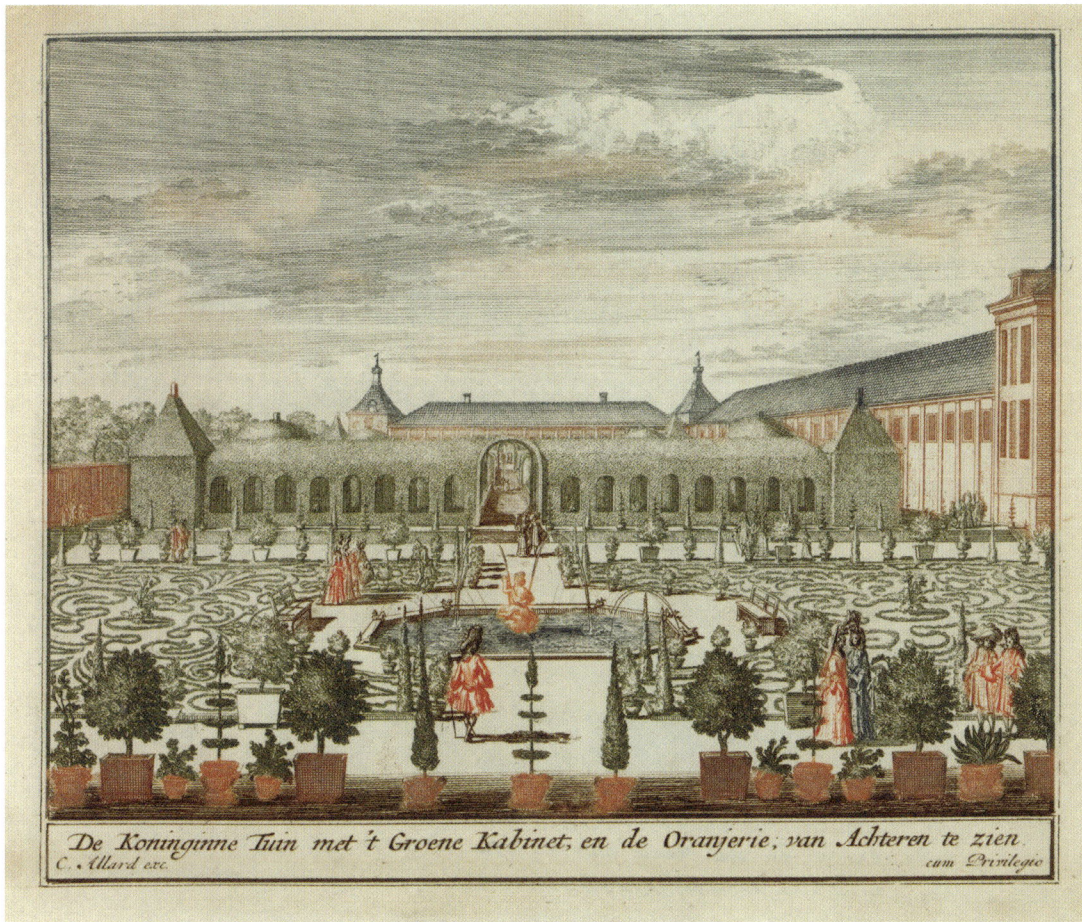

De Koninginne Tuin met 't Groene Kabinet, en de Oranjerie, van Achteren te zien.

C. Allard exc.   cum Privilegio

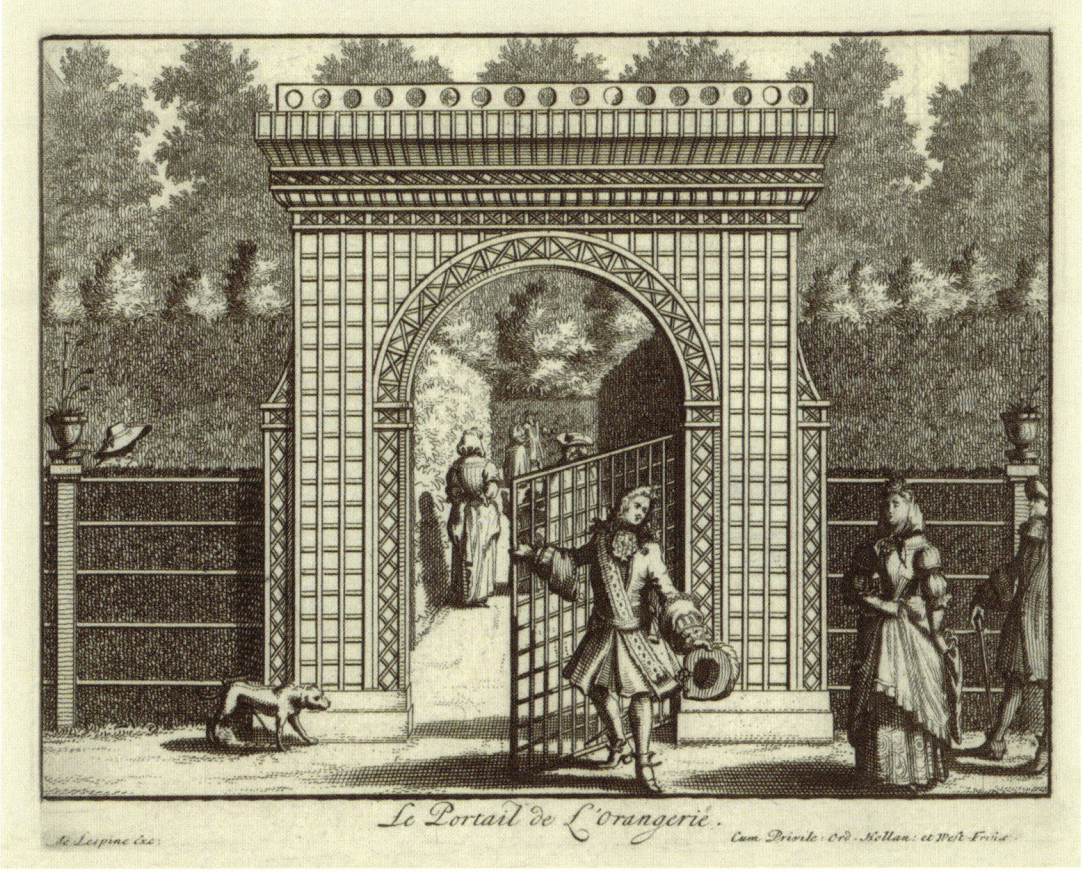

Le Portail de L'orangerié.

de Lespine exc.   Cum Privile. Ord. Hollan: et West-Frisie

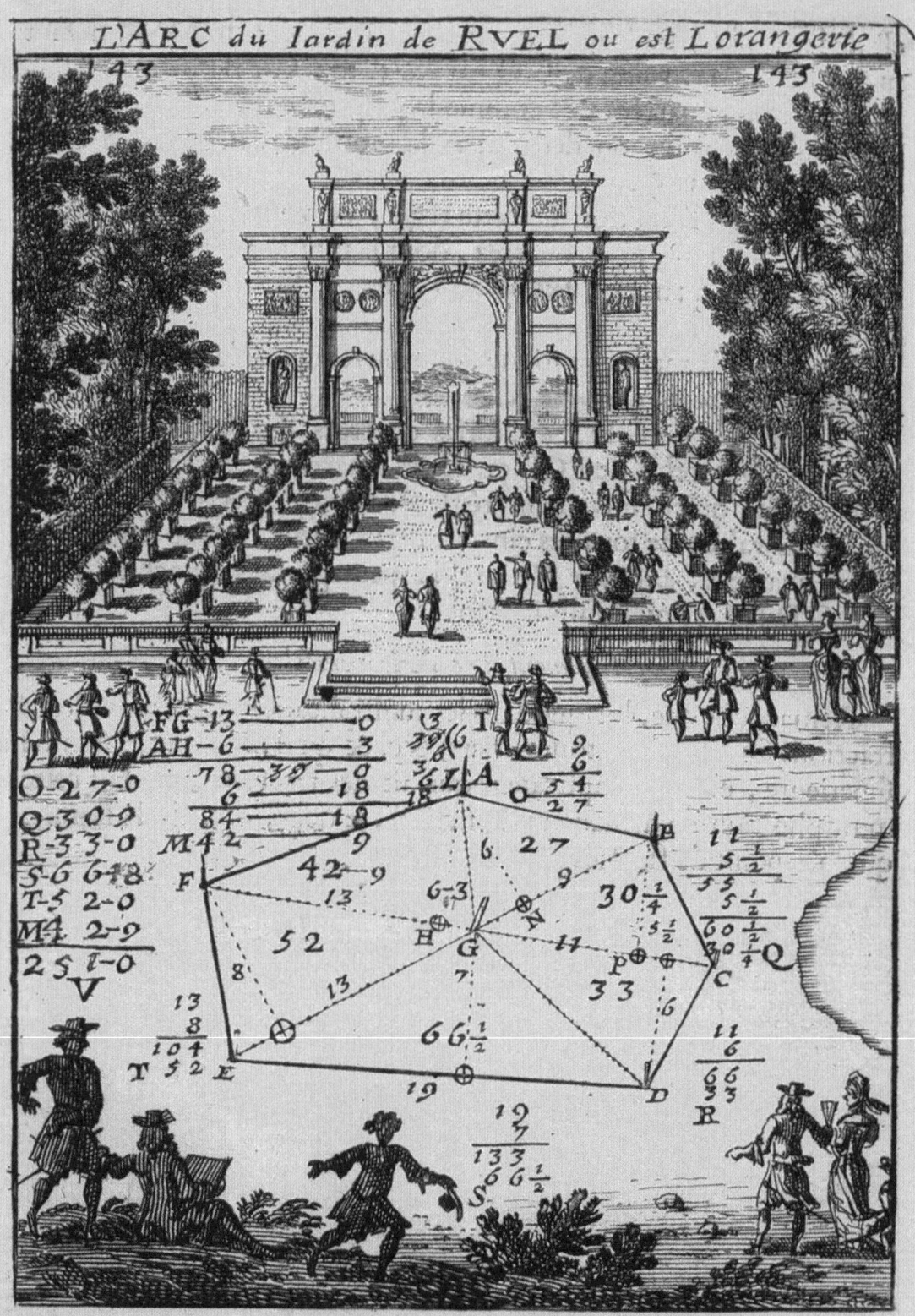

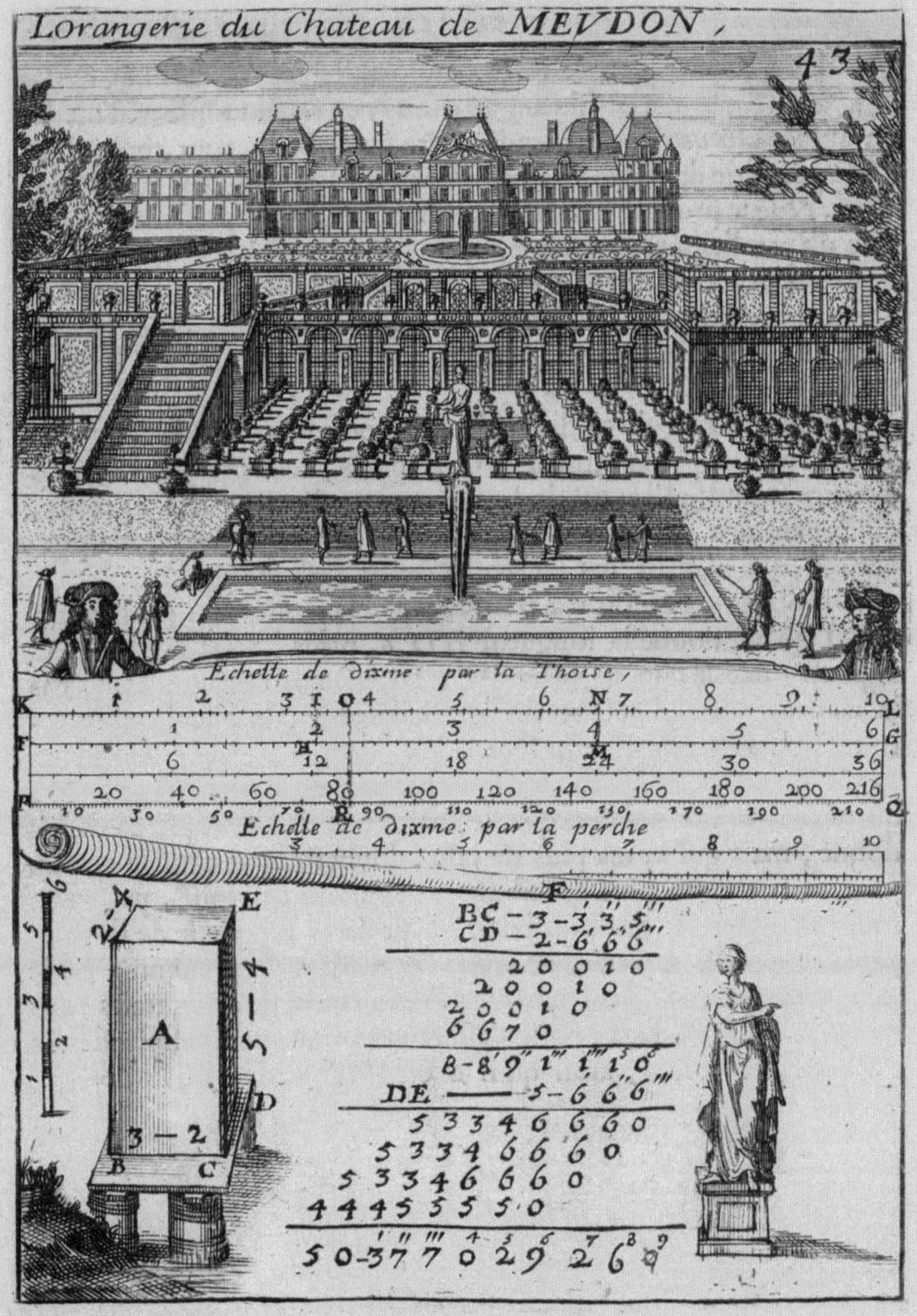

Lorangerie du Chateau de MEVDON,

Echelle de dixme par la Thoise,

Echelle de dixme par la perche

BC — 3 - 3'3''5'''
CD — 2 - 6'6''6'''

                    2 0 0 1 0
                  2 0 0 1 0
                2 0 0 1 0
              6 6 7 0

B - 8'9''1''1''1'0'''
DE ——— 5 - 6'6''6'''

5 3 3 4 6 6 6 0
5 3 3 4 6 6 6 0
5 3 3 4 6 6 6 0
4 4 4 5 5 5 5 0

5 0 - 3'7''7''' 0' 2'9'' 2'6''' 9

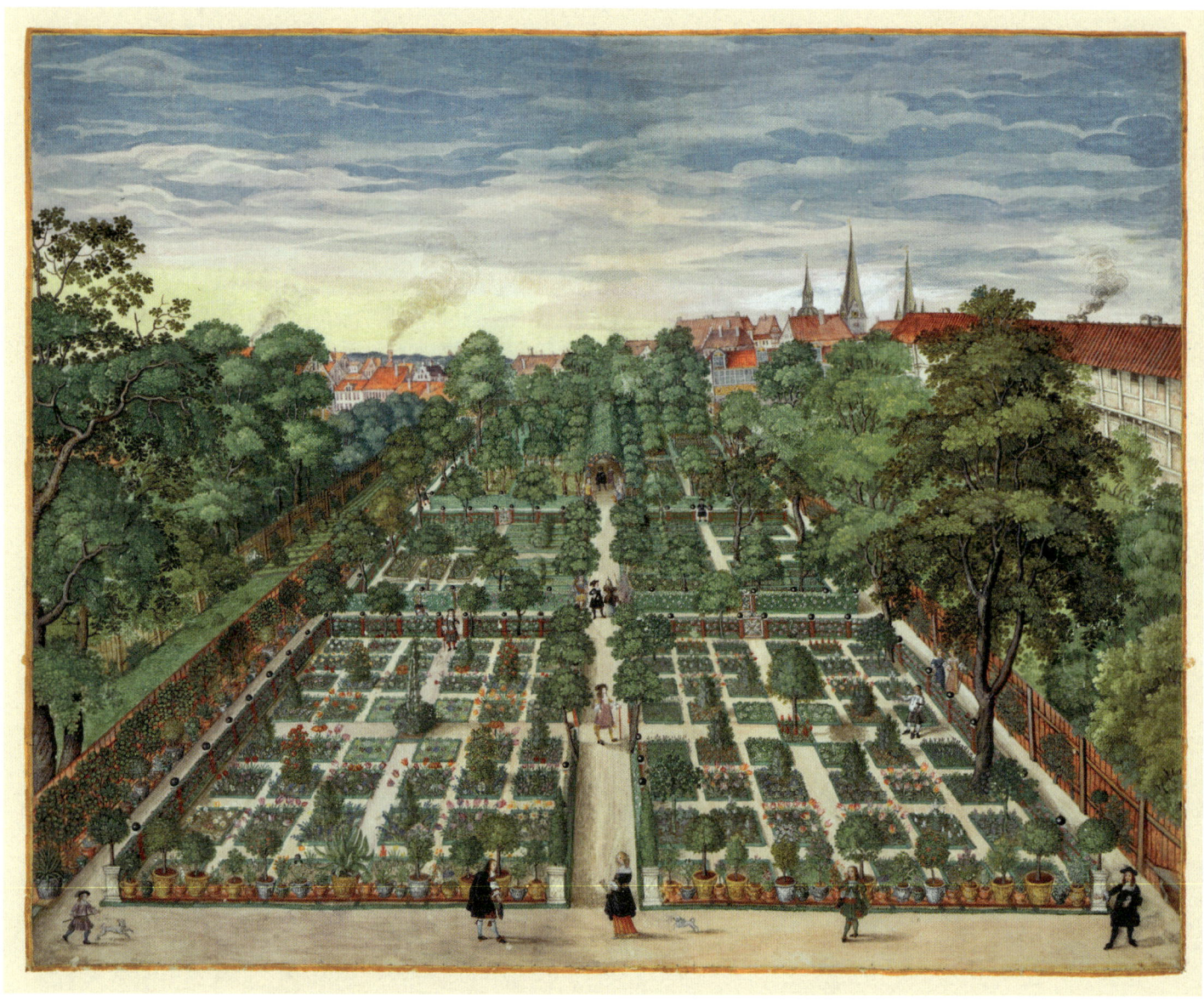

PREVIOUS PAGES: Orangeries at the chateaux of Ruel (now Rueil) and Meudon, France, engravings from *La géométrie pratique* by Alain Manesson-Mallet, 1702. **ABOVE:** Garden of Caspar Anckelmann in Hamburg, Hans Simon Holtzbecker, 1664–71. **OPPOSITE, ABOVE:** The orangery in the park of Zorgvliet, Jan van den Aveelen, engraving from *Toneel van Nederlandse Lusthoven*, 1718. **OPPOSITE, BELOW:** Orangery gardens at the Palace of Versailles, 18th century.

The fashion for orangeries and the intricately laid-out gardens that surrounded them spread throughout northern Europe. Such complex garden design required familiarity with geometry, such that engravings of formal garden layouts appeared in the work of French mathematician Alain Manesson-Mallet (1630–1706, see previous pages).

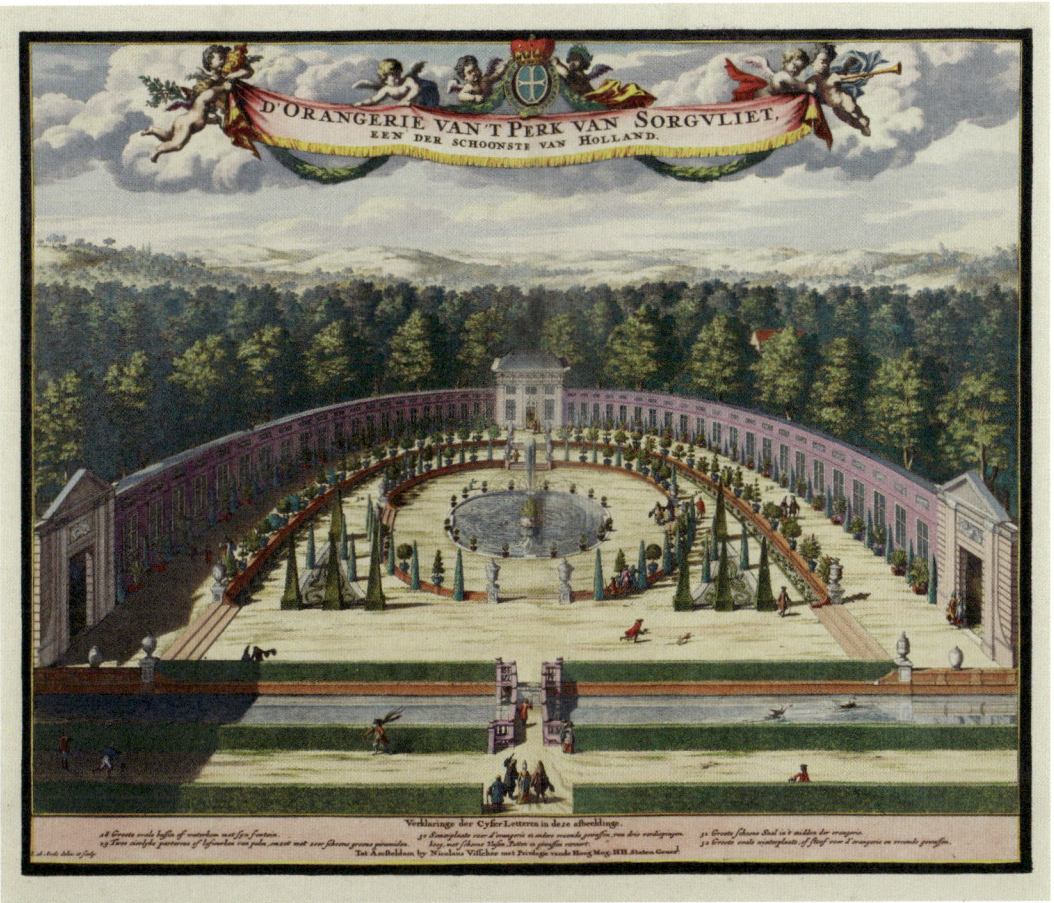

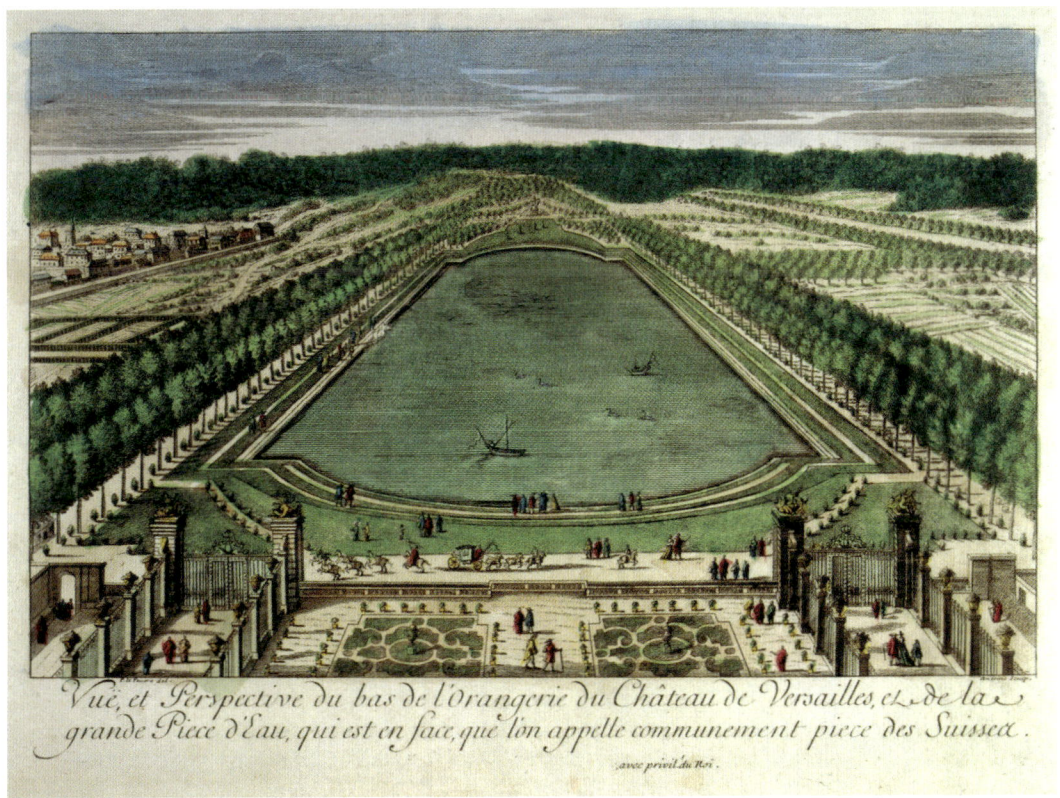

De Caus had a grove of thirty 60-year-old orange trees moved to the Hortus Palatinus, a remarkable achievement in itself, and successfully established using the temporary heated wooden structure system. The garden was largely destroyed in the Thirty Years' War (1618–38), after which it was not rebuilt, though elsewhere the architecture of permanent stone orangeries became increasingly monumental. In Fontainebleau, France, Louis XIV (1638–1715) converted Henry IV's aviary into an orangery, then took over Vaux-le-Vicomte, where his minister of finance had made a collection of 190 orange trees over a century old, and ordered that the trees be removed to a new orangery at his hunting lodge in Versailles.[20] One of the sour orange trees, known as Grand-Bourbon, had been grown from a seed planted in 1421 and was still thriving in the 1820s,[21] reportedly dying only in 1894.[22] In 1686, the orangery was demolished and replaced with the current building,

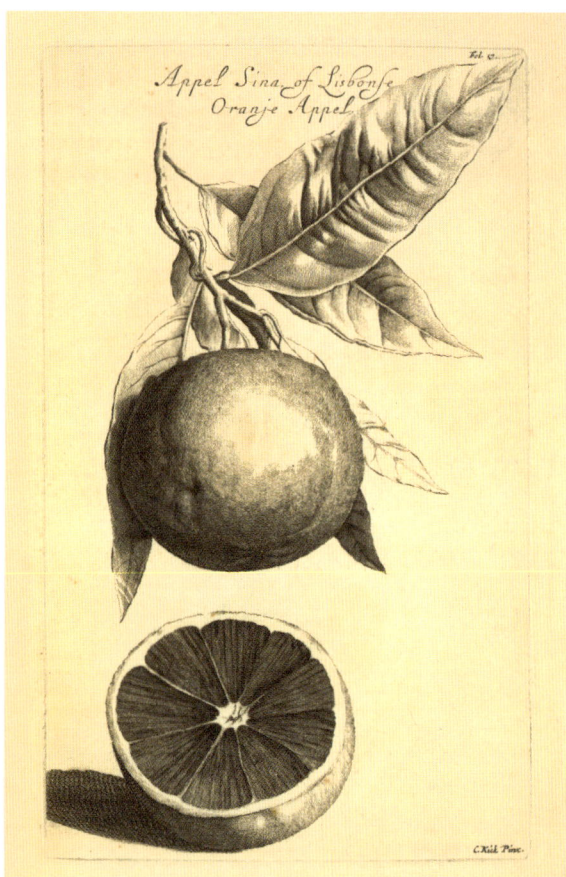

A sweet orange, engraving after Cornelis Kick from *Nederlantze Hesperides* by Jan Commelin, 1676. The fashion for growing citrus in Europe created a market for lavish books like this.

soon to become famous for its flourishing collection of citrus, likely the best in Europe at the time. There is a central gallery 150 m (492 ft) long with two side galleries under the staircases. It has never been surpassed. Indeed, Louis XIV seems to have been obsessed with citrus:

Whenever the king gave in his gardens one of those brilliant parties which contributed almost as much to his fame amongst foreign nations as his conquests, orange trees were used to decorate the porticoes, the halls of green, and the other similar embellishments. Orange trees were also amongst the principal ornaments of the Great Gallery of the Palace of Versailles.... Even in his own apartments did the king insist on keeping these trees; and his gardeners, in order to satisfy his passion in this regard, had even discovered the secret of how to obtain orange blossoms all the year round.[23]

Not to be outdone after visiting Versailles, the Elector of Saxony, Augustus II (1670–1733), had his own citrus collection placed in a magnificent orangery opened at his Zwinger Palace in Dresden in 1709. Meanwhile, in the Netherlands, not surprisingly, orangery-building became even more popular with the construction of an orangery at Het Loo, a Baroque palace of 1684–86, built for William and Mary. As the increasingly competitive 'citrusmania' gripped royal houses and affluent society in general, citrus-inspired designs became more and more common motifs in embroidery, sculpture and paintings, while the plants, flowers and fruits featured in poems and songs, making citrus familiar to ordinary people.

After the Restoration of the monarchy in England in 1660,[24] conservatories (or greenhouses or orangeries) had become fashionable there too, and enabled the growing of more and more tender species of plants as they became available from expanding commerce in tropical countries. There was also a rising demand for books on citrus, as well as other tender plants. In 1683 there appeared in London an octavo English edition of Jan Commelin's *Nederlantze Hesperides* of 1676, called *The Belgick, or Netherlandish Hesperides. That is: the management, ordering, and use of the limon and orange trees, fitted to the ... climate of the Netherlands.* Commelin (1629–1692), a Leiden-born dealer in pharmaceuticals, government official and botanist, relied heavily on Ferrari and to a lesser extent J. J. Pontanus's *Horti Hesperidium libri II* of 1514 as well as the Jesuit priest Frans van Sterbeeck's (1631–1693) manuscript book, *Citricultura*, which was not published until 1682. However, Commelin had his own orangery so could add his own personal observations, including the rather strange advice that:

*The place whence they must be fetched, is St Remo,
Situated by or upon the River Nervi, whence they are
brought to Genoa. We must bring no trees out of any
other climates to Plant or Order here in our Netherlands,
because the Condition of that place, doth, above all other
places in Italy, agree best with our climate.… If Trees
be brought out of any other Countries, as Spain, or
Portugal, it will certainly be in Vain and to no Purpose;
because they cannot well endure our cold changeable
and uncertain Air.*[25]

Besides describing the oranges and lemons then
available, their cultivation and protection in winter,
Commelin added recipes for orange brandy, candied
orange flowers, orange-flower water, preserves and
ointments, as well as lemonade.[26] Van Sterbeeck's
citrus notes were returned to him by Commelin
in 1677. He began his *Citricultura* with five
congratulatory poems by prominent Antwerp citizens
and included a chapter on the identity of Adam's
apple, the forbidden fruit, discussing apples, figs
and even bananas in that regard. The bulk of Van
Sterbeeck's book again deals with oranges, citrons
and similar fruits, lemons and, fourthly, other exotic
trees, grafting and pests. The book ends with a
discussion on other exotics, including pomegranates,
passionflowers and oleanders.

Among other books on citrus, reinforcing the
importance of these plants in raising the status of their
growers, one of the most magnificent of all appeared
early in the eighteenth century, the *Nürnbergische
Hesperides* by Johann Christoph Volkamer (1644–
1720), who modelled it on Ferrari's *Hesperides* (see
p. 98).[27] Volkamer's father, Johann Georg, had visited
Italy, studied in Padova and, on his return, built a
substantial greenhouse in his garden at Gostenhof.
Johann Christoph also travelled in Italy and began to
study citrus-growing in gardens across the country.

He and others prepared drawings for a sumptuous
book as a worthy successor to Ferrari, but his plates
are enhanced by charming perspective drawings
of gardens where the citrus grew and other scenes:
the German ones in the Nuremberg area in the first
volume, those in Italy, most from near Venice, the
Brenta, the Euganean Hills and near Bologna (but also
his own garden in Nuremberg as well as Schönbrunn
Palace in Vienna) in the second. Although striking,
this was not an original format, as it was used by
Abraham Munting (1626–1683) in his *Naauwkeurige
Beschryving der Aardgewassen* of 1696, in which
uprooted plants hover similarly, but over classical or
pastoral landscapes. A third volume to include citrus
in gardens and greenhouses elsewhere in Europe was

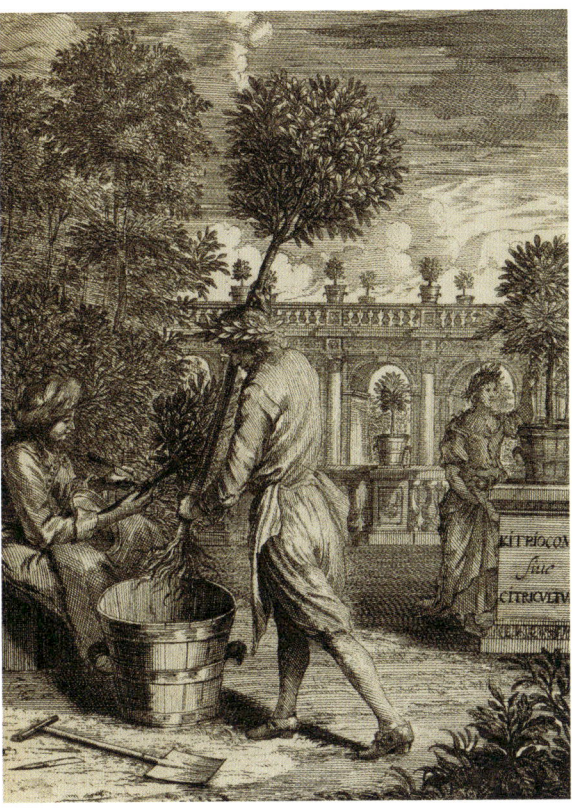

Frontispiece, Franz Ertinger, engraving after C. E. Biset from
Citricultura by Frans van Sterbeeck, 1682. Van Sterbeeck's book
included descriptions of citrus, but also dealt with pests and
propagation techniques such as grafting and budding.

never completed.[28] Although the number of cultivars
(and many other rare and fashionable plants) covered
in the book is greatly increased from Ferrari, discussion
of their cultivation is not really further advanced.

At this time, an orangery was begun in the gardens
of the Belvedere in Vienna by Prince Eugene of
Savoy (1663–1736).[29] Completed in 1714, this
Pomeranzenhaus was much more than a winter home
for citrus. It was an elaborate portmanteau building
with two large rooms for citrus year-round, flanking
a double-storey ballroom, with an apartment for the
prince. Shortly afterwards, an orangery resembling
a French chateau was built by garden designer Batty
Langley (1696–1751) at Wrest Park in Bedfordshire,
England. But all these were to be outdone by that
built by Emperor Franz I Stephan (1708–1765) in
1754 at the summer palace of Schönbrunn in Austria,
probably to the designs of French architect

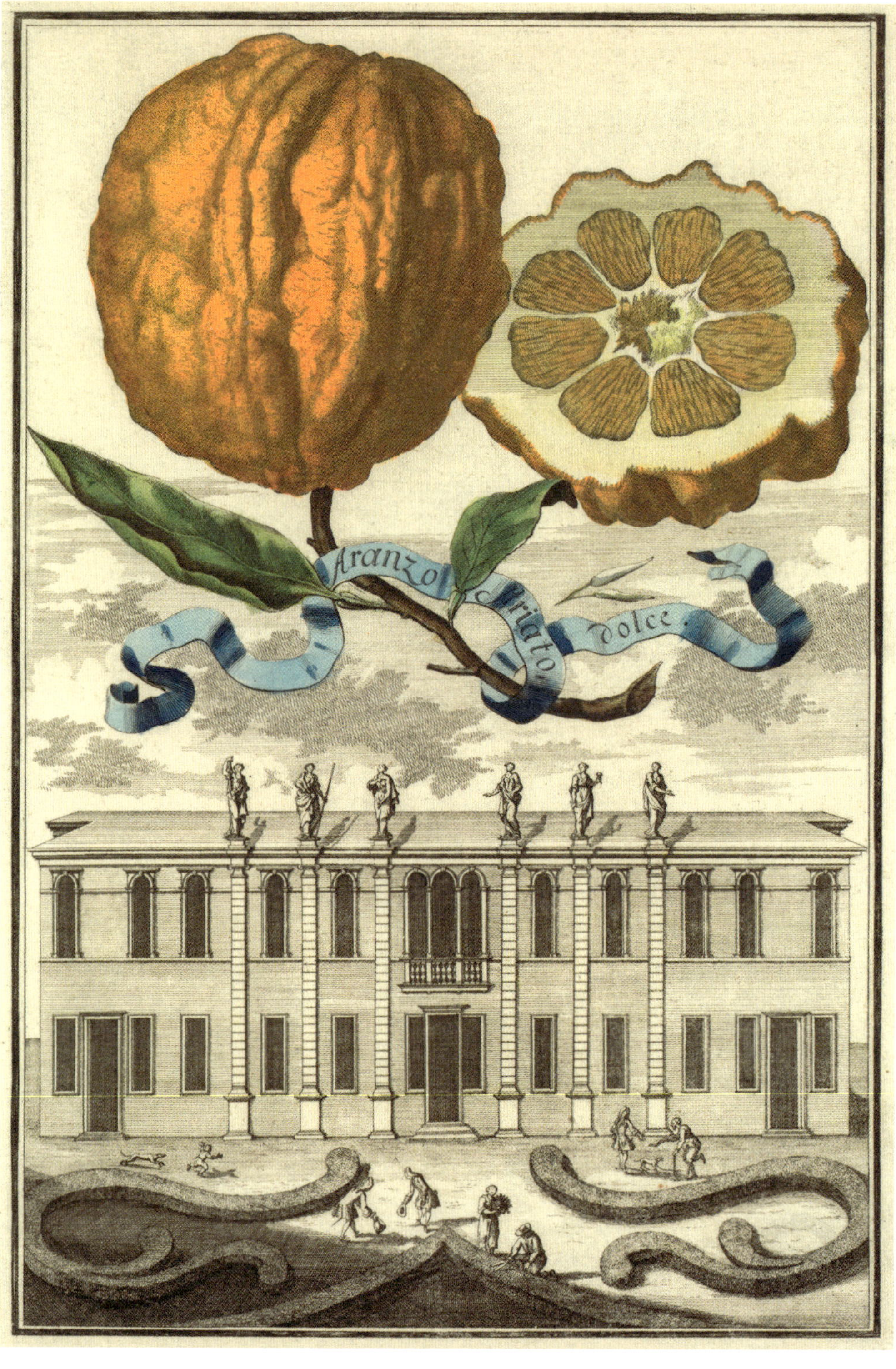

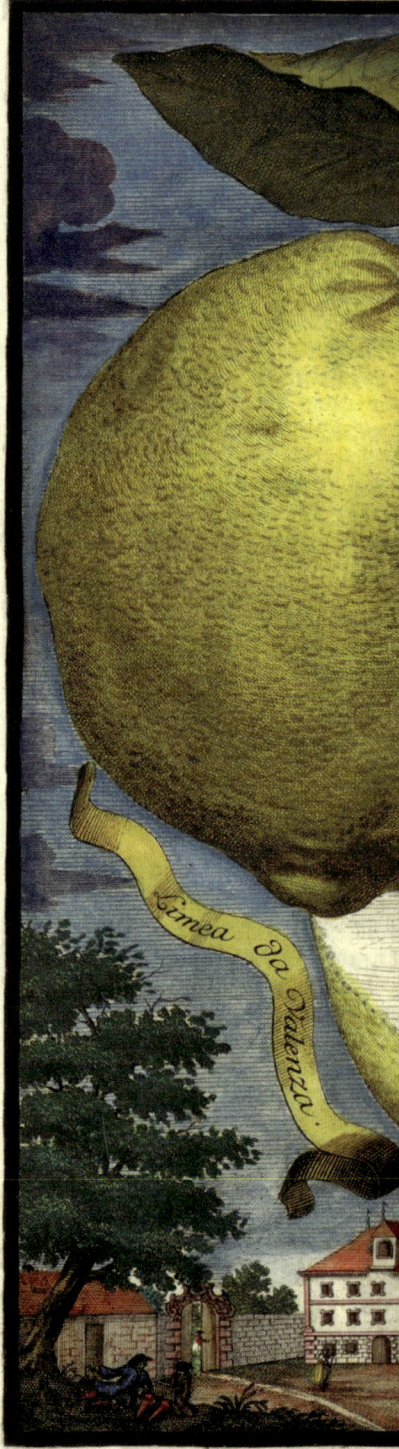

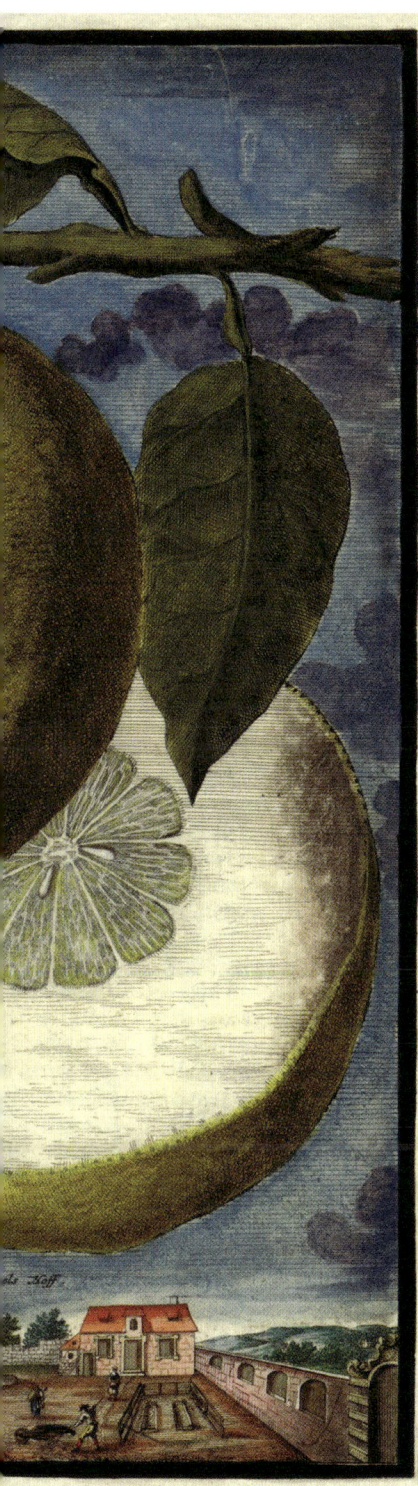

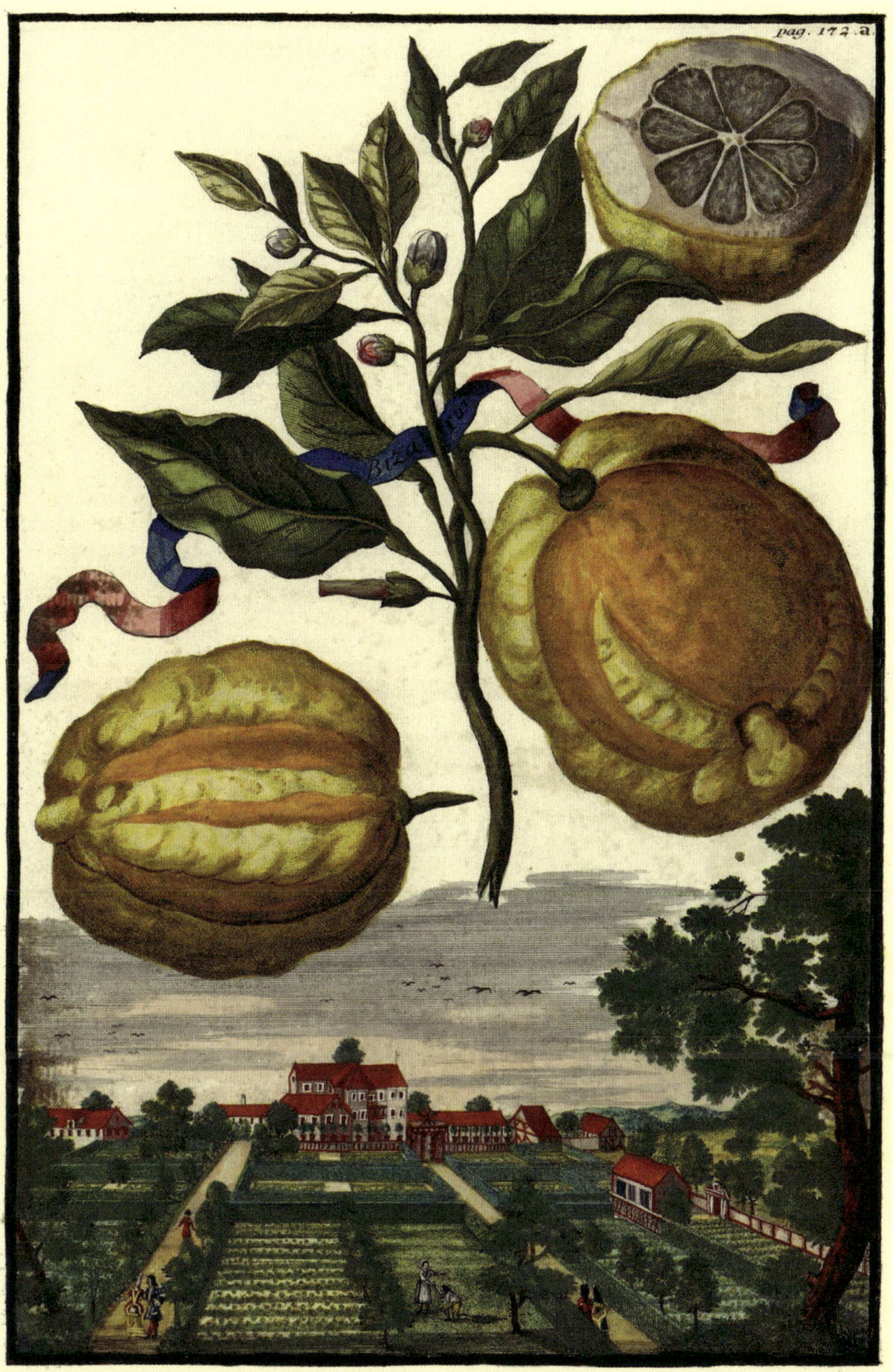

'Aranzo striato dolce', 'Limea da Valenza' and 'Bizaria', engravings from *Nürnbergische Hesperides* by Johann Christoph Volkamer, *c*. 1708.

In the tradition of Ferrari, German botanist Volkamer connected his citrus monograph with the mythological Hesperides. Volkamer discussed many more kinds of citrus than Ferrari and augmented his plates with vignettes of gardens beneath the fruits themselves. Nevertheless, many of the citrus fruits depicted in his work cannot be reliably identified.

Jean-Nicolas Jadot de Ville-Issey (1710–1761). It is 189 m (620 ft) long and 10 m (33 ft) wide – second only to the orangery at Versailles.

Some growers managed to succeed spectacularly with much more modest arrangements. The gardeners of Archibald Campbell, 3rd Duke of Argyll (1682–1761), at Whitton, west of London, coaxed citron trees to produce fruits 'as large and perfectly ripe as they are in Italy and Spain' by training the plants against a south wall with flues and glass covers in winter.[30]

By comparison with all these monumental continental orangeries, the orangery at the Royal Botanic Gardens, Kew, near London, built in 1761 to the design of William Chambers (1723–1796), is rather modest, being only 28 by 10 m (92 by 33 ft). Unlike that at the Belvedere, but like most of the surviving significant orangeries (including English examples at Mount Edgcumbe, Cornwall, completed c. 1760, and Burghley House, Lincolnshire, built during the late 1760s), it is now a restaurant, as light levels inside are insufficient to sustain healthy citrus trees in the long term. This was always a problem in the first orangeries, as they had rather small windows, although English writer John Evelyn (1620–1706) realized as early as 1693 the need for large windows 'so high as almost to touch the timbers which support the ceiling'. Despite this, small windows were the norm, owing, no doubt, to the cost of glass. The result was that citrus trees deteriorated if they were not moved outside regularly. By 1860, there was a report on 'the dying orange trees at Versailles.... many of the orange trees are very ill and many will die'.[31] The last of Kew's 'unhappy trees', many of which had already been moved to Kensington Palace, were transferred to Kew's glasshouses while the orangery was reopened in 1863 as the Timber Museum.

CATALOGUES AND COLLECTIONS

By the early 1700s, the number of kinds of citrus available in commerce had greatly increased. The catalogue of Polish grower Caspar Wilhelm Scultetus, published in 1731, listed 162 different kinds of citrus plants, with 35 'Citronate', 85 'Citronien ... Limonien and Lumien' and 42 'Pomeranzen'. The heyday of the orangery meant that there was a plentiful supply of good fresh material for scientific work, so that a number of taxonomic treatises, as opposed to practical horticultural works larded with mythology and recipes, began to appear. Some of these were beautifully illustrated using the new techniques of colour printing. A pioneer in this was the Italian

Giorgio Gallesio (1772–1839), a politician and botanist living near Genoa. One of his books, *Traité du citrus*, was first published in 1811 and, showing how popular it became, was reprinted in 1826; so significant was it that an English translation was issued in the USA under the title *Orange Culture* as late as 1876. Gallesio was not merely an observer; he was an experimentalist, carrying out trials and making hybrids at his estate, now the Villa Gallesio-Sanguineti, on the eastern slopes of the Valle dell'Aquila near Finalborgo. In his book, he argued that his experimental findings showed that permanent variation in plants was due neither to environmental factors nor to grafting, which were common misapprehensions among growers at the time. Instead he proposed that (in modern terms) variation occurred through outcrossing different kinds of citrus. He was very Darwinian in his thinking. Gallesio prepared a new classification of the genus *Citrus*, building on that of Linnaeus, but wisely writing:

Citrus is a genus whose species are greatly disposed to blend together, and whose flower shows great facility for receiving extraordinary fecundation; it hence offers an infinite number of different races which ornament our gardens, and whose vague and indefinite names fill the catalogues.... This is perhaps one of the most difficult portions of our work, first, because the botanists or agriculturists who have described the varieties have not always done so with the exactness requisite to enable us to recognize them among so many different names; and, secondly, because in the course of centuries several of these varieties have disappeared, from frosts or other influences, and been replaced by a quantity of new varieties which resemble them, and which, by means of some slight differences, create confusion in the application and comparison of these descriptions.[32]

Without the advantages of modern techniques, Gallesio recognized as species then in Europe only the citron, the lemon (actually a hybrid involving the citron), the sour orange and the sweet orange (both the same hybrid species); the lime he incorrectly considered to be a hybrid between the lemon and the orange.[33] Tellingly, he wrote of the citron, 'for several centuries a constant species [but] it produced hybrids ... so soon as it was placed in the same soil with lemon and orange trees'.[34] He reported that at that time the

---

'Cedro ordinario' (citron, *Citrus medica*), engraving from *Nürnbergische Hesperides* by Johann Christoph Volkamer, c. 1708.

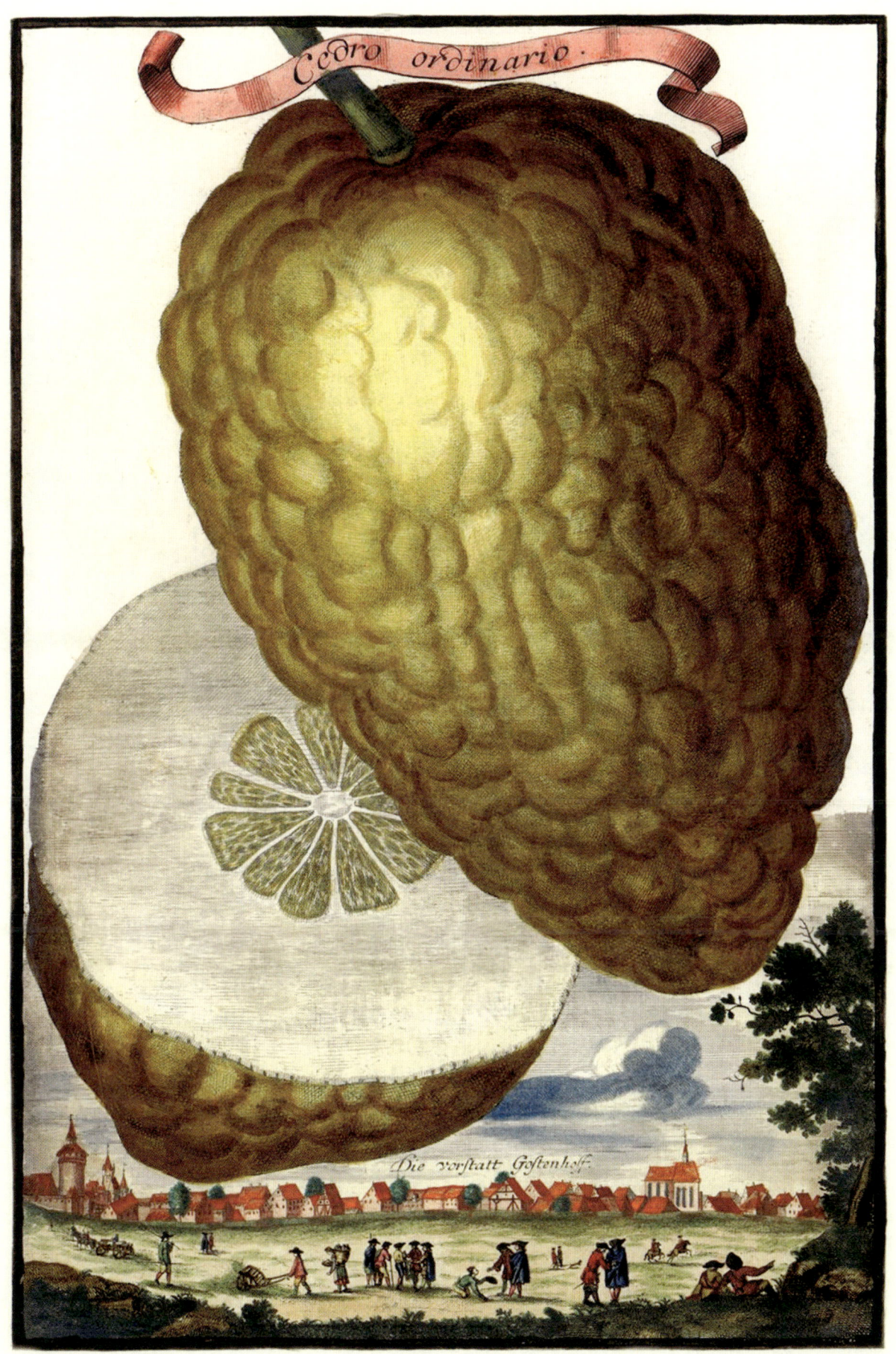

Cedro ordinario.

Die vorstatt Goftenhoff.

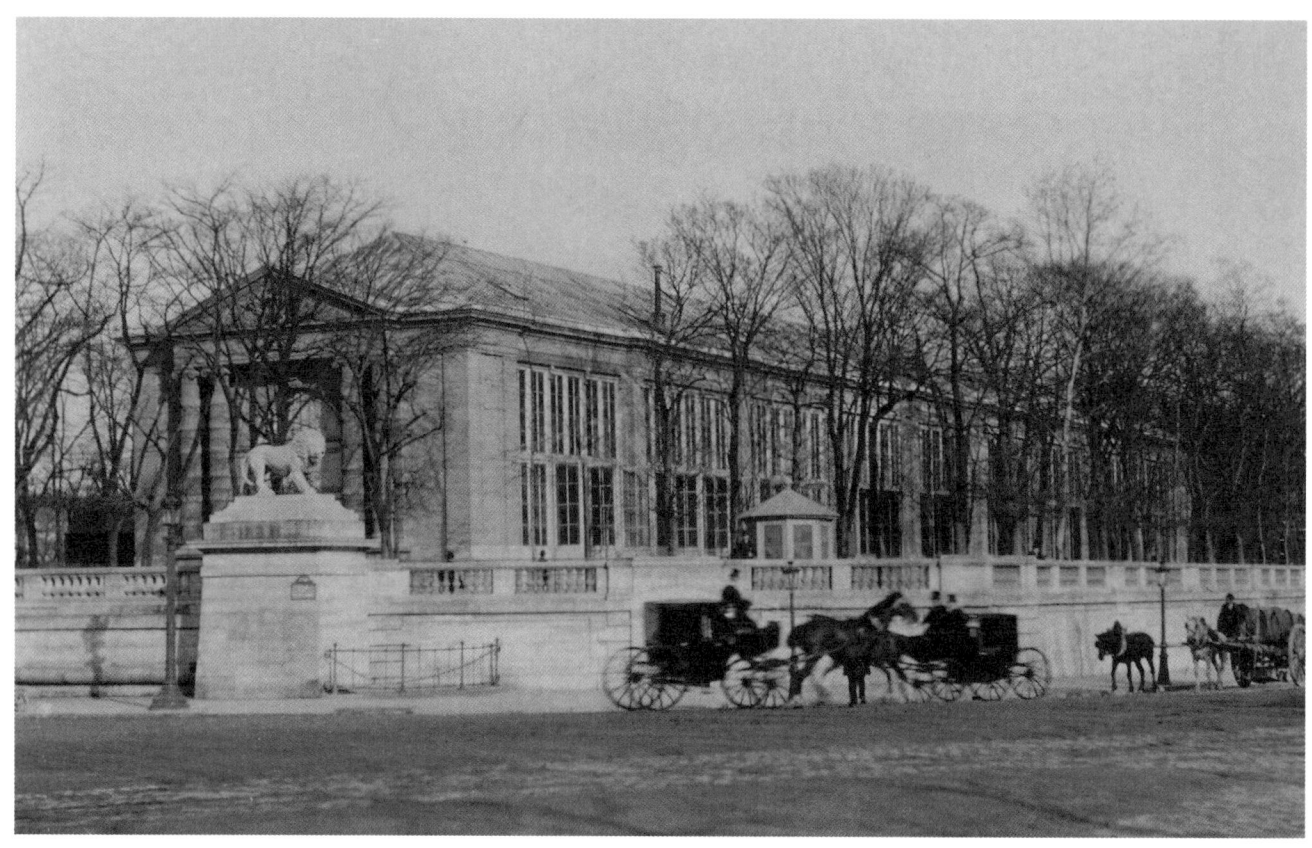

Did you ever sleep in a field of orange trees in bloom? The air which one inhales deliciously is a quintessence of perfumes.... It is like opium prepared by fairy hands and not by chemists.

GUY DE MAUPASSANT, 'EN VOYAGE', 1882

L'Orangerie au jardin des Tuileries, Hippolyte Blancard, *c.* 1890
One of modern-day Paris's major museums started life as the winter home for the citrus collections of the Jardin des Tuileries. Opened in 1852, the orangery was built for Napoleon III.

sour orange (bigarade) was considered more strongly scented than any other citrus and that it was grown merely for its flowers near Paris, while in southern France, notably at Grasse and Nice, besides St Remo in Italy, it was being cultivated solely for the distillation of them for scent.[35] As for the 'Adam's apple', which he considered to be a hybrid of the orange and citron (in other words a lemon, in which he was probably right), it was widely planted at Versailles, in the Jardin des Plantes and elsewhere in Paris, though the fruit was 'good for nothing, and is sought for its beauty only, as it is neither edible when raw, nor agreeable for confits'.[36]

Contemporary with Gallesio was Antoine Risso (1777–1845), a pharmacist with a broad interest in many aspects of natural history who ran the Jardin de Naturalisation in Nice.[37] In his *Essai sur l'histoire naturelle des orangers* of 1813, he too made a classification of citrus, describing many new cultivars with practical information including much on pests and diseases besides uses. More importantly, this account was a precursor to perhaps the most magnificent citrus work of the nineteenth century, the *Histoire naturelle des orangers*, issued in nineteen parts in Paris from 1818 to 1822 (see pp. 172–73), with illustrations by Pierre-Antoine Poiteau (1766–1854). There were later English translations and a second edition, called *Histoire et culture des orangers*, prepared by Rouen botanist Alphonse du Breuil (1811–1890) and published in 1872.

Before Risso and Poiteau's monumental book came out, Étienne Michel's *Traité du citronier* of 1816 was published in Paris. It was an early issue of a section of the 1819 edition of one of the folio volumes of Henri-Louis Duhamel du Monceau's (1780–1782) monumental *Traité des arbres et arbustes*, largely adopting Risso's 1813 classification with some magnificent colour-printed illustrations by Pierre-Joseph Redouté (1759–1840).

At almost the same time yet another account, in German, appeared. After spending six years in Italy as tutor for the family of Wilhelm von Humboldt (brother of the even more celebrated Alexander von Humboldt), Friedrich Sickler (1773–1836)[38] published his folio work *Der vollkommene Orangerie-Gärtner, oder Vollständige Beschreibung der Limonen, Citronen und Pomeranzen, oder die Agrumi in Italien und ihrer Cultur* (1815), which was heavily reliant on Gallesio's work. In it, he described one of the most famous Italian citrus collections, namely that of the Borghese in Rome, who had some seventy kinds in cultivation, and discussed citron trees with fruits over 40 cm

(16 in.) in length and over 7 kg (16 lb) in weight. Sickler's account of what were known in Italy as *agrumi* (bitter fruits) had little scientific merit, but he made watercolour drawings of 'the most remarkable sorts';[39] these went to the Horticultural Society of London in 1818, but are now lost.

While Sickler and other naturalists were trying to document citrus in Europe and bring order to their classification, orangery-building was reaching its most extravagant heights, particularly in Sickler's own homeland – what is now Germany. The biggest orangery in northern Germany is that built at Potsdam to the designs of the Prussian King Frederick William IV (1795–1861), who commissioned the architect Friedrich Ludwig Persius (1803–1845). Begun as early as 1826,[40] construction continued until the late 1860s. The building replaced the old orangery at Frederick the Great's (1712–1786) Sanssouci and was really a small palace with royal suites, including reception rooms, servants' quarters, lots of statuary and two greenhouses 103 m (338 ft) long by 16 m (52 ft) wide for some one thousand plants. The complete ensemble survives and is over 300 m (984 ft) long, with the original floor-heating system still in good working order. Built in Italian Renaissance style, it incorporated elements inspired by the Villa Medici in Rome and the Uffizi in Florence. Sadly, Frederick William died, severely incapacitated, in 1861, so did not live to see his extravaganza completed.

At the same time, in Paris , an orangery was being built for Napoleon III (1808–1873), to hold the citrus collection of the Jardin des Tuileries, the tubbed trees formerly overwintering in the Grande Galerie of the Louvre. The orangery was opened in 1852 and today is perhaps the most internationally known of all orangeries because, as the Musée de l'Orangerie, it houses Impressionist art, notably the murals *The Water Lilies* by Claude Monet (1840–1926).

By the end of the nineteenth century, orangeries in general were becoming showcases for a variety of tender plants, a concept far distant from what has been considered their original humanist intent (see p. 98). Their decline in that allegorical sense had already begun a hundred years before, when the idea of reviving the ideals of Classical antiquity was being replaced by a fascination with collecting the exotic for its own sake.[41] In short, the ideals of the Renaissance were replaced by materialistic ostentation, though paradoxically sometimes with architecture aping the Renaissance, the rich using the latest technology to display what were in effect the spoils of empire.

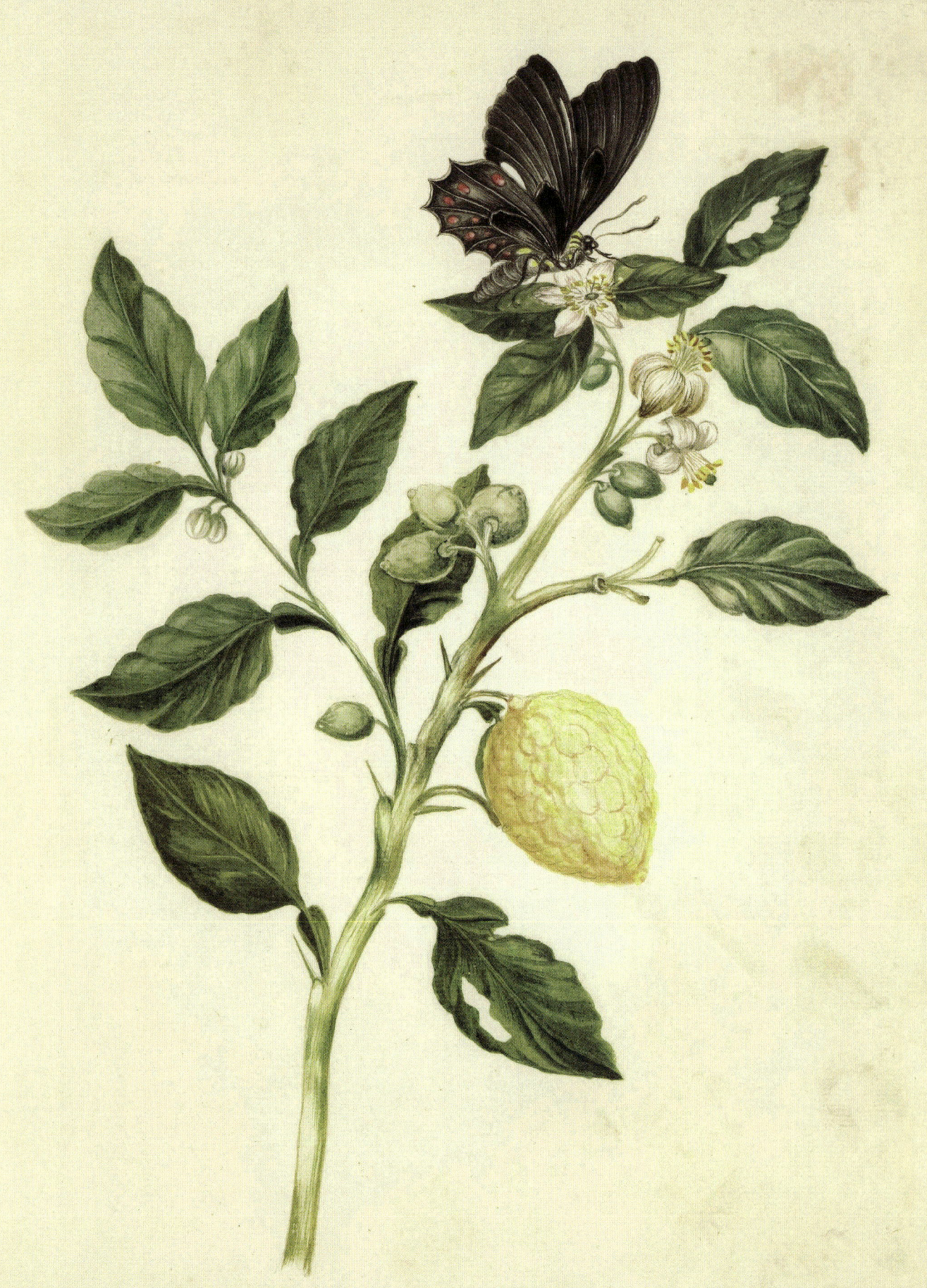

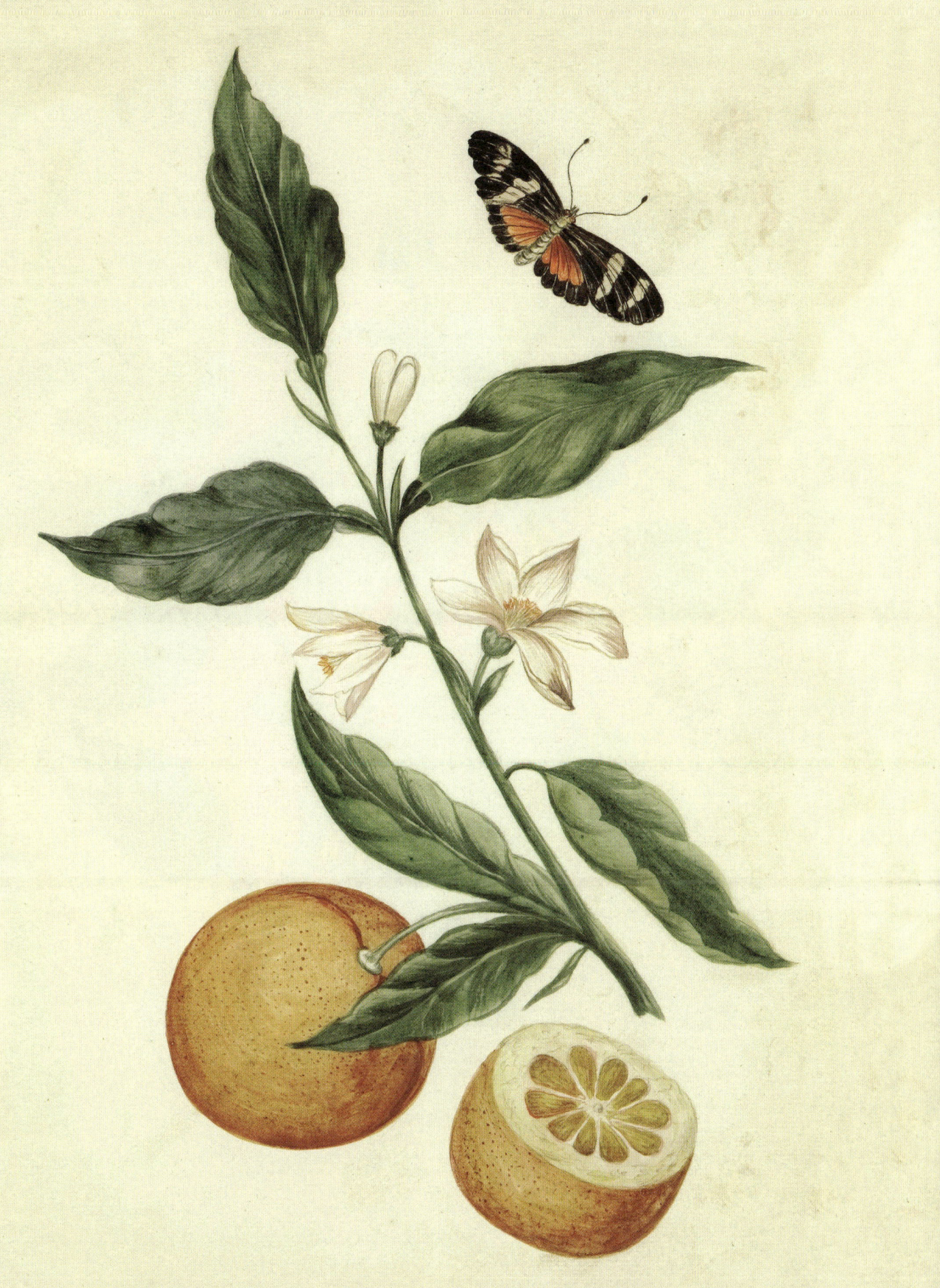

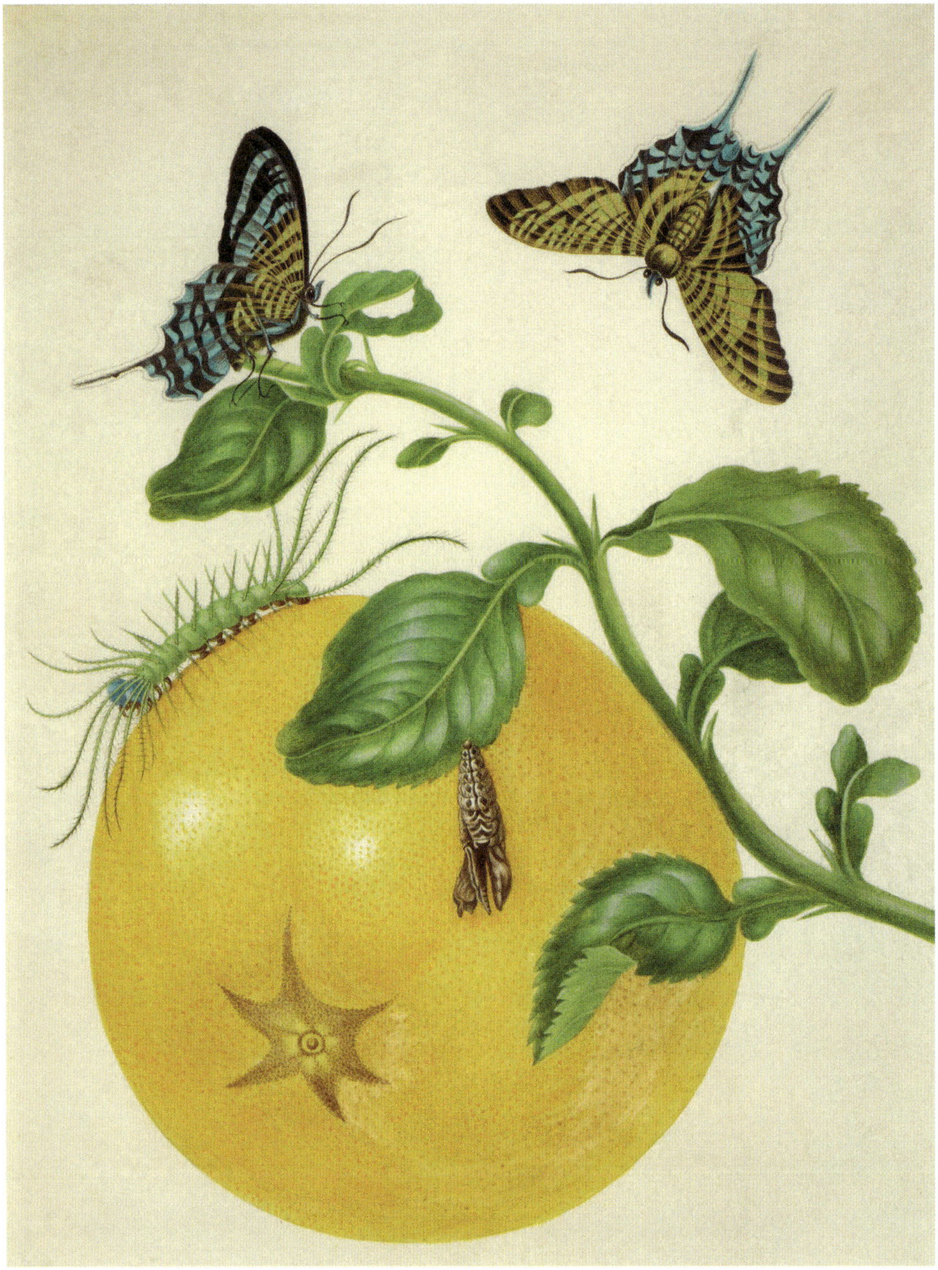

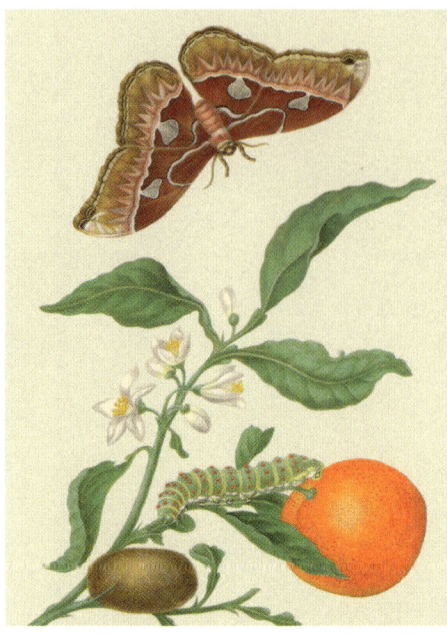

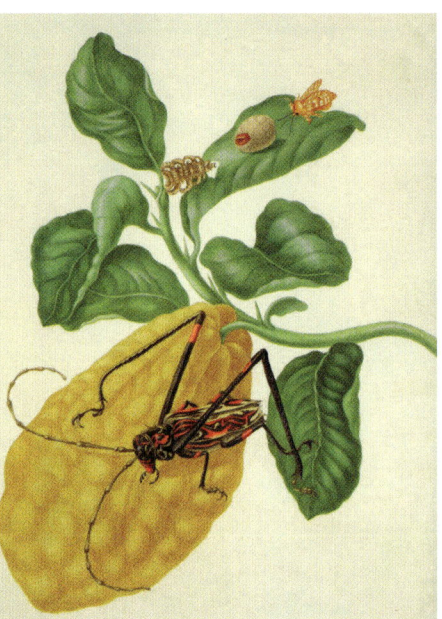

PREVIOUS PAGES: 'Flowering branch of lemon tree, fruit and butterfly' (left) and 'Branch with blossoming orange blossom, oranges and butterfly' (right), Cornelis Markée, *c.* 1763. **ABOVE, CLOCKWISE FROM LEFT:** 'Branch of Pomelo with Green-Banded Urania Moth', 'Branch of Seville Orange with Rothschildia Moth' and 'Citron with Monkey Slug Moth and Harlequin Beetle' from *Metamorphosis insectorum Surinamensium* by Maria Sibylla Merian, 1702–3. **OPPOSITE:** Johann Jakob Haid (attrib.), plates from *Phytanthoza iconographia* by Johann Wilhelm Weinmann, 1737.

Merian (1647–1717) travelled to Suriname, then a Dutch colony, in 1699 to study the insects there. Her watercolours and observations were later published as *Metamorphosis insectorum Surinamensium*. Some of her studies were later copied by Markée (*c.* 1709–1769; see previous pages). Weinmann (1683–1741) was a German apothecary who established a botanic garden in Regensburg and prepared an important survey of known plants in his *Phytanthoza iconographia*.

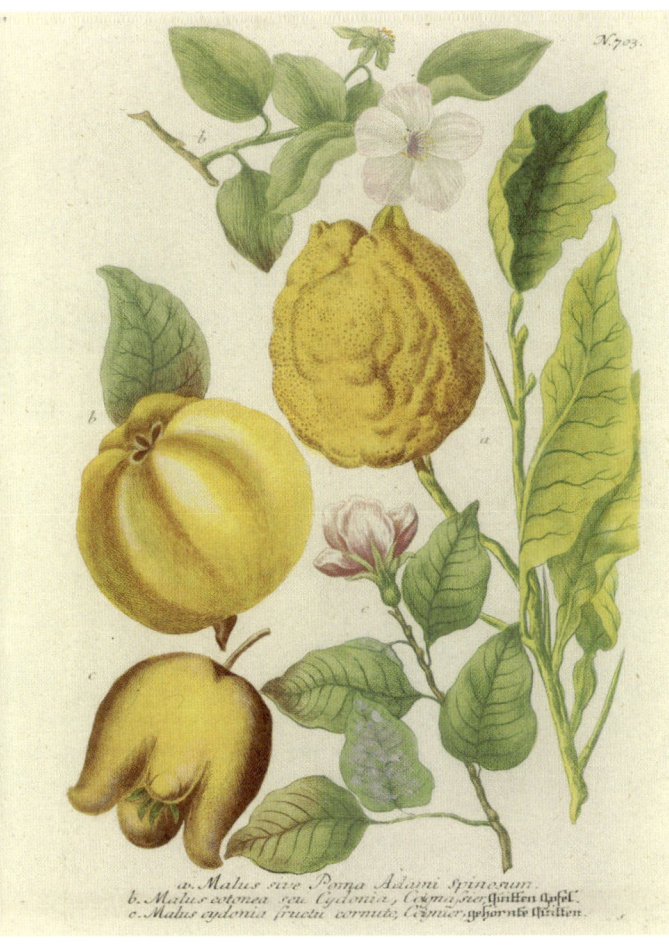

a. *Malus sive Poma Adami spinosum.*
b. *Malus cotonea seu Cydonia, Coignassier, Quitten Apfel.*
c. *Malus cydonia fructu cornuto, Coignier, gehörnte Quitten.*

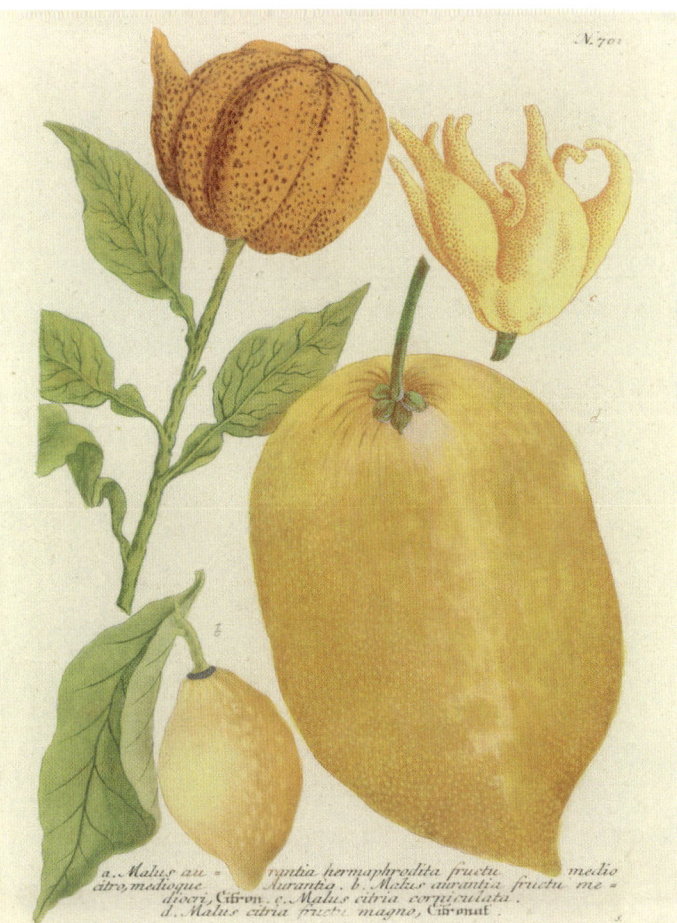

a. *Malus au rantia hermaphrodita fructu medio*
citro, medioque Aurantia. b. Malus aurantia fructu me -
diocri, Citron. c. Malus citria corniculata.
d. *Malus citria fructu magno, Cuironat.*

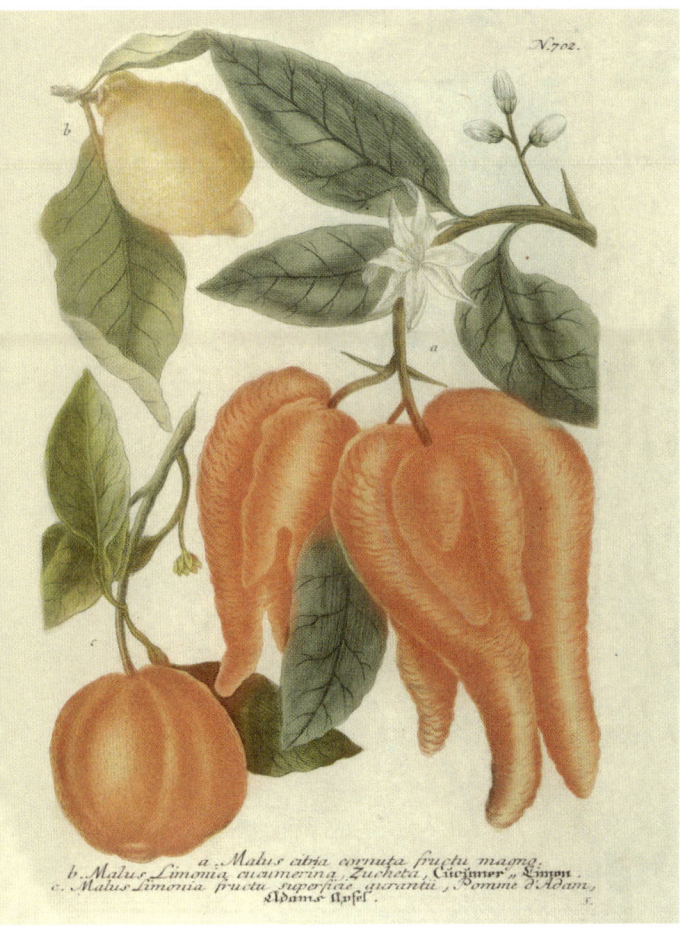

a. *Malus citria cornuta fructu magno.*
b. *Malus Limonia cucumerina, Zucheta, Cucumer Limon.*
c. *Malus Limonia fructu superficie aurantii, Pomme d'Adam,*
Adams Apfel.

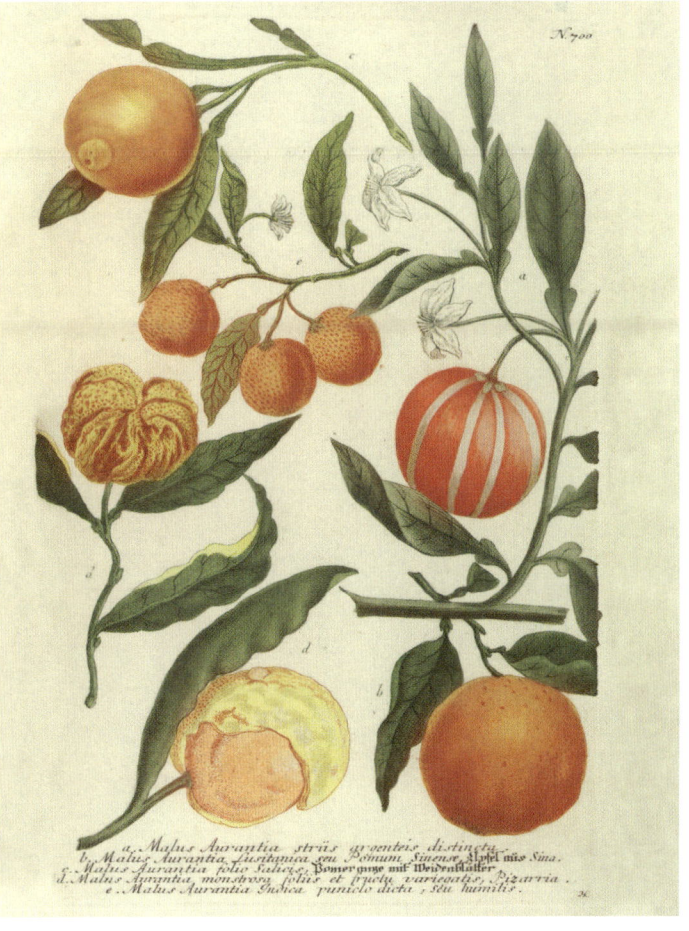

a. *Malus Aurantia striis viventis distincta.*
b. *Malus Aurantia Lusitanica seu Pomum Sinense, Apfel aus Sina.*
c. *Malus Aurantia folio Salicis, Pomerantze mit Weidenblätter.*
d. *Malus Aurantia monstrosa, foliis et fructu variegatis, Bizarria.*
e. *Malus Aurantia Indica, pusilla dicta, seu humilis.*

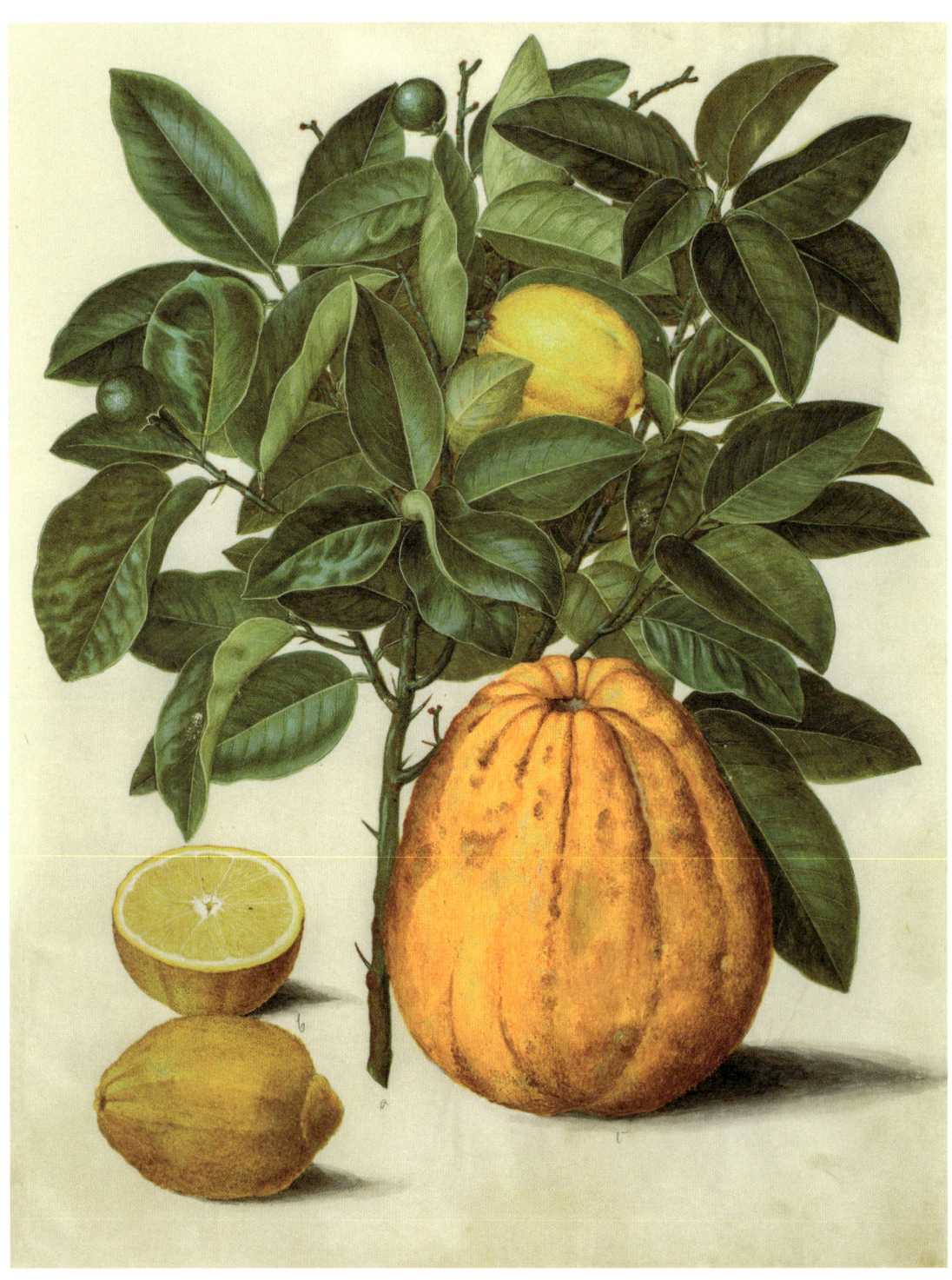

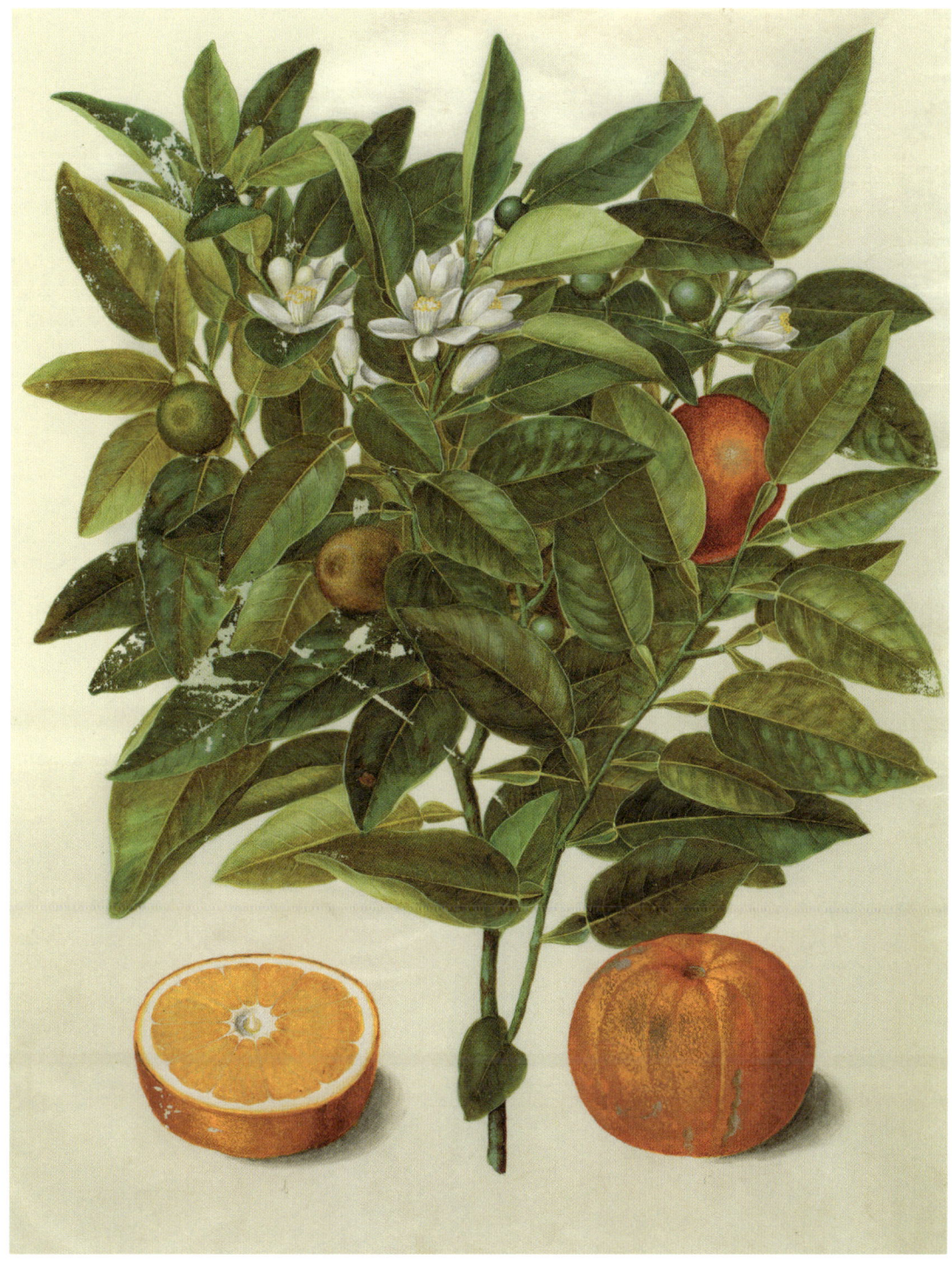

Lemon and pomelo (opposite) and sweet orange (above), Hans Simon Holtzbecker, 1649–59.

Holtzbecker (1610/20–1671), from Hamburg in northern Germany, was one of the most renowned botanical artists of the 17th century. Much of his work was attributed to Maria Sibylla Merian (see previous pages), who was also German and lived only slightly later than him. Holtzbecker was commissioned by Friedrich III, Duke of Holstein-Gottorp, to record plants in his new orangery, reputedly the first of its kind in Germany.

Pl. CXV.

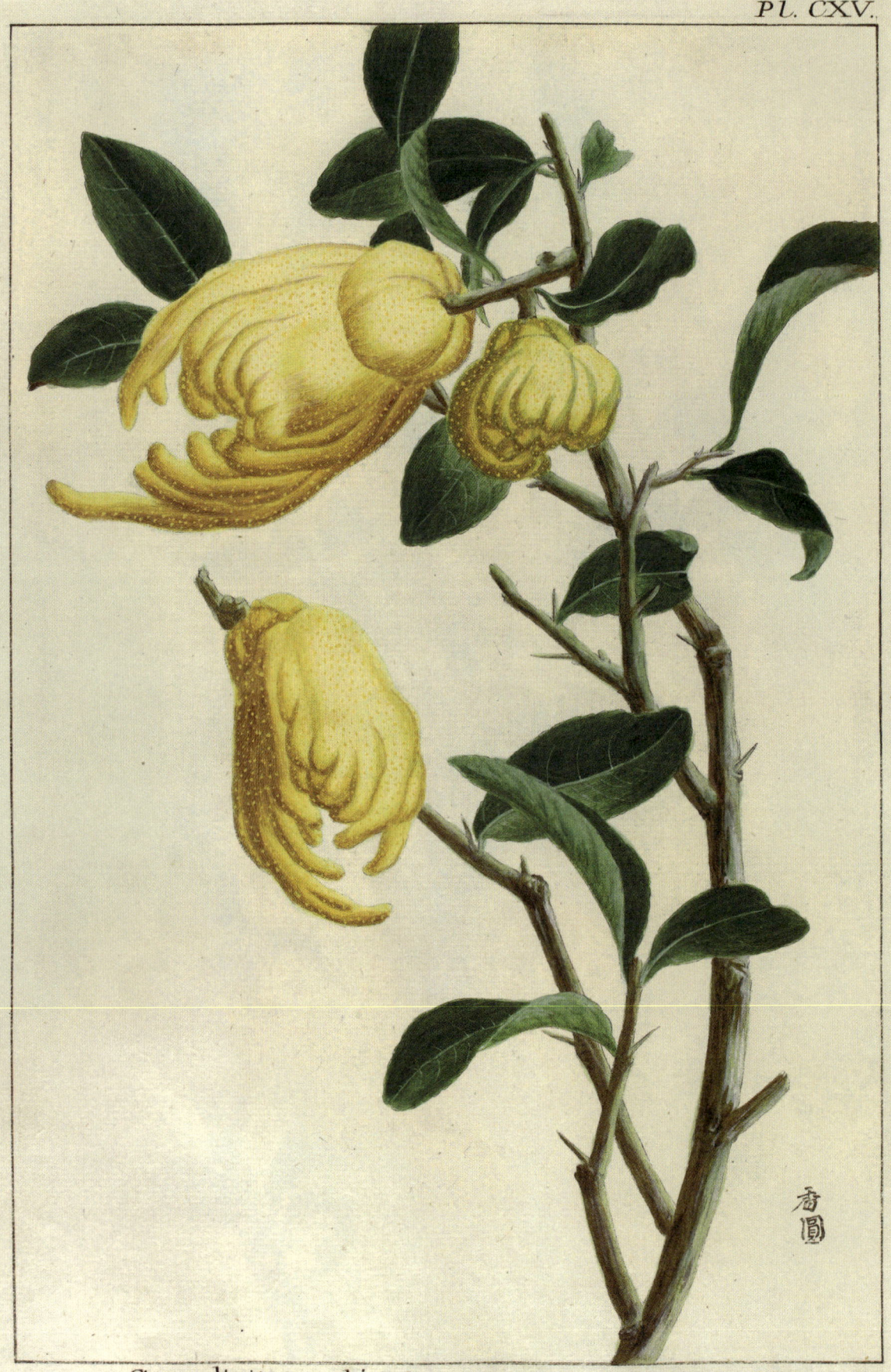

Citrus digitata nobis. *heung-yune Monstres doigtiés.*

Still offered in Buddhist temples and presented as a New Year gift in China today, is the fruit of an ancient group of cultivars, in which the hesperidium comprises partially separate, finger-like carpels and resembles a fruit attacked by bud mites. In China fingered citrons are said to promote good fortune (fu shou), happiness and longevity; in the West they are sold by florists as Buddha's hand.

SEE P. 39

Citrus digitata nobis (fingered citron, a famous cultivar of *Citrus medica*) from *Le grand jardin de l'univers* by Pierre-Joseph Buc'hoz, 1785.

Buc'hoz (1731–1807) was a French lawyer and doctor, who recommended music as therapy for depression ('melancholy'), but also a botany teacher. He poured out natural history publications, over 500 works, mostly conspicuously plagiarized from the work of others. Today he is very well known in the art trade because of his popular, and increasingly prized, coloured engravings, much influenced by Chinese painting.

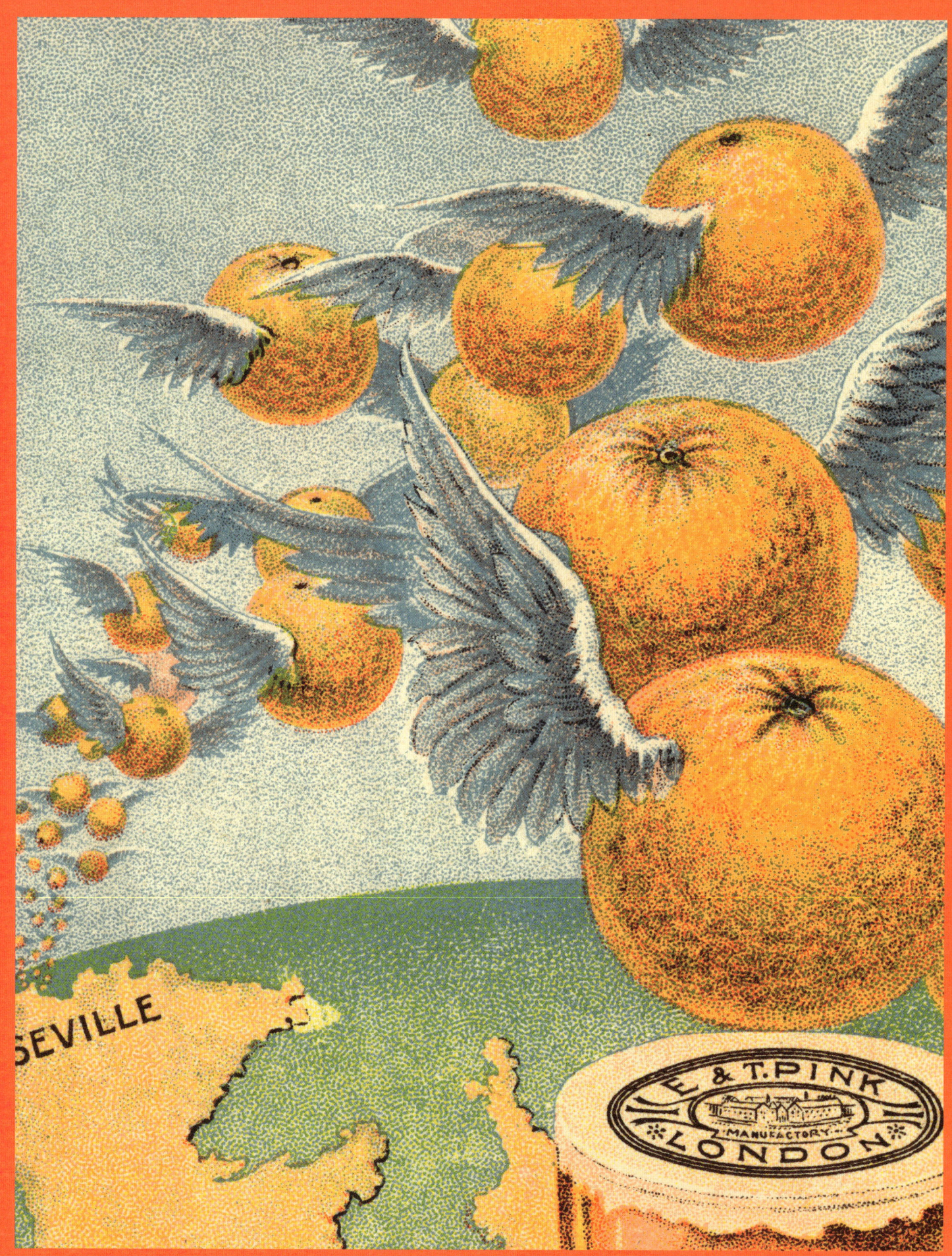

FOUR

Empire, Exports and Vitamin C

THE FIRST EUROPEAN COLONIES were in the Atlantic islands - the Canary Islands overrun by Spain from 1402, Madeira settled by Portugal beginning in 1420.[1] These were reached after rather short sea voyages, but economic and then imperial ambitions further afield - to Africa, Asia and the Americas - meant longer voyages with often catastrophic consequences. In 1577, a Spanish galleon was found adrift in the Sargasso Sea; all on board were dead.[2] A surgeon on an English ship wrote of his experience of the affliction that had caused that disaster:

It rotted all my gums, which gave out a black and putrid blood. My thighs and lower legs were black and gangrenous, and I was forced to use my knife each day to cut into the flesh in order to release this black and foul blood. I also used my knife on my gums, which were livid and growing over my teeth.... And the unfortunate thing was that I could not eat, desiring more to swallow than to chew.... Many of our people died of it every day, and we saw bodies thrown into the sea constantly, three or four at a time.[3]

This was scurvy.

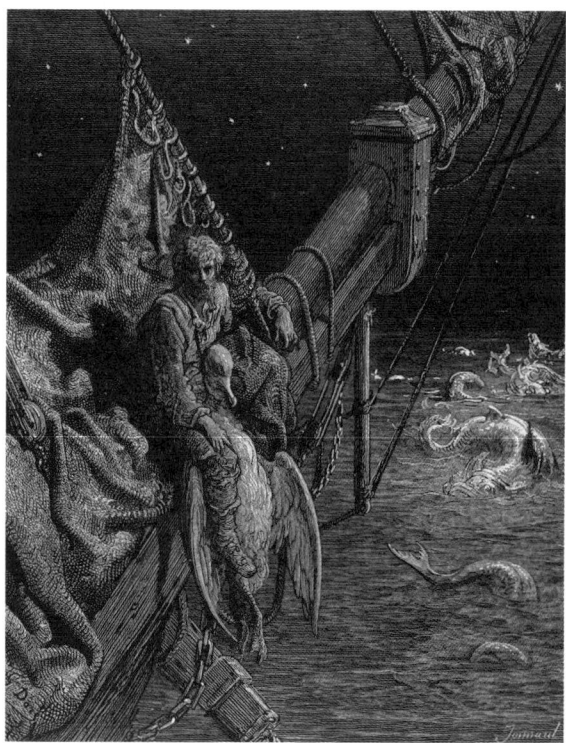

Plate after Gustave Doré from 'The Rime of the Ancient Mariner' by Samuel Taylor Coleridge, 1876. The poem gives a vivid account of the hallucinations, suffering and death associated with scurvy.

SCURVY

Scurvy was known to the ancient Greeks. Hippocrates wrote of Alexander the Great's 329 BCE voyage from the Indus to the head of the Persian Gulf that a large number of men suffered from 'pains in the legs, foetid breath, gangrene of the gums and eventually lost their teeth'.[4] Later, Roman soldiers stationed in northern Europe also suffered from it.[5]

The disease is a condition now known to result from a deficiency of vitamin C, which is crucial to the synthesis of collagen in humans and certain other animals. A healthy human has 900–1,500 mg in the body and uses *c.* 50 mg a day.[6] After around sixty to ninety days of inadequate intake, the vitamin C level in blood plasma falls to a quarter of healthy levels. The body then ceases collagen production, resulting in the progressive disappearance of cartilage: teeth and hair fall out, bones become brittle and the skin forms ulcers. Early symptoms are lethargy and depression, then the appearance of spots on the skin, especially on the lower limbs; mucous membranes begin to rot and become blackened by seeping blood, the gums become spongy. Other symptoms include halitosis, diarrhoea, internal bleeding and then lung and kidney failure leading to coma and death.

In scurvy patients, the brain survives longer than other organs because transport of vitamin C into the cerebrospinal fluid is the last to give way. The Scottish naval physician Thomas Trotter (1760–1832) reported in *Observations on the Scurvy*, 'In dreams they are tantalised by the favourite idea; and on waking, the mortifying disappointment is expressed with the utmost regret, with groans, and weeping, altogether childish.'[7] Sensitivity to sounds is enhanced such that the report of a gun could kill scorbutic sailors, and even fresh water or the much-anticipated taste of fruit could result in a seizure.

Most plants and animals synthesize vitamin C from monosaccharide sugars, but some mammals, for example humans and other 'higher' primates, cannot produce it, though the more 'primitive' primates can. The failure is due to a malformed enzyme produced in the liver. It must be concluded that such a failure was not an evolutionary disadvantage among animals getting adequate supplies from their food - in other words, our ancestors were fruit eaters, gaining their vitamin C through their diet.

The ancient Chinese carried fresh food in the form of live ginger (12 mg vitamin C per 100 g) on their voyages,[8] and, in the West, it has long been known that citrus fruits can reverse the symptoms of scurvy, leading to a full recovery.[9] As early as 1227 CE,[10] in his *Compendium medicinae*, the most important British medical book of the Middle Ages, Gilbertus Anglicus (Gilbert 'the Englishman', 1180–1250) recommended the use of citrons and lemons in Palestine.[11] In the 1497 voyage to India of the Portuguese Vasco da Gama, the use of citrus fruits was understood, in that, on the East African coast, sick crew seemed to recover after eating them, but da Gama still lost well over half of his men, largely to the disease. The Portuguese Ferdinand Magellan (1480–1521) lost 208 men out of 230, again largely to scurvy, during the first circumnavigation of the world, and, by the 1590s, Sir Richard Hawkins (c. 1562–1622), who claimed to have seen over 10,000 scurvy cases in his life, 'wished some learned man would write of it, for it is the plague of the sea and the Spoyle of Mariners'.[12]

Hawkins's contemporary Sir James Lancaster (1554–1618), commanding the first English East India Company's fleet to the Malay Archipelago in 1601 on his flagship *Red Dragon*, had 'bottles of the juice of lemons, which he gave to each one, as long as it would last, three spoonfuls every morning.... the general cured many of his men and preserved the rest'.[13]

Mariners planted (or left behind in discarded fruits) seeds of citrus at landfalls during their voyages. The Portuguese planted fruit trees and vegetables on St Helena in the south Atlantic as a source of fresh food on the journey to India; by 1583, there were so many citrons, lemons and oranges growing 'without planting or setting, that all the valleys are full of them, which is a great pleasure to behold, for that it seems to be an earthly Paradise'.[14] In April 1593, bottled lemon juice from the island kept scurvy under control in one ship until it reached Madagascar and more fresh food. Oranges, lemons and citrons were well established in coastal Sierra Leone, such that by 1606 one traveller wrote, 'Of orange trees I shall make no special mention, since the forests are full of them.'[15]

Lemons were carried to combat scurvy on Dutch East India Company ships from their first voyage in 1602 onwards. The Dutch also established a garden in their colony at the Cape of Good Hope for provision of fresh food, thereby becoming independent of St Helena, where British ships often attacked them.[16] Commander Jan van Riebeeck (1619–1677) arrived in April 1652 and, after a false start and much privation, laid out the garden: in June 1654 citrus (including lemons from St Helena, first fruiting in 1661)[17] reached South Africa, in effect beginning today's major export industry.

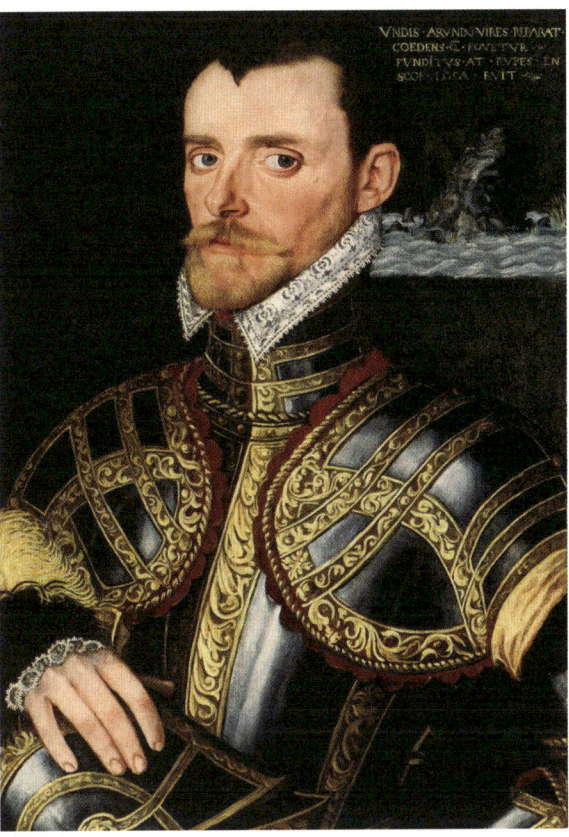

Portrait of Admiral Sir Richard Hawkins, unknown artist, c. 1590. Hawkins wrote down his observations of the effects of citrus on sufferers from scurvy, 'to be a certaine remedie for this infirmitie'.

The British East India Company had been advised by John Woodall (1570–1643), who supplied its medicine chests, of the importance of fresh food in fending off scurvy. In his book, *The Surgeon's Mate*, of 1617, he recommended lemon juice as both prophylactic and curative. Yet, despite all the obvious successes of citrus with the Dutch, the British and indeed others, it fell out of favour by the end of the century in the face of the fashionable 'imbalance of bodily humours' theory of Western medicine – though perhaps cost was a more pressing factor. Fresh fruit was expensive, and although boiling the juice down to 'rob' allowed easy storage, it destroyed the vitamin (especially if boiled in copper kettles). Lemon juice ceased to be standard issue and was replaced by other 'remedies', including purging with seawater, bleeding, taking acids or even making sufferers work harder.

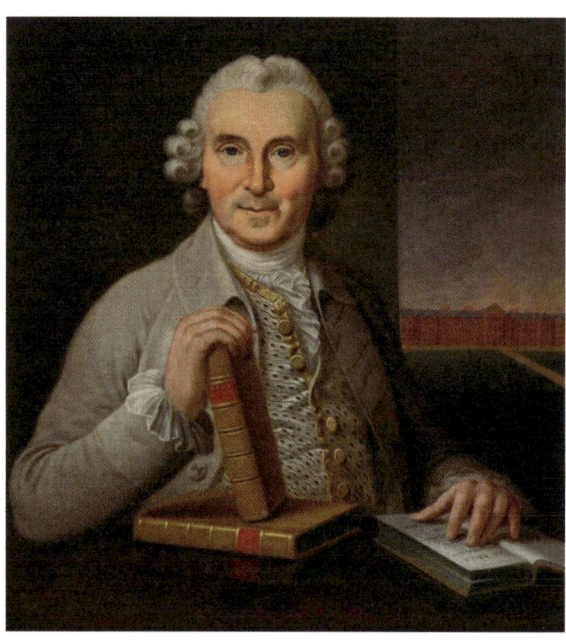

Portrait of James Lind, Sir George Chalmers, 1783. Lind established that citrus fruits could prevent and cure scurvy, but his treatise on the disease, published in 1753, was mostly ignored. Only in 1796 was lemon juice routinely issued to sailors of the British Royal Navy.

By 1712, it was held by one authority that fresh food caused inflammation of the gut: 'one must, when ships reach countries abounding in oranges, lemons, pineapples etc., ensure that the crew eat very little of them since they are the commonest cause of fevers and obstructions of the vital organs.'[18] Of the 184,899 men recruited or press-ganged for navy service in the Seven Years' War with France (1756–63), 133,708 died from disease, including scurvy, while only 1,512 were killed in action.[19]

According to Captain William Bligh (1754–1817) of the *Bounty*, outbreaks of scurvy on a ship were the results of bad management, so when his surgeon diagnosed scurvy on his ship, he insisted on its being logged as rheumatism. Much later, during his first expedition of 1911–12 in the Antarctic, Captain Robert Scott (1868–1912) was beside himself when Ernest Shackleton (1874–1922) developed scurvy and sent him home (perversely, a tin of Cooper's Oxford marmalade taken on the expedition was in perfect condition when found in 1980).[20] On the second expedition, Scott even considered omitting from his log any mention of scurvy.

All this, in retrospect, seems quite extraordinary. However, no one really knew what it was in food that prevented or cured scurvy. Another issue was that the disease was often confused with other nutritional deficiency illnesses, including the blindness attributed to scurvy that was actually due to a lack of vitamin A.

Despite the work of the Scottish physician James Lind (1716–1794),[21] who identified citrus as a promising treatment in 1753, and James Cook's (1729–1779) use of citrus on his Pacific voyages leading to physicians arguing its importance to the British Admiralty, only in 1796 did the British navy adopt lemon juice as standard issue at sea. Thenceforth there were daily draughts of lemon juice, usually added to the rum ration.[22] The consequently improved health of the British navy gave Nelson a strategic advantage over Napoleon's fleet at the Battle of Trafalgar in 1805.

Lind was a naval surgeon during the War of the Austrian Succession (1740–48). His 1753 work *Treatise of the Scurvy* described 'this foul and fatal mischief'. In May 1747, Lind performed one of the very first of what are now called controlled experiments. He divided twelve scurvied seamen into pairs, each receiving different treatments, including seawater, cider, vinegar, and oranges and lemons.

Those given citrus quickly recovered. Lind believed the cause of the disease was tainted food and faulty digestion and excretion, the skin thus likely to acquire 'the most poisonous and noxious qualities, and a very high degree of putrefaction'. It took six years for Lind to publish his findings.[23] He had the cure, but could neither see the cause nor couch his findings in the framework of contemporary science. Somewhat tactlessly, he also made criticisms of the theories of more influential men.[24] However, his advice seems to have had some effect on Cook, who lost no crew to scurvy.

Cook was apparently unconcerned to victual his ships with citrus in particular: he carried sauerkraut and fresh onions (up to 30 mg vitamin C per 100g), and he would put into land wherever possible, as he was aware of the value of fresh leaf vegetables gathered by his crews.[25] However, on approaching Tahiti in April 1769, the botanist Joseph Banks (1743–1820) began to succumb to scurvy despite following Cook's orders for all of the company to take a pint of malt wort each day. Thankfully he had a supply of lemon juice, as prescribed in Nathaniel Hulme's *A Proposal for Preventing the Scurvy* (1768), and drank six ounces a day.[27]

Another Scottish doctor, Gilbert Blane (1749–1834, Physician to the Fleet 1779–83),issued a pamphlet, *On the most effective means for preserving the health of seamen, particularly in the Royal Navy*, in 1780.[26] He prescribed citrus juice as a prophylactic as well as a cure; later, as Commissioner of the Sick and Wounded Board, he pushed the Admiralty to introduce lemon juice as a daily ration. Following this development, citrus-growing became a strategic necessity in maintaining crews' health, in the same way that oak was strategic for shipbuilding and hemp for rigging and cordage. Between 1795 and 1815 the navy used 1.6 million gallons (7.3 million litres). As one historian opined, 'one might say the British Empire blossomed from the seeds of citrus fruits'.[28]

Only in 1911 was it clearly recognized that scurvy was caused by a dietary deficiency.[29] The next year, a Polish-American biochemist, Casimir Funk (1884–1967), developed the concept of vitamins, one of which was thought to be the antiscorbutic factor. Albert Szent-Györgyi (1893–1986), a Hungarian physiologist, completed a PhD under Frederick Gowland Hopkins (1861–1947) at Cambridge, where he isolated 'hexuronic acid' from adrenal glands. Returning to Hungary he found that this acid was the hitherto uncharacterized antiscorbutic factor: vitamin C. Once its structure had been determined, it was renamed L-ascorbic acid in view of its qualities. In 1933 vitamin C was synthesized from glucose, making it the first vitamin to be artificially produced. Soon it could be made in bulk and, in 1934, Hoffman-La Roche, which bought the patent, was the first pharmaceutical company to mass-produce and sell synthetic vitamin C, under the name Redoxon.

The last chapter in the citrus and vitamin C story provides a strange twist to the life of one of the world's greatest scientists, the American Linus Pauling (1901–1994). Remarkable for a prodigious research output and winner of both a Nobel Prize for Chemistry and the Nobel Peace Prize, Pauling was a pioneer in quantum chemistry and molecular biology – besides advocating for nuclear disarmament. In his later career he promoted vitamin C megadosage, namely the use of doses comparable to or even higher than levels produced in the livers of mammals able to synthesize vitamin C. Initially he argued that vitamin C was efficacious in dealing with the common cold, which involves the inflammation of the upper respiratory tract by viruses, subsequently infected by bacteria; he later argued that it would prevent cardiovascular disease and cure late-stage cancer, but so far there is no hard scientific evidence to support any of his assertions.

THE LIME

In 1845, ships in the West Indies were provided with lime juice from Britain's Caribbean colonies, substituting for lemon juice from the Mediterranean, no doubt a cost-cutting exercise. Lime juice, though, has only half the vitamin C that either lemons or oranges have and is therefore less effective in preventing scurvy. During the Crimean War (1853–56), the French army ate many vegetables, the British far fewer. This led to the British suffering more deaths from scurvy than any other single cause. The fact that lime juice had been sent out in 1854, but never used, was one of the scandals raised by no less a critic than Florence Nightingale (1820–1910) in her evidence on the conduct of the war.[30] Under the 1855 Carriage of Passengers Act in the USA, lime juice was given to emigrants – one gill (*c.* 120 ml) a week; limes were preferred to lemons, as cheap supplies came from

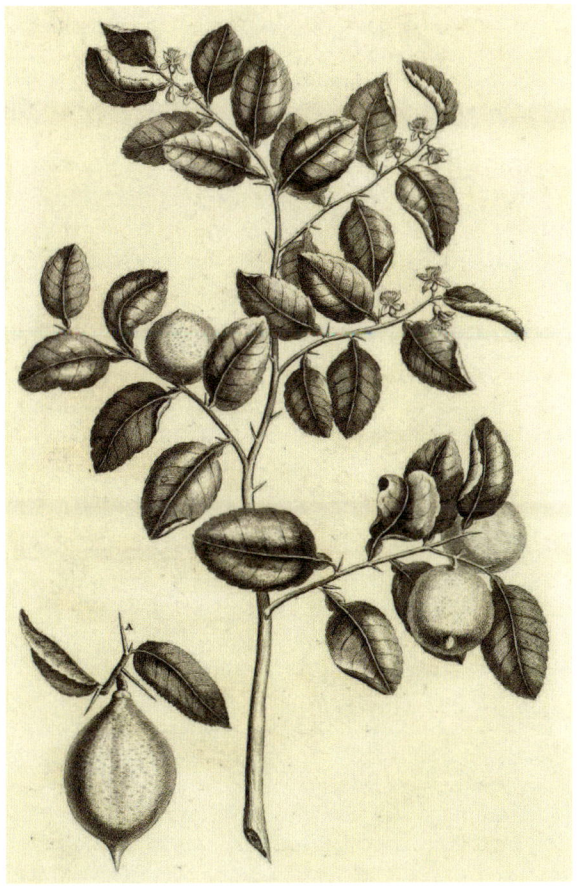

Limao nipis, from *Herbarium Amboinense* by Georg Eberhard Rumpf, 1741. Rumpf noted the antiseptic qualities of the lime as well as its ability to stay fresh on long sea voyages.

the West Indies, where the sugar industry was failing and limes were promoted as exports.[31] In 1860 lime juice was in use throughout the Royal Navy and in 1865 the merchant navy too, hence Americans calling British sailors 'limeys' or 'limers'.

The first recorded mention of the lime is that of 'Abd al-Latif al-Baghdadi (1162–1231) in the thirteenth century, who described the fruit as 'balm lemon of smooth skin the size of a pigeon's egg.'[32] The Arabic word *limoon* comes from the Hindi *lime* or *limbu*, no doubt ultimately from the Malay name *limau*; the English 'lime' (originally 'lyme', a generic word for citrus) derives from *lemon*, with the same etymology.

In a voyage begun in 1626, written up in his *Travels in Africa, Persia and Asia the Great* (1677), a courtier of Charles I of England, Sir Thomas Herbert (1606–1682), wrote of 'oranges, lemons and limes' on Mohéli (now part of Comoros) in the western Indian Ocean. Robert Knox (1641–1720), an East India Company ship's captain who was held captive on Sri Lanka for nineteen years (and brought cannabis to scientific notice in England), added in his *An historical relation of the island Ceylon* (1681):

At the time of year that there is most plenty of Lemmons [i.e. limes], they take them and squeez the juyce into an earthen Pot, and set over the fire, and boil it so long, till it becomes thick and black like Tar. This they set for their use, and it will keep as long as they please. A very small quantity of it will suffice for sawce. They call it Annego.[33]

As *anuga*, this was still being prepared until at least the middle of the last century in the Bibile area in Uva Province, formerly famous for the seedless 'Bibile Sweet' sweet orange. *Anuga* was used in Ayurvedic medicines as well as in cooking.[34]

But where had these limes in the Indian Ocean originated? They were well known in what is today's Indonesia. Georg Eberhard Rumpf (see p. 106), writing on *limao nipis* (*nipis* = thin, i.e. the peel) in his *Herbarium Amboinense* commented of limes that,

The juice ... is used daily to give a tart taste to all kinds of food, since all Indian Nations prefer [its] juice in their food over any kind of Vinegar. And this juice is by far the better tasting. One is also wont to make a comfit out of the green peel.... Such pickled [limes], mixed with Sulphur, clean and dry all kinds of Scabies and creeping sores.[35]

Rumpf also described how limes were used to clean weapons, stating 'if one plucks these [limes] by hand (when they are ripe), then wraps string around them

and hangs them up in such a way that they do not touch one another, one can keep them for a long time on Sea journeys'.[36]

In the early nineteenth century, lime juice was imported from the West Indies to Scotland for 'Glasgow punch', a drink based on rum. Lime and lemon were the sources of commercial citric acid at the end of the nineteenth century. Concentrated lime juice, prepared like *anuga*, was exported, its citric acid used in calico printing. In 1867, a ship chandler from Leith, Scotland, Lauchlan Rose (1829–1885), patented a method for preserving lime juice not with alcohol, but with sugar. The limes largely came from Dominica, where Rose bought plantations of them. From 1903 until the 1920s, the industry, which extended to citric-acid production, gave the island its greatest prosperity, and Rose's company did not cease its activities there until 1978.[37] From the 1930s the British company also manufactured a lime marmalade. Rose's lime juice cordial, the world's first concentrated fruit drink, is now made and distributed in the USA by Coca-Cola, with most of the limes now grown in Peru and Mexico and using high-fructose corn syrup instead of sugar.

THE MAKRUT (*CITRUS HYSTRIX*)

As with both oranges and lemons, it turns out that there is no such thing as a wild lime. When visiting Sri Lanka (then Ceylon) in the 1880s, the ever perspicacious Emanuel Bonavia was being shown around the Royal Botanic Gardens Peradeniya by the director when he had a revelation:

While going over the ground his foreman placed in my hands a Citrus fruit, which I had never seen before, and which he said he had obtained from a neighbouring Cocoa estate ... it flashed across my mind that I held in my hand the wild form of the ancestor of all the cultivated true Limes. Dr Trimen and myself then visited the tree. The immense winged petiole of the leaf rather staggered me. The large and small spines and the blade of the leaf were like those of the cultivated Lime.[38]

This was the makrut (*Citrus hystrix*) and, indeed, it is one of the parents of the culinary or Key lime (*Citrus × aurantiifolia*), so called due to its association with the Florida Keys, USA; the other being the citron (*C. medica*). That this original hybrid lime, unlike the lemon, has no orange (mandarin × pomelo) in its ancestry likely explains its lower levels of vitamin C. The makrut is not native to India and it is unclear whether the lime arose there after the makrut had been introduced to the citron, or when the citron had been introduced to the makrut's homelands in the Malay Archipelago. The makrut is apparently wild from Burma and Thailand to Sumatra and eastwards to New Guinea, though its natural distribution has been obscured due to its subsequent spread far into the Pacific.[39]

Although mentioned in Western literature from 1726, it was not scientifically named until 1813, by Augustin Pyramus de Candolle (1778–1841), a banker-botanist then based in Montpellier, France, in his account of the plants growing in the botanic garden there. The fruit's characteristic spiny form (hence the name *hystrix*, from the ancient Greek word for porcupine) convinced de Candolle that it was a new species. At the end of the eighteenth century, seeds from introduced plants growing on Mauritius had reached a Nîmes merchant, one Rolland-Compland. Through him a plant reached the botanic garden in 1808 and, by 1816, was being offered in the French nursery trade.

The fruits of the makrut are not eaten because of acrid oil in the vesicles surrounding the seeds. They have been used medicinally and in many places for washing clothes; it is possible that it was spread by Polynesians for washing hair as well as clothes, as the macerated leaves form a lather in water. The most familiar forms in Western cultivation with bumpy fruits were early called kaffir lime, a term to be avoided as it derives from the Arabic *kafara*, meaning an infidel, and by the end of the nineteenth century was in western India a term of abuse before being applied derogatorily by white South Africans to Black people. In Sri Lanka it is widely called leech lime because of its efficacy in deterring leeches when the juice is smeared on boots and legs, while in other places the Thai word *makrut* is used. Its foliage is perhaps most familiar in the West as the lime leaves characteristic of Thai cooking.

Although the Key lime is widely grown throughout the tropics and was the original lime for margaritas and pisco sours, the more commonly seen sort in supermarkets of temperate countries is the seedless 'Tahiti' lime (*Citrus × latifolia*), which is a cross between the original lime and the lemon, therefore having the 'blood' of four species: the citron,

---

Advertisement for Rose's Lime Juice Cordial, 1898. Patented by Lauchlan Rose, a Scottish ship's chandler, in 1867, the first-ever fruit concentrate was preserved by the addition of sugar, rather than alcohol.

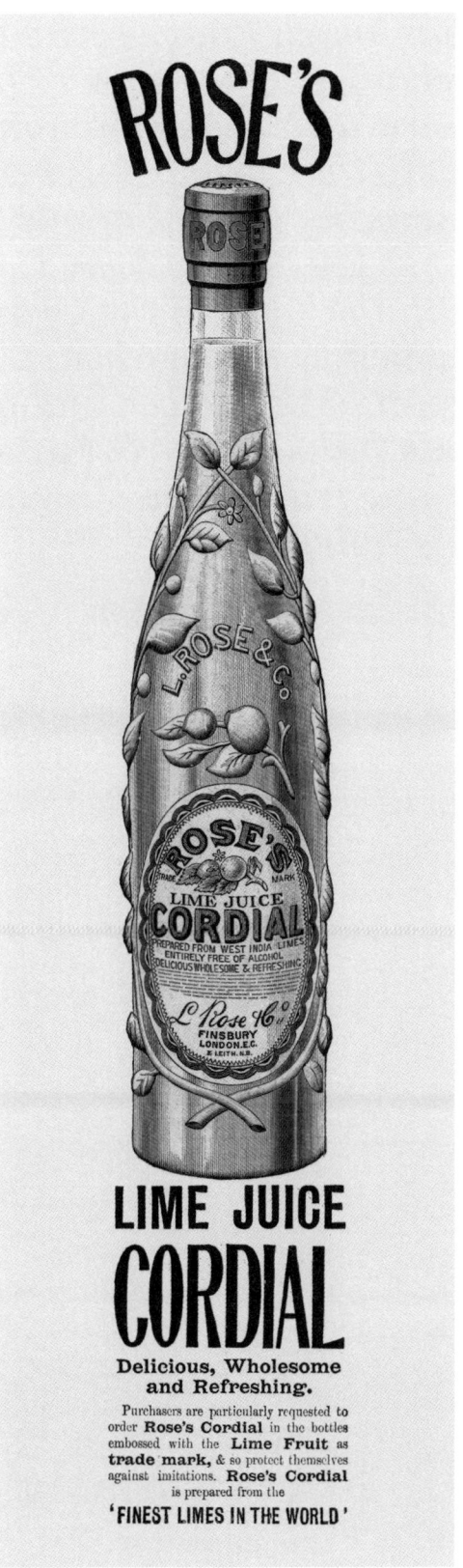

That which I have seene most fruitfull for this sicknesse, is the sower Oranges and Lemmons, and a water which amongst others (for my particular provision) I carryed to the Sea.... Comming aboord of our Shippes, there was great joy amongst my Company; and many with the sight of the Oranges and Lemmons, seemed to recover heart. This is a wonderfull secret of the power and wisdome of God that hath hidden so great and unknowne vertue in this fruit, to be a certaine remedie for this infirmitie.

RICHARD HAWKINS, *OBSERVATIONS IN HIS VOYAGE INTO THE SOUTH SEA*, 1593

Der Doctor Schnabel von Rom, published by Paul Fürst, 1656.

A plague doctor from Rome is shown in a satirical engraving wearing a hat, coat and mask with a beak (*Schnabel* in German) filled with linen soaked in aromatic citrus oils and spices. Because diseases were thought to be caused by bad air and noxious smells, counteracting them with strong aromas was thought to protect the wearer from infection. In Germany's Black Forest, mourners still carry lemons at funerals. Long associated with health, citrus fruits had also been recommended for the treatment of scurvy (dubbed 'the plague of the sea' by Richard Hawkins) since the 13th century, though it was not until the 20th century that the relationship between scurvy, citrus and vitamin C was understood.

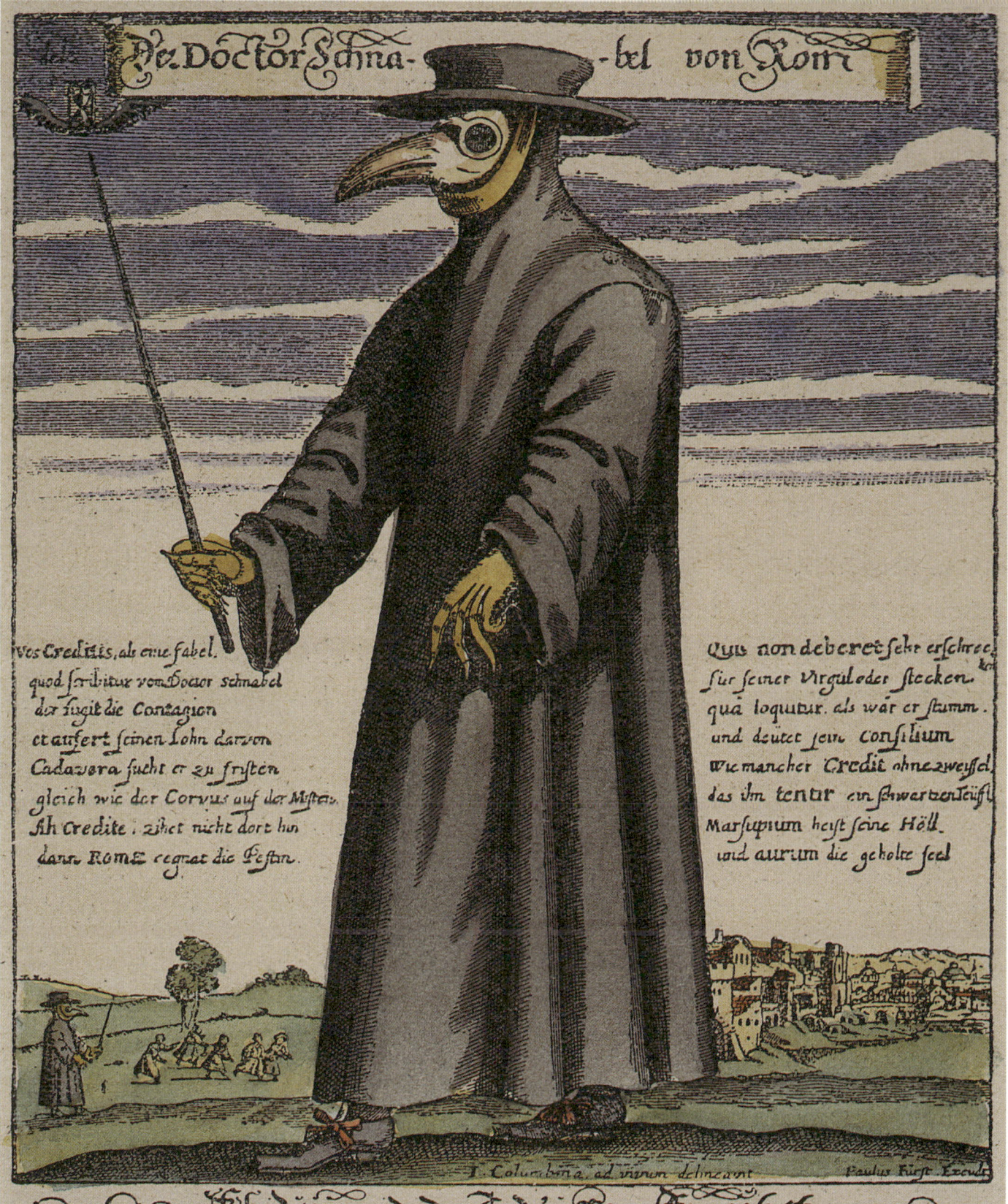

Der Doctor Schnabel von Rom

Vos Creditis, als eine fabel.
quod scribitur von Doctor schnabel
der fugit die Contagion
et aufert seinen Lohn darvon
Cadavera sucht er zu fristen
gleich wie der Corvus auf der Mistia.
Ah Credite, ziehet nicht dort hin
dann ROM regnat die Pestin.

Qui non deberet sehr erschrecken
für seiner Virgul oder stecken.
qua loquitur als wär er stumm.
und deutet sein Consilium
Wie mancher Credit ohne zweiffel
das ihm tentir ein schwerzer Teuffl
Marsupium heist seine Höll.
und aurium die geholte seel

I. Columbina ad vivum delineavit   Paulus Fürst Excud:

Kleidung wider den Tod zu Rom. Anno 1656.
Also gehen die Doctores Medici daher zu Rom, wann sie die an der Pest erkrankte Personen besuchen, sie zu curiren und fragen, sich wider Gifft zu sichern, ein langes Kleid von gewaxtem Tuch ihr Angesicht ist verlarot, für den Augen haben sie grosse Crystalline Brillen, wie die Nasen einen langen Schnabel voll wolriechender Specerey, in der Hände, welche mit Handschuhen wol versehen ist, eine lange Ruthe und darmit deuten sie, was man thun, und gebrauchen soll

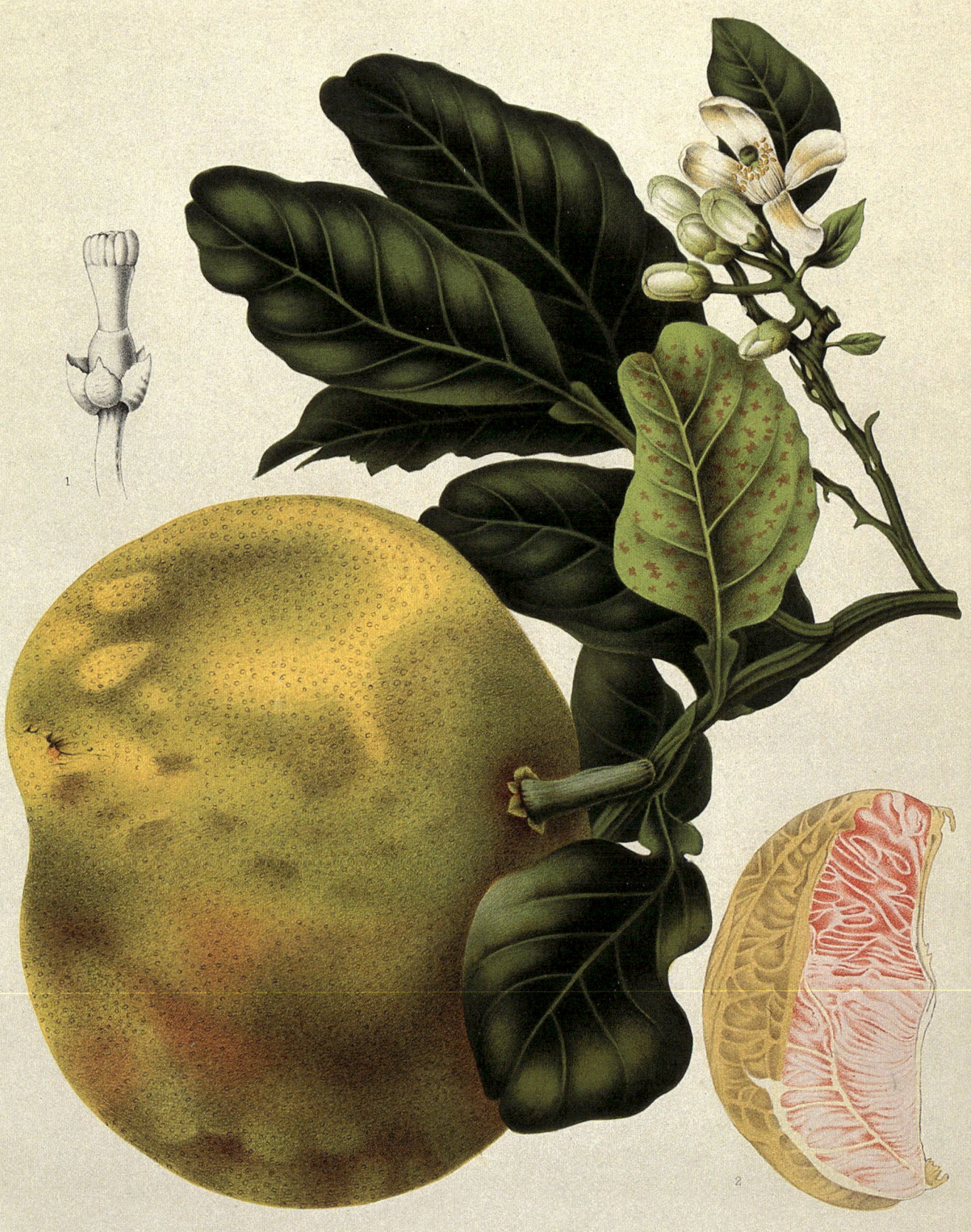

Point d'après nature par Mme Berthe Hoola van Nooten, à Batavia.　　　Chromolith. par P. Depannemaeker, à Ledeberg-lez-Gand.[Belgique].

CITRUS DECUMANA.

Librairie C. Muquardt, éditeur Bruxelles.

the pomelo, the mandarin and the makrut, the citron as a 'double dose' via both parental hybrids. To most people, especially in the USA, 'Key lime' means 'the official Florida state pie', being made with limes, sweetened condensed milk and egg yolk, though the first recipe dates from only 1940. Today, it is usually made with concentrate from Tahiti lime,[40] despite efforts to impose fines on those selling Key lime pies not made with Key limes.[41]

COLONIZATION

On 22 November 1493, during his second voyage to the New World, the Genoese Christopher Columbus (1451–1506) introduced Old World novelties from Gomera in the Canary Islands to what is now Haiti, on the island of Hispaniola: seeds of citrons, lemons and oranges, as well as cattle, goats and sheep.[42] The first citrus on the American continent was planted on 12 July 1518 on the coast of present-day Mexico.[43]

By 1525, sour and sweet oranges, citrons and lemons grew throughout the Caribbean, and by the 1570s to 1580s they were common in Brazil, likely introduced by Jesuits in the state of Bahia as early as 1530.[44] The deplorable transportation of enslaved African people, beginning in 1538,[45] provided the workforce and did so until 1869. The British, too, heavily exploited enslaved people in their overseas colonies until 1833.

Citrus appeared in Florida, which had been colonized in 1539, probably at St Augustine in 1565.[46] In a letter of 1577 from Santa Elena (Parris Island, south-east South Carolina), Bartolomé Martínez informed the Spanish king, 'I planted with my own hands grape vines, pomegranate trees, orange and fig trees'.[47] However, it was in the West Indies that one of the most important modern citrus crops arose: the grapefruit. Its name alludes to seventeenth-century reports of the pomelo 'tasting like unripe grapes. If properly hung to avoid bruising, they will remain good during a four or five month voyage. Not only do they taste good, but they are healthful for you'.[48] Captain Philip Chaddock (see p. 33) is said to have introduced pomelos (*Citrus maxima*) to Barbados, probably directly from Asia.[49] By 1830 an apomictic hybrid between a pomelo and a sweet orange was

being called 'Barbadoes grapefruit' and by 1684 it was the most valuable citrus in Barbados.

Grapefruit was introduced to Florida by the Spanish Odet Philippe (1787–1869), who, in 1823, settled in Tampa Bay, having brought this plant and other citrus from the Bahamas (or perhaps Cuba). Some of his original trees survived to at least 1932.[50] A seedling from his trees was the original 'Duncan' grapefruit, distributed by insurer A. L. Duncan from 1892 and planted widely. It has now been largely replaced by cultivars like 'Marsh',[51] the first seedless one, which arose in 1860 near Lakeland, Florida, followed by the the pink-fleshed 'Thompson', a sport (somatic mutant) on a branch of 'Marsh' at Oneco, Florida, in 1913.[52]

Before this significant innovation in the Caribbean, it is likely that the Spanish introduced citrus to Vanuatu in the early seventeenth century, because when James Cook arrived there in 1774, 'peppers and oranges' were being grown.[53] It was Cook himself who introduced oranges to Tahiti, those near Cape Venus being still alive in 1880. However, oranges soon became invasive, spreading across the island and, without grafting or any care whatsoever, producing good fruit. Traders from California used to arrive around February and pack so many boxes of oranges that some 5–7 million a year were exported to the USA. Around 1870 production declined due to fungal disease, insect attack and mineral deficiencies, so that by 1900 they had disappeared on all land below 300 m (984 ft). Although the Spanish had visited what is now the Cook Islands in the early seventeenth century, it seems likely that it was Cook himself who introduced citrus there, in 1767; in the 1860s it was an important export.[54] Supplies of limes went to New Zealand to be used in cordials, skin cleansers and pomades, until cheaper sources eclipsed the trade by about 1920.

Elsewhere in the Pacific, sweet oranges were established in Hawaii as early as 1792, when Scottish doctor-naturalist Archibald Menzies (1754–1842) germinated, aboard ship, seeds collected in South Africa and gave the resulting seedlings to Hawaiian chiefs. Although they were widely grown in the archipelago, commercial cultivation of oranges ended with the introduction of the Mediterranean fruit fly at the beginning of the twentieth century.

HEALTH AND WEALTH

Citrus was in Arizona by 1707 and, in 1769 (likely introduced from Mexico), in California, which

Pomelo (*Citrus maxima*), Pierre-Joseph Depannemaeker after Berthe Hoola van Nooten, *c.* 1885. The pomelo was introduced to the West Indies by European colonizers. It hybridized with the sweet orange to produce the grapefruit.

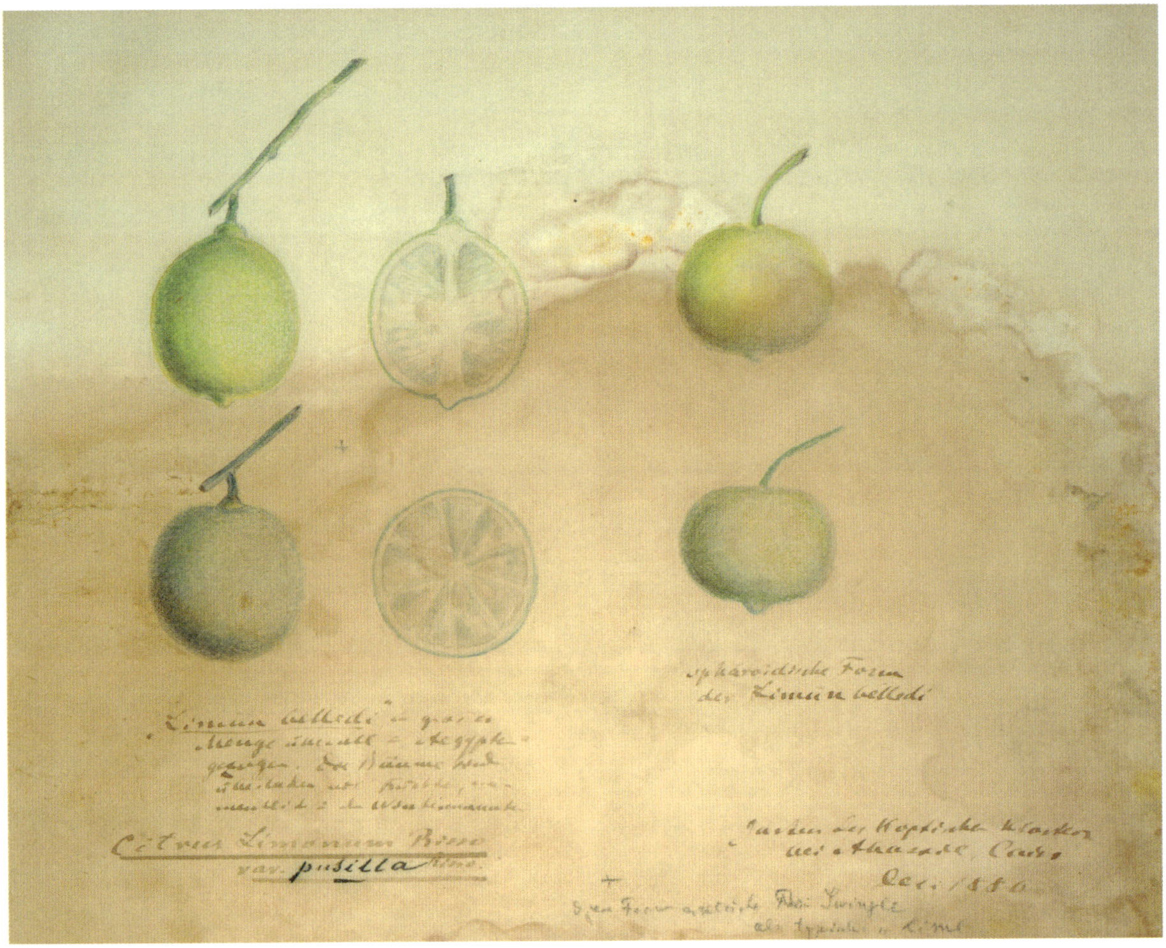

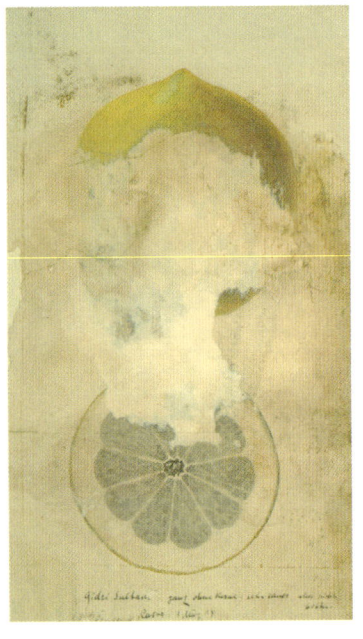

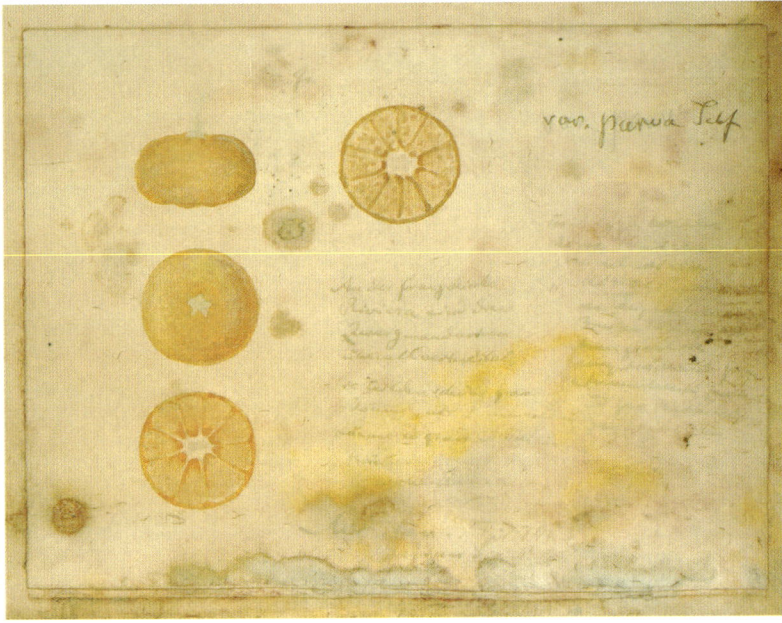

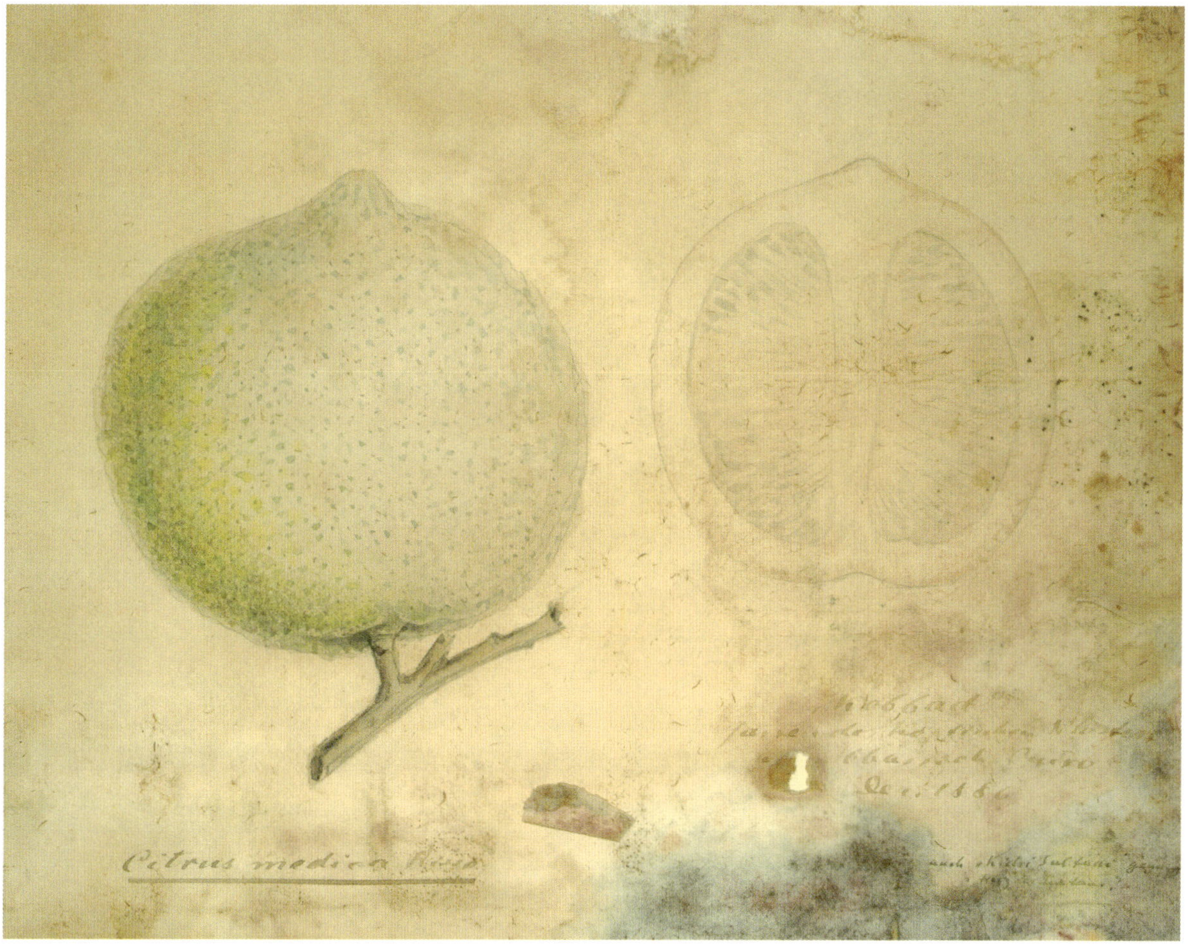

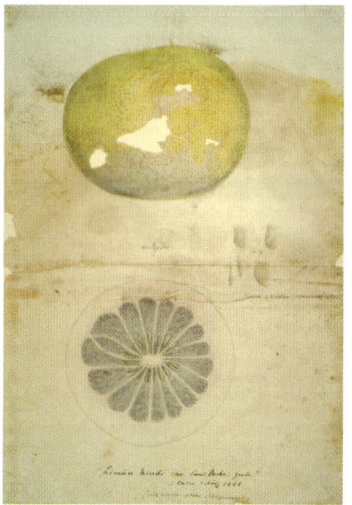

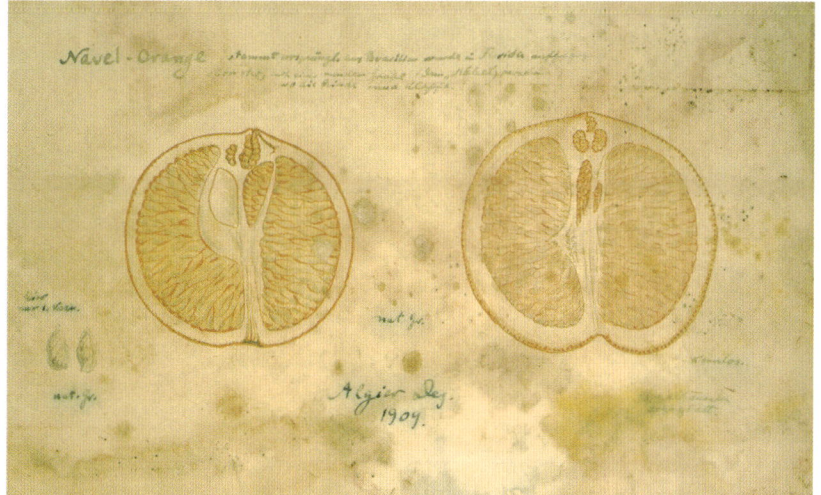

Citrus fruits in cultivation, Georg Schweinfurth, 1880–1913.

Schweinfurth (1836–1925) was best known for his expeditions in tropical Africa (1869–70) and Libya (1873–74). In 1875, he settled in Cairo, where some of these unpublished sketches were made. In 1888, he moved to Berlin but continued visiting the Mediterranean until at least 1909, when the last of these drawings was made in Algiers.

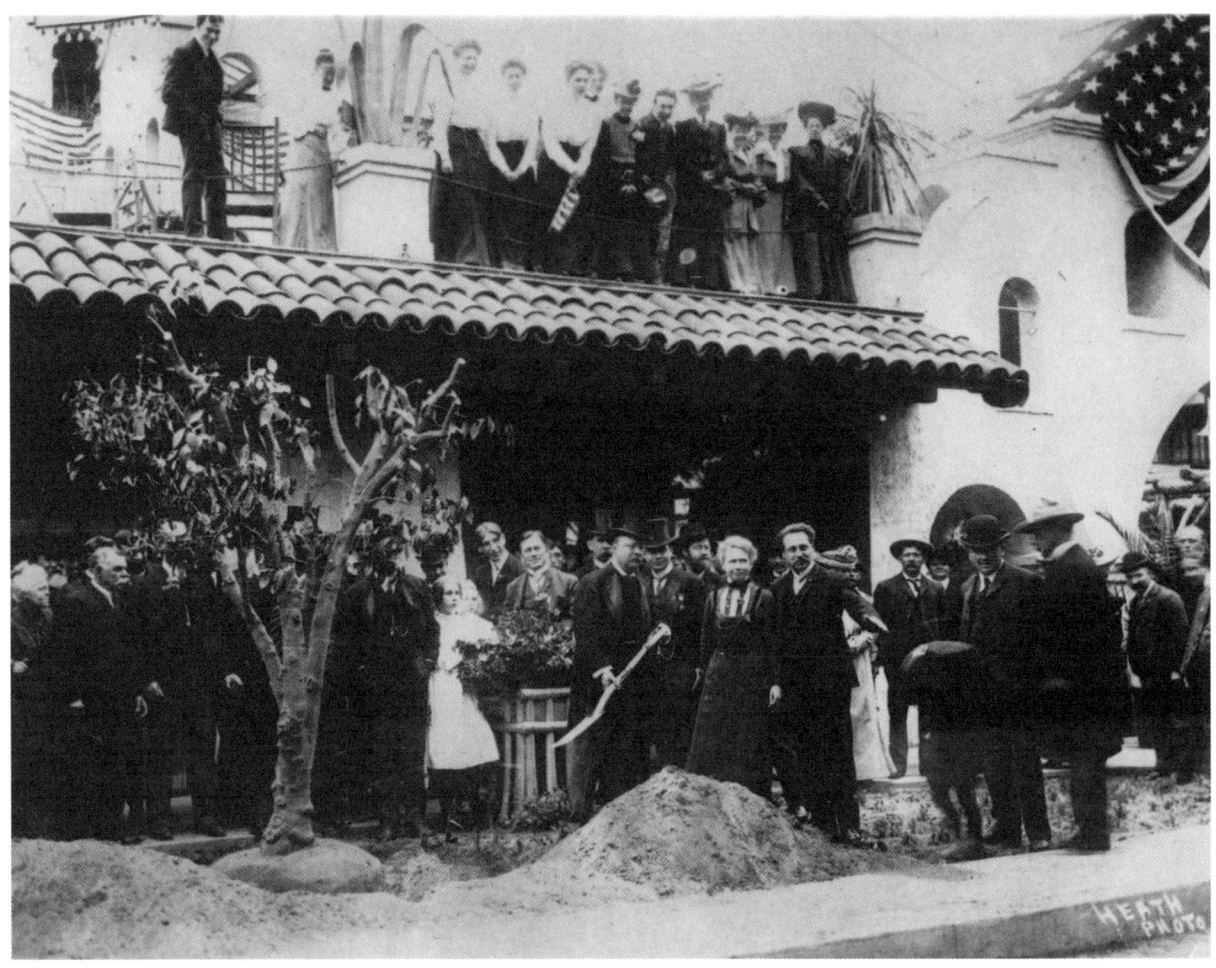

President Theodore Roosevelt replants California's 'first navel orange tree' at the Mission Inn, Riverside, California, 1903. This tree was one of two planted in California in 1873 that were the source of all the state's navel oranges.

had been colonized by Jesuits in 1634. The first commercial orchard there was started in 1841 on land later occupied by the Arcade Depot railway station in Los Angeles. The proprietor, William Wolfskill (1798–1866), shipped oranges east by train from 1877, effectively beginning today's major industry.[55]

Oranges and, particularly, the railway were key factors driving migration to California. With one-way tickets reduced to $1, the population increased by 345,000 between 1880 and 1890. By 1886 in Los Angeles County alone 5,000 hectares (12,000 acres) of groves were bearing oranges,[56] though by then there was competition for the eastern US market from Mexico, and in fact exports from Jamaica to the USA have increased almost tenfold between 1875 and 1884. Riverside, California, was founded in 1871, principally as a citrus operation; by 1893 it was the richest city per capita in the country. Drought put paid to grazing and the age of irrigated citrus took over, until urbanization led, by the 1950s, to up to

1,200 hectares (3,000 acres) of orange groves being grubbed up every day in Los Angeles County alone. Southern California and even the eponymous Orange County are no longer home to commercial orange groves.

In 1873, two trees of 'Bahia', a seedless navel orange from Bahia, Brazil, were planted in Riverside by Eliza Lovell Tibbets (1823–1898). She declared that, as they had come via Washington DC, the trees were 'Washington navels', and the superfluous name has stuck. Long before, by 1828, the botanic garden in Sydney, Australia, was growing 'Bahia' introduced in 1824 and, as early as 1835, it was introduced to

Florida from Brazil,[57] where it had apparently arisen at the end of the eighteenth century. It was likely a mutant of 'Selecta' propagated by the Portuguese from about 1822[58] – but frost killed it in Florida. One of Tibbets's two trees, which were the source of a major part of the California citrus industry, is still standing at Riverside.[59]

By 1900, this winter-ripening cultivar was the most planted orange in California. Around the world it rose to be second only to the summer-ripening 'Valencia' in commerce, while many mutants of 'Bahia', notably 'Bahianinha', have become important too, especially in Brazil, now a major exporter. 'Valencia' was said to have been introduced to California via Thomas Rivers' English nursery in 1876 and then called 'Rivers Late'[60] (they had supplied Florida in 1870, where it was first called 'Hart's Tardiff'[61]). Others argue that it was developed in California by William Wolfskill, who sold his patented hybrid to the Irvine Ranch owners, who planted nearly half their lands with

it. From one million boxes of oranges in 1887 the California industry, increasingly competitive and then oligopolistic, produced more than 65.5 million boxes of oranges, lemons and grapefruit in 1944, driving the southern Californian economy in the way cotton did in the north-eastern USA.[62] Indeed, it has been cogently argued that, long before, the orange industry had been pivotal to the reconstruction of the South, particularly Florida, after the Civil War.[63]

On the other hand, by 1885, the American trade in 'green' oranges had had a catastrophic effect in Sicily, where the British Vice-Consul in Catania wrote:

Orange pickers, Ormond, Florida, William Henry Jackson, 1890s. Jackson (1843–1942) specialized in geographical surveys and commercial landscape photography, and later sold his work as postcards. His photographs are a testament to the hard lives of agricultural labourers.

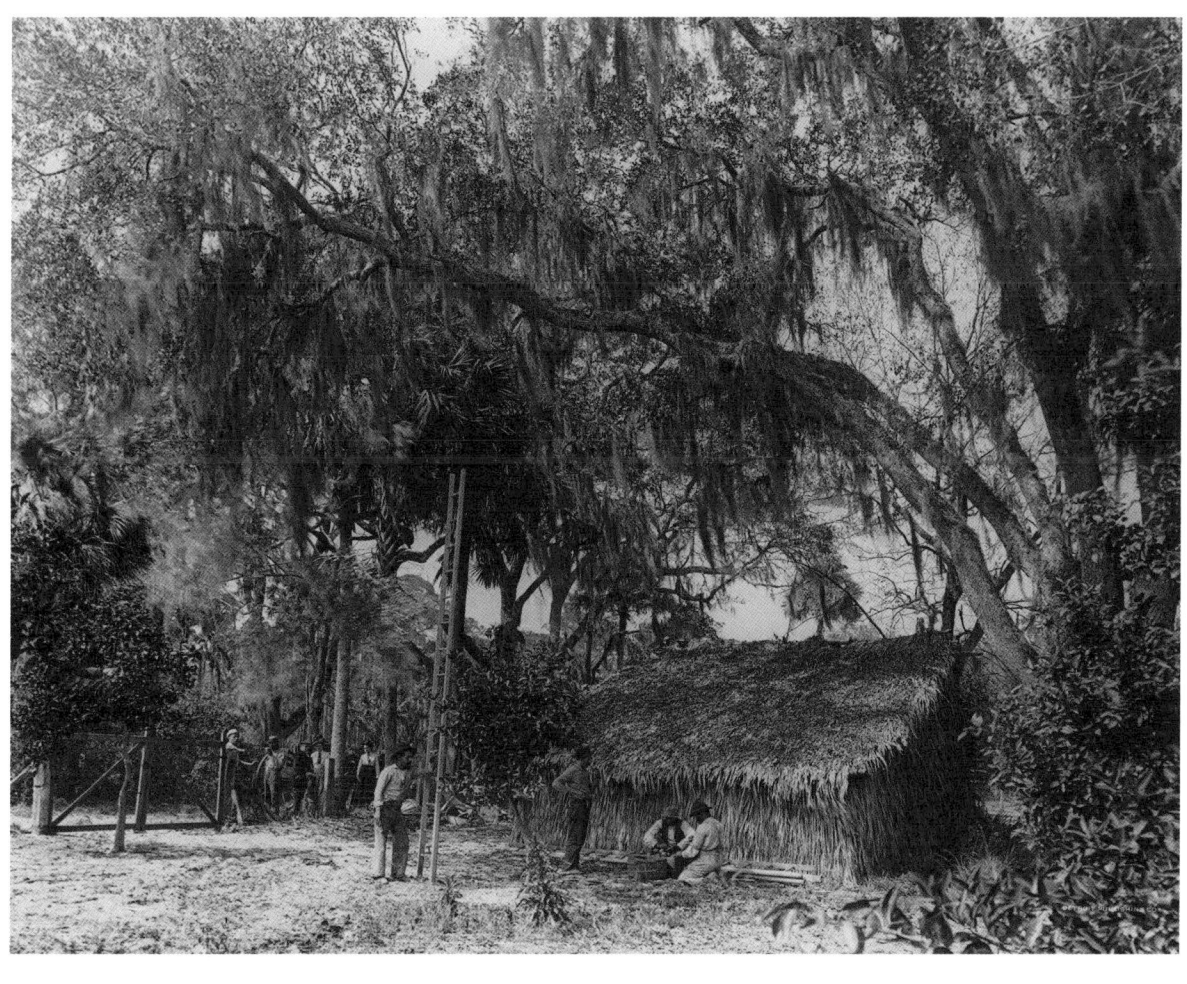

These important products, which a few years ago promised to become a branch of commercial resource to the island, have proved instead a failure to all those who have turned their fields and grounds into orange and lemon groves. Prices in America (the chief country where exported) are so very low, that it is more convenient for the producers to let the fruit rot on the trees than to go to the expense of packing it up for exportation.[64]

From 1886 to 1887 the Florida citrus industry produced 1.2 million boxes of oranges, seven years later 5 million, the next year around 6 million and in 1930–31 some 35 million, of which 27.2 million were exported from the state.[65] With the rise of the Florida and California industries, the Valencia region of Spain lost its American market by 1886. However, in that same year, 2 million boxes (3 billion oranges) in a single season were exported from Spain, largely to northern Europe, while in the previous five years Cordoba had tripled its export of sour oranges to London and Liverpool for marmalade-making.[66] Oranges were also exported from Portuguese territories, for example, in 1856 São Miguel, one of the Azores, exported to England 200 shiploads amounting to 200 million oranges, 25 million of which were sold in theatres and in the streets. In 1881–82 nearly a million packages, each containing on average 400 oranges, were landed in London.

NATIVE CITRUS OF AUSTRALASIA

When Australia, home to native species of *Citrus*, was first settled by Europeans in 1788, the British 'First Fleet' arrived with oranges, limes and lemons picked up from Rio de Janeiro on the way, as instructed by Joseph Banks. After over 250 days at sea, the settlers planted fruit trees with seeds from Rio and the Cape of Good Hope. The first to bear fruit were those grown by the Reverend Richard Johnson (1756–1827), the colony's chaplain, in the two acres allotted to him in Bridge Street, Sydney. In 1807 George Suttor (1774–1859) was selling oranges at 2/6 a dozen at a Sydney market[67] and, by 1830, a grove at Kissing Point was producing 12,000 dozen oranges a year. By 1834, Suttor was selling '2,000 dozen oranges at 8 pence, and 1,000 dozen lemons at 6 pence'.[68]

By World War I, the demand for lemon juice in Australia had not abated and a lemon industry developed on Norfolk Island.[69] In 1917 alone, 901 casks of lemon juice and 1,121 of peel – products of about four million lemons – were exported and, by 1921, the industry accounted for 57 per cent of the island's exports. But as the mainland Australian citrus

industry revived in the 1920s, Norfolk Island's lemon industry fizzled out.

As early as 1828, the botanic garden in Sydney was growing sixteen cultivars of sweet orange, including the 'Bahia' navel introduced from Brazil in 1824; four kinds of 'mandarin'; three of pomelo and two of citron; besides lemons and limes. In the same year, the colony was exporting oranges and lemons to Tasmania. From about 1839 to 1860, Richard Hill (1810–1895), a prosperous pastoralist and politician, owned a large plantation on the Lane Cove River, a branch of the Paramatta River flowing into Sydney Harbour. He exported oranges to the Victorian goldfields and at its peak made profits of £50 a day. In 1856, his 800 trees yielded 56,000 dozen oranges for the Sydney market;[70] in 1858 his yield was some 60,000 dozen oranges.[71]

Unlike the Americas and Africa, Australia already had native *Citrus* species, some of which had long been used by Aboriginal people, but it was not until some years after European colonization that they were scientifically described. Their documenting was at least in part a way of assessing what of the 'natural productions' would be of commercial significance in the economies of the colonies that were to make up the Commonwealth of Australia in 1901. These species are now of great interest because of their disease-resistance, yet are little known. They are here dealt with in some detail for the first time.

The Australian species are part of the easternmost subgrouping of the genus *Citrus*. Although closely related to one another according to DNA analyses, between them they have much greater morphological diversity and ecological tolerance than the rest of the genus put together. New Caledonia has two or three native species, Papua New Guinea three, but Australia has six, found in a wide range of habitats, from tropical rainforest to semi-arid grassland. The first known to Western science was *Citrus australis*, the *dooja*, which is restricted to the subtropical rainforests of south-east Queensland and can reach 18 m (59 ft) tall, spreading by suckering roots. It produces round, rough-skinned green to yellow fruits up to 8 cm (3 in.) in diameter, but was not recognized as a species of *Citrus* until 1858.

Poster for the 'Eat More Fruit' campaign, Victoria, Australia, *c.* 1924. The Victorian Railway Commissioners promoted the health-giving properties of citrus, reasoning that a demand for fruit would lead to growth for the railway's freight business.

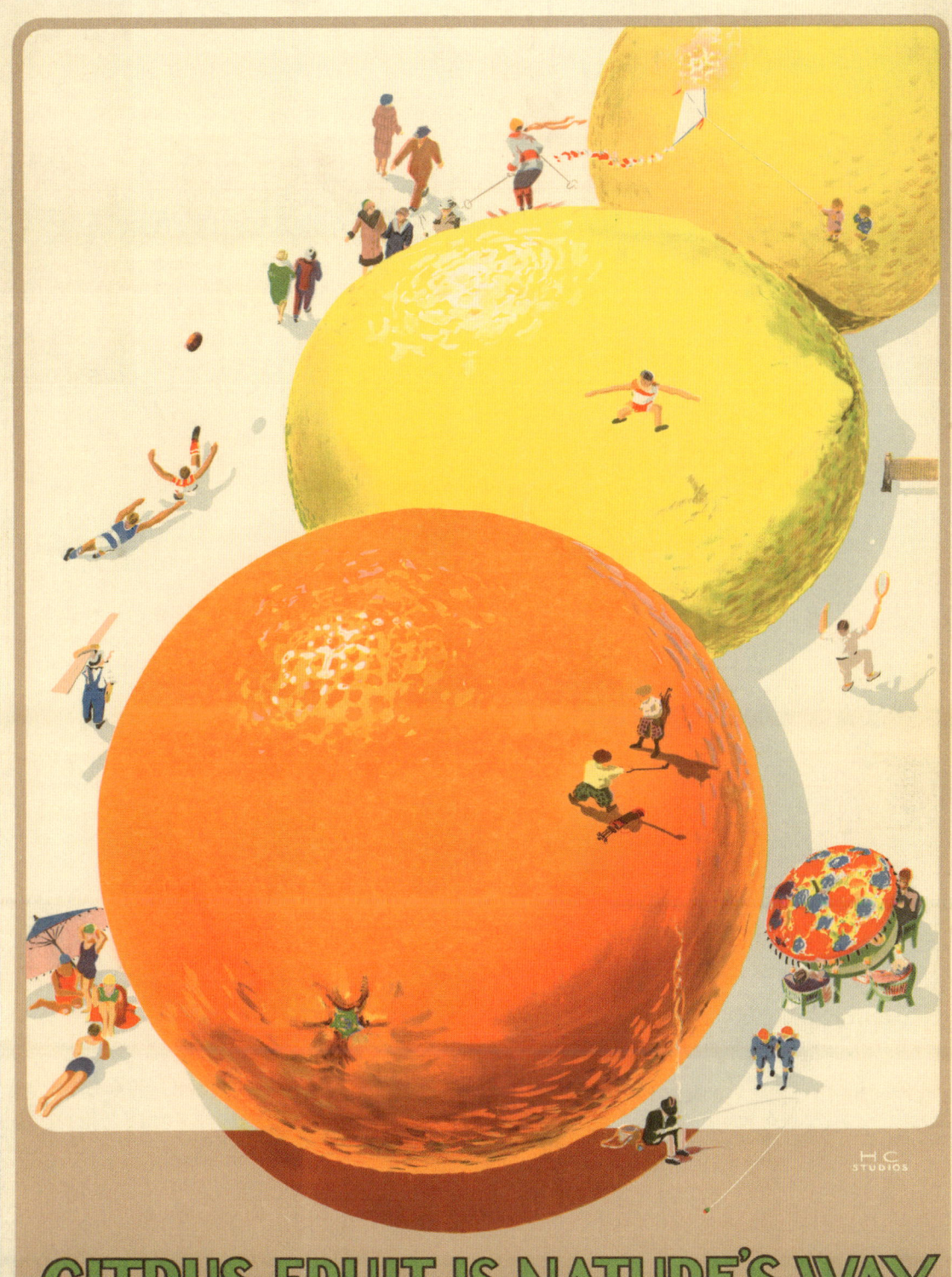

CITRUS FRUIT IS NATURE'S WAY TO KEEP YOU FIT FOR WORK AND PLAY!

J.E.HACKETT. Print. Melb.   Victorian Railways Poster Nº48

Meanwhile, the desert lime, lime bush or desert kumquat (*Citrus glauca*), had been known to science from the collections of Major Thomas Mitchell (see p. 116). He collected it in 1846 on his fourth expedition into what is now Queensland, but only in 1932 was it recognized as a true *Citrus*. It grows in the semi-arid regions of eastern Australia and is able to tolerate temperatures up to 45 and down to −24 °C (113 to −11 °F). It can be found as a tree with a distinct trunk up to 12 m (39 ft) tall or form a dense spiny thicket, spreading by heavily armed root suckers such that it can be a pest in pasture land. The fruit has three times the vitamin C found in oranges.

The finger lime (*Citrus australasica*), so called because of its elongated fruits, grows wild in subtropical

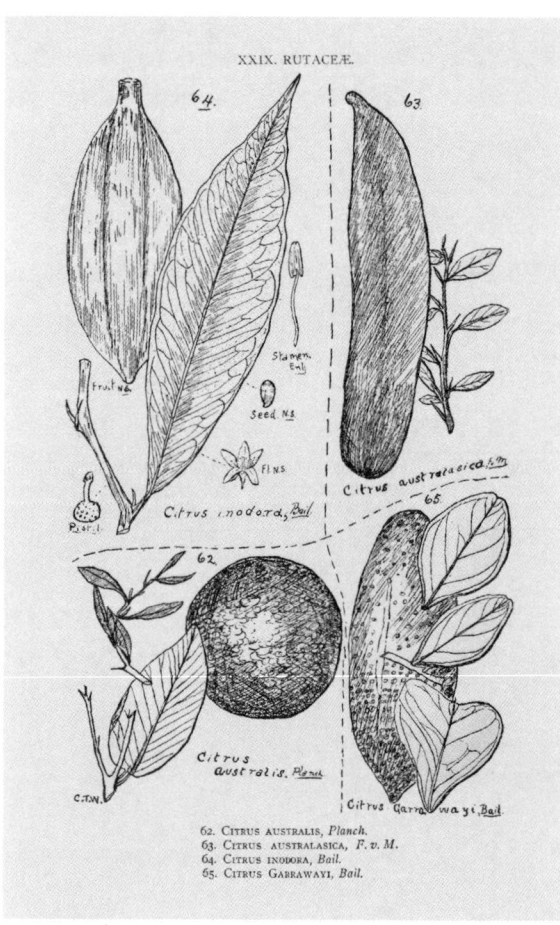

Australasian citrus in *Comprehensive catalogue of Queensland plants, both indigenous and naturalised* by Frederick Manson Bailey, 1909. Bailey described several indigenous citrus species previously unrecorded in Western science including the Russell River lime (*Citrus inodora*) and the Mount White lime (*Citrus garrawayi*).

rainforests in the north-east corner of Australia, from Brisbane in Queensland south to the Clarence River in northern New South Wales. It was named by Baron Ferdinand von Mueller (1825–1896), Director of the Botanic Garden in Melbourne, in 1858, based on materials sent from Moreton Bay by Walter Hill (1820–1904), a Scot and superintendent at the Brisbane Botanic Garden. It is an understorey tree reaching up to 10 m (33 ft) high when grown from seed. The fruits range from yellow or green, through red and purple, to almost black. They contain acid juice and yield up to 82 mg of vitamin C per 100 g (3½ oz) of pulp and are now a significant 'bush tucker' commercial proposition.

Shortly after the formal description of *Citrus australasica*, and perhaps collected before it in the French colony of New Caledonia, *Citrus oxanthera* was first described by the Montpellier-born missionary Abbé Jean Xavier Hyacinthe Montrouzier (1820–1897), a Marist priest and naturalist. Up to five unarmed species with large, strongly scented orange-like flowers have been recognized as restricted to New Caledonia.

No new *Citrus* species were found in the Pacific for more than twenty years. *Citrus warburgiana* was to be described in 1902 by a Queensland colonial botanist, the English-born Frederick Manson Bailey (1827–1915), from material collected in the Milne Bay Province of what is now Papua New Guinea in 1884 by the Belgian-born William Edington de Margrat Armit (1848–1901). Armit had migrated to Australia and became a sub-inspector in the Queensland Mounted Native Police in 1872. Despite being involved in punitive expeditions against First Australians, in 1880 he proposed that a work on Aboriginal culture should be written to help white people understand them better. He later settled in north Queensland, making a living as a writer (as 'a Queensland Police Officer') and naturalist, returning to New Guinea in 1884 and writing *Notes on the philology of the islands adjacent to the south eastern extremity of New Guinea* (1886). Nevertheless, the citrus name commemorates the German botanist and Zionist Otto Warburg (1859–1938), who in 1891 had described from his own collections in Ceram and the Aru Islands south-west of New Guinea a *Citrus medica* var. *aruensis*. Bailey thought this close to his new plant, but Warburg's specimens were destroyed during the Allied bombing of Berlin in 1944, so it can never be proved definitively.

Before publishing *Citrus warburgiana*, though, Bailey described another new species, this time from

Australia, based on material he himself had collected in July 1889.[72] Bailey encountered his new *Citrus* on the Russell River in north-east Queensland during the Bellender-Ker Range Expedition of Archibald Meston (1851–1924), a politician and naturalist. The Russell River lime (*Citrus inodora*), a vulnerable species, is found only in the lowland rainforest between Innisfail and Cairns in northern Queensland, the wettest part of Australia, with over 430 cm (169 in.) of rain a year. Much of its original habitat was cleared long ago by Europeans for banana and sugar-cane cultivation. This small tree differs from all other Australian citrus in having (to us at least) scentless flowers. It somewhat resembles *Citrus polyandra* from New Ireland, Papua New Guinea, which was collected early in the twentieth century by the German missionary Father Gerhard Peekel (1876–1949), but is now also known from mainland Papua New Guinea.

Peekel was based at Lamekot Mission/Namatanai in southern New Ireland, where he found *Citrus polyandra* 'in the gardens of the natives and at waysides; abundant, but nowhere truly wild',[73] suggesting it may have been imported from the interior or the mainland. It is an unarmed tree up to 8 m (26 ft) high with bright *yellow* flowers and sweetish edible yellow fruits about 7 cm (2¾ in.) across. Peekel was working on a local flora, until the outbreak of World War II. In January 1942, the Japanese bombed Kavieng, New Ireland, and most Europeans were evacuated before they captured the island. Peekel, aged sixty-six and in ill health, was the only one of the fourteen priests there to survive, no doubt because he was of potential use to the aggressors for his knowledge of the local plants recorded in his manuscript, by now in 12 volumes with 1,200 plates. Peekel died in 1949 without seeing his remarkable work published. An English translation, *Illustrated Flora of the Bismarck Archipelago for Naturalists*, was the only version to be printed, in Papua New Guinea in 1984.

Bailey described a third Australian species, resembling the finger lime. The Mount White lime (*Citrus garrawayi*) is named after Roland Walter Garraway (1859–1942), an Englishman who, like Armit before him, migrated to Australia and became a sub-inspector in the Queensland Native Mounted Police. As a police cadet in October 1885, he was part of a detachment that allegedly killed at least six Aboriginal people at Irvinebank near Cairns, but he was later to become 'protector' of Aborigines at Coen.[74] Famous as a good 'bush man', Garraway was interested enough in the native flora to send specimens, including a citrus in 1904, to Bailey, who wrote to Garraway that

the specimen was 'of particular interest as it adds another to our Queensland Citrus'.[75] The species, found in monsoon forest, is protected under the Queensland Nature Conservation Act 1992.

The remaining *Citrus* species of the western Pacific were not to be scientifically described until the last quarter of the twentieth century and the early years of the twenty-first. Three species were still to be collected and named by botanists – two in Papua New Guinea, the other in the 'Top End' of Australia. The Brown River finger lime (*Citrus wintersii*), found only in the Central Province of Papua New Guinea, was first collected in May 1960[76] by Kevin Joseph White (1924–2012), a Queenslander who did much to develop forestry in New Guinea. The name commemorates Harold F. Winters (1913–2019) of the Department of Agriculture's Plant Introduction Station, Beltsville, Maryland, USA, who not only collected it in 1970, but first described the plant (as *Microcitrus papuana*), his only botanical christening.

The Humpty Doo lime (*Citrus gracilis*) was named in 1998, having been first collected in 1971 by John L. McKean (1941–1996), an ardent ornithologist of the Flora and Fauna Unit of the Northern Territory Conservation Commission. *Citrus gracilis*, named for the form of its rather graceful foliage, is a straggly tree growing up to 8 m (26 ft) tall in eucalypt woodland. Suckering and coppicing like *Citrus glauca*, it has globose fruits up to 8 cm (3 in.) in diameter.

Last to be named, though first collected in October 1953 by Australian-born botanist Leonard John Brass (1900–1971), is *Citrus wakonai*, known only so far from Goodenough Island (and named for a village there). In 2000, using Brass's field-notes, Malcolm Smith, a citrus breeder at the Bundaberg Research Station in Queensland, working with local people, was able to find the very populations Brass had seen over fifty years before. *Citrus wakonai* is a rainforest tree up to 6 m (20 ft) tall with rather straggling habit, often coppicing and suckering from the base, like *Citrus glauca* and *Citrus gracilis*, and it has the shortest juvenile phase of any known *Citrus*, as little as 144 days from germination.[77]

The consequence of the Age of Empire, besides generating the mighty citrus industry of America and elsewhere, was the encountering of the extremely diverse, yet still relatively little known, Australasian species. Yet, with somewhat intact wild populations, these may well help save the industry as we have come to know it – besides benefiting the Indigenous peoples on whose lands they grow.

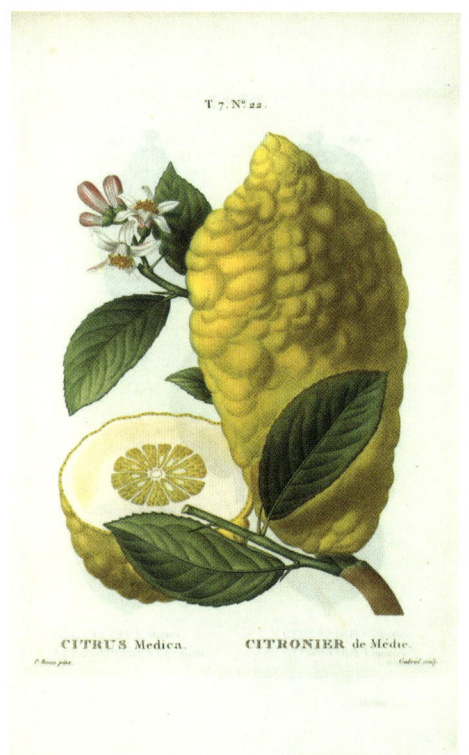

T. 7. N° 22.

CITRUS Medica.   **CITRONIER** de Médic.

P. Bessa pinx.   Gabriel sculp.

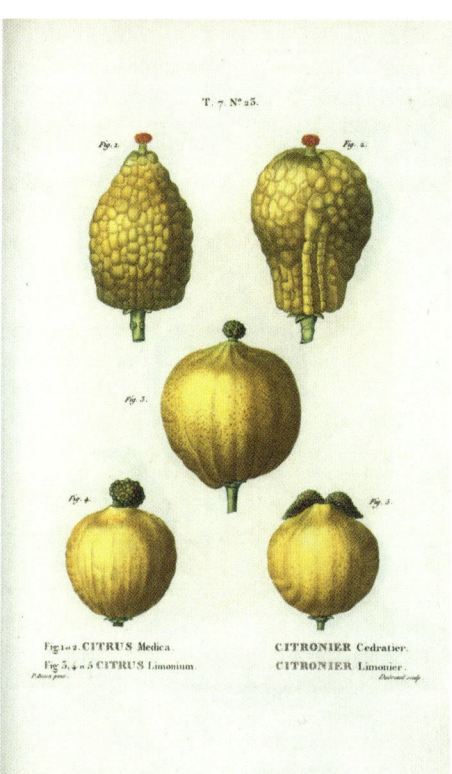

T. 7. N° 23.

Fig. 1.   Fig. 2.

Fig. 3.

Fig. 4.   Fig. 5.

Fig 1 et 2. **CITRUS** Medica.   **CITRONIER** Cedratier.
Fig 3, 4 et 5 **CITRUS** Limonium.   **CITRONIER** Limonier.

P. Bessa pinx.   Dubreuil sculp.

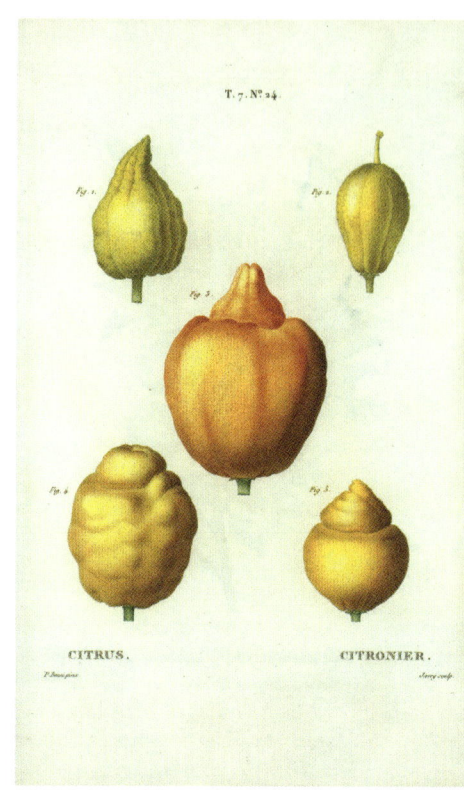

T. 7. N° 24.

Fig. 1.   Fig. 2.

Fig. 3.

Fig. 6.   Fig. 5.

CITRUS.   **CITRONIER**.

P. Bessa pinx.   Savy sculp.

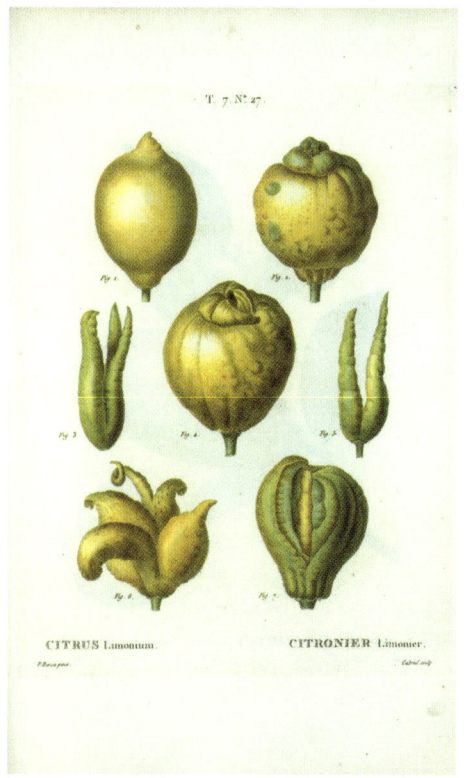

T. 7. N° 27.

Fig. 1.   Fig. 2.

Fig. 3.   Fig. 4.   Fig. 5.

Fig. 6.   Fig. 7.

CITRUS Limonium.   **CITRONIER** Limonier.

P. Bessa pinx.   Gabriel sculp.

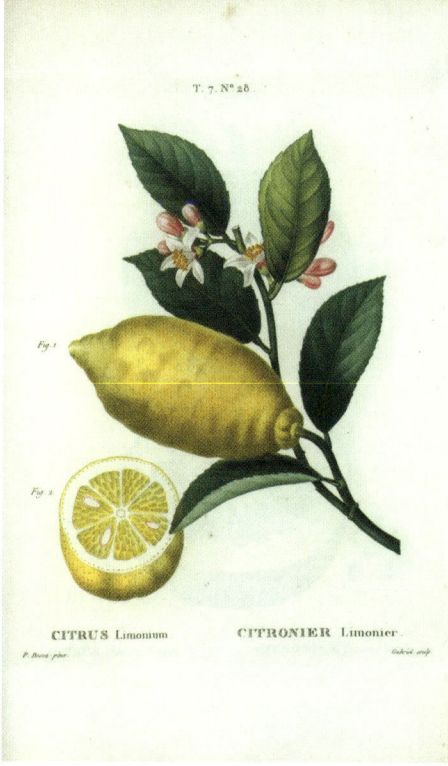

T. 7. N° 28.

Fig. 1.

Fig. 2.

CITRUS Limonium.   **CITRONIER** Limonier.

P. Bessa pinx.   Gabriel sculp.

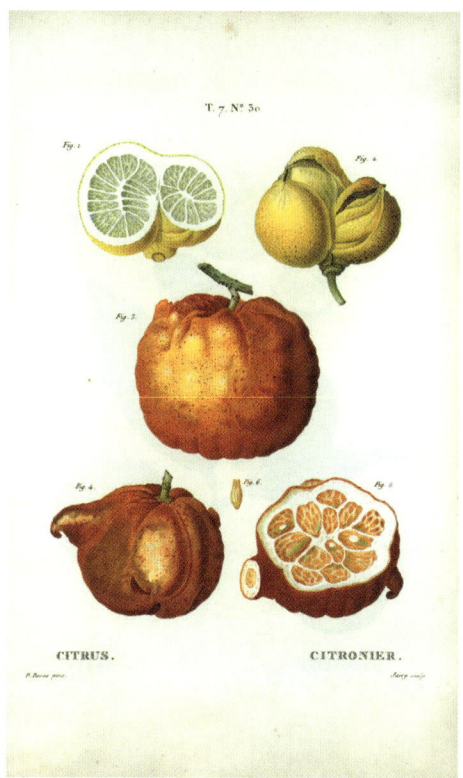

T. 7. N° 30.

Fig. 1.   Fig. 2.

Fig. 3.

Fig. 4.   Fig. 6.   Fig. 5.

CITRUS.   **CITRONIER**.

P. Bessa pinx.   Savy sculp.

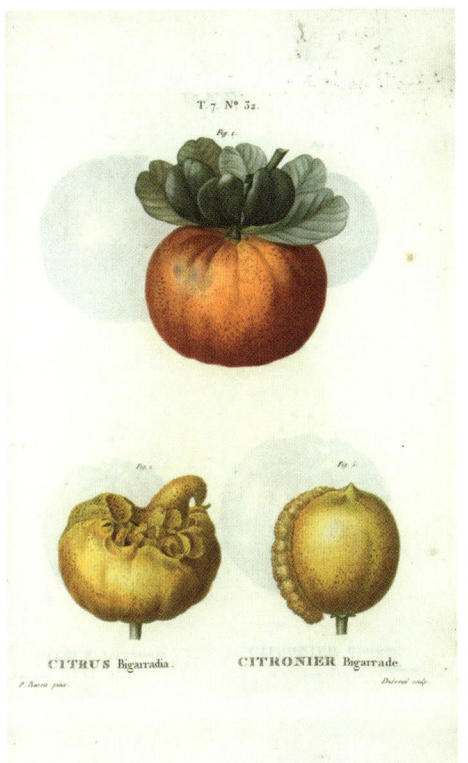

T. 7. Nº 32.

Fig 1.

Fig 2.   Fig 3.

CITRUS Bigarradia.   **CITRONIER** Bigarrade.

P. Bessa pinx.   Dubreuil sculp.

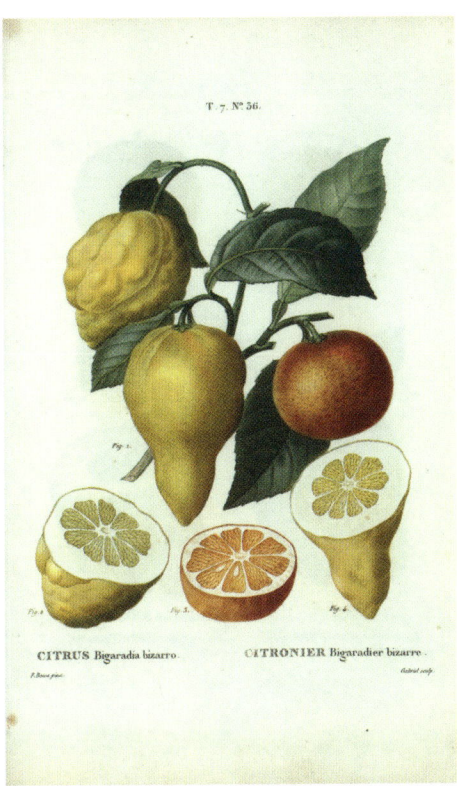

T. 7. Nº 36.

Fig 1.

Fig 2.   Fig 3.   Fig 4.

CITRUS Bigaradia bizarro.   **CITRONIER** Bigaradier bizarre.

P. Bessa pinx.   Gabriel sculp.

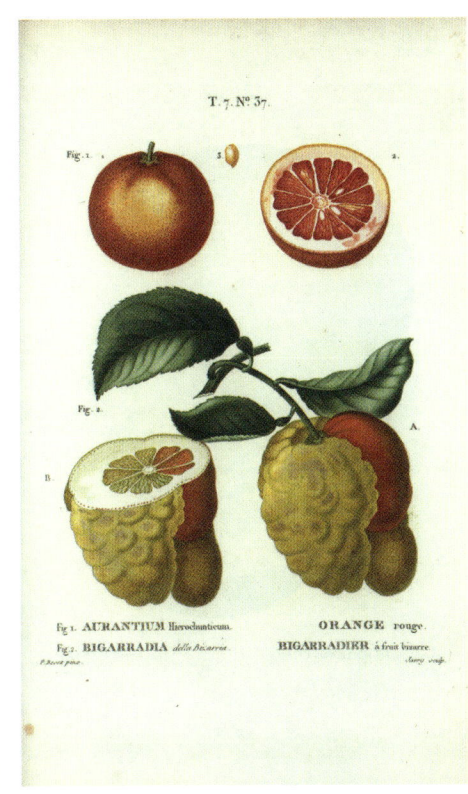

T. 7. Nº 37.

Fig 1.   5.   2.

Fig 2.   A.

B.

Fig 1. **AURANTIUM** Hierochunticum.   **ORANGE** rouge.
Fig 2. **BIGARRADIA** della Bizarra.   **BIGARRADIER** à fruit bizarre.

P. Bessa pinx.   Savry sculp.

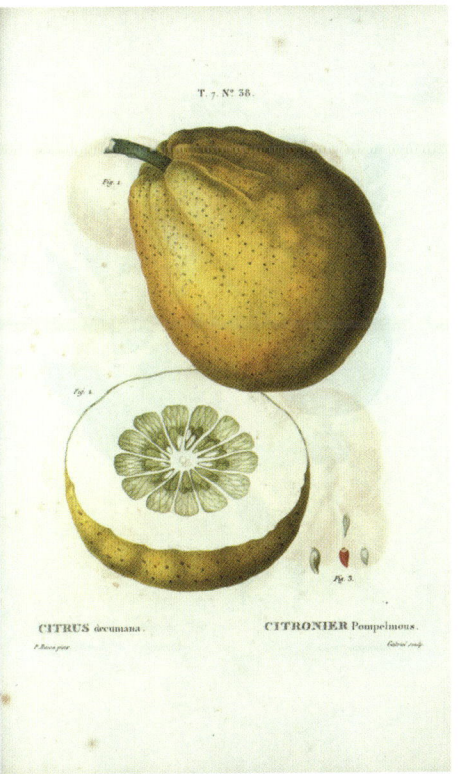

T. 7. Nº 38.

Fig 1.

Fig 2.

Fig 3.

CITRUS decumana.   **CITRONIER** Pompelmous.

P. Bessa pinx.   Gabriel sculp.

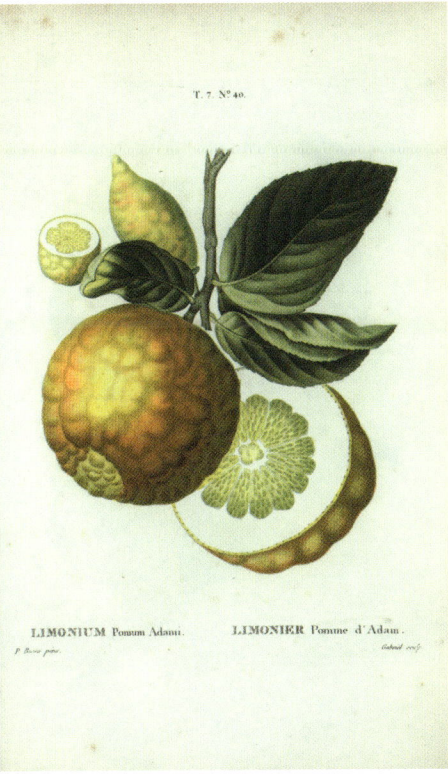

T. 7. Nº 40.

LIMONIUM Pomum Adami.   **LIMONIER** Pomme d'Adam.

P. Bessa pinx.   Gabriel sculp.

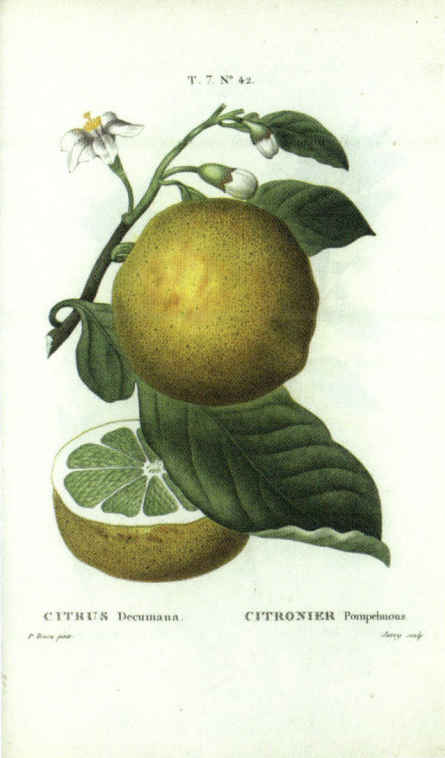

T. 7. Nº 42.

CITRUS Decumana.   **CITRONIER** Pompelmous.

P. Bessa pinx.   Savry sculp.

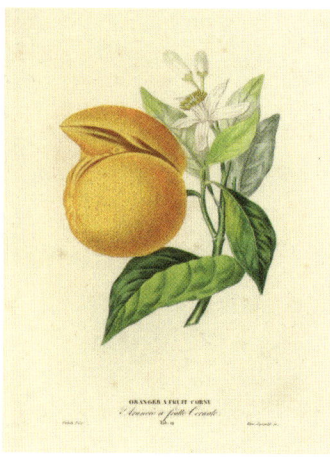

ORANGER À FRUIT CORNU
Aranciu à fiutti Cornutu

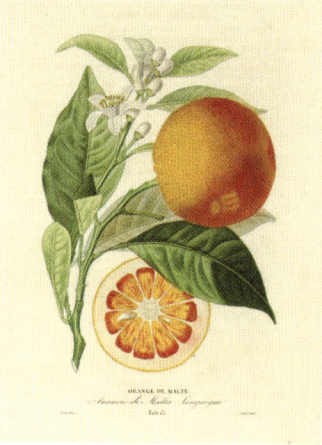

ORANGE DE MALTE.
Aranciu de Malta Sanguignu

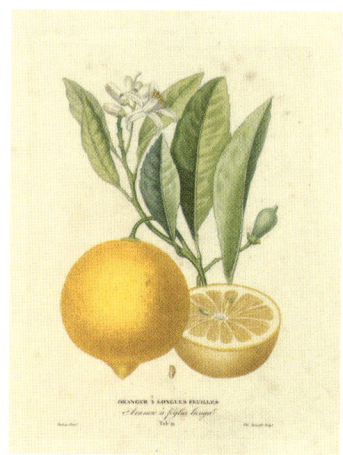

ORANGER À LONGUES FEUILLES
Aranciu à foglia longa

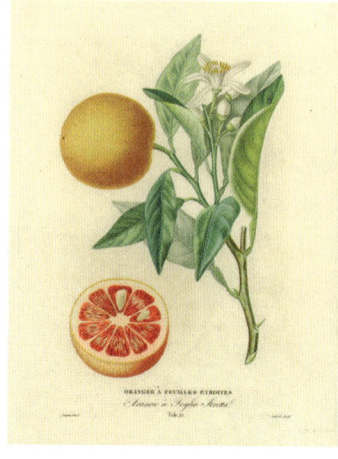

ORANGER À FEUILLES ÉTROITES
Aranciu à foglia Stretta

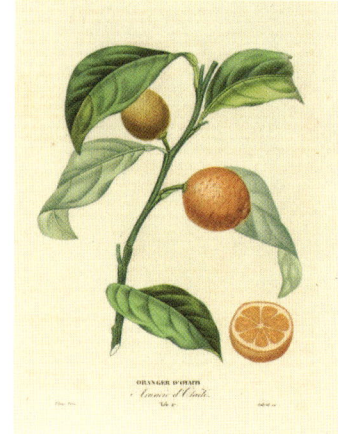

ORANGER D'OTAITI
Aranciu d'Otaiti

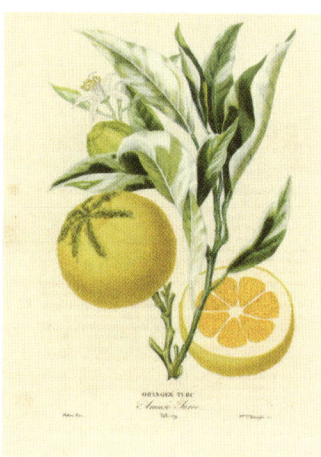

ORANGER TURC
Aranciu Turcu

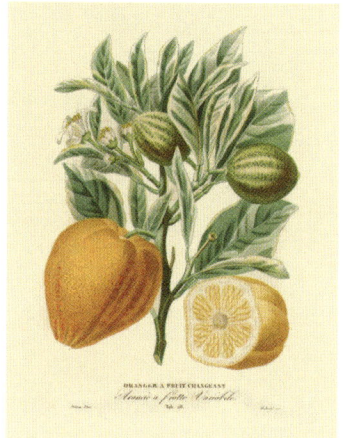

ORANGER À PETIT CHAGEANT
Aranciu à frutti d'ambla

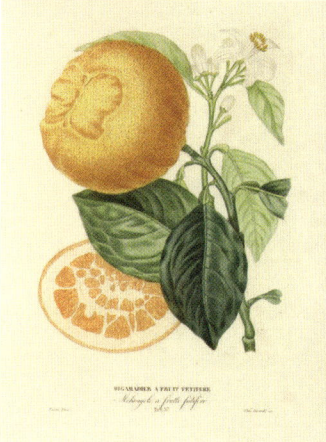

BIGARADIER À FRUIT PETITE
Melangolo à frutti piffiu

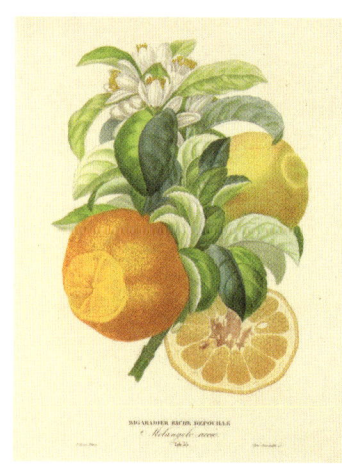

BIGARADIER BUCH. DOPOUILLÉ
Melangolo secco

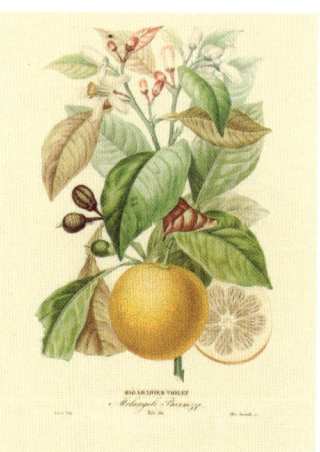

BIGARADIER VIOLET
Melangolo Pavonazzo

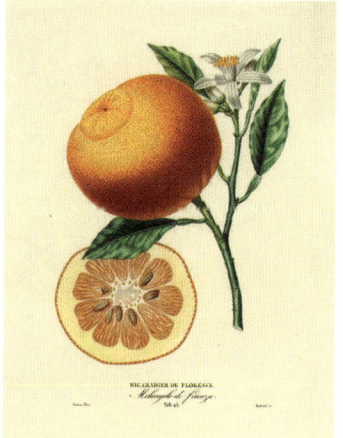

BIGARADIER DE FLORENCE
Melangolo de Firenze

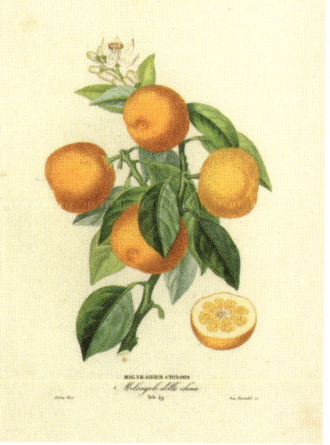

BIGARADIER CHINOIS
Melangolo della china

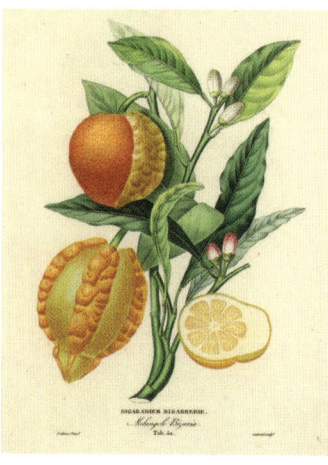

BIGARADIER BIZARRERIE
Melangolo Bizzaria

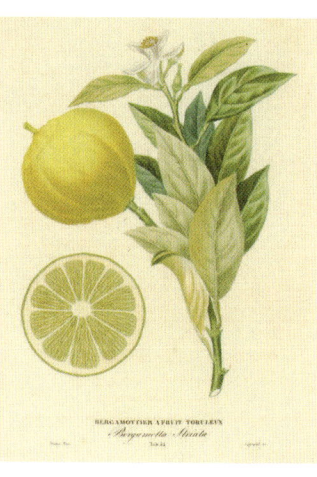

BERGAMOTTIER À FRUIT TORULEUX
Bergamotta Striata

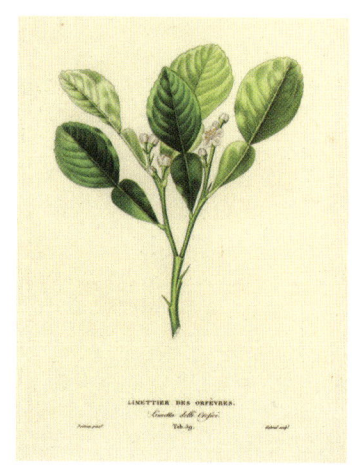

LIMETTIER DES ORFÈVRES.
Limetta delle Orefici

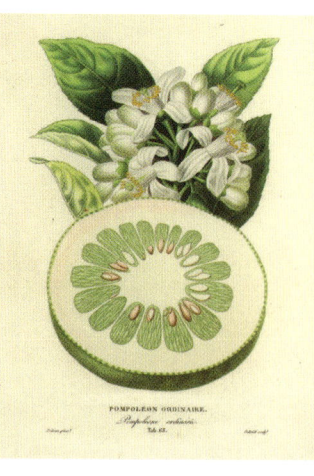

POMPOLÉON ORDINAIRE.
Pompleone ordinario

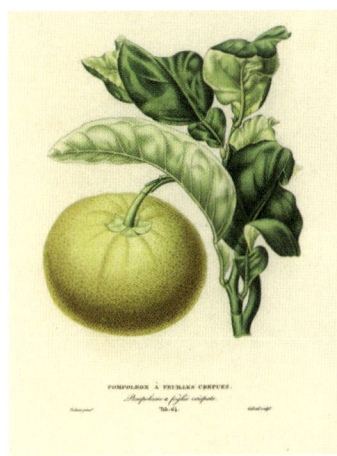

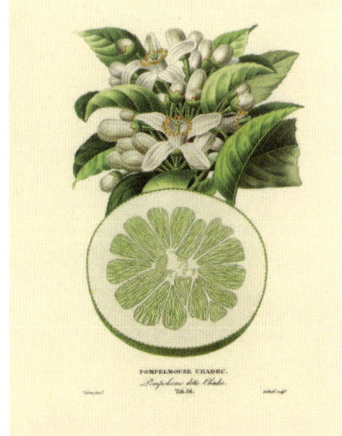

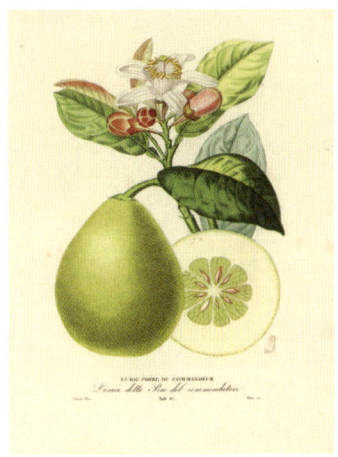

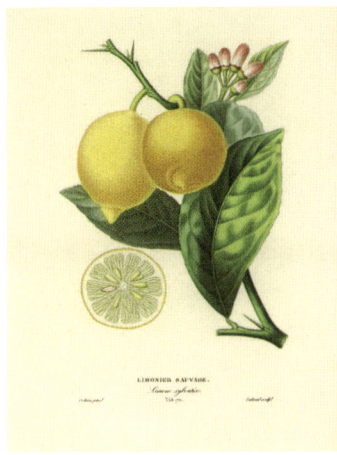

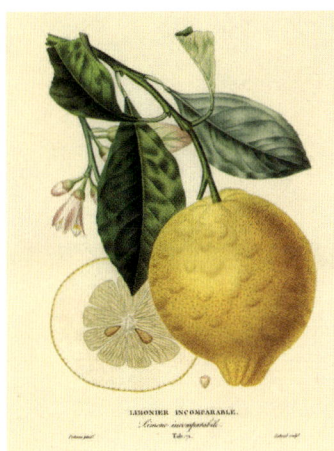

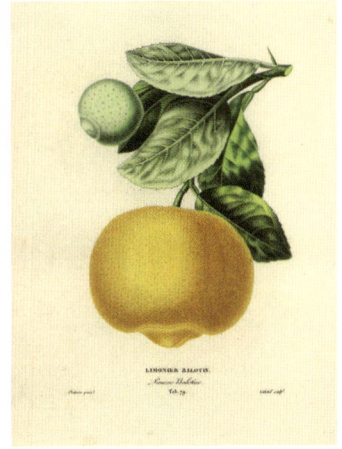

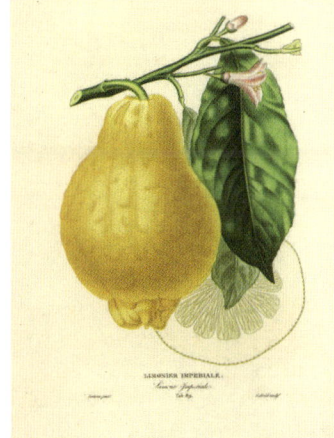

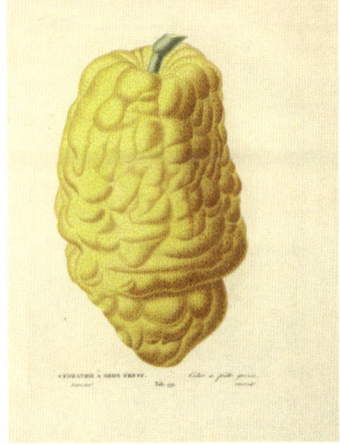

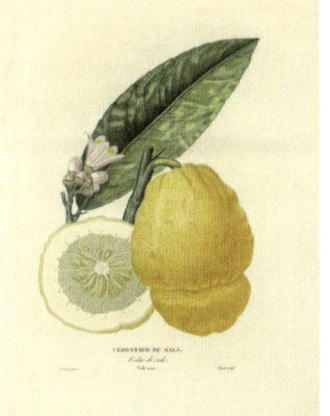

PREVIOUS PAGES: Hand-coloured plates, Pierre-Joseph Redouté, from *Traité des arbres et arbustes* by Henri-Louis Duhamel du Monceau, 1801–19.

OPPOSITE AND ABOVE: Plates from *Histoire naturelle des orangers* by Antoine Risso and Pierre-Antoine Poiteau, 1818–22.

Risso, a pharmacist and naturalist, and Poiteau, who trained and practised as a gardener and botanist, together described 169 different kinds of citrus: over 70 kinds of orange, nearly 50 lemons, as well as limes, citrons and other citrus fruits. They also discussed the orange liqueur they called 'curasow' (curaçao, see p. 197).

BELOW: *Citrus aurantium* from *Flora medica, oder, Abbildung der wichtigsten officinellen Pflanzen* by David Dietrich, 1831.
OPPOSITE: *Citrus limonum* Risso from *Dictionnaire universel d'histoire naturelle* by Charles Henry Dessalines d'Orbigny, 1849.

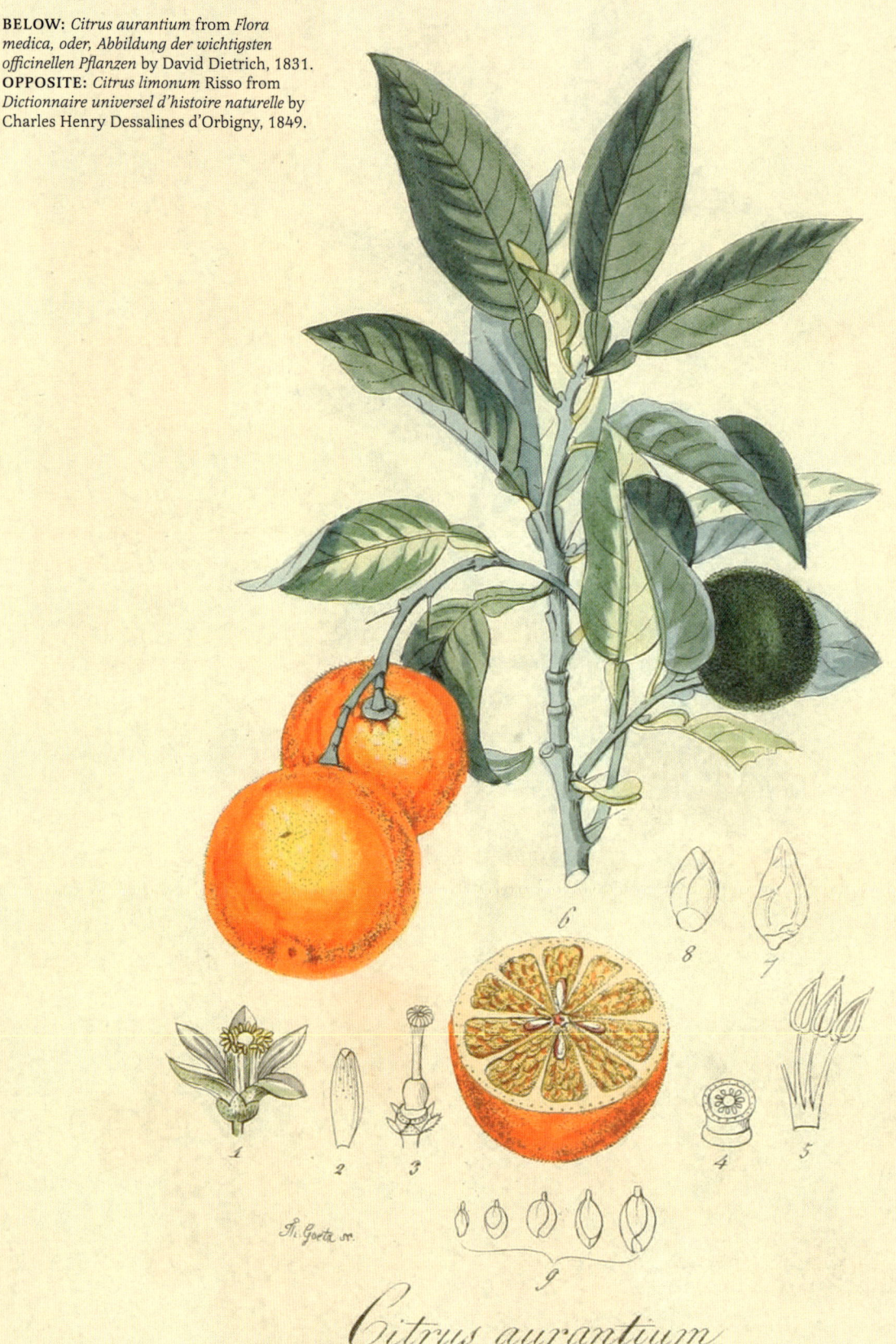

Citrus aurantium.

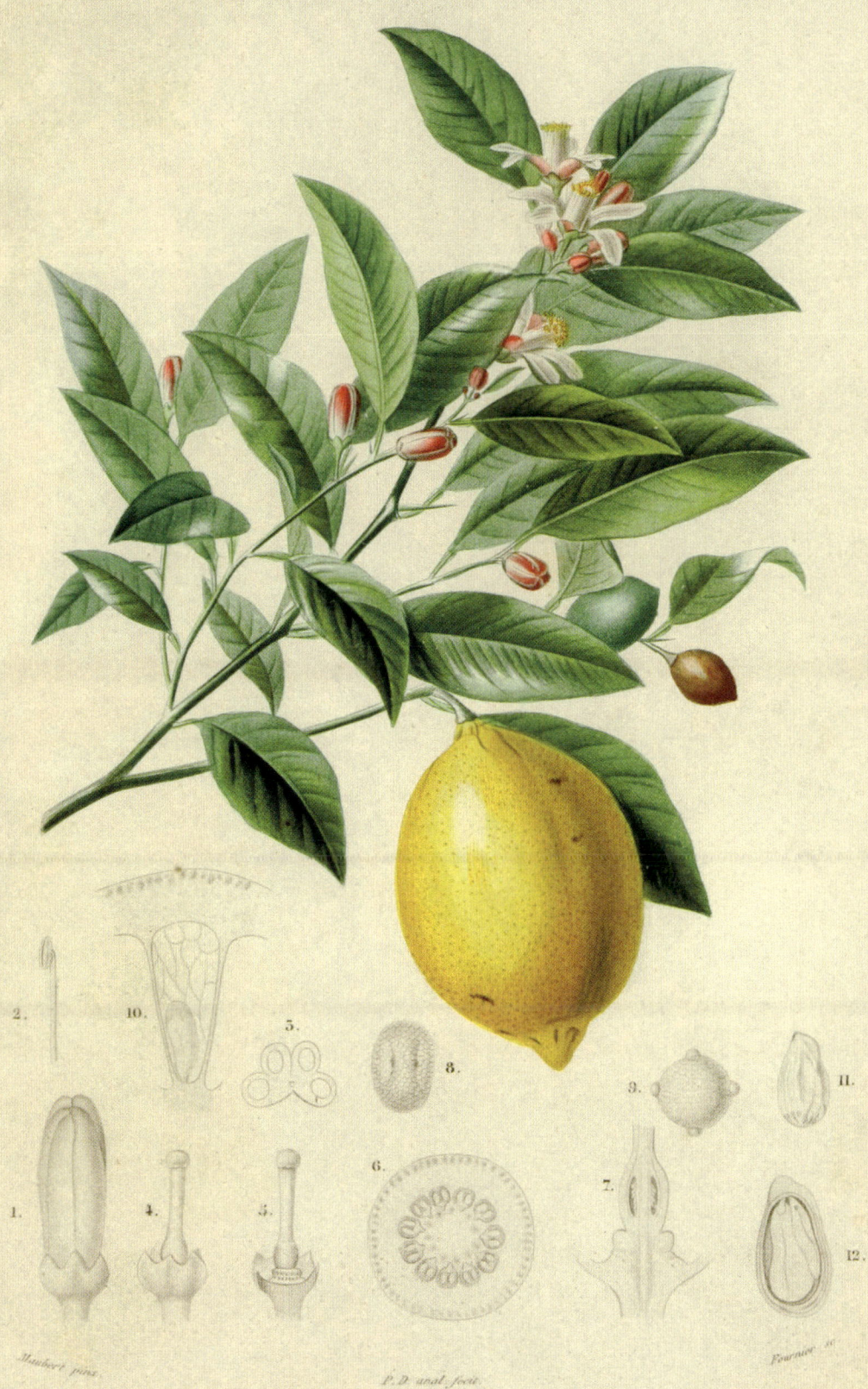

AURANTIACÉES. *Oranger Limonier.* (**Citrus** Limonum, *Risso.*)

Maubert pinx.   P. D. anal. fecit.   Fournier sc.

Fallou imp.

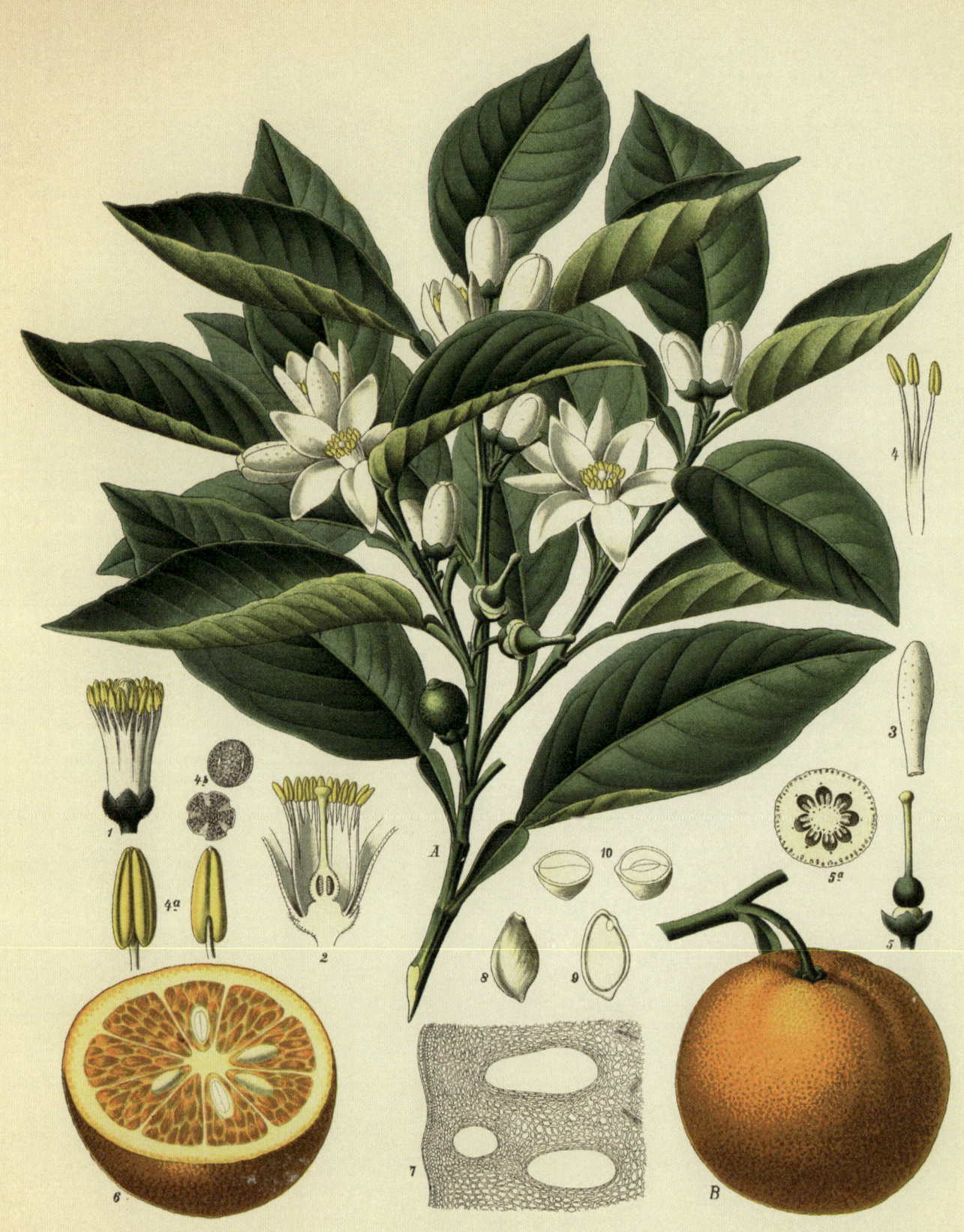

Citrus vulgaris Risso.

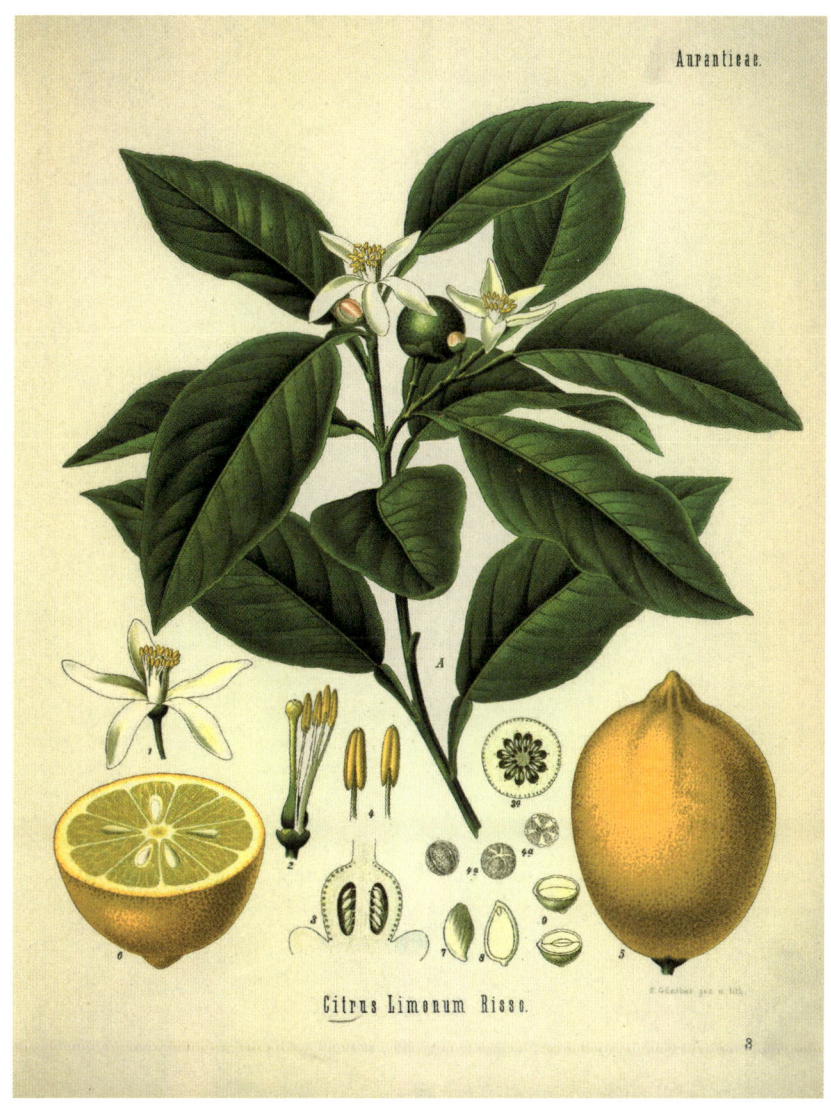

Aurantieae.

Citrus Limonum Risso.

OPPOSITE AND ABOVE: *Citrus vulgaris* Risso (sour orange, *Citrus × aurantium* Sour Orange Group, opposite) and *Citrus limonum* Risso (lemon, *Citrus × limon*, above), plates from *Medizinal-Pflanzen* by Hermann Köhler, 1883.

Published after his death, Köhler's (1834–1879) German herbal gathered together plants of medicinal interest from many European countries. The Latin names on the plates are those of Antoine Risso (see pp. 141 and 172–73).

OVERLEAF: *Citrus medica* L and *Citrus aurantium* from *Flora Lekarska* by Edward Winkler, 1852.

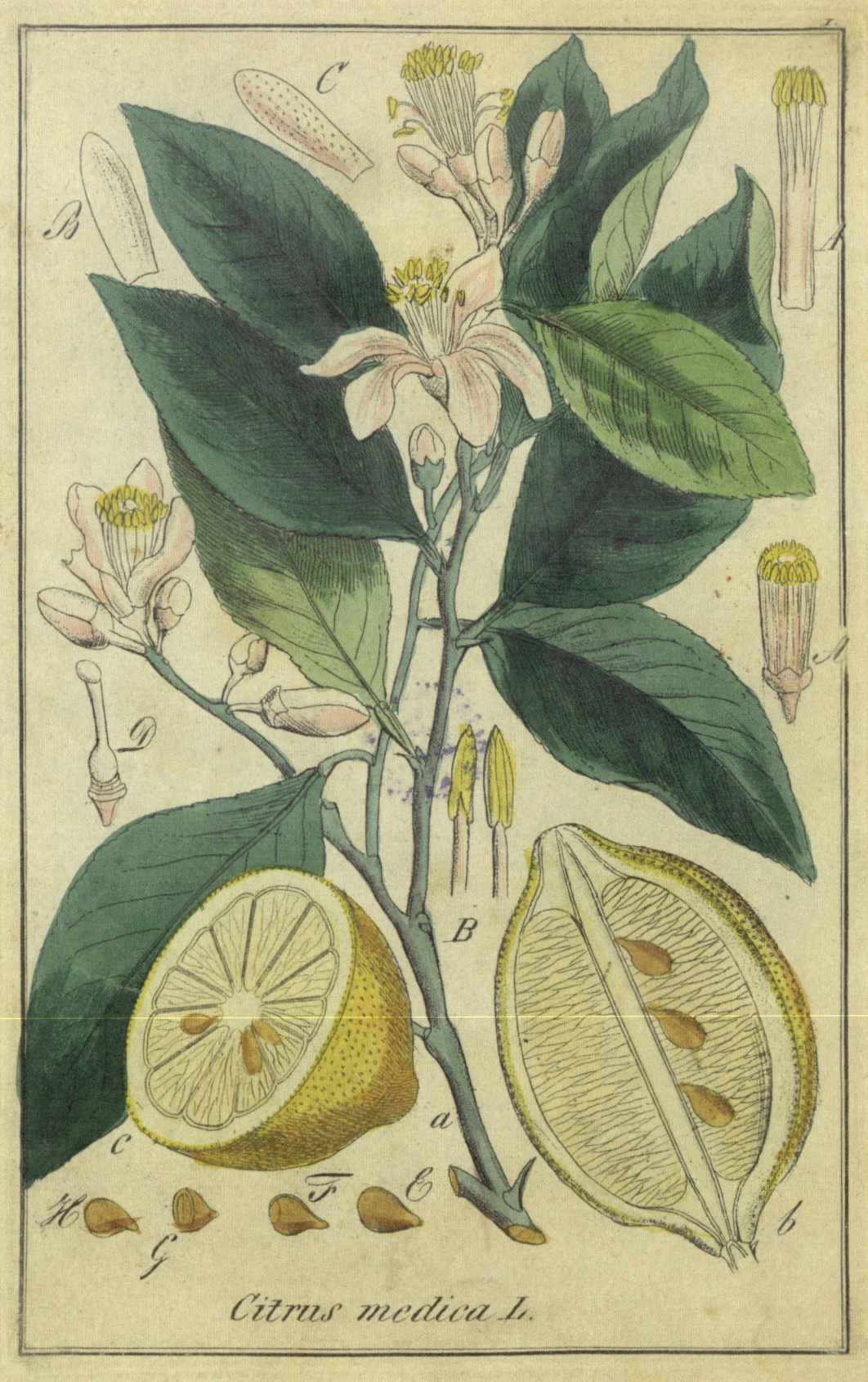

Citrus medica L.

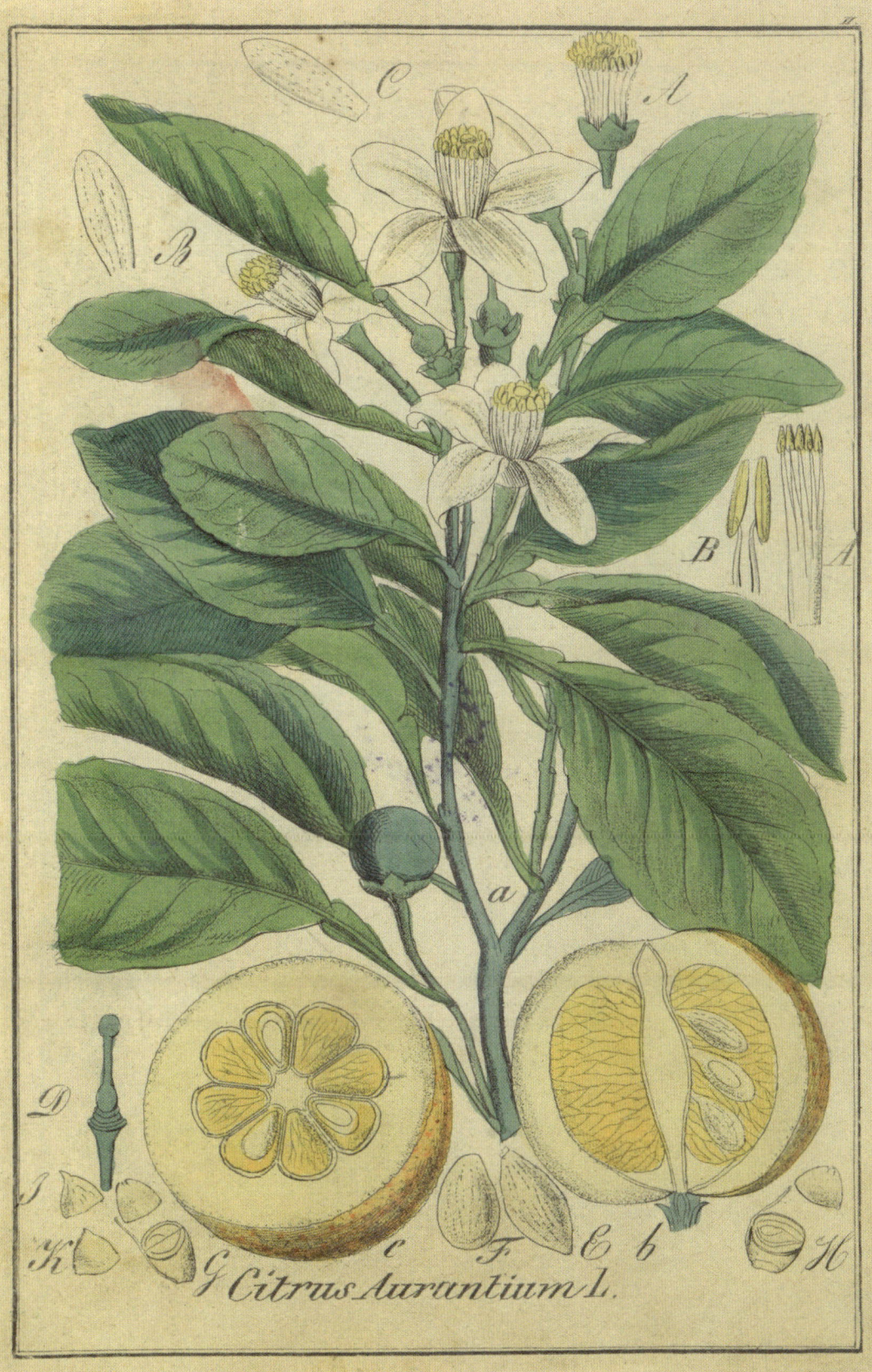

G. Citrus Aurantium L.

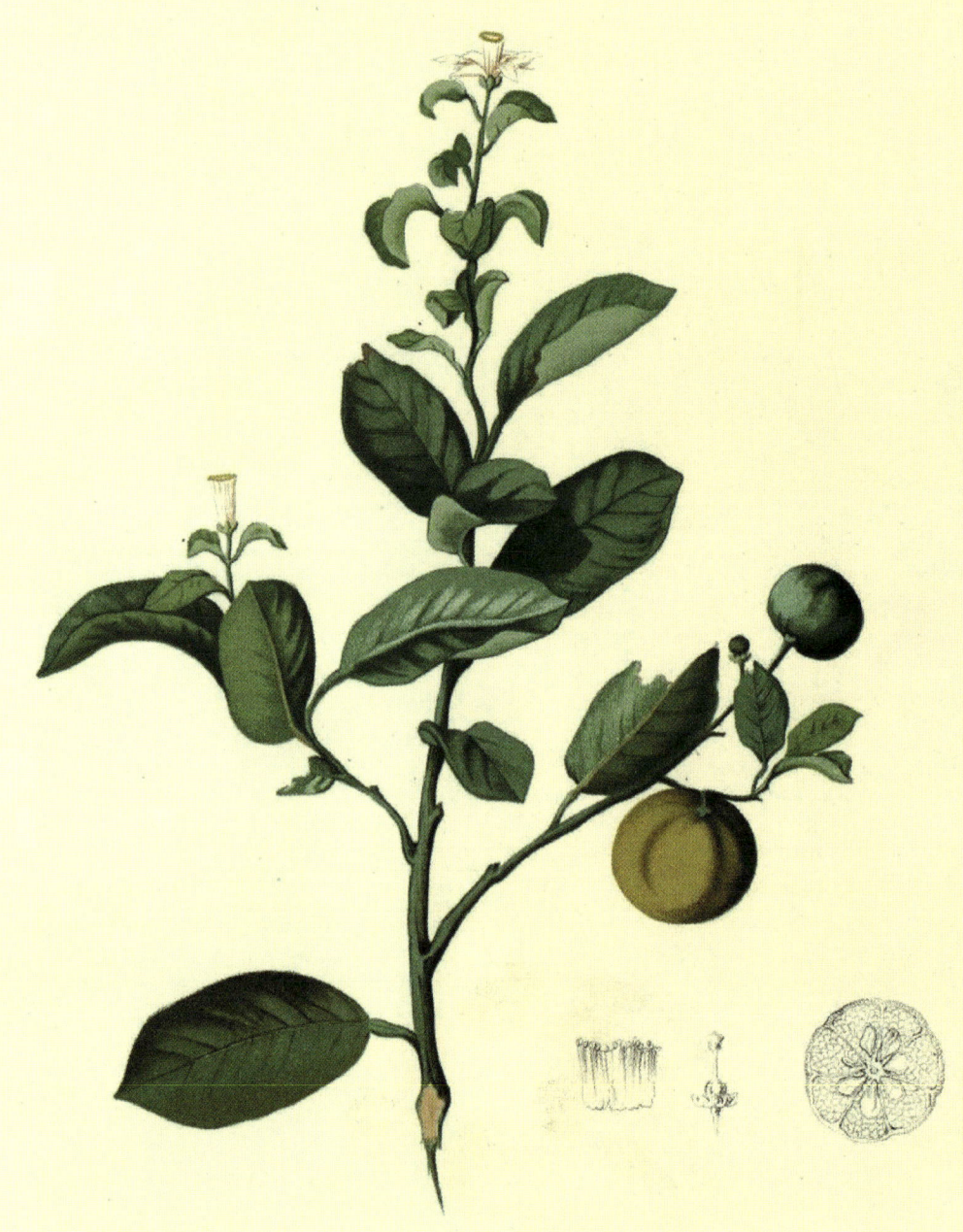

CITRUS MITIS.—Blanco.

R. Garcia B.

Lit C. Verdaguer. Barcelona.

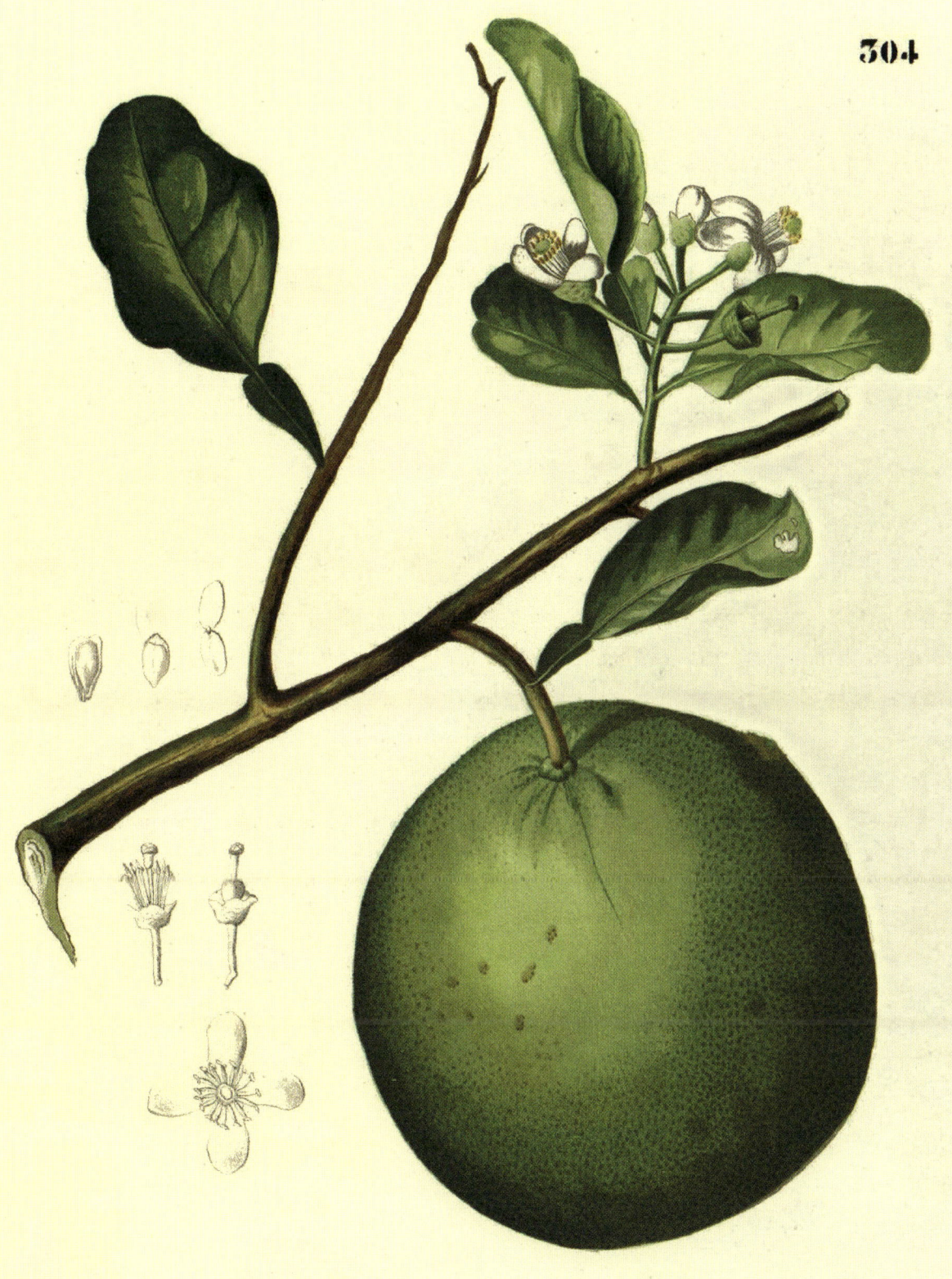

CITRUS DECUMANA.—Linn.—Blanco.—DC.

C. Verdaguer Barcelona.

Certain animals, especially parrots, are fond of the citrus pulp whether sweet or sour, and are quite equal to carrying its seeds to long distances ... carried with the pulp to their nests for their young, and dropping it on the way.

EMANUEL BONAVIA, *CULTIVATED ORANGES AND LEMONS*, 1888

PREVIOUS PAGES: *Citrus mitis* Blanco (calamondin, *Citrus × microcarpa*, p. 180) and *Citrus decumana* Blanco (pomelo, *Citrus maxima*, p. 181) from *Flora de Filipinas* by Francisco Manuel Blanco, 1880–83.

OPPOSITE: *A Branch of Orange Blossom with a Bee-Eater*, Pancrace Bessa, 1816.

The Parisian Bessa (1772–1835) was one of the most prolific botanical painters in France the first half of the 19th century. He was a pupil of the famous Dutch painter and engraver Gerard van Spaendonck (1746–1822), as was Pierre-Joseph Redouté (see p. 141) with whom Bessa worked on a number of important projects. Bessa was appointed flower painter and drawing master to the Duchesse de Berry in 1820 and then worked on the royal collection of watercolours on vellum until his death.

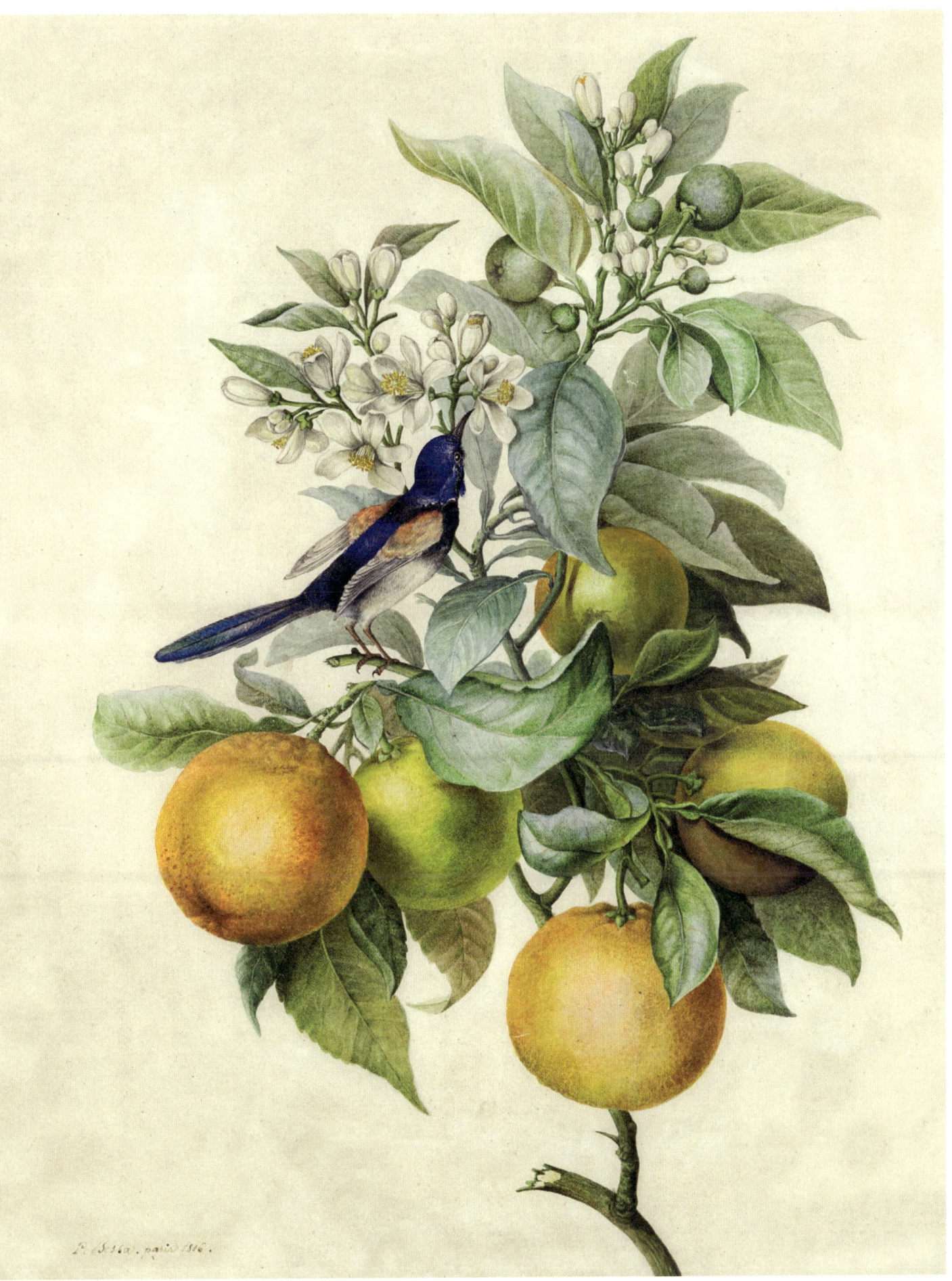

P. Bessa, paris 1816.

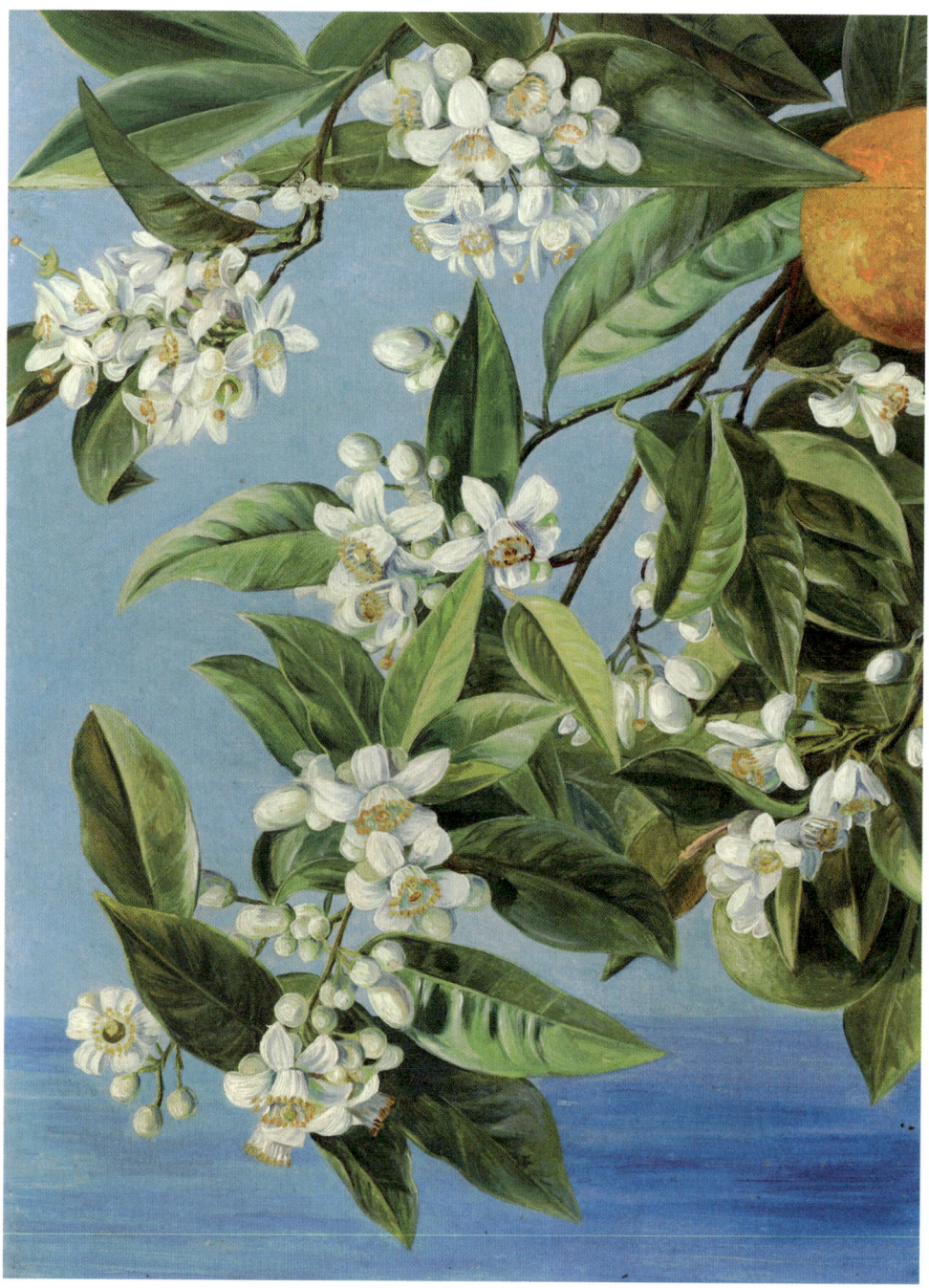

Orange Flowers and Fruits, Painted in Teneriffe, 1875 (above), *Foliage, Flowers and Fruit of the Citron, and Butterfly; painted in Brazil*, n.d. (opposite) and *Pomelo, henna branch and flying lizard*, 1876 (overleaf), Marianne North.

North (1830–1890) was a prolific English botanical artist, who endowed an art gallery at Royal Botanic Gardens, Kew, to house her paintings.

Did I not paint? And wander and wonder at everything? Every rock bore a botanical collection fit to furnish any hot-house in England.

MARIANNE NORTH, *RECOLLECTIONS OF A HAPPY LIFE*, 1892

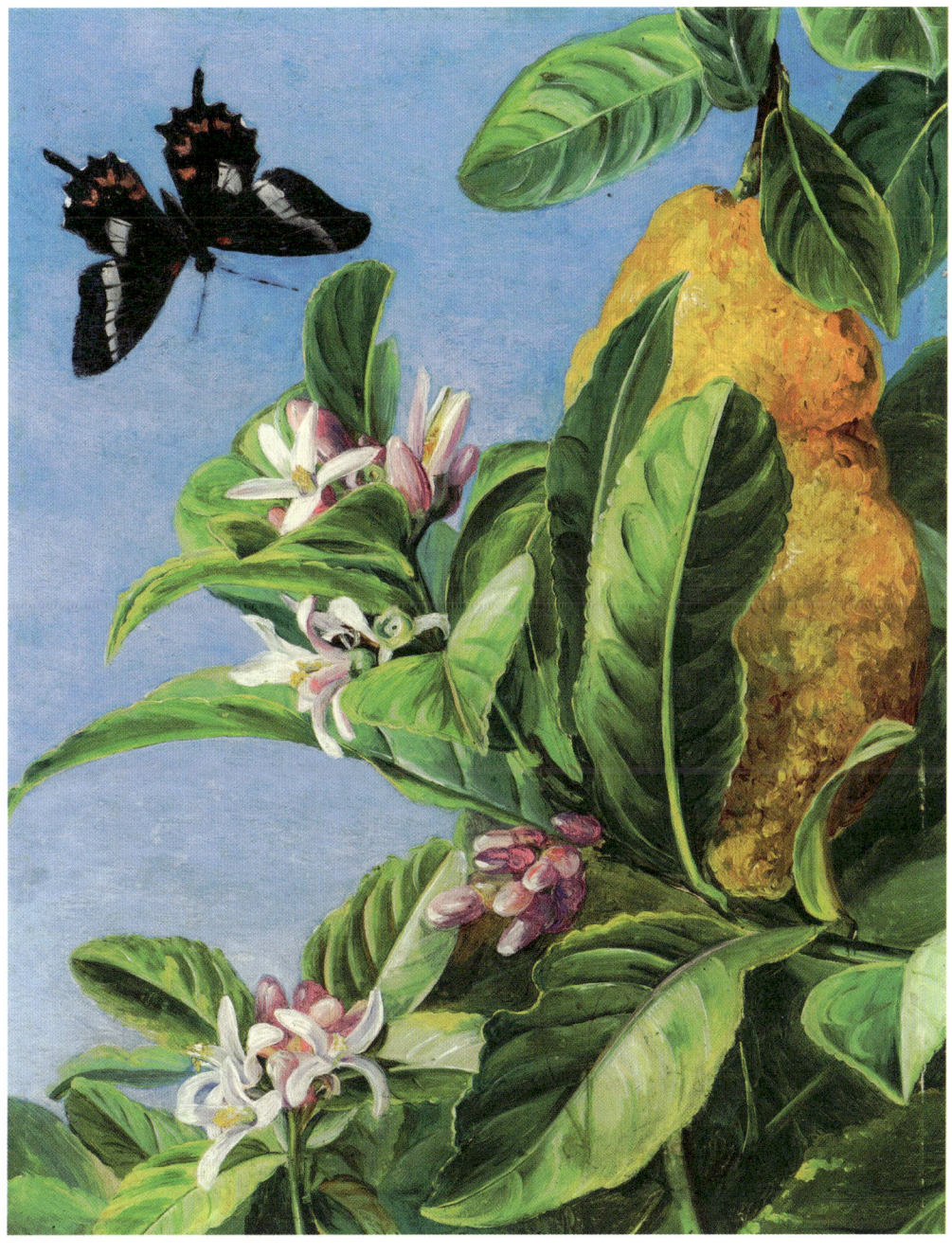

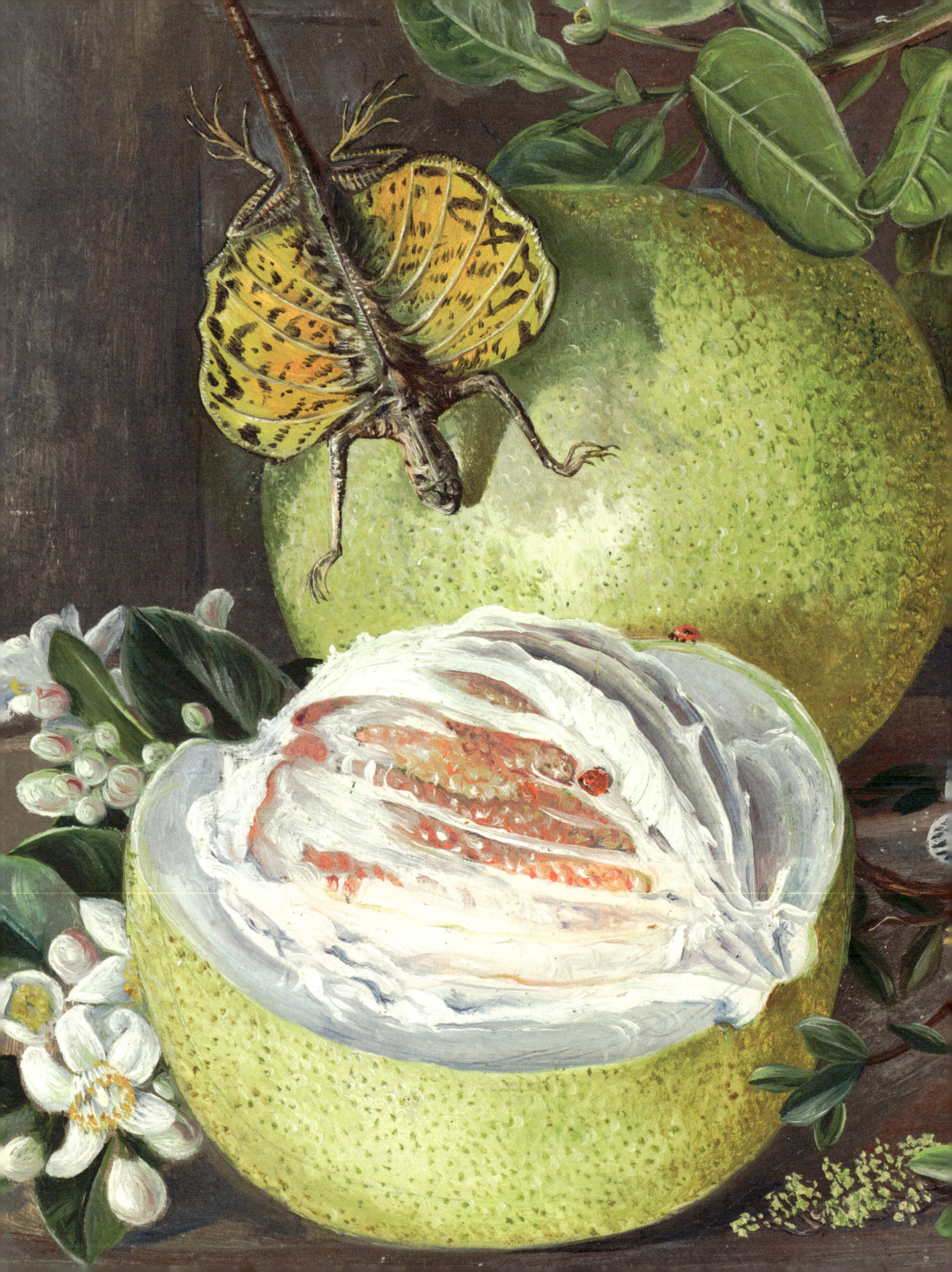

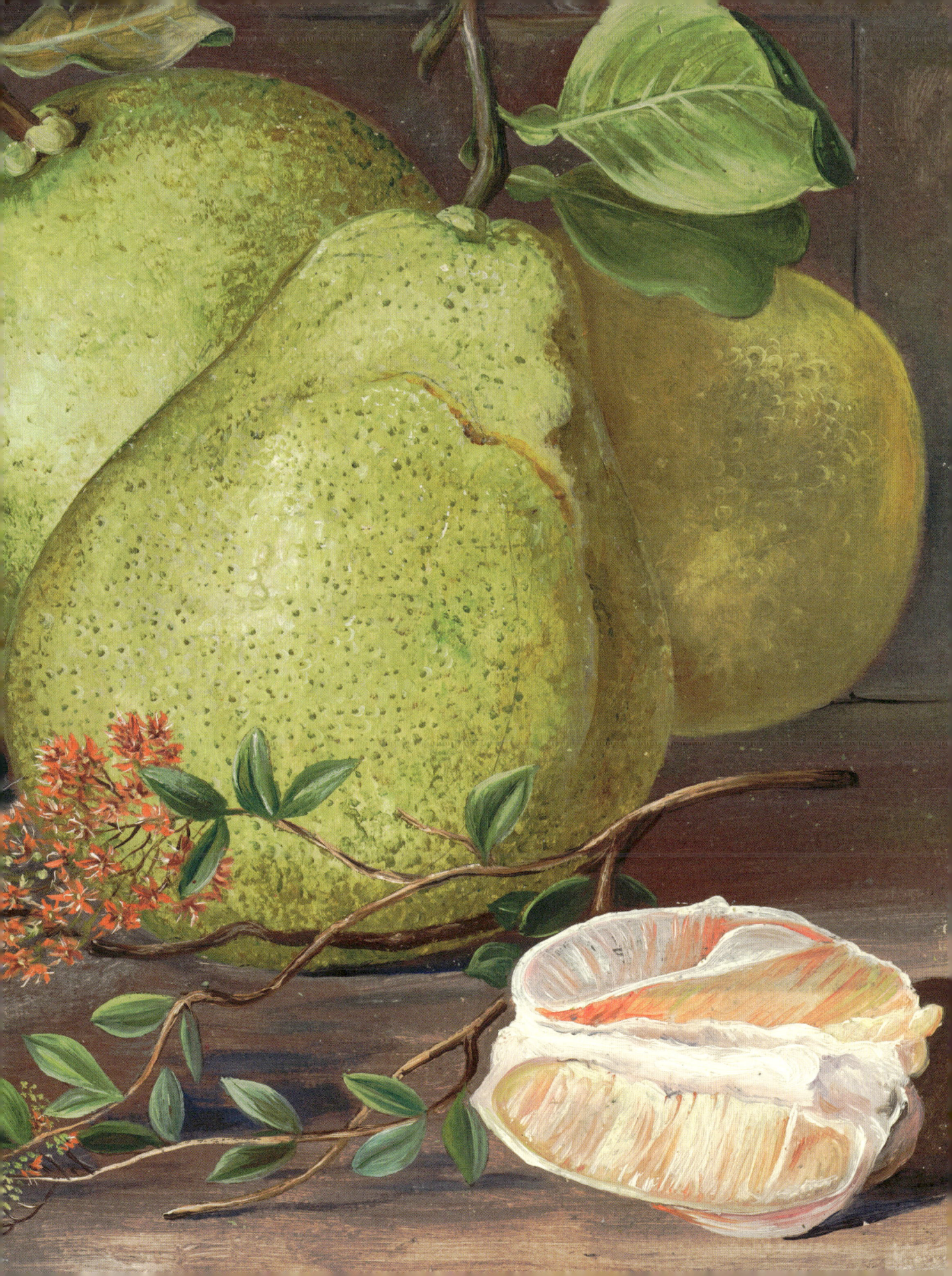

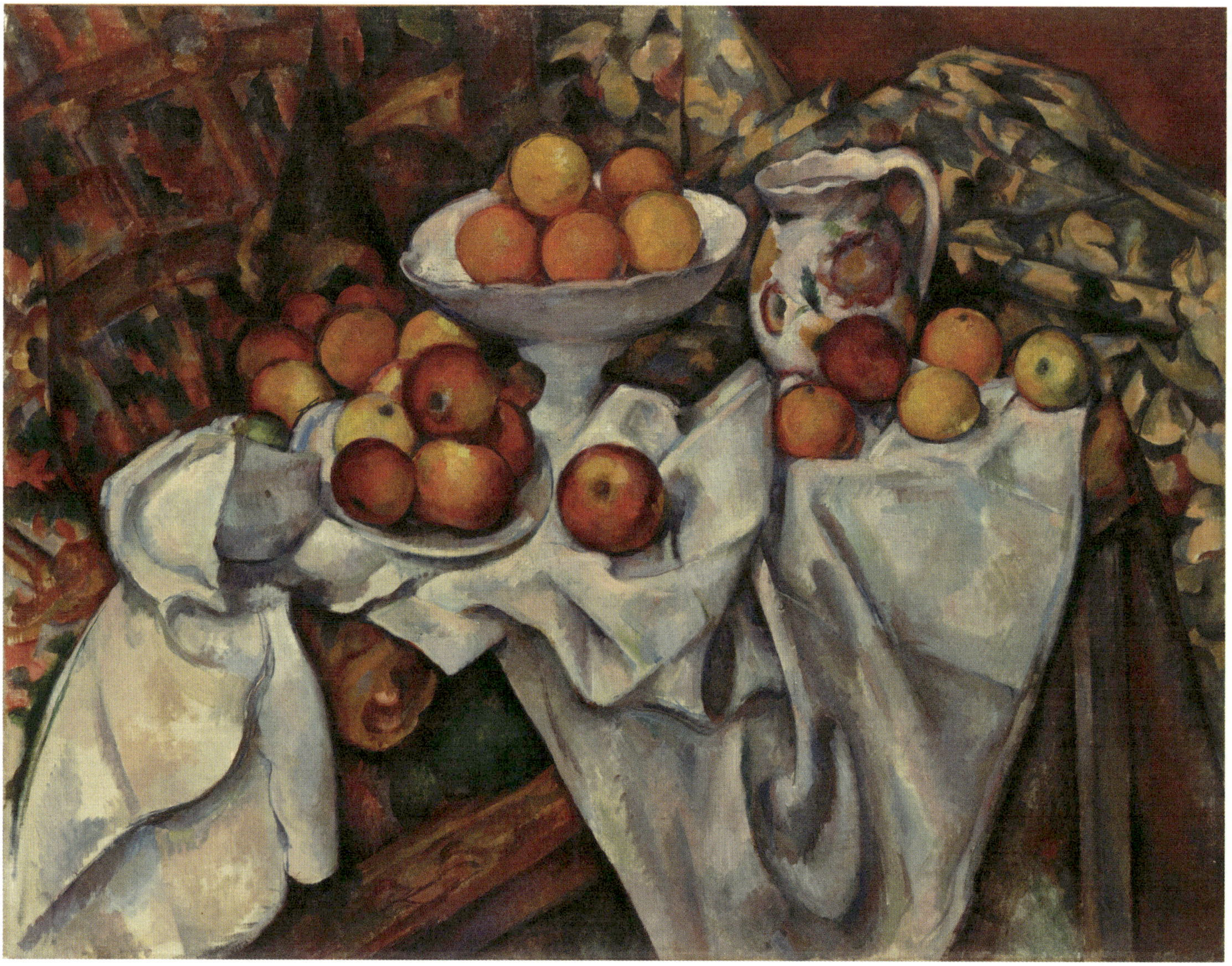

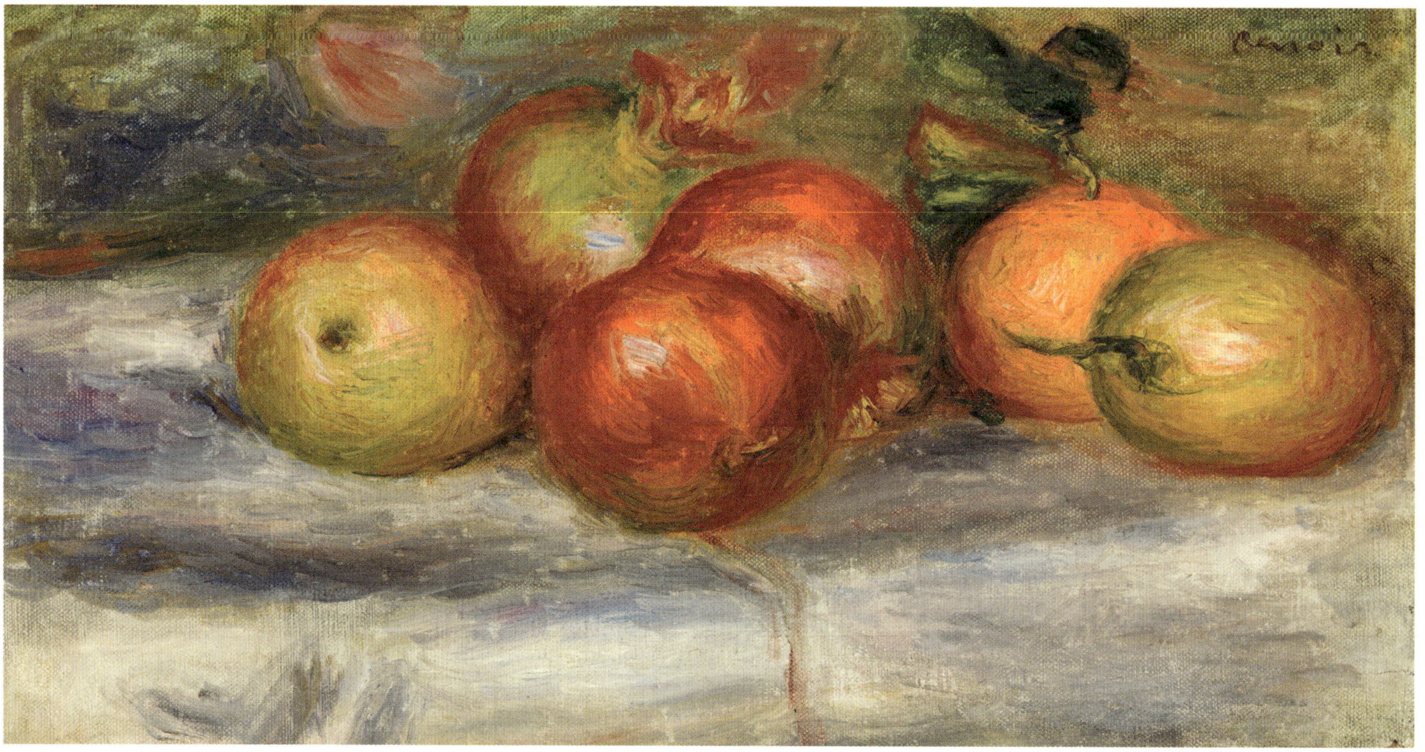

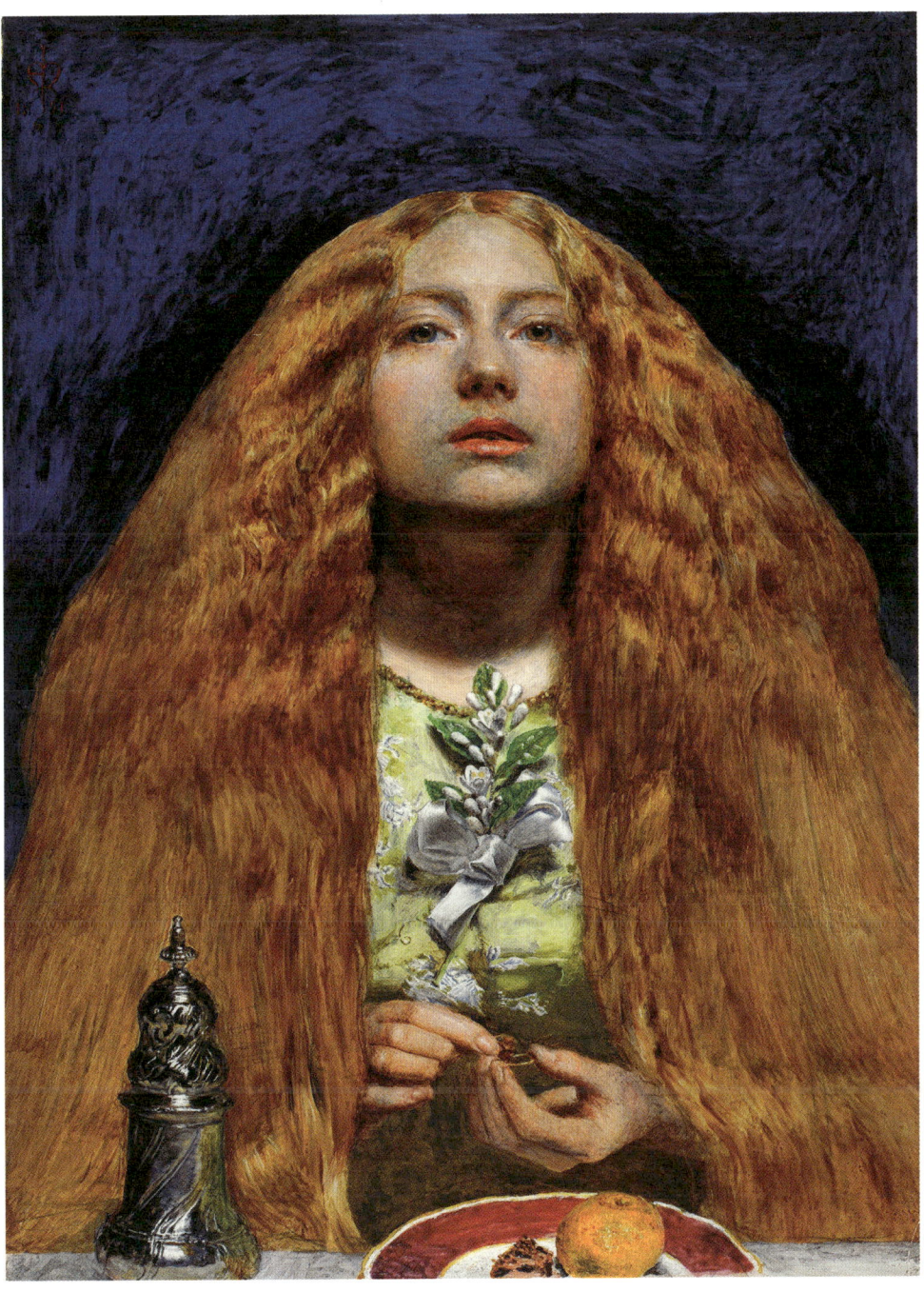

OPPOSITE, ABOVE: *Still Life with Apples and Oranges*, Paul Cézanne, *c.* 1899. **OPPOSITE, BELOW:** *Apples, Oranges and Lemons*, Pierre-Auguste Renoir, *c.* 1911. **ABOVE:** *The Bridesmaid*, John Everett Millais, 1851.

Citrus has remained a favourite subject throughout all the movements in fine art, as the modern masterpieces here show. In Millais's portrait the orange blossom signifies virginity. In the Victorian era, it was held that a bridesmaid would have a vision of her true love if she passed a piece of wedding cake through a ring nine times, as depicted here. Meanwhile, Cézanne wrote to a friend, 'they [fruits] love having their portraits done ... They exhale their message with their scent. They reach you with all their smells and tell you about the fields they've left, the rain that made them grow, the dawns they watched.'

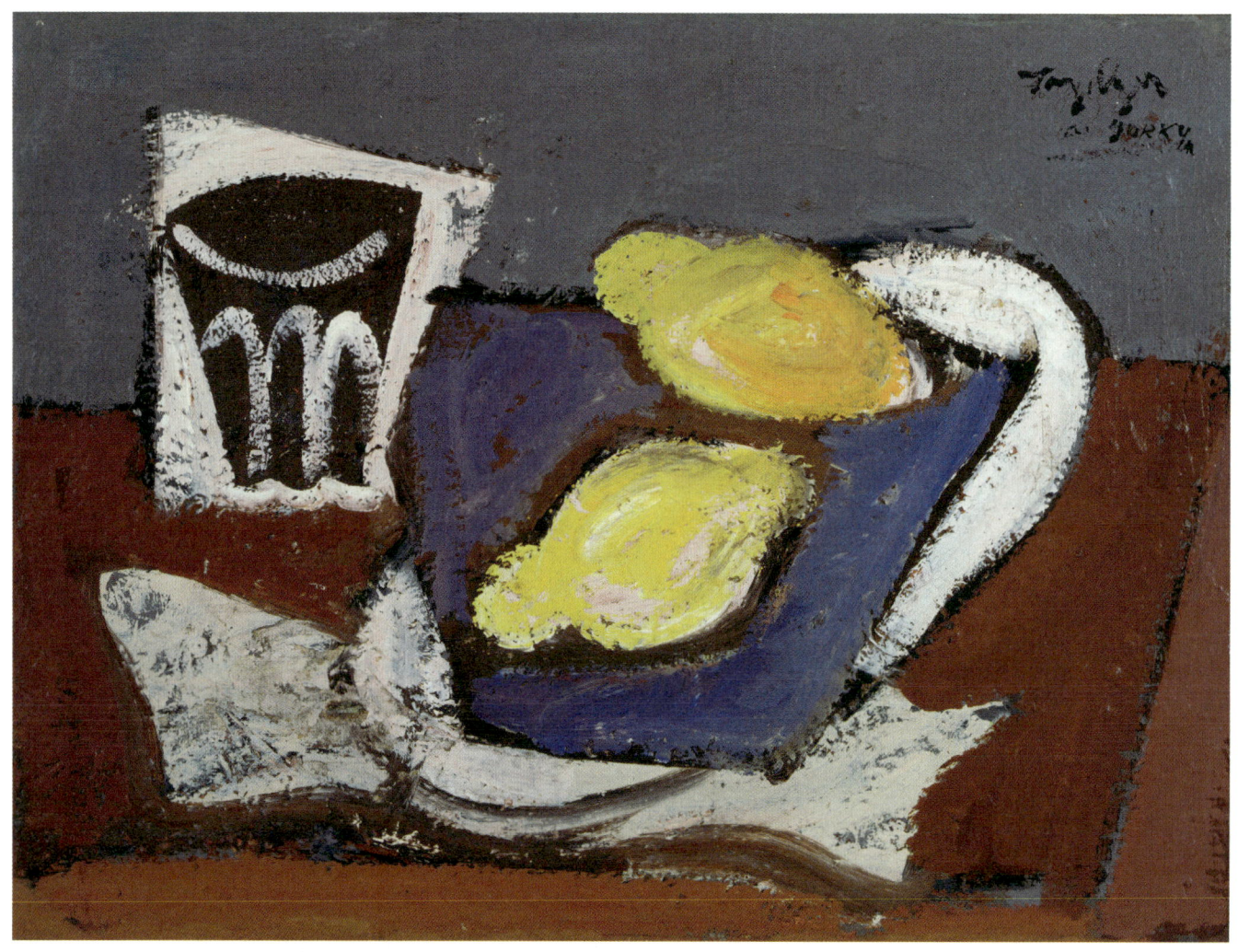

ABOVE: *Still Life with Lemons*, Arshile Gorky, early 1930s. **OPPOSITE, ABOVE:**
Still Life with White Jar, Orange and Book, Vilhelm Lundstrøm, 1932–33.
OPPOSITE, BELOW: *Untitled* (still life with parrot and fruit), Frida Kahlo, 1951.

Gorky (1904–1948), an Armenian painter who settled in the USA, followed in the footsteps
of Poussin and Cézanne in his still-life depictions of fruit, while evoking the Cubism of Picasso.
Danish artist Lundstrøm (1893–1950) also followed Cézanne and Picasso, his still-life paintings
however pursuing a formal ideal geometry. Kahlo (1907–1954) painted rich, colourful
still lifes with subjects that evoked natural sensuality and political symbolism.

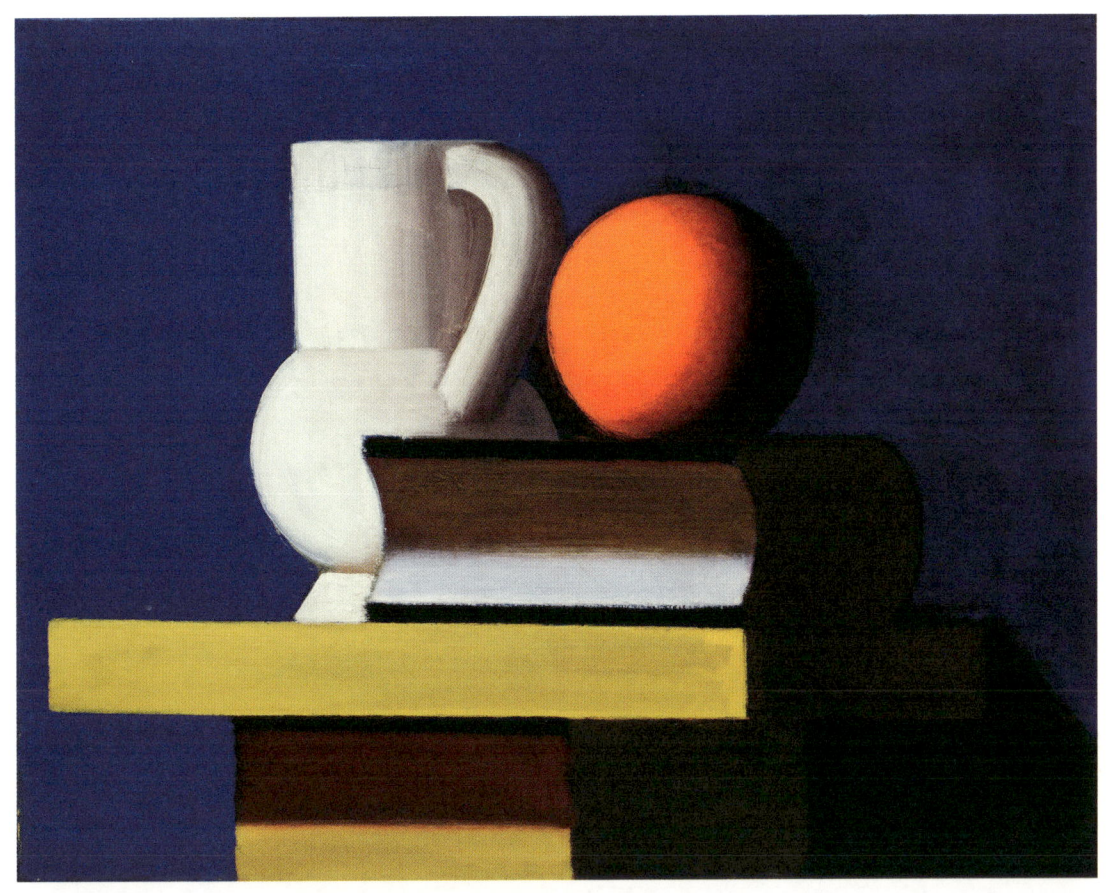

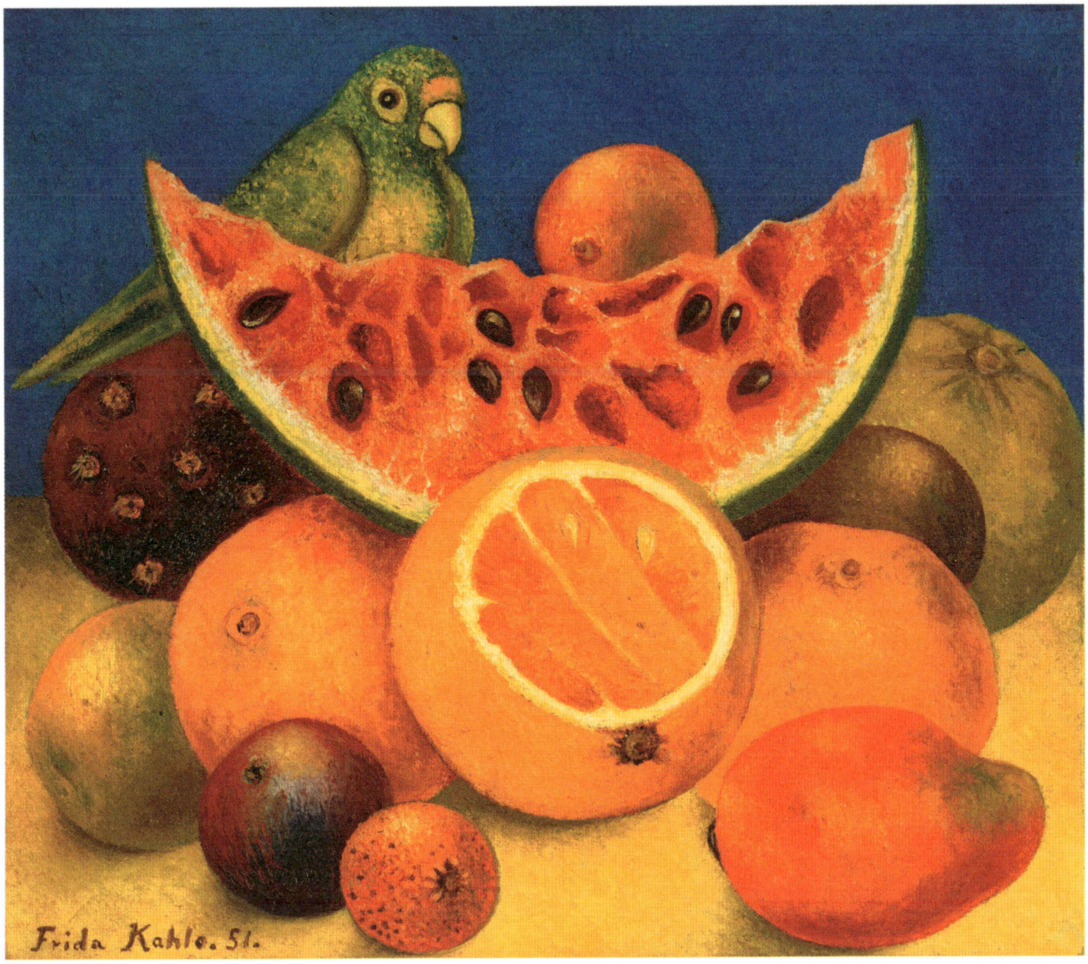

Frida Kahlo. 51.

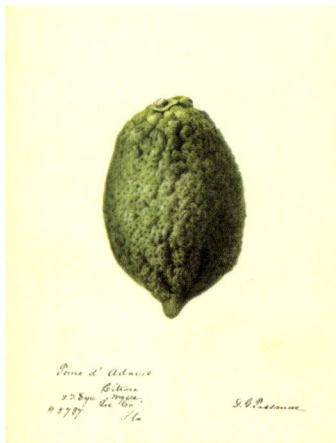 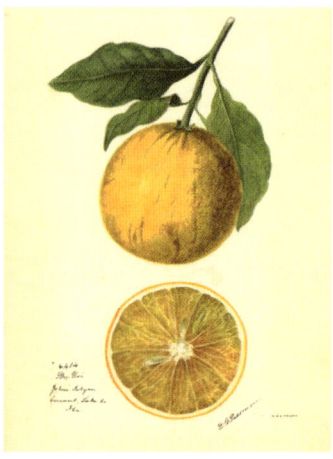 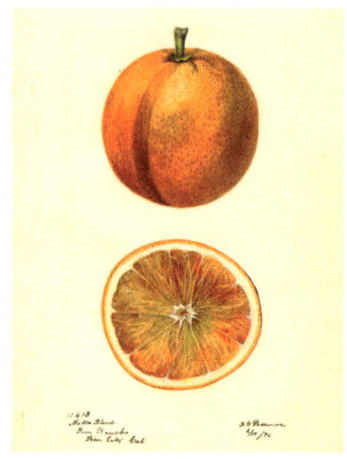 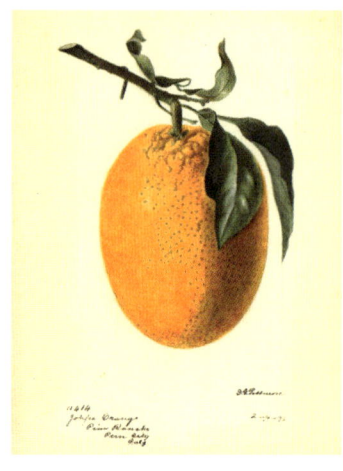

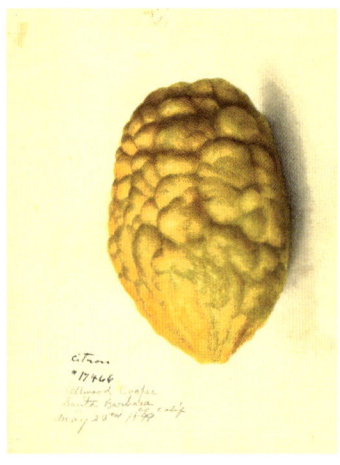 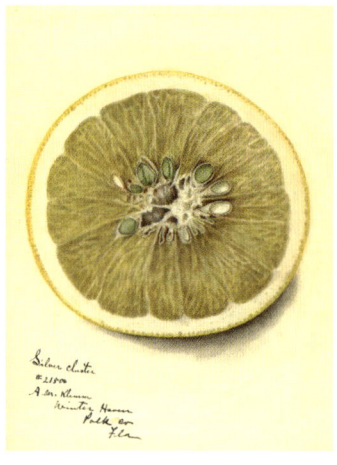 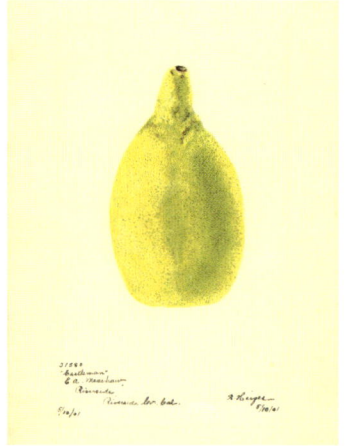 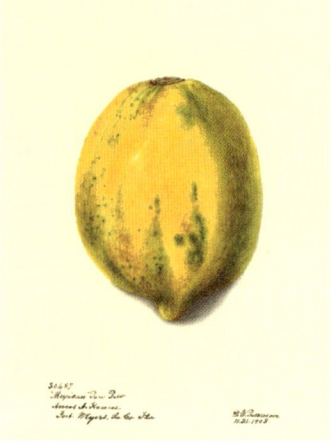

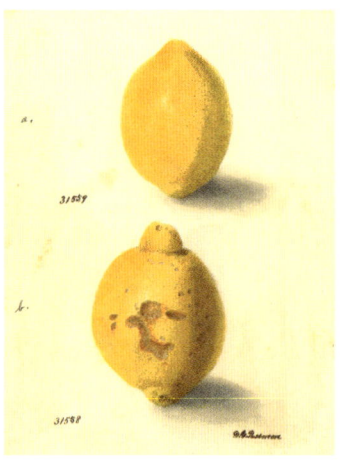 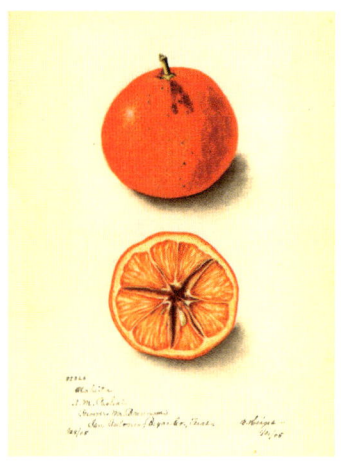 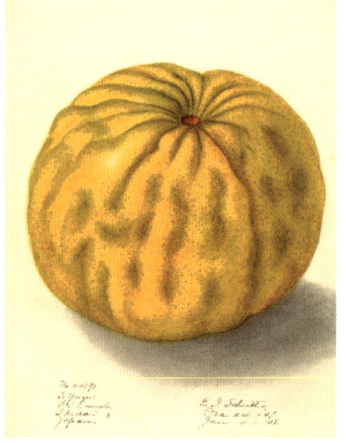 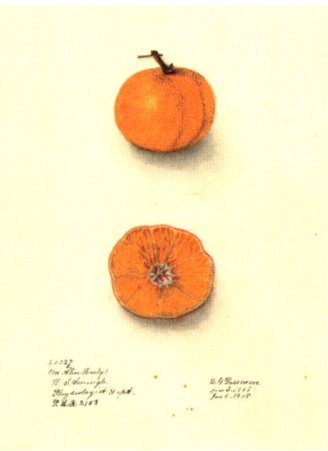

'Pomological Watercolours', commissioned by the US Department of Agriculture (USDA) Division of Pomology, late 19th to early 20th century.

From 1886 to 1942, the USDA Division of Pomology commissioned more than 7,500 watercolours, line drawings and lithographs from 66 artists of fruits growing in the US. Among the most prolific contributors were Amanda Newton (c. 1860–1943), Deborah Griscom Passmore (1840–1911), Ellen Isham Schutt (1873–1955) and Royal Charles Steadman (1875–1964).

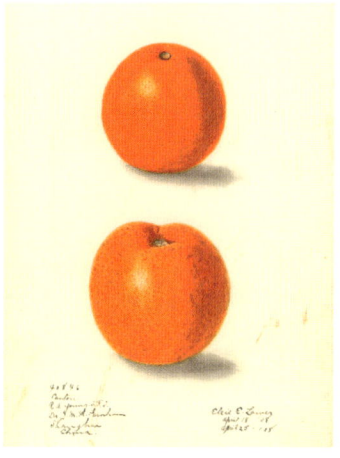

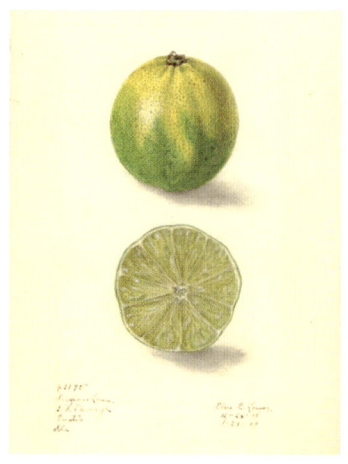

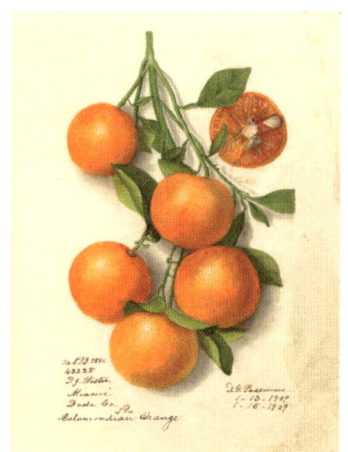

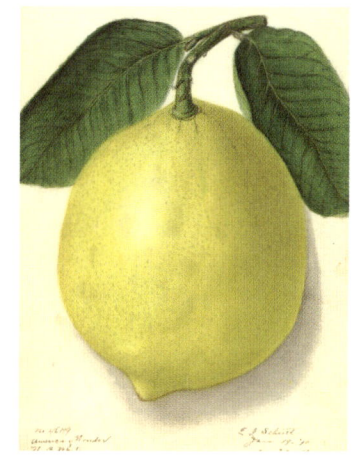

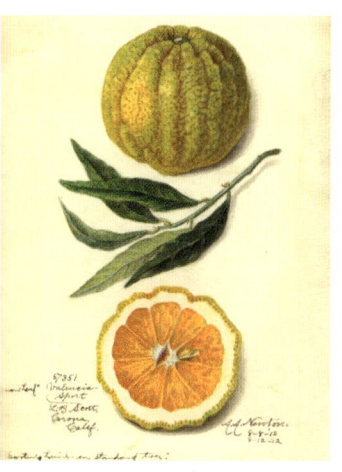

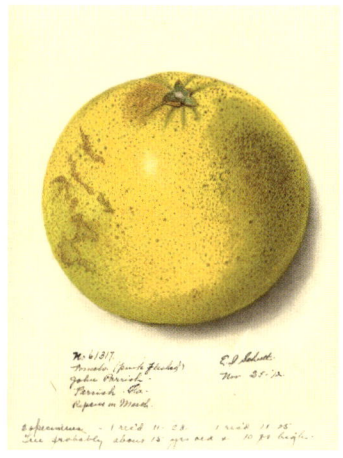

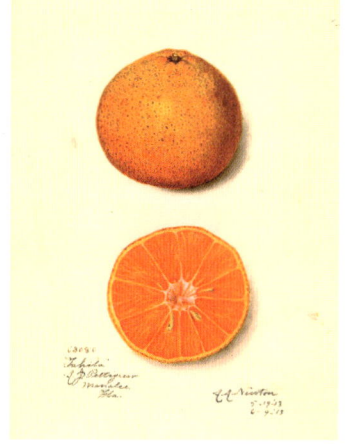

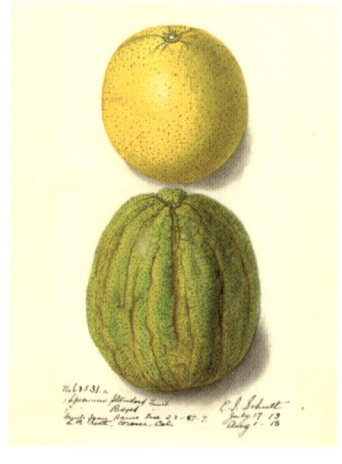

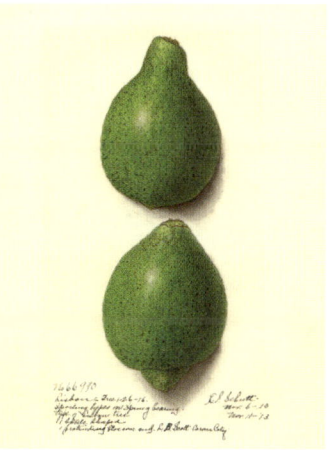

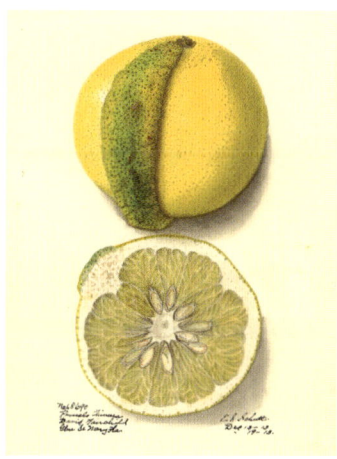

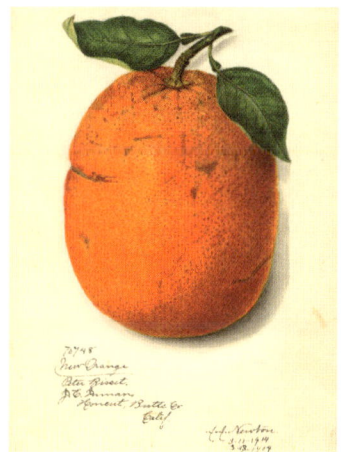

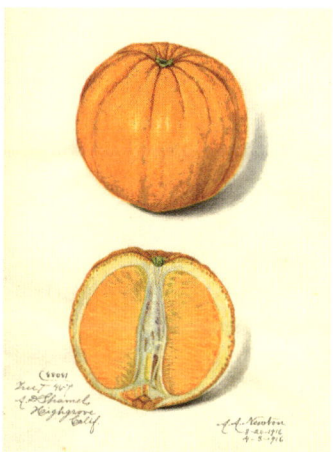

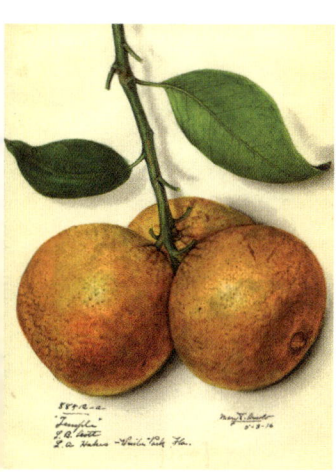

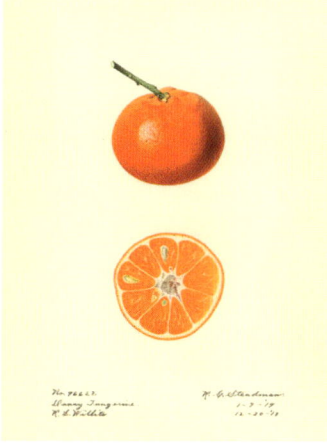

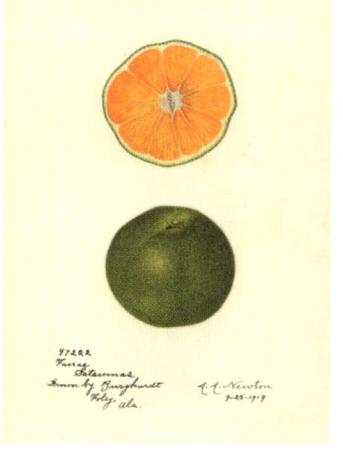

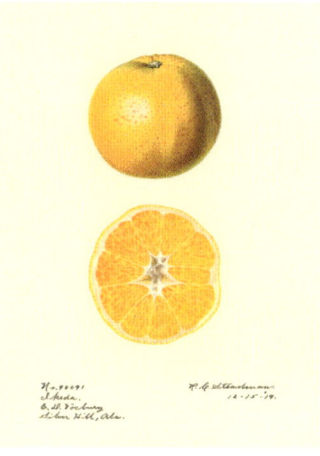

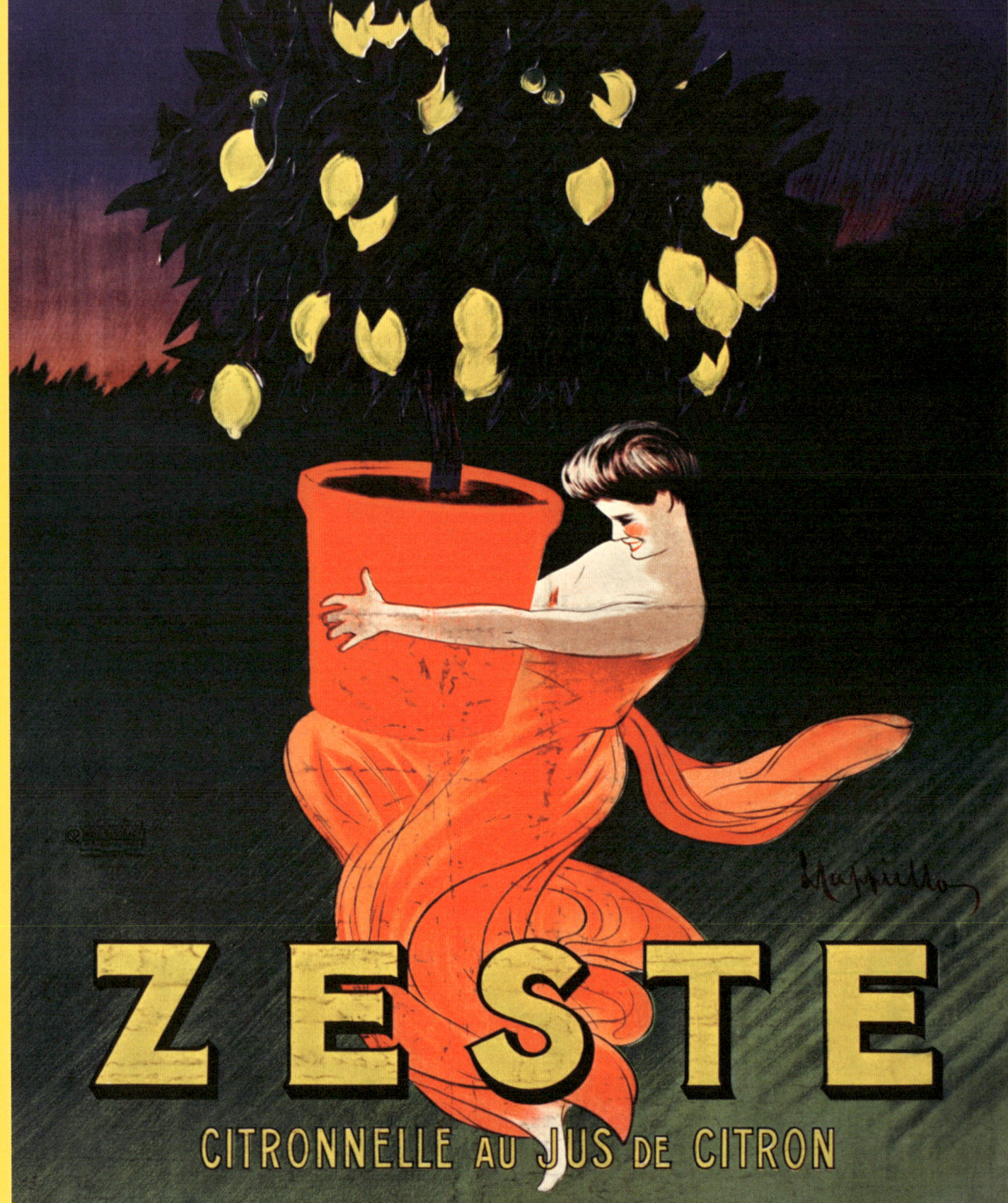

ZESTE

CITRONNELLE AU JUS DE CITRON

SEUL
FABRICANT FOURNIER-DEMARS Sᵗ AMAND
(CHER)

FIVE

Progress
and Perils

THE WORLD'S MOST FAMOUS LOVER, Giacomo Girolamo Casanova (1725–1798), wrote an autobiography that has scandalized and entertained readers for two centuries. An account of his many sexual escapades, the work prompted conversations both public and private about sex. Despite his licentious life, Casanova lived to old age and, although picking up (and no doubt passing on) venereal disease, his use of prophylactics was well known. Among various methods, he used a watery solution of lemon juice, now shown to be an effective spermicide.[1]

This technique was known in eighteenth-century Istanbul, and in the late nineteenth century in tropical America a solution of lemon juice 'mixed with a decoction of the husks of mahogany nut' was employed for the same purpose. Still in use as a contraceptive in the 1960s, lemon juice was judged as unsurpassed by 'any modern clinical contraceptive', and up until the twenty-first century sex workers in Kano, Nigeria, used the same method. Experimental work has shown that sperm motility is reduced by citric acid, which destroys proteins of the mitochondria, to zero in just a few minutes.[2]

At the beginning of the twenty-first century, experiments were carried out in Australia, through the LemonAIDS programme, to ascertain whether lemon juice was able to inhibit cellular infection with HIV. At concentrations of 5 per cent and above, infection seemed to be inhibited; a 50 per cent solution (pH 2.7) neutralized all virus within two minutes. It was concluded that the citric acid causes

surface proteins in the envelope of the virus to be altered such that no binding and subsequent infection can occur.[3] Lemon juice was more efficacious than lime juice, though both are better than commercially available microbicides. It is interesting to note that human ejaculate already contains citric acid, so that such applications may be merely reinforcing a natural microbicide. It is estimated that a microbicide with only 40 per cent efficacy against HIV transmission, at 30 per cent coverage of the sexually active population, would avert more than 5.6 million HIV infections.

FOLK MEDICINE

Theophrastus (see p. 39) wrote of the use of the citron as both a mouthwash[4] and an antidote to poison.[5] Even before that, the pre-Socratic philosopher Democritus (c. 460–370 BCE) told the story of some Egyptian criminals who were sentenced to be thrown into a snake-pit but, having eaten citron, were unharmed. The citron's qualities were extended to all poisons and extolled for this well into the sixteenth century CE; even today, citrus extracts are used to treat the effects of venomous snake bites in Colombia.

In the first century CE, citron was suggested as a remedy for gout and, two centuries later, a medical book prescribed citron syrup to stop a cough. By the twelfth century, lemon juice was said to relieve stomach problems and inebriation; to lower temperature and cure acne, boils and abscesses; among many other properties. In the Middle Ages in Europe, peel of the rare and expensive sour orange was being recommended not merely as a flavouring but, when powdered and dissolved in wine, to prevent worms as well as the Black Death (bubonic plague, 1346–1353).[6] It has been shown that lime oil is indeed effective against a number of bacteria, even cholera (*Vibrio cholerae*), if taken with food.

By the end of the sixteenth century, forms of citron were included in the category of 'heart plants' through the now discredited Doctrine of Signatures, which held that the medicinal qualities of plants were signalled by their shape and other features, and appeared thus in Giambattista della Porta's *Phytognomica* published in Naples in 1588. From then onwards citrons figured conspicuously in European pharmacopoeias, describing many uses beyond the treatment of scurvy.[7] In Haiti, oranges are still used to treat colds, fevers, liver and gallbladder problems, rheumatism and skin conditions. In Colombia, quaffing copious amounts of lime juice is said to counteract the psychedelic effects of ayahuasca.[8]

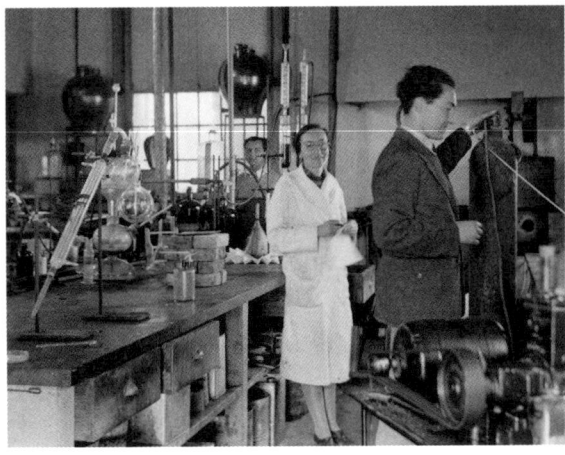

Natural and synthetic essential oils from citrus juice being analysed at a laboratory at the Frutarom Company, Haifa, Israel, 1939, for food and pharmaceutical applications.

MODERN MEDICINE

As early as 1948, it was realized that the antiseptic potency of citrus essential oils was greater than that of the pioneering surgical antiseptic phenol.[9] Some of the constituent compounds active against bacteria, fungi, protozoa and insects are phytoalexins, produced by the plant in response to attack. The insect-repellent qualities mean that citrus deters the mosquitoes carrying malaria as well as removing fire ants, wasps, and even fleas on a cat, but does not harm mammals.[10]

Citrus fruits have important antioxidant qualities that are beneficial in the treatment of diseases such as Alzheimer's, where neuroreceptors are damaged by free radicals. Differences between citrus fruits in their efficacy may be at least partly due to their different parentages. The most efficacious tested is the citrange (*Citrus × insitorum*), a cross between the bitter orange (*Citrus trifoliata*) from northern China and an orange (*Citrus × aurantium*).

Citron extracts have analgesic, antibiotic, antiparasitic and antimicrobial properties.[11] They are also helpful in the management of diabetes and effective in lowering cholesterol. Citrus is the major source of a group of chemical compounds known as coumarins, considered effective against cancer.[12]

FOOD AND DRINK

Aside from these medicinal qualities, citrus fruits are nutritionally low in protein and fat, offering mainly carbohydrates such as sucrose, glucose and fructose. They also provide dietary fibre, helping to lower cholesterol, and are sources of minerals and many B vitamins besides vitamin C; their carotenoids can be converted into vitamin A.[13] Commercially valuable industrial products are the anti-sweetening agents hesperidin, naringin and their by-products, used instead of quinine in some tonic waters for example. Peel oil is used to make carvone, a synthetic spearmint oil for chewing gum,[14] while there is limonene in coke, so that the Coca-Cola Company is one of the biggest users of citrus waste.

In the eighteenth century, a number of citrus-flavoured liqueurs became fashionable in continental Europe as *digestifs* or in cocktails. A sour-orange liqueur was developed by the Bols distillery in Amsterdam after the discovery that aromatic oil could be extracted from the unripe peel of otherwise useless fruits from the Dutch possession of Curaçao in the

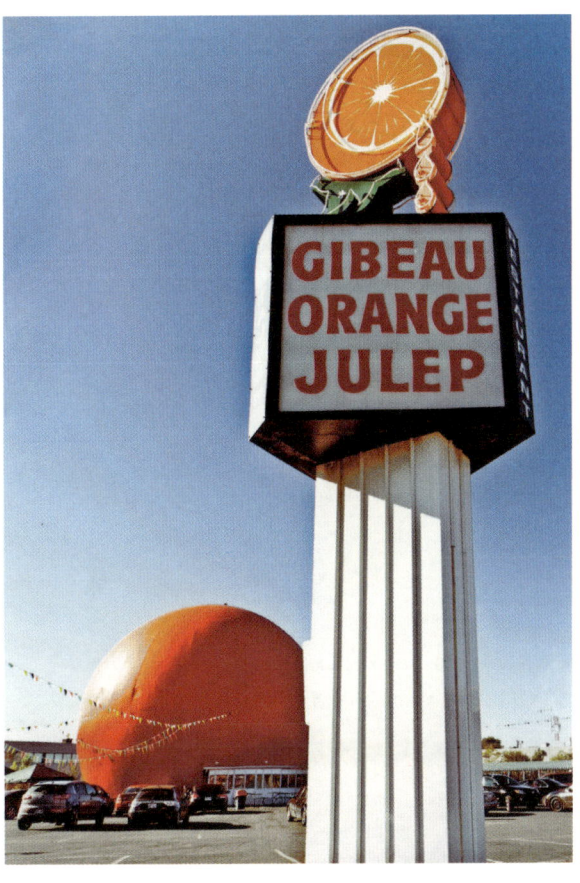

Gibeau Orange Julep restaurant, Montreal, built in 1966. At the three-storey destination, diners can accompany their meal with an orange 'julep' – a variation on a sweet, occasionally alcoholic drink. Such 'citrus architecture is also in the USA and Australia.

Caribbean. The characteristic colour of today's blue curaçao is due to the synthetic dye E133 (Brilliant Blue). The drink became fashionable in the 1960s, likely because it was a component of the popular cocktail Blue Hawaii, which gave the title to one of Elvis Presley's feature films.

French confectioners Jean-Baptiste Combier and his wife Joséphine are credited with the formulation of 'triple sec' to flavour chocolates sold in their sweet shop at Saumur in the Loire, distilling a mash of sun-dried peel of sour oranges from Haiti with sweet 'Valencia' oranges. Cointreau and Grand Marnier survive today as French orange-based alcohol brands; while in Italy lemon-based limoncello and *chinotto*-flavoured Campari are popular.

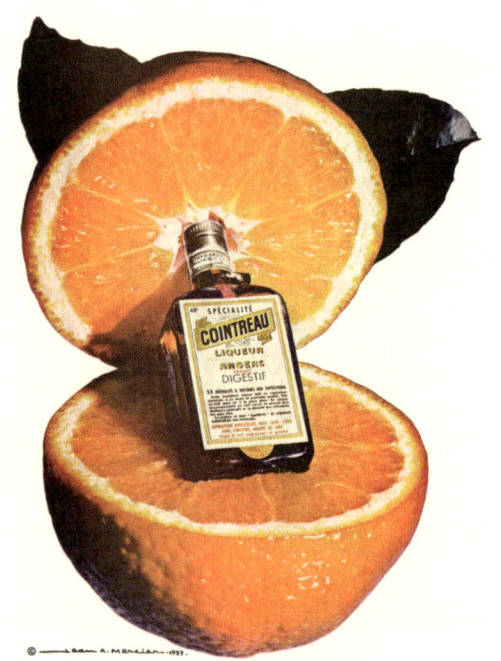

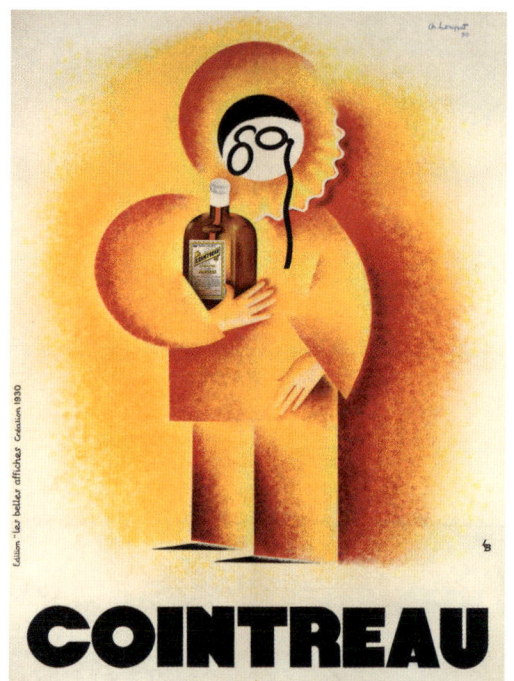

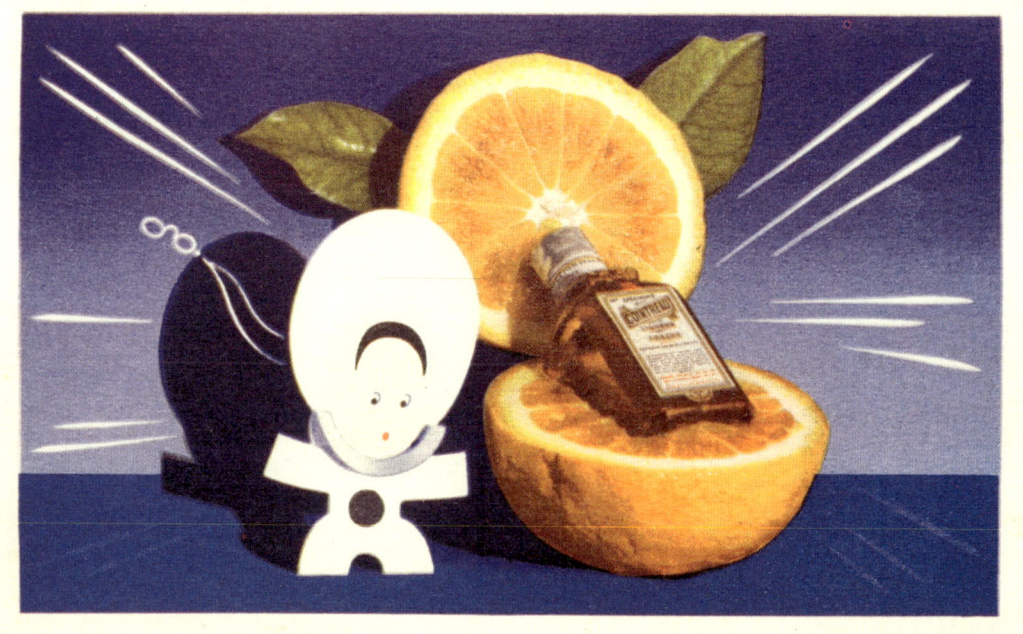

Advertisements for Cointreau liqueur by Charles Loupot, 1930 (top right) and Jean-Adrien Mercier, 1957 (top left), 1950 (above) and 1940 (opposite).

This popular French *digestif*, originally formulated to a secret recipe by brothers Adolphe and Édouard-Jean Cointreau in Saint-Barthélemy-d'Anjou, France, is triple-distilled and blends sweet and sour orange peel with alcohol from fermented sugar-beet. Variations on the popular clown character 'Pierrot Cointreau' featured in advertising for many years. The drink was first sold in 1875 in a four-sided bottle to set it apart from the usual rounded bottle shape, and remains a classic cocktail ingredient.

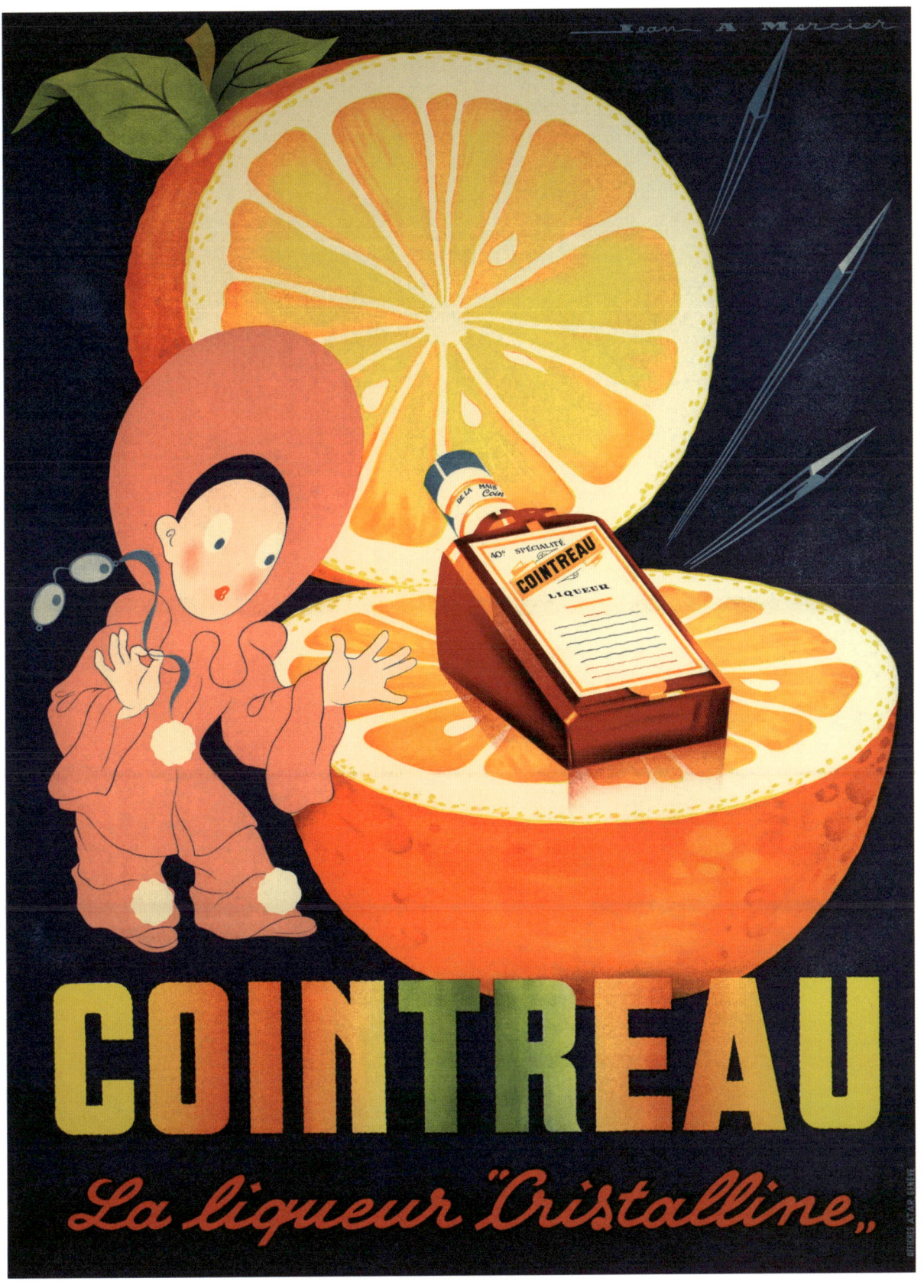

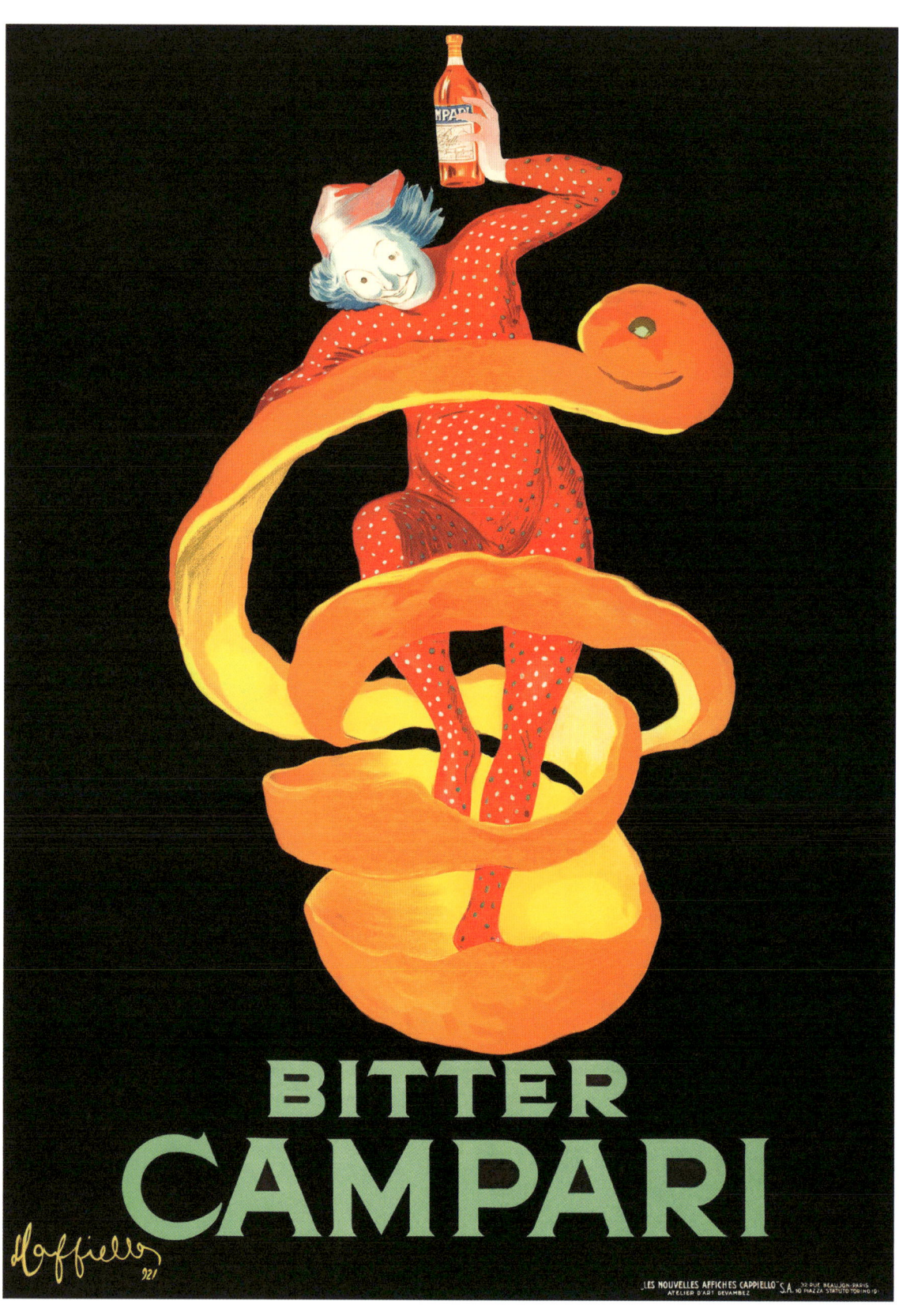

Recent food fashions in Western cooking include the use of yuzu and of Australian 'bush tucker' citrus, especially finger limes, whose vesicles are the 'vegetable caviar' paired with oysters, for instance. As with so many edible plants, fermented citrus products include wines, notably those made since the 1970s from citrus waste (oranges, Key limes, grapefruit, tangerines and tangelos) by Florida Orange Groves Winery in St Petersburg, Florida.

HEALTH AND HYGIENE

Non-dietary uses of citrus were noted early in Europe. The anonymous *Dat Batement van Recepten* (1549), based on an Italian work of about 1525,[15] included a recipe for removing stains from wool and linen using citrus, especially lemon juice. In Britain, lemon juice is still promoted to remove sweat and rust stains from clothes as well as iron stains from concrete, to bleach chopping boards, shine brass and polish hardwood furniture, besides combatting dandruff, relieving prickly heat, and treating vomiting, travel sickness, stings and burns.[16] Increasingly, commercial cleaning products and air-fresheners incorporate citrus oils, not merely to mask but to dissolve airborne odours. Limonene from citrus-peel waste combined with sulphur, a waste product of the petrol industry, may be used to remove toxic metals from soil and water.

Citrus oils are now commonplace ingredients in beauty products including soap, shampoo, suntan products and hand creams, besides finding uses in aromatherapy. The centuries-old association of citrus with health extends to diet, as seen in the use of anti-scorbutic regimes for sailors in the eighteenth century. In the twentieth century, fashion and science combined to make new dietary fads, such as the 'grapefruit diet'; various studies have cited evidence of effective weight management, as well as regulation of insulin and glucose.

MODERN HYBRIDS

The increasing importance of oranges in the nineteenth century led to selections of bud mutants and later to breeding programmes to provide novelty and variation. This was, of course, merely a deliberate

extension of what had been happening 'naturally' for centuries in Asia, where spontaneous mutants like the 'Buddha's Hand' had been selected and where previously geographically isolated species were brought together, leading to hybridization.

Most of the selections and deliberate hybridizations were made in the USA and in the commercially most significant *Citrus × aurantium* (mandarin × pomelo) complex of oranges, tangerines, commercial mandarins and grapefruit. The name 'tangerine' came into use in the mid-nineteenth century for a group of small-fruited cultivars with easily removed peel, initially exported from Tangier, Morocco. These have more features of the original wild mandarin. Some commercial mandarin cultivars are said to have originated in India and were introduced to Florida via Jamaica in the mid-nineteenth century, but it may be the same as the *ponki* from China, long used as a rootstock in Japan.[17] Also in this group are the readily peeled cultivars called satsumas, introduced to the West around 1878 from the Satsuma area (now Kagoshima Prefecture) of southern Japan.

Spontaneous crosses between commercial mandarins and sweet oranges are known as tangors. One such is the King orange, first reported by Europeans from Hue in Vietnam in 1791, though no doubt introduced from Chinese gardens.[18] It was exported to the USA as 'King of Siam' (now known as 'King') around 1882 and is still much grown in Vietnam. Others include the 'Temple' tangor, noticed in Jamaica in 1896,[19] and the ortanique ('or' from orange, 'tan' from tangerine and 'ique' from unique), also found in Jamaica *c.* 1920. The popular 'Clementine'[20] is yet another tangor, named after Brother Marie-Clément Rodier (1839–1904), a missionary who ran an orphanage near Oran, Algeria, where he selected the fruit in 1902.

WALTER SWINGLE

In 1909, the clementine was introduced to the USA by the nation's greatest ever citrologist, Walter Tennyson Swingle (1871–1952).[21] Born on a farm, Swingle had little formal schooling, but attended classes at what is now Kansas State University, where he met David Fairchild (1869–1954), son of the head of the institution.[22] At seventeen Swingle was Assistant Botanist in the school's experimental station; he trained as a mycologist and published twenty-seven scientific papers by the time he was twenty. On the recommendation of Fairchild, he was appointed to the newly established United States Department of Agriculture (USDA) in Washington DC in 1891,

Advertising poster for Campari, Leonetto Cappiello, 1921. Gaspare Campari (1828–1882) created his eponymous beverage, flavoured with *chinotto*, in Novara, north-west Italy, in 1860.

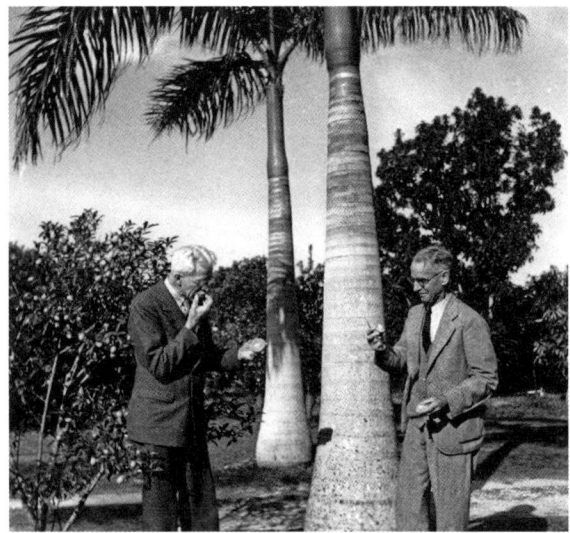

working on the diseases of citrus in Florida. He introduced the pollinating fig wasp from Algeria to the US, thereby allowing Smyrna figs to be grown there, and trained himself as a geneticist.[23] As a 'scout', he helped the Library of Congress acquire over 100,000 Chinese books on plants. Perhaps most significantly, he was responsible for making more deliberate citrus hybrids than anyone before him.[24] By 1904, some 1,780 were established at the USDA's subtropical garden in Miami.[25]

Although citrus-grafting had long been practised in China (see p. 29), citrus trees in the West were usually grown as seedlings until the mid-1800s. Grafting became much more common only after phytophthora root rot spread (it was first recorded in the Azores in 1842, with sour orange being somewhat resistant). Seedling rootstocks chosen for taking grafts were bitter orange (*Citrus trifoliata*, very widely used today), mandarins, sour orange, sweet orange and rough lemon (*Citrus* × *otaitensis*, discarded by the 1960s as not very resistant to phytophthora). It was Swingle who bred the citrange rootstocks (crosses between the bitter orange and the orange, *Citrus* × *insitorum*, meaning 'of the grafters'), which are resistant to phytophthora, tolerant of citrus tristeza virus and have some resistance to parasitic worms. His 'Troyer' cultivar – named after Albert Melville Troyer (b. 1866), a satsuma-grower of Fairhope, Alabama, who first fruited it – is now the most widely used rootstock worldwide. Troyer travelled to the Black Sea coast in 1934, invited by the Soviets to help establish satsuma plantations there, and soon became a Soviet citizen, renouncing his American citizenship. Shortly thereafter, he was accused of sabotage and sentenced

to ten years imprisonment in a forced labour camp, where he no doubt died.[26]

Swingle's team also made the first deliberate tangerine–grapefruit hybrids, which he called tangelos, in Eustis, Florida, in 1897. The first marketed, from about 1915, were 'Sampson' and 'Thornton'[27] but they were too puffy and thin-skinned for successful packing and shipment, so were used in 'orange' juice instead. Swingle's 1908–12 crosses between 'Bowen' grapefruit and 'Dancy' tangerine yielded 'Seminole' and 'Minneola', with better shipping qualities and released in the 1930s. However, Swingle himself had to admit that 'tangelo-like' fruits were not uncommon in Japan long before this[28] – and indeed spontaneously derived citranges were growing in the Yangtze Valley early last century.

Swingle synthesized tangors, including 'Murcott' (especially important today), raised in the USDA citrus-breeding programme in Florida and now sold as 'honey tangerine'. Other important tangors include 'Shiranui', developed by the Japanese Department of Agriculture in 1972 and now grown in Brazil as well as the Far East.[29] It is also known as 'Sumo citrus' in the USA and *hallabong* in South Korea, because the base of the fruit has a broad 'neck', and so resembles Mount Halla (Hallasan) on Jeju Island. Swingle also used kumquats in his breeding programmes: hybrids with West Indian lime gave the limequat (*Citrus* × *floridana*), first made in 1909, while crosses with citranges gave citrangequats (*Citrus* × *georgiana*). Citranges crossed with calamondin (itself a hybrid of the kumquat and mandarin) gave rise to the citrangedins, e.g. 'Glen'.

Swingle understood the importance of genetic diversity and the dangers of monoculture when most of the citrus industry was (and still is) genetically very limited, most of the crop comprising a very small number of clones covering a very narrow genetic range. He was an early advocate of establishing permanent living collections of crop wild relatives so as to ensure there was a reliable source of variable germplasm for breeding new cultivars in the face of pests and disease.

ABOVE: T. Ralph Robinson and Walter T. Swingle, Florida, 1930s. The two men worked closely together at the Office of Horticultural Disease and Crops, USDA, creating tens of thousands of controlled crosses. **OPPOSITE:** Swingle's team created the tangelo, a new type of orange-like citrus fruit made by hybridizing tangerines and grapefruit. Other creations included the citrange and limequat.

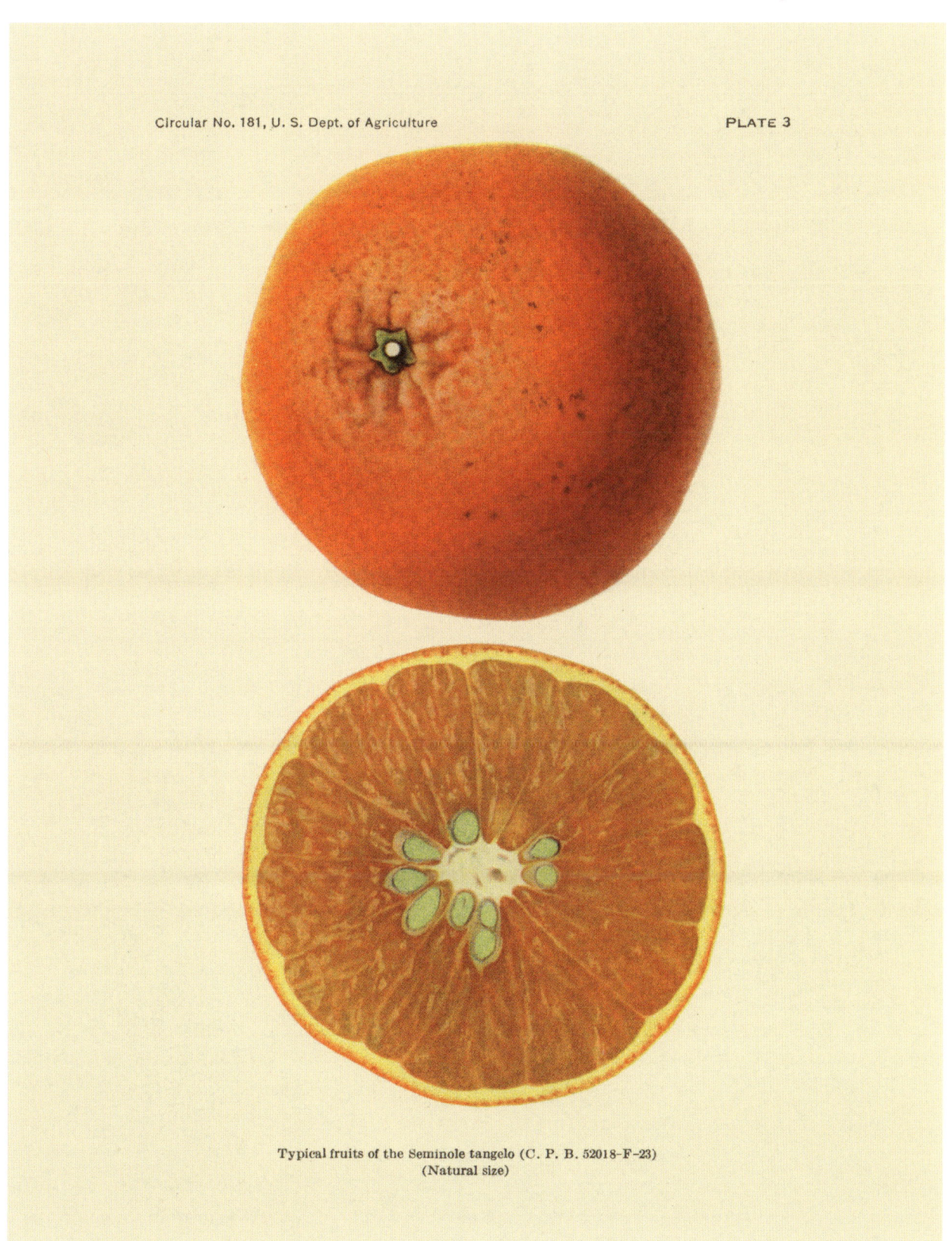

Typical fruits of the Seminole tangelo (C. P. B. 52018-F-23)
(Natural size)

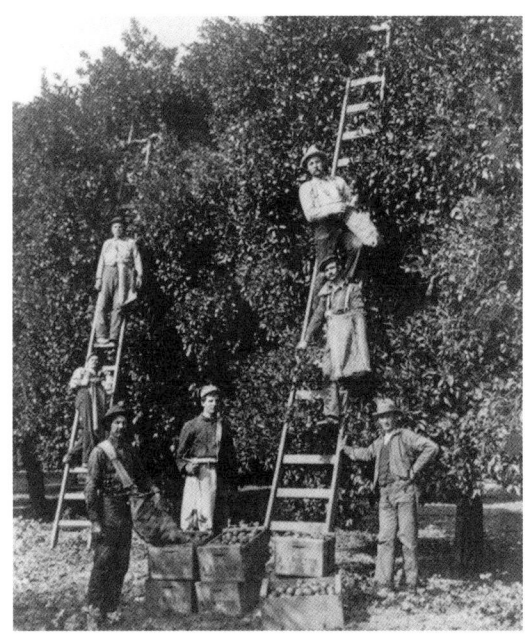

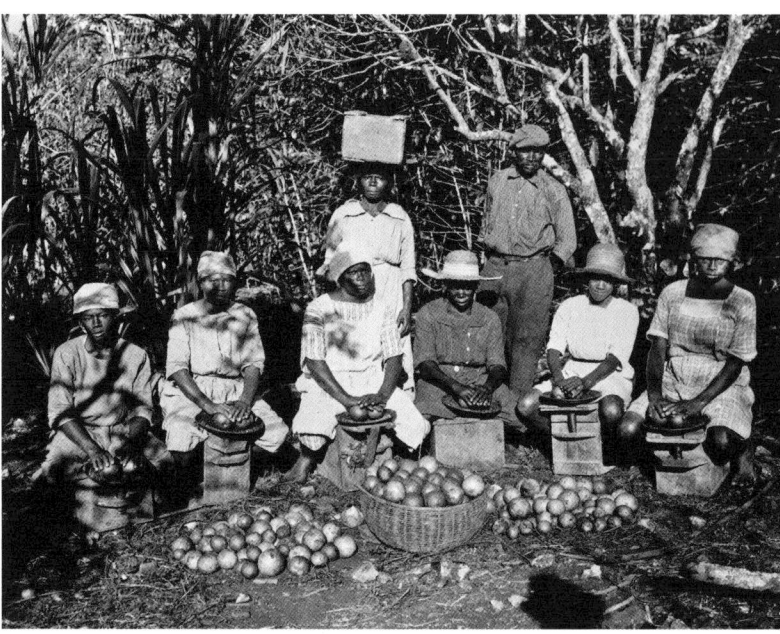

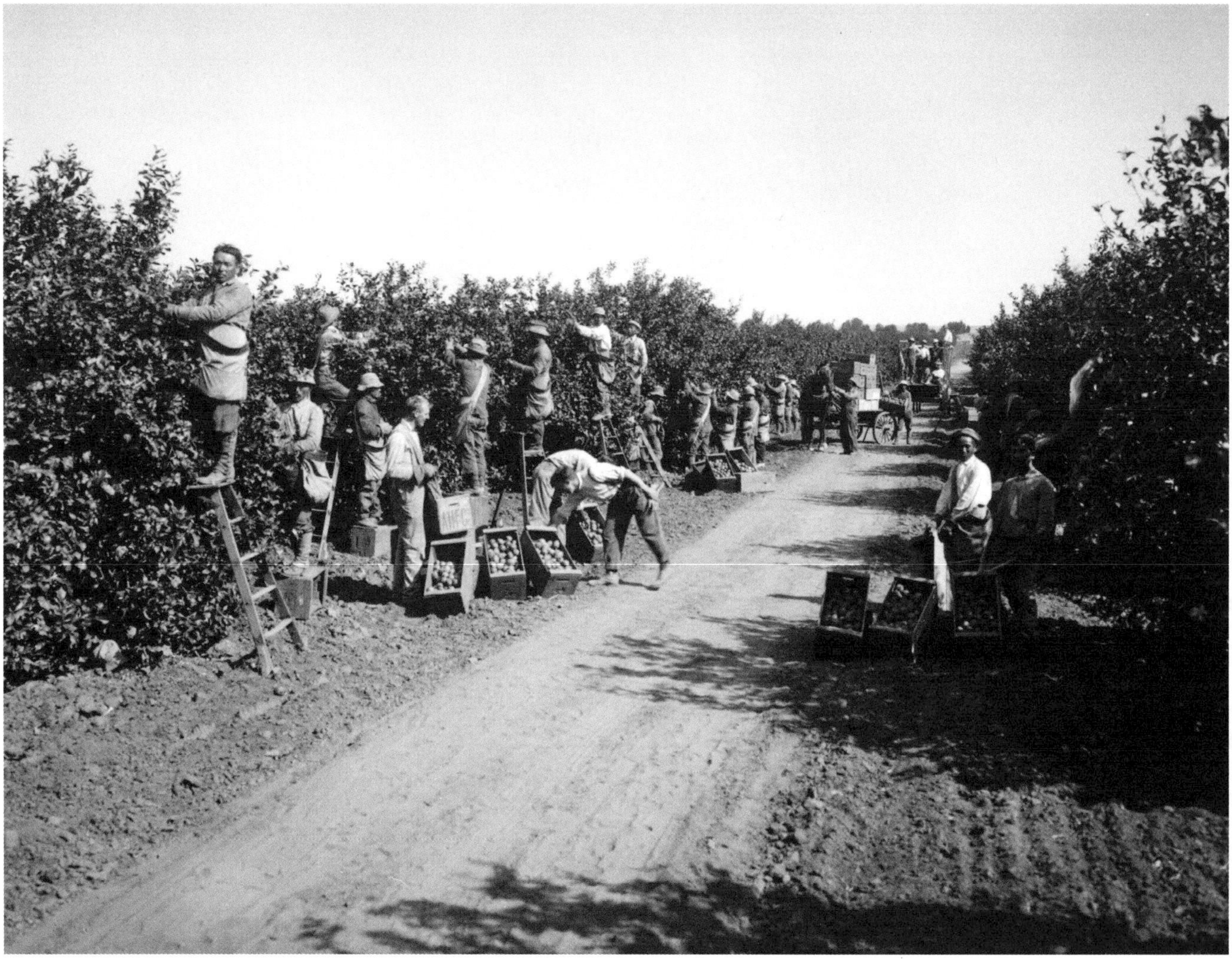

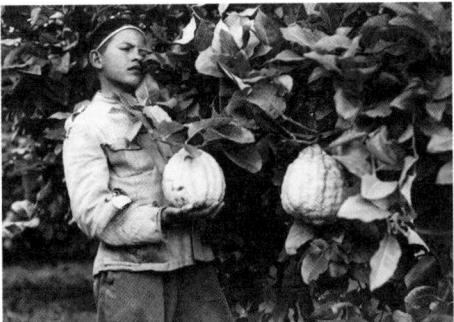

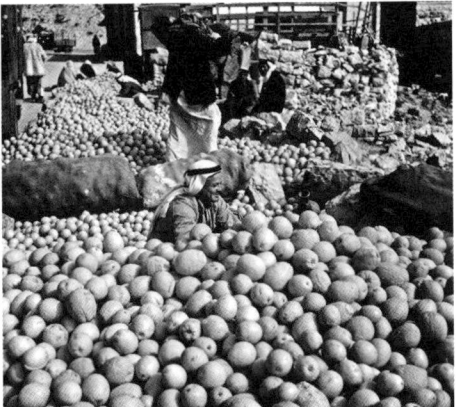

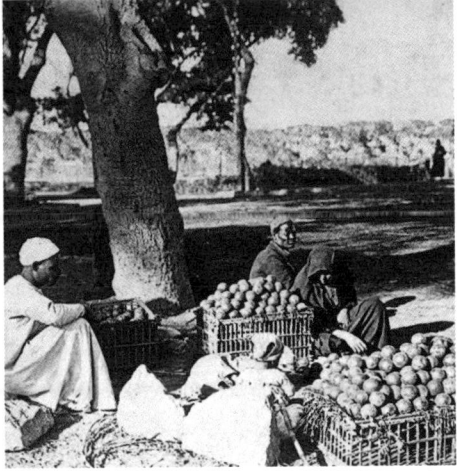

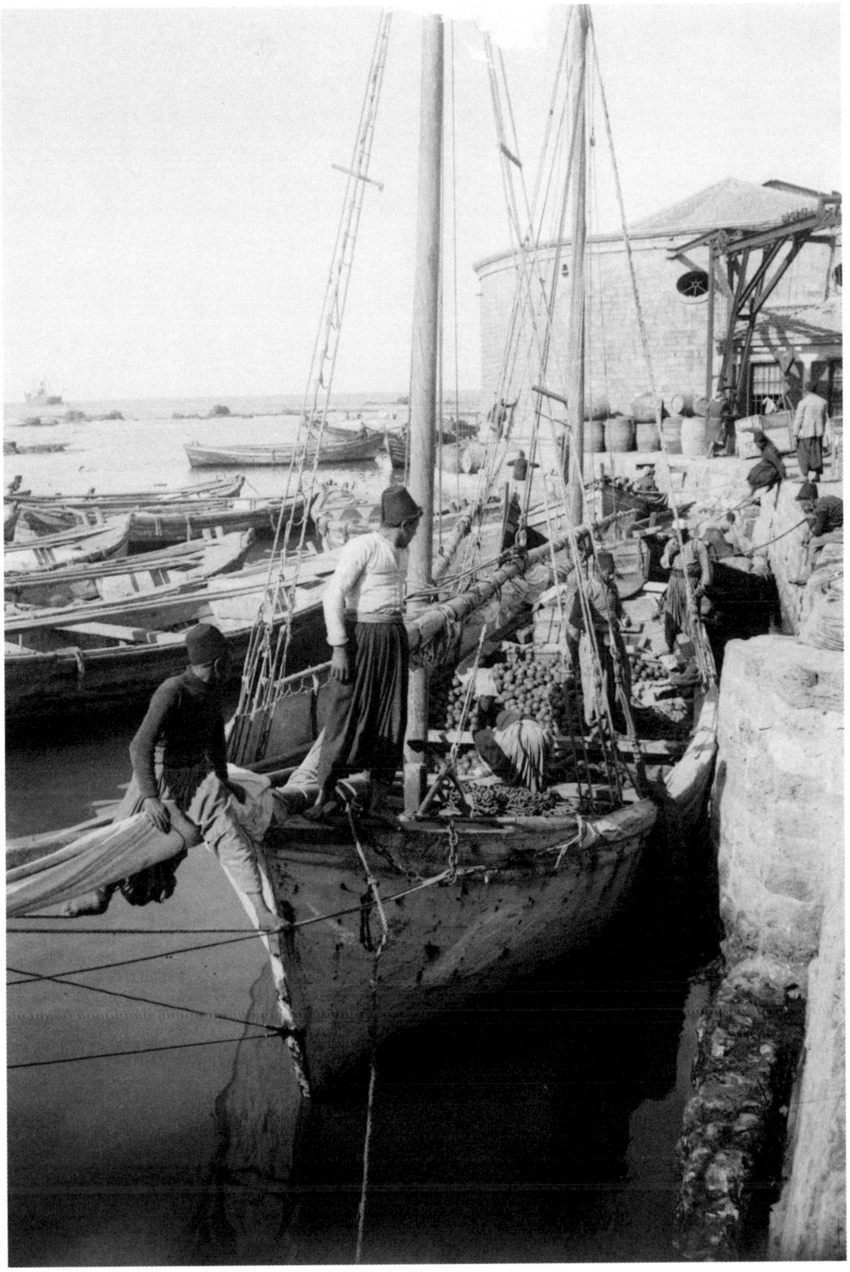

OPPOSITE, CLOCKWISE FROM TOP LEFT: Picking and packing oranges in the orchards near Los Angeles, California, 1905; orange rinders extract orange oil to be sold for flavouring, Jamaica, c. 1925; workers harvesting oranges in groves, near Riverside, California, n.d.

ABOVE, CLOCKWISE FROM TOP LEFT: Boy with giant citron, Libya, 1935; coasting boat being loaded with oranges, Jaffa, c. 1900; Cairo orange sellers, Egypt, c. 1896; orange market, Palestine, 1940.

The global citrus industry boomed in the late 19th and early 20th centuries; fresh orange production is still significant in several countries including Egypt and the western USA. Jaffa was Palestine's cultural and commercial centre; its orange groves – famed for 'Shamouti' (thick-skinned, seedless; also known as Jaffa oranges, selected around 1844) and 'Baladi' (thinner-skinned, seeded) – once produced up to 1.5 million crates a year (each crate with 150 oranges). After the formation of the state of Israel in 1948, however, many groves were cleared, because the trees used precious water and they were a reminder of the Palestinian society then disappearing.

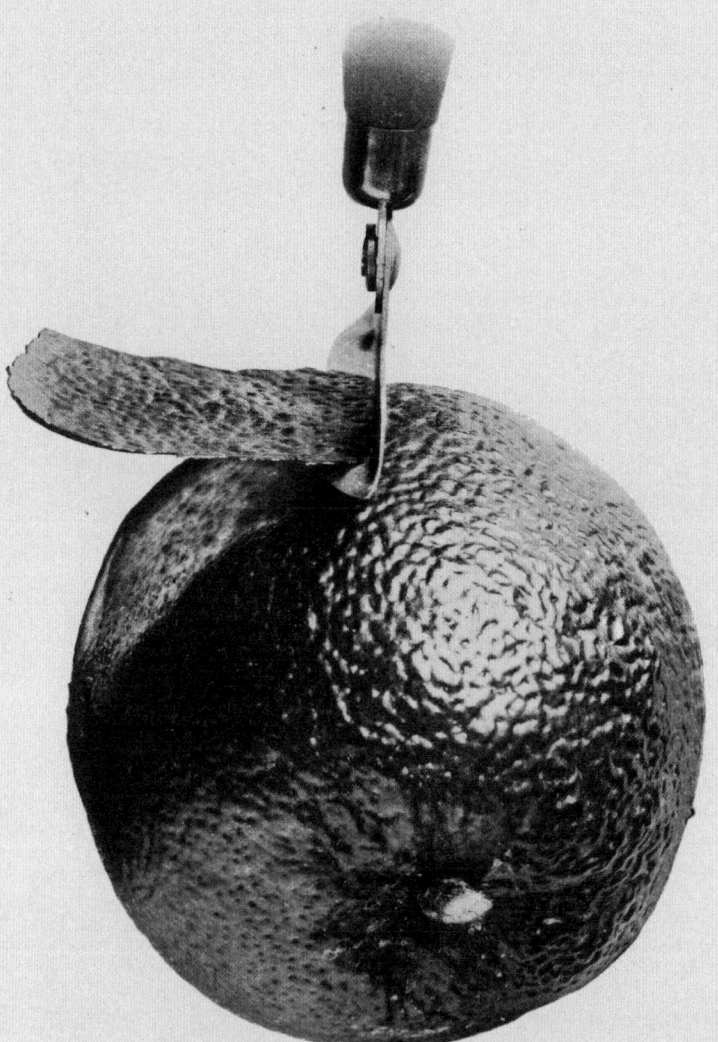

MORTON CITRANGE, NATURAL SIZE.

A remarkable hybrid between the Japanese orange, hardy as far north as Washington, and the ordinary orange. To avoid the strong oil in its glands, the skin should be peeled before cutting the fruit. (Plate 1.)

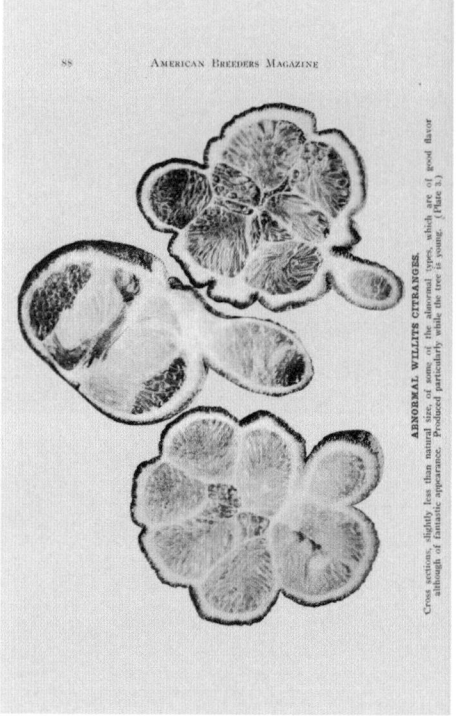

ABNORMAL WILLITS CITRANGES.

Cross sections, slightly less than natural size, of some of the abnormal types, which are of good flavor although of fantastic appearance. Produced particularly while the tree is young. (Plate 3.)

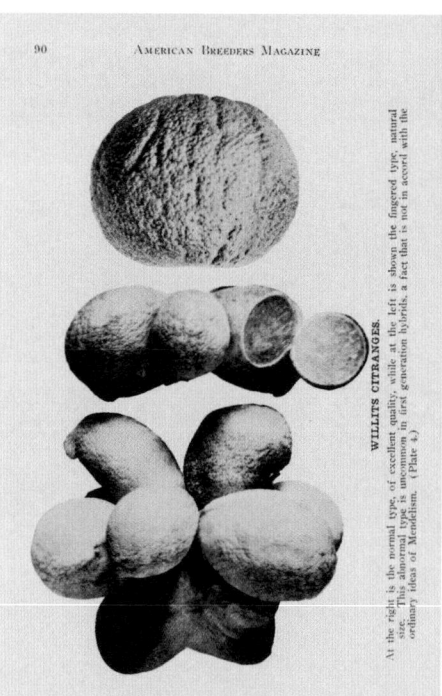

WILLITS CITRANGES.

At the right is the normal type, of excellent quality, while at the left is shown the fingered type, natural size. This abnormal type is uncommon in first generation hybrids, a fact that is not in accord with the ordinary ideas of Mendelism. (Plate 4.)

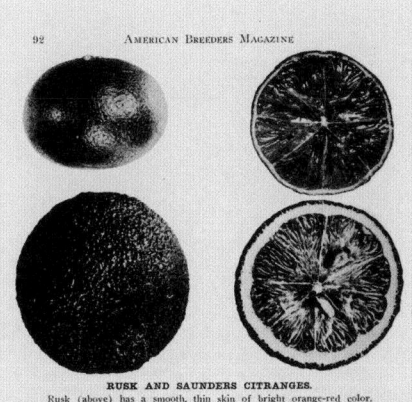

RUSK AND SAUNDERS CITRANGES.
Rusk (above) has a smooth, thin skin of bright orange-red color, while the Saunders (below) has a thick skin and very large, prominent oil glands, making it a good keeper. Both natural size. (Plate 5.)

Citrus, these African cherry oranges have compound leaves composed of from three to seven very large leaflets. It is not uncommon for a single compound leaf of an African cherry orange to have a surface ten times as large as that of a common orange leaf. When we reflect that the sugar that sweetens the fruit and the aromatic substances that give it flavor and perfume are formed in the leaves we realize how important it is to secure large-leaved forms of our cultivated plants.

Through a new system of grafting it has been possible to force some of these African cherry oranges to flower when they were less than two years old. In this way it has been possible to make a few

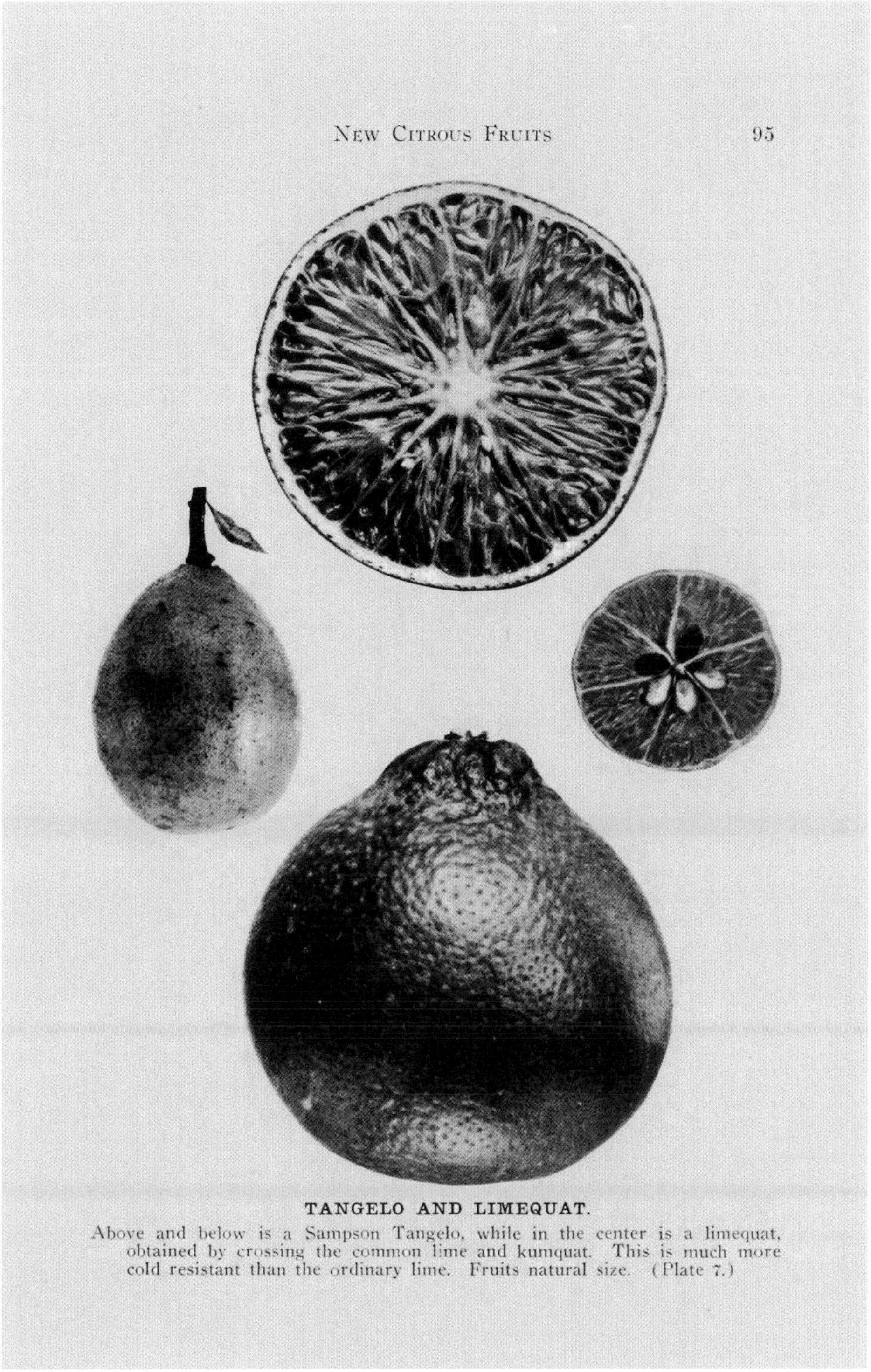

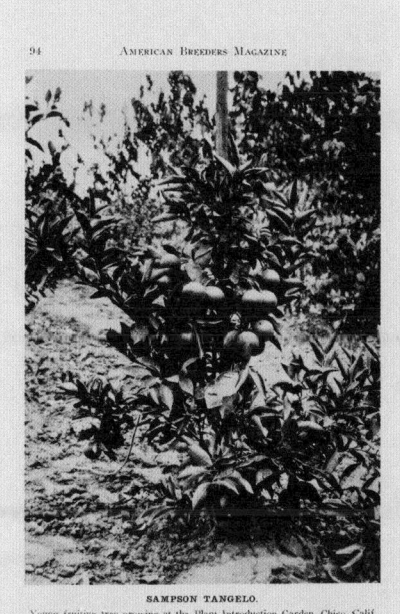

SAMPSON TANGELO.
Young fruiting tree growing at the Plant Introduction Garden, Chico, Calif. This hybrid is now cultivated commercially, and trees often fruit abundantly while still very young. (Plate 6.)

TANGELO AND LIMEQUAT.
Above and below is a Sampson Tangelo, while in the center is a limequat, obtained by crossing the common lime and kumquat. This is much more cold resistant than the ordinary lime. Fruits natural size. (Plate 7.)

Pages from Walter Swingle's article 'New Citrous Fruits: Successful Hybrids – The Citrange, Tangelo and Limequat – Cold-Resistant Substitutes for the Lemon and Lime – Future Possibilities', *American Breeders Magazine*, 1913.

Citranges (*Citrus ×insitorum*) are crosses between a bitter orange (*Citrus trifoliata*) and sweet oranges. The very hardy 'Morton' (1897) commemorates Julius Sterling Morton (1832–1902), Secretary of Agriculture; it is little grown nowadays. 'Willits' (before 1895) commemorates Edwin Willits (1830–1896), first Assistant Secretary of Agriculture; in 1909 Swingle crossed it with a kumquat, giving the citrangequat (*Citrus ×georgiana*), which therefore has contributions from four wild *Citrus* species in its genome.

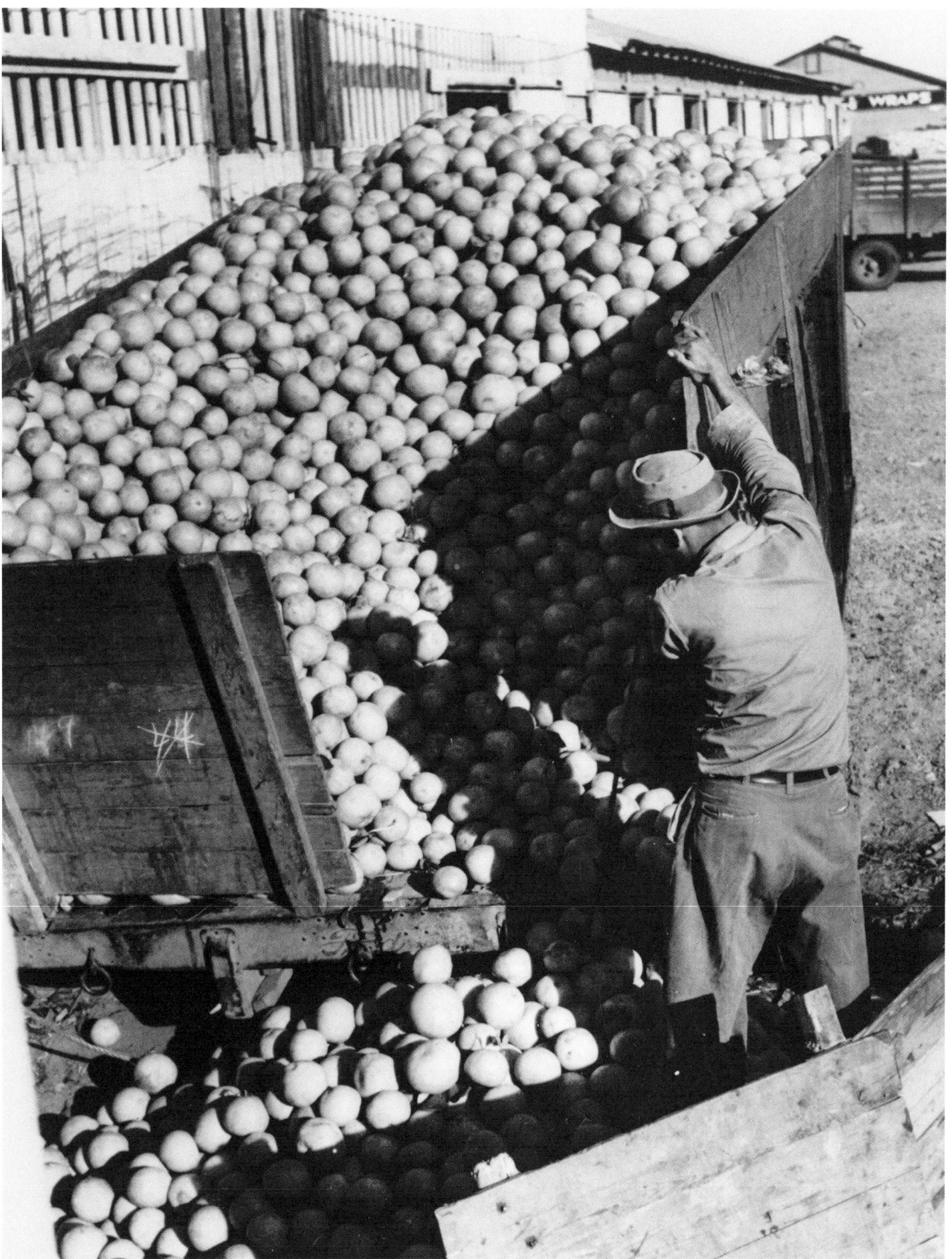

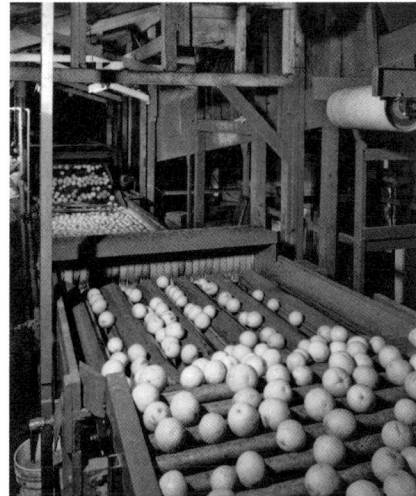

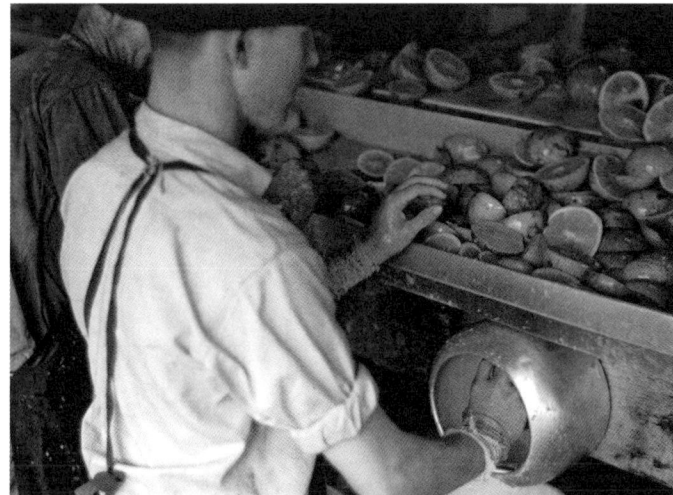

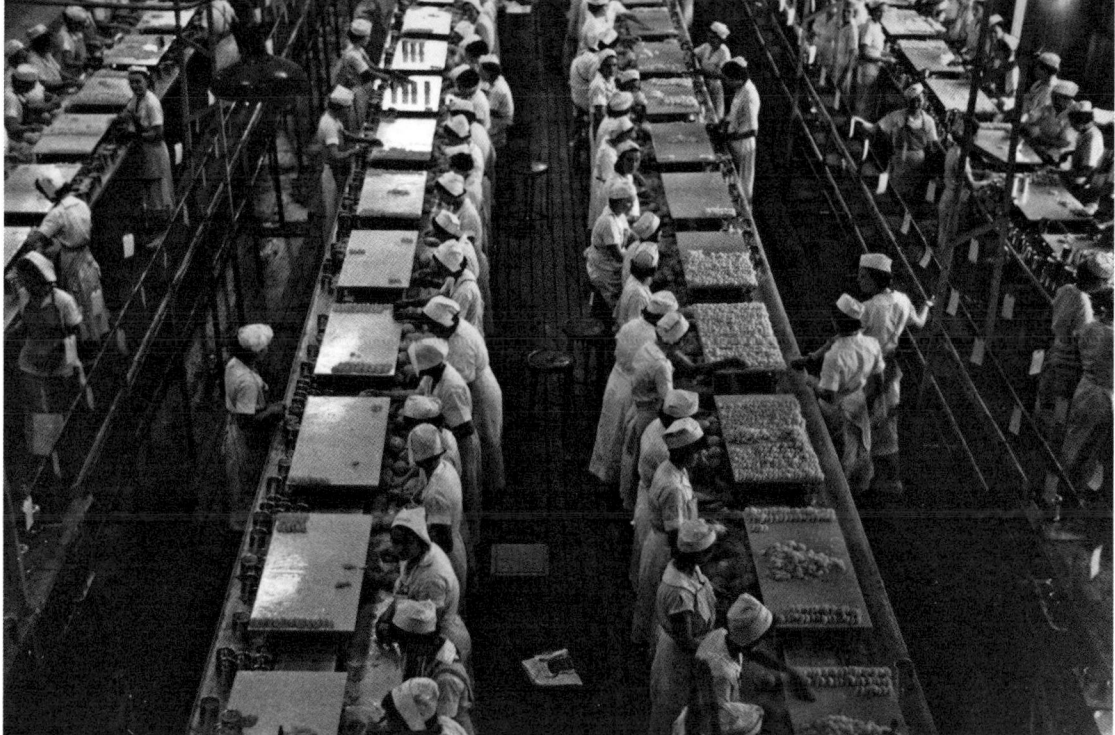

OPPOSITE: Unloading a grapefruit truck at a juice plant, Weslaco, Texas, Arthur Rothstein, 1942. **TOP LEFT:** National Orange Company Packing House, Riverside, California, Brian Grogan, 1968. **TOP RIGHT AND ABOVE:** At a grapefruit canning plant, Winter Haven, Florida, Arthur Rothstein, 1937.

Arthur Rothstein (1915–1985) was one of several distinguished photographers commissioned by the Farm Security Administration (1937–42) and the US Office of War Information (1942–44), originally to document the lives of the rural poor and later to record the lives of migratory agricultural workers in the western and midwestern states of the US.

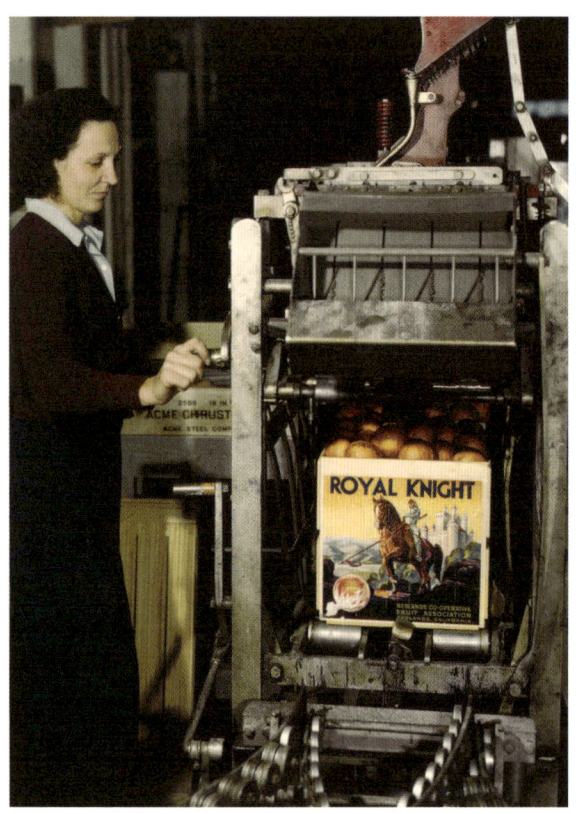

Operating a machine for putting lids on crates at a co-op orange packing plant, Redlands, California, Jack Delano, 1943.

Swingle established labs and collections in Florida, but in 1939 USDA citrus work was moved to Orlando. Swingle's trees were replanted there but most subsequently died. He retired but was given facilities at the University of Miami, where he wrote 'The Botany of Citrus and its Relatives of the Orange Subfamily', a fundamental scientific text, in many senses still unsurpassed, published in 1943 in Herbert Webber (1865–1946) and Leon Batchelor's (1884–1958) monumental *The Citrus Industry*.

Swingle's account was the first modern synthesis of the citrus group, and to frame it he had to tackle the problem of classification, which had become very confused, largely because the hybrid origins of almost all the cultivated citrus were not understood and because some authors insisted on giving species names to the hybrids, apomictic clones and bud sports. Indeed, the matter is only now being resolved[30] and, in retrospect, it is remarkable to read what Linnaeus had written about the two species

he recognized in 1753, noting of the citron (*Citrus medica*), 'The varieties of this [including the lemon], which are very numerous at the present day, are mixed together with the following [the orange]'.[31] With his and other taxonomic botanists' general disdain for working on cultivated, as opposed to 'wild', plants, Daniel Oliver (1830–1916), later Keeper of the Herbarium at the Royal Botanic Gardens, Kew, wrote, 'Feeling it utterly hopeless usefully to define [the species of *Citrus*], I have thought it best to leave the Oranges, Lemons, Limes and their allies as I found them.'[32] By Swingle's time, some botanists recognized up to 256 'species', others just a few – or even merely a very variable single one.

The chief culprit was the Japanese Chozaburo Tanaka (1885–1976), whose intimate knowledge of cultivated citrus had led him to recognize as species what we would now call cultivars. Very diplomatically[33] – even as rival systematists were from countries at war – Swingle, partly using as cover the publications of the little-known Russian botanist Arkadii Efimovich Kozhin (1884–1957), who worked in the Institute of Plant Breeding in Leningrad (now Saint Petersburg),[34] drastically reduced the numbers, but not to the extent it has now been possible to do as a result of DNA and other studies. On the other hand, many of the new genera Swingle rather optimistically described, notably *Fortunella* (kumquats), *Eremocitrus* and *Microcitrus* (Australian 'limes'), have now been taken back into *Citrus*. Webber prepared an invaluable account of citrus cultivars for *The Citrus Industry* and advocated the use of 'cultivar groups' rather than formal botanical names – the approach used today.[35]

BREEDING AND SELECTION

In his breeding programme, Swingle selected for hardiness and peelability, among other traits. These and the vogue for different peel colour and reduced thorniness are still being pursued, as is the dwarfness favoured by the pot-plant trade. The ornamental citrus industry took off in Italy in the 1960s, though a dwarf cultivar of rough lemon (*Citrus × otaitensis*), the winter-fruiting 'Otaheite', was well known in the USA in the 1940s and the Hong Kong kumquat (*Citrus japonica* 'Chintou'), known as *kinkan* in Japan, was grown there then. Modern ornamental citrus grown as houseplants include 'Buddha's Hand' (cultivars in the *Citrus medica* Fingered Group – see p. 39),[36] besides hybrids of those, 'Meyer' lemons, kumquats and the *chinotto*.[37] The current most sought-after trait is a reduced number of seeds, as well as the more familiar ones leading to disease resistance, improved yields, fruit quality and long shelf-life.[38] Manipulation of the chromosome complement has led to seedless mandarins, lemons, limes, pomelos and grapefruit. Gamma radiation has also been used, as in the development of the seedless lemon 'Ayelet' derived from 'Villa Franca'. The almost-seedless and red-fleshed 'Star Ruby' grapefruit was generated in Texas between 1959 and 1963 – the first citrus cultivar to be patented in the US[39] – using irradiated seeds and buds from 'Ruby Red' (a name used for natural mutants of 'Thompson Pink', first recorded in 1929) and released in the 1970s.[40]

Seedless cultivars are preferred by consumers even if flavour is compromised, as is the case with commonly grown seedless 'Marsh' grapefruit, a chance seedling arising in Florida c. 1860, by comparison with the seeded 'Duncan'. So significant is seedlessness that in 2007 Californian citrus-growers lobbied the State Legislature for a Seedless Mandarin Protection Act in an attempt to establish 'no-fly zones' around seedless groves, because seeded fruits were produced after cross-pollination by bees, thereby making the crop a quarter to a third of the value of seedless ones.[41]

ORANGE JUICE

After the 1918–19 Spanish Flu pandemic,[42] milk or juice was recommended for disease prevention, providing a major commercial opportunity for orange juice. The aggressive promotion of orange juice led to the ubiquitous OJ of today's American breakfast. In 1945, orange-juice powder was developed, and in 1954 an evaporator was perfected to remove water from the juice, though concentrate loses most of the flavour. This drawback was overcome by Louis Gardner McDowell (1912–1986) and others who proposed over-concentrating the juice and then adding 'cutback', which is mainly fresh orange juice but also includes peel oil, pulp and so forth.[43] By the 1960s, a single Florida plant run by Minute Maid used eight million oranges a day in juice production.

The use of pasteurization led to the success of the mighty Tropicana brand.[44] Italian-born American Anthony Rossi (1900–1993) founded Tropicana in the sunshine state of Florida and was 'the first to bring fresh juice to the masses',[45] and by 1957 he was shipping 5.7 million litres (1.5 million gallons) from Florida to New York every week. After switching to transport by rail,[46] as early as 1971 the company was running two trains a week, each carrying around a million gallons of juice to New York. In 1993, Tropicana used 20 per cent of all Florida's oranges.

Synthetic juice is increasingly taking the market.[47] In 2000–1 Americans drank more than 19 litres (5 gallons) of orange juice per head per annum, the peak recorded consumption; by 2021–22 this had fallen to less than 7.5 litres (2 gallons) each.[48] In that year the USA produced 190,000 metric tonnes of orange juice, but this was far outstripped by Brazil at 1.14 million.[49]

Citrus in the USA seem to be declining in favour generally, because, despite the per capita consumption of fresh berries (strawberries, blueberries and cranberries) doubling from 1990 to 2007, and that of pineapples and papayas increasing by more than five times, sales of fresh citrus declined from 12 kg (26.6 lb) per person in 1998 to 9.3 kg (20.6 lb) in 2008.[50] Oranges, the historically prominent choice, showed a continuous decline in per capita consumption from 5.6 kg (12.4 lb) in 1990 to 4.5 kg (9.9 lb) in 2008 – though tangerines increased from 0.6 to 1.4 kg (1.4 to 3.1 lb).[51]

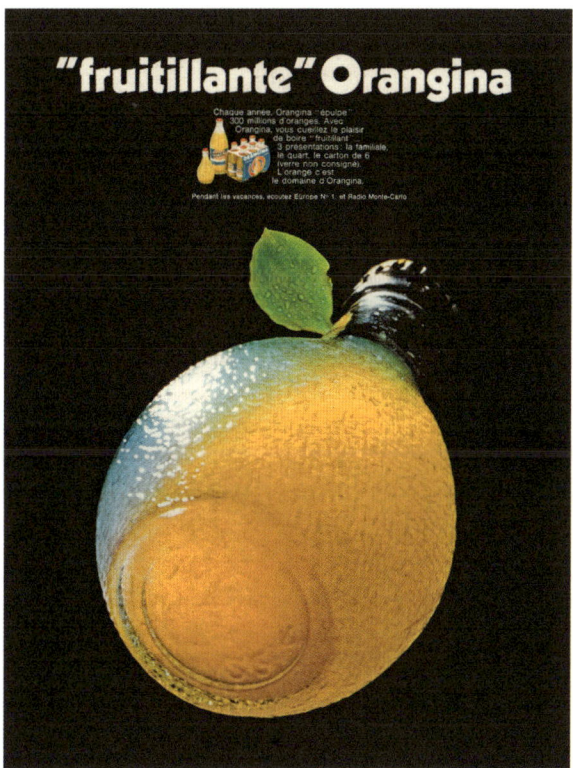

Advertisement for Orangina, 1960s. 'Every year, Orangina pulps 300 million oranges.' The soft drink, a mixture of blended concentrated citrus juice, sugar and carbonated water, first appeared in France in 1936. From 1951 it was sold in a distinctive rounded bottle with a surface textured like citrus peel.

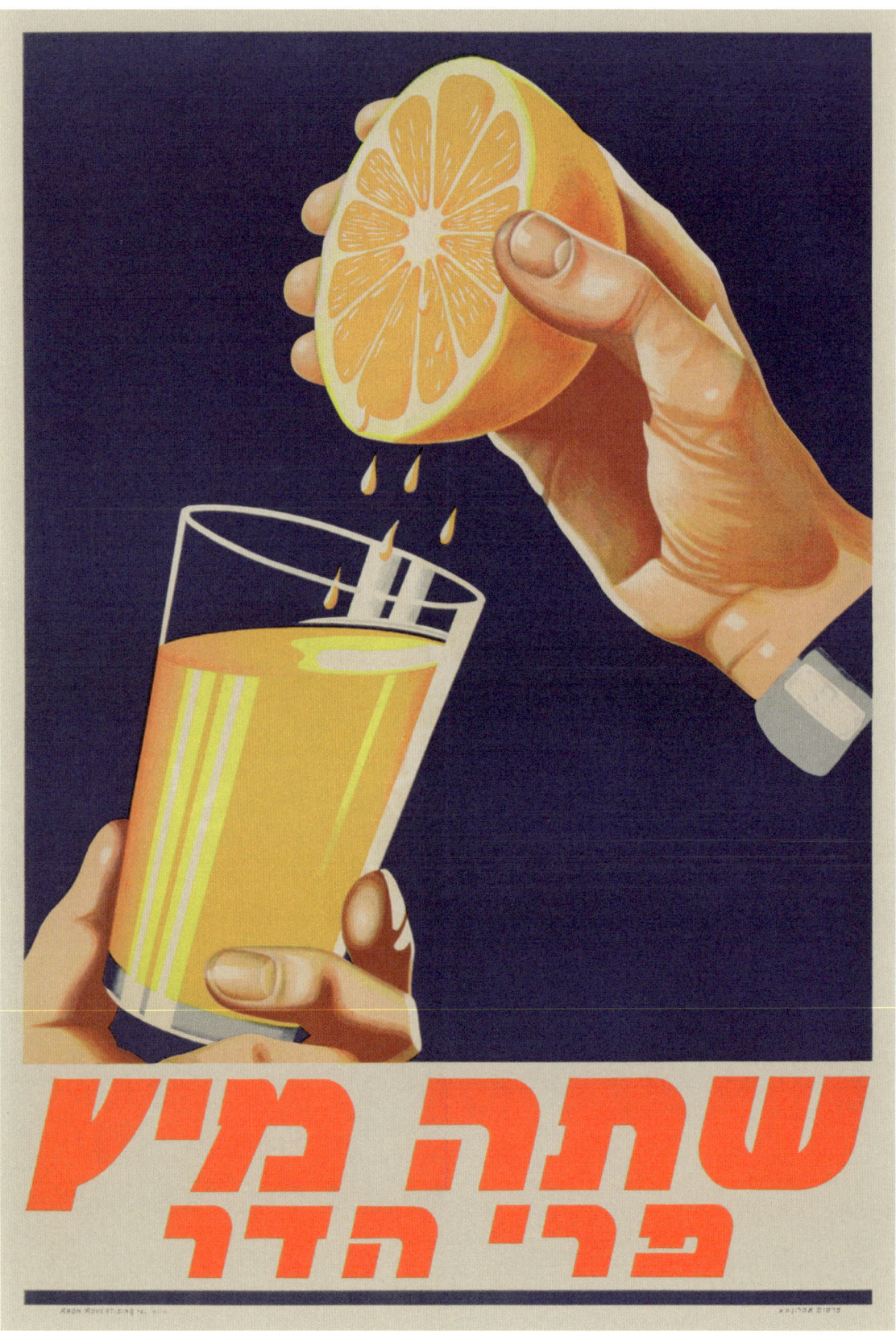

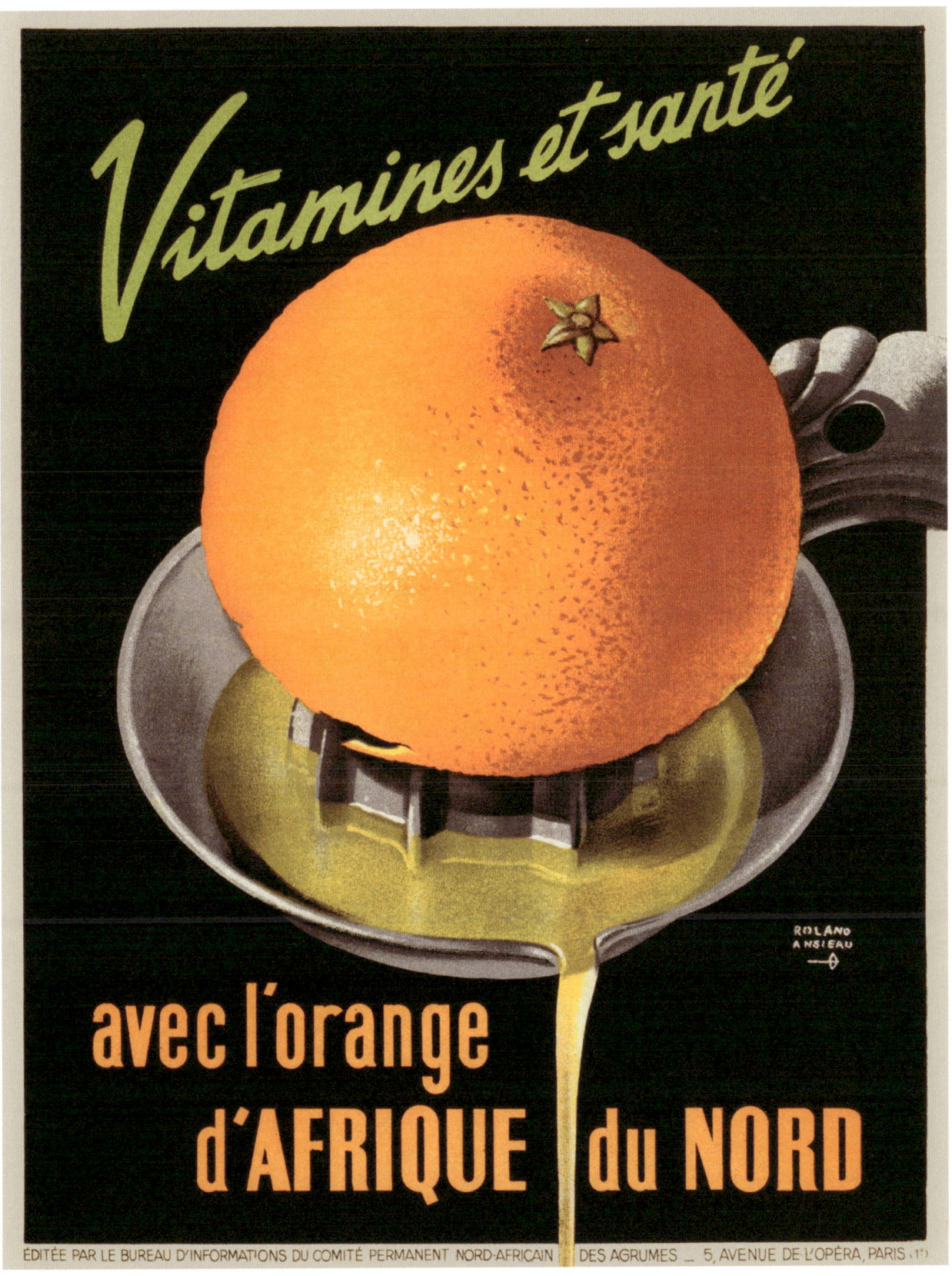

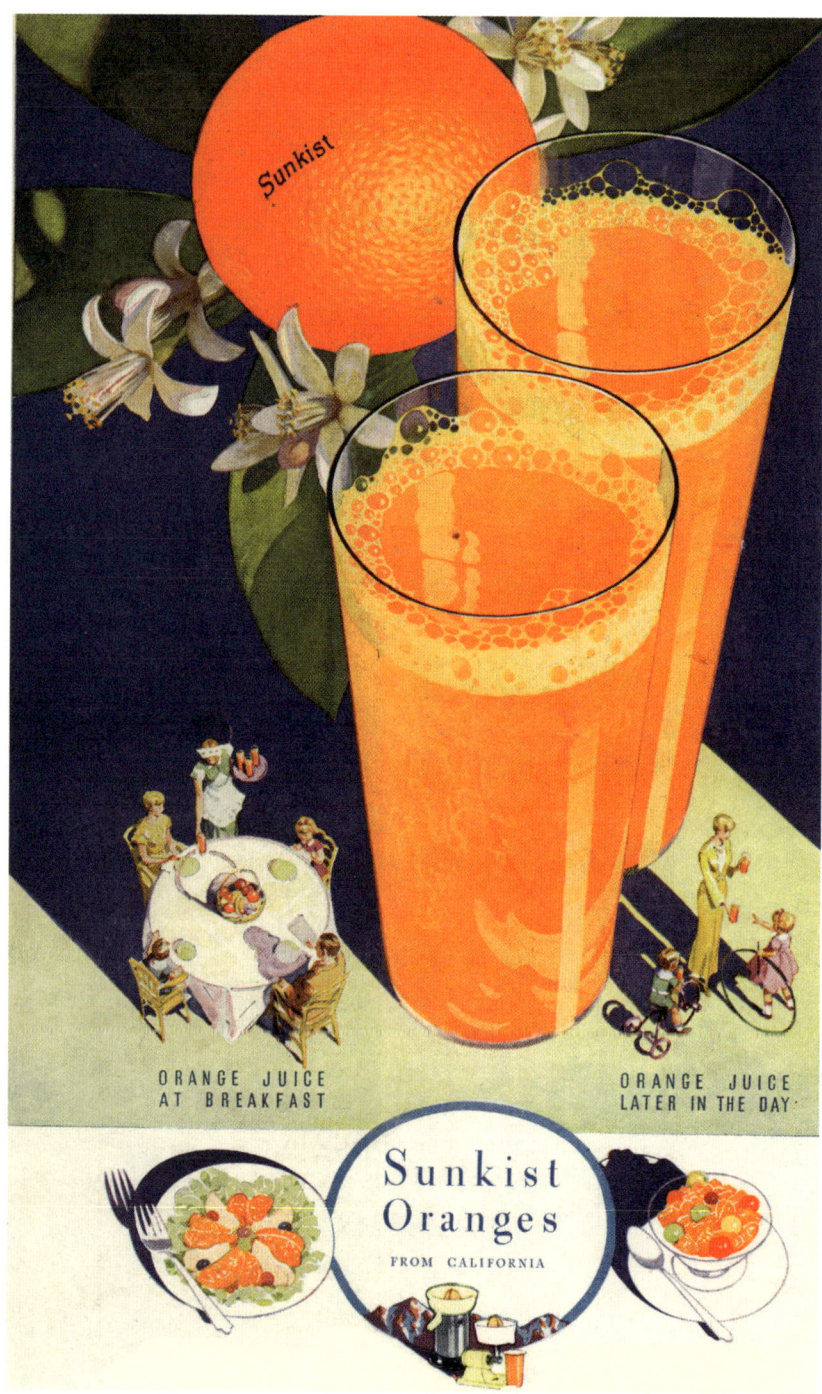

PREVIOUS PAGES: 'Drink citrus juice', Franz Krausz, *c.* 1947 (left); 'Vitamins and health', Roland Ansieau, *c.* 1950 (right). **ABOVE:** Sunkist orange juice, 1936. **OPPOSITE, CLOCKWISE FROM TOP LEFT:** 'Drink Belorange', Toni Bernat, *c.* 1949; 'Bali – as much vitamin C as from freshly squeezed orange juice', Bernard Villemot, 1975; 'Orangina – with orange pulp', Bernard Villemot, 1965; 'L'Orade luxury orangeade', *c.* 1950.

Posters advertising orange juice all over the world in the mid-20th century evoked sunshine, luxury, leisure, health and fun.

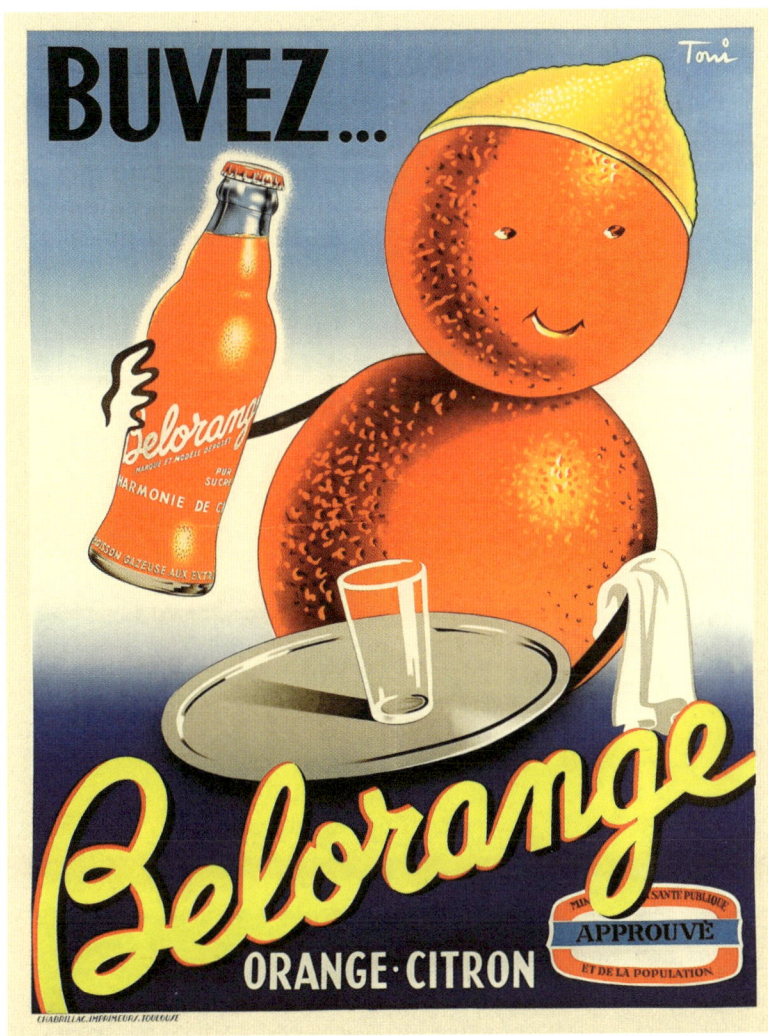

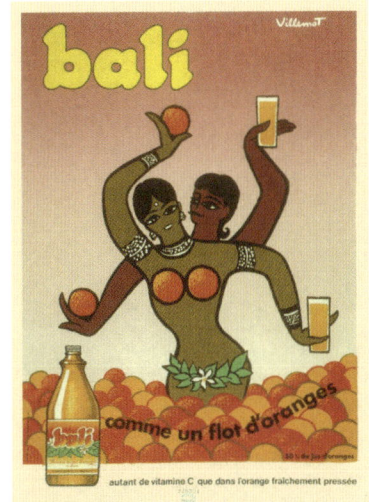

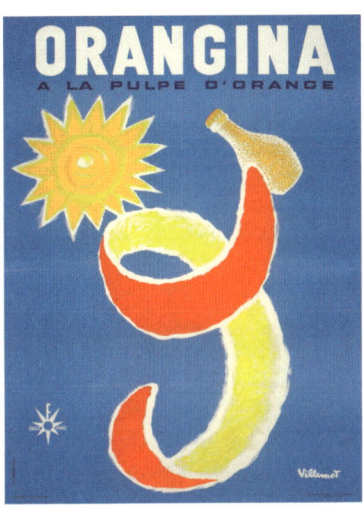

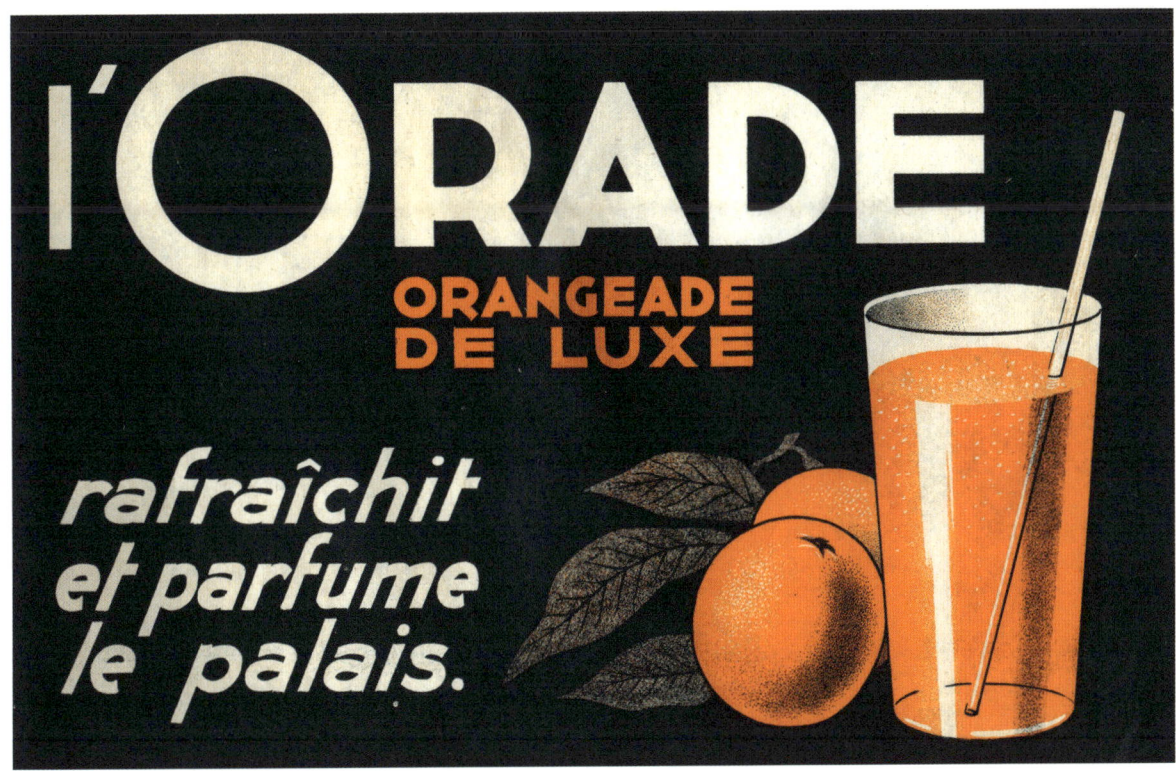

CITRUS IN PERIL

Citrus in cultivation has long been beset by many challenges, including climate. In February 1895, frost killed to the ground almost all Florida orange trees, so that production was reduced to 3 per cent of what it had been and took sixteen years to get back to previous levels. There were earlier major freezes there in 1747, 1766 and 1774, with 1825 perhaps the worst, while that in 1835 also killed almost all oranges in the state. The groves of Douglas Dummett (1806–1873) on Merritt Island off the Atlantic coast survived and, from them, the industry could restart.[52]

There were severe frosts in 1934 and in 1962, when, once again, almost all Florida oranges were killed. The Brazilian José Cutrale (1926–2004) saw an opportunity and sold his concentrate to the Atalanta Corporation for Minute Maid juice in the USA.[53] He ran his company with his son and successor, José Luis Cutrale (1946–2022), the billionaire 'Orange King'. The company, Sucocítrico Cutrale, acquired estates in Florida and, at one time, accounted for 90 per cent of world orange juice production, the Brazilian operation some 60 per cent of concentrate.

In 2016, the company admitted engaging in anti-competitive practices between 1999 and 2006 and paid an administrative fine of more than £70 million for participation in an illegal cartel that suppressed the prices growers could get for their oranges, forcing thousands of them out of business and into financial ruin.[54] None of the victims was ever compensated and, after failing to get redress in Brazil, pursued court proceedings in Europe.

Nowadays, when there are frost warnings, oil heaters are moved into the groves – at great expense. In 2017 in Scalea in southern Italy, a severe frost damaged most of the citrons, except those sheltered by photovoltaic panels. The owners thereafter distributed scions to growers who had lost their plants, helping to protect the region's traditional crop.[55] The practice of 'agrivoltaics', where crops are grown beneath solar panels, not only produces energy but also provides shade and shelter for the citrons. Furthermore, it allows some economic clout for citrus-growers, who need not sell at low prices but are able to leave fruit hanging until the market price lifts. The cover provided by the panels reduces water requirements by 70 per cent and the fruits are not only larger, of a better colour and with fewer defects from exposure to the weather, but also the rind has a higher concentration of essential oils.

Other perils are human-induced. In 1959, Minute Maid extended operations into the swampy savannahs of Florida near the Indian River, erected dykes and pumped out the native habitat to plant 600,000 orange trees on 7,000 acres.[56] The company has done more since, including establishing the largest lemon grove in the world. From the early 1970s in the Algarve, Portugal, which has been providing oranges to northern Europe since 1580, the area began to greatly increase citrus planting, to the detriment of the traditional dry agri-silvicultural system.[57] The countryside became transformed as more reservoirs and sinking of boreholes increased the water supply. The result has been the loss of the traditional way of life (perjoratively called 'peasant farming'), with the drift of young people to the towns, while the intensive irrigation is bringing up salts from far below ground, thereby compromising the very cash crops that the European Union has promoted.

Citrus trees suffer many pests,[58] including mites, aphids, mealy bugs, whitefly, fruit flies, but especially scale insects, notably red scale (*Aonidiella aurantii*, commonly known as California red scale but probably native to China), which infests all above-ground parts of the plant. Although still a major pest in Australia, biological control there is successful thanks to Integrated Pest Management programmes begun in the 1970s. Minute parasitic wasps are released and their larvae feed on the body of the scale insect and pupate under the scale cover; the first exotic parasitic wasp was introduced from China as early as 1902.

Biological control has a long history in China, where, since at least 304 CE (the oldest record of such control anywhere), weaver ants (*Oecophylla smaragdina*) have been used to capture other insect pests, their nests sold for the purpose. In Trinidad, oranges and grapefruit were attacked by the Asian citrus blackfly (*Aleurocanthus woglumi*), which was also controlled by the introduction of parasitic wasps, *Amitus hesperidum* and *Encarsia perplexa*, native to India and Vietnam. The wasps grow inside the blackfly larva, consuming it alive from within. However, *Encarsia perplexa* can itself be parasitized by another wasp, *Encarsia smithi*, which was accidentally introduced into Florida in the 1970s.[59]

Major diseases of citrus include certain fungi, notably species of *Phytophthora* (water moulds) leading to collar and root rot, especially in waterlogged conditions. Much more serious is citrus tristeza virus (the name derived from 'sadness' in Spanish and Portuguese for the devastation it wrought in South America in the 1930s). Spread by aphids, it has led to colossal losses of citrus trees all over the world and put millions more out of production. Seeds, in general, are free of viruses, but growing citrus from seeds leads to the long juvenile phase before flowering. This has been overcome by grafting *in vitro*, using minute shoot tips on seedling rootstock, resulting in virus-free mature plants without any reversion to the juvenile phase.[60] Other approaches to the problem include breeding in resistance from naturally resistant species; especially promising is *Citrus wakonai* (see p. 169), as it flowers and fruits in its first year.

Other serious diseases are caused by bacterial infections. Citrus canker is caused by *Xanthomonas citri*, leading to leaves and fruits dropping prematurely.[61] The disease spread from south-east Asia, reaching the USA in 1910. Biotechnological approaches to resistance include CRISPR-based genome editing, but studies have concluded that no resistant genes have been characterized so far. After a series of outbreaks and successful elimination in Florida, it was introduced again in 1995 and is now impossible to eradicate – containment management is the only option.

HUANGLONGBING

Huanglongbing (HLB)[62] is the official name for the bacterial disease also known as citrus greening. The Chinese term literally translates as 'yellow dragon disease', but the meaning 'yellow shoot disease' (*huangshaobing*) is intended.[63] Yellow refers to the colour of infected foliage, while 'citrus greening' refers to the fact that the fruits of infected plants remain green and therefore unsaleable. The symptoms of HLB include a blotchy mottling of the leaves and misshapen fruits of poor colour and worse flavour, being salty and bitter.

Where the disease is prevalent, citrus trees may not live more than four years, and those surviving for five to eight may never bear marketable fruits. Mandarins and other cultivars with that genome in their make-up, especially sweet oranges, generally have the severest symptoms; those such as lemon, grapefruit and sour orange, with more pomelo and other species in their ancestry, often less so. The pomelo itself and also limes are more tolerant, though in southern Florida even these now have severe symptoms. It seems that no citrus, hybrid or wild, is completely resistant to it, though a degree of tolerance has been found in some Australian species. The evolutionary explanation for such resistance is elusive.

The disease is the worst thing that has ever happened to the citrus industry. The bacteria are spread by insects known as psyllids. In feeding, the psyllids inject bacteria into the plant, where the bacteria multiply and fill the transport system such that materials cannot pass from the leaves to the rest of the plant. Infection can follow as little as fifteen minutes' feeding time, and 100 per cent infection can take place after as little as an hour's feeding. The disease is spread not only by psyllids but also by marcotting (air layering), and grafting using infected scions. Because plants have a system of diverting materials around a diseased or damaged area until the last few working sieve-tubes cease functioning, an infected citrus tree can look reasonably fine. This latent phase can last several years, during which time the tree can act as a source of bacteria to be spread to yet other plants.

The most widespread vector, the Asiatic citrus psyllid (*Diaphorina citri*), is now a worldwide pest of citrus. The adult insects are only 3–4 mm (approximately ⅛ in.) long and can occur in large numbers on the undersides of citrus leaves, being most active as the trees produce new growth. The whole lifecycle can last as little as fifteen days in Florida, where there are nine or ten generations a year.

Although HLB was long considered to have originated in China, it seems highly unlikely that this was the case. The very fact that there is no resistance in edible Chinese species of *Citrus* suggests it has jumped from another kind of plant altogether. It has been argued that the bacteria associated with HLB (*Candidatus* Liberibacter) originated in Gondwanan Africa in association with indigenous trees in the family Rutaceae and speciated in Africa and on the Indian plate after the break-up of Gondwana.

The African form of the bacterium (*CL*af) evolved to be associated with the African citrus psyllid (*Trioza erytreae*) and Rutaceae such as species of *Calodendrum*, *Teclea*, *Vepris* and *Zanthoxylum*. The Asian form (*CL*as) and associated citrus psyllid would appear to have originated in India, with perhaps curry leaf (*Bergera koenigii*) as its original host, and spread almost worldwide from there since the 1800s. Its dispersal to south-east Asia in the nineteenth century and to the Americas in the 1900s was followed by Africa in the 2010s. The spread of *CL*as means that the most severe form of the disease now occurs in the Americas and throughout much of subtropical and tropical Asia, east to New Guinea and west to the islands of the Indian Ocean, and now Africa. Ironically, it has been argued that the spread can be attributed to the well-meaning plan to reduce scurvy

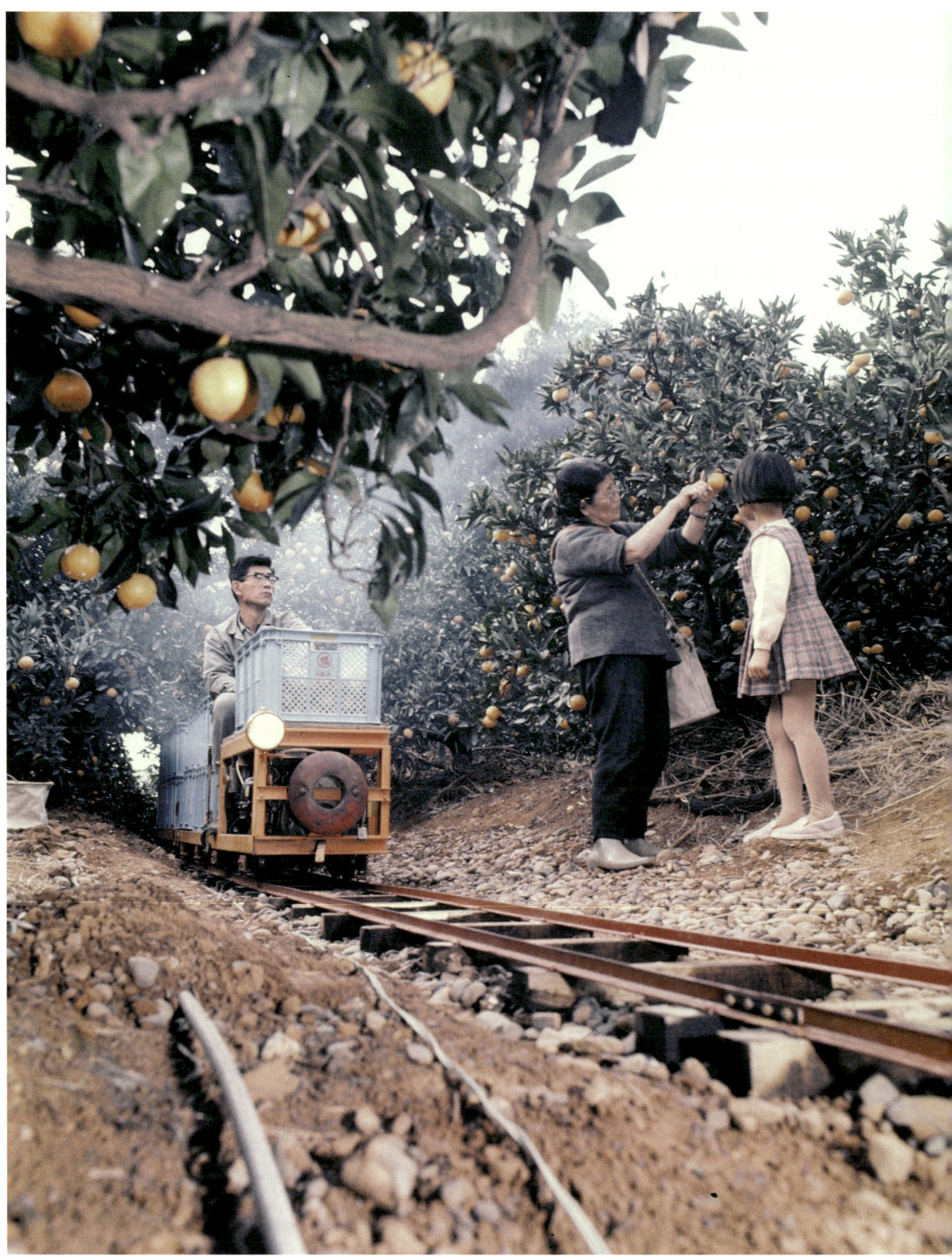

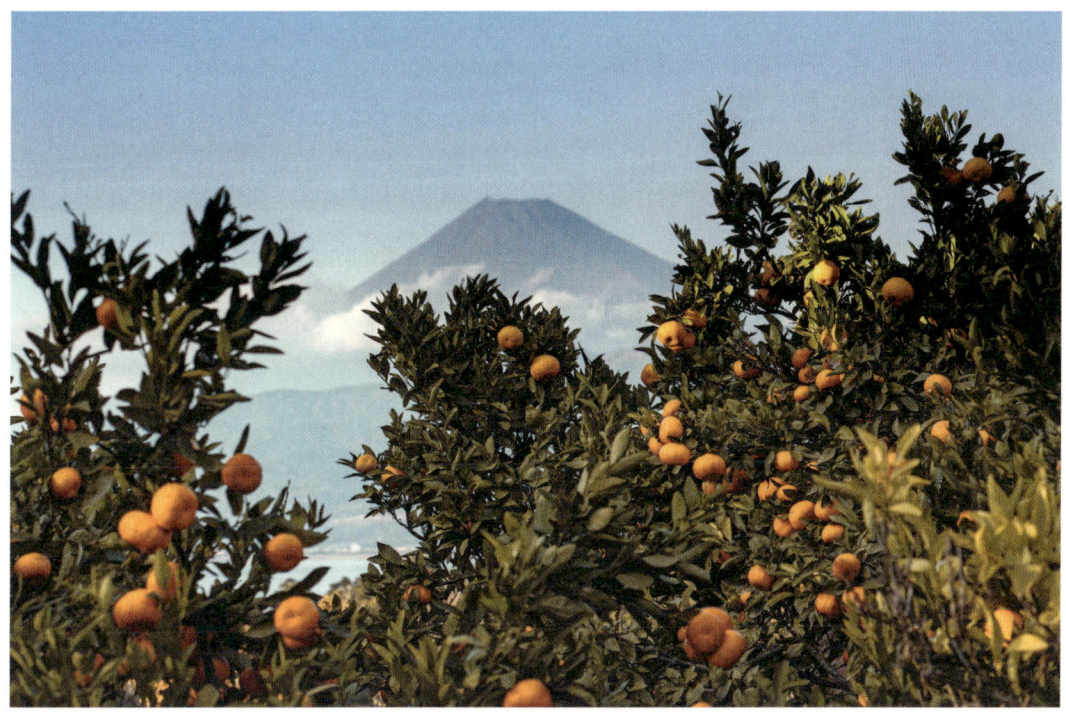

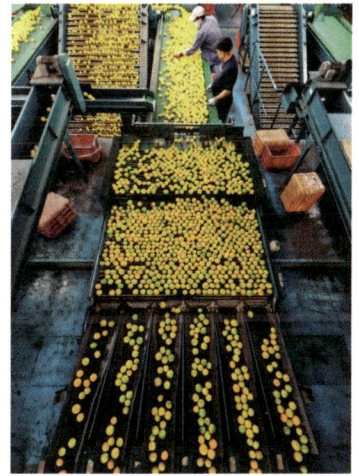

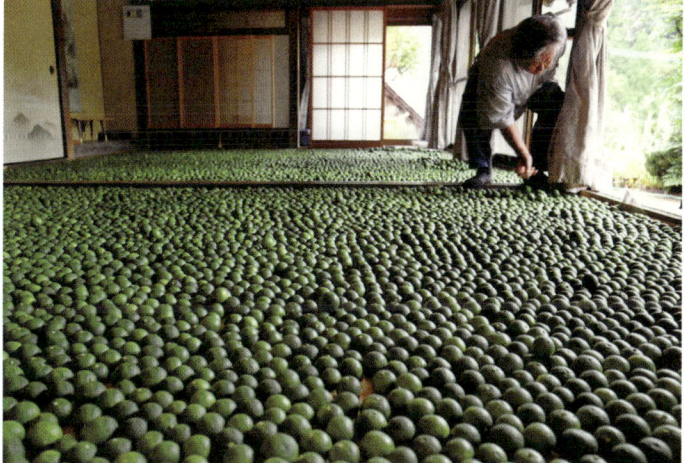

OPPOSITE: Orange harvest, Shizuoka, Japan, 1969. **ABOVE, CLOCKWISE FROM TOP:** Satsumas, Japan, n.d.; a farmer airs sudachi, Kamiyama, Tokushima, Japan, 2015; satsumas at a fruit-sorting facility, Tanabe City, Wakayama, Japan, 2023.

Part of the *Citrus × aurantium* complex, 'Citrus unshiu' or the satsuma is an easily peelable fruit that derives its name from the former Satsuma province of Japan. By 1922, seedless satsumas accounted for about 70 per cent of the country's citrus. The sudachi (*Citrus × sudachi*) is an Ichang papeda/mandarin cross (see p. 36) used in Japan in place of vinegar or as a flavouring for ice cream, soft drinks and liqueurs.

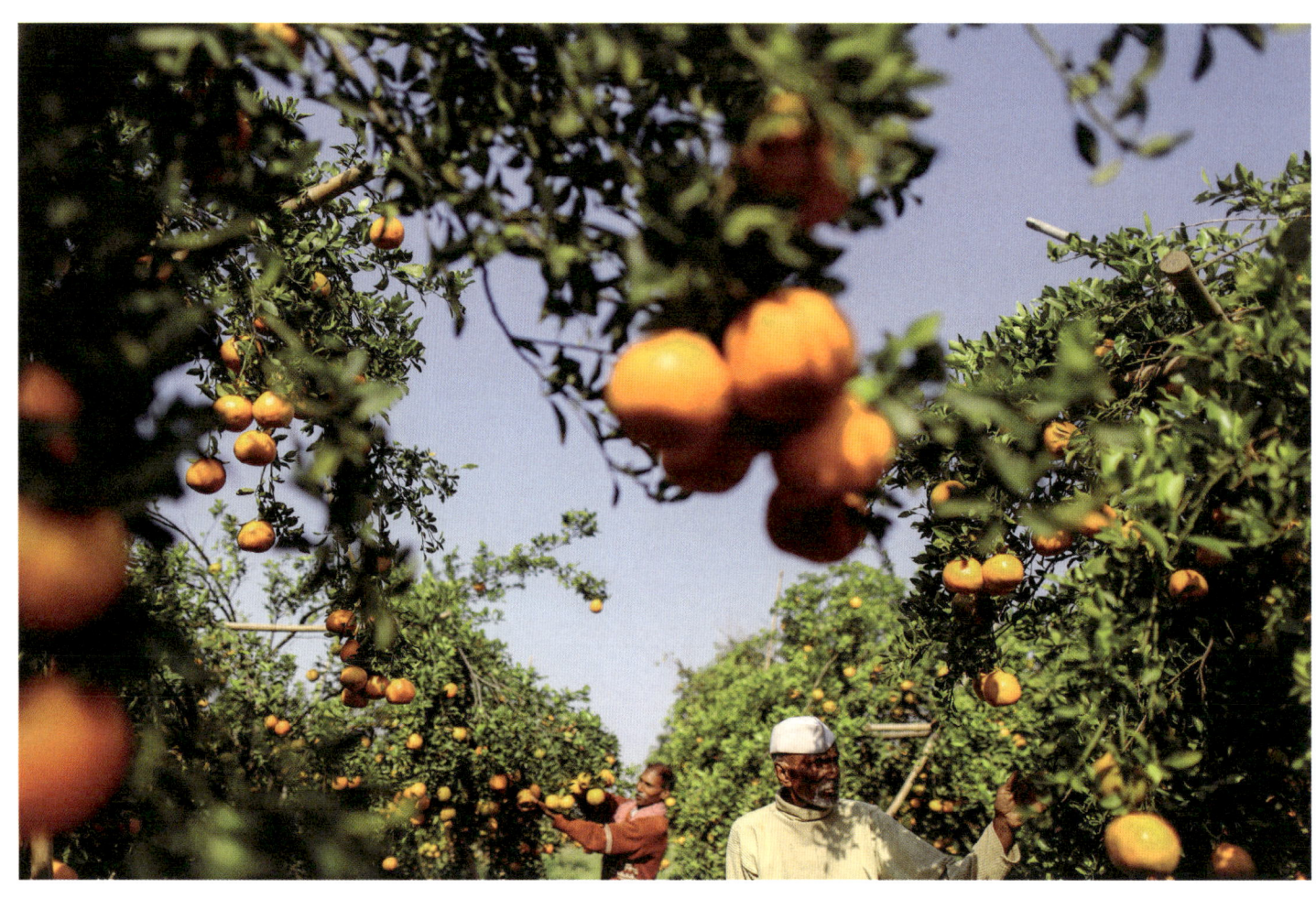

ABOVE: Farm workers inspect oranges during a harvest in Maharashtra, India, 2014. **OPPOSITE:** Sorting oranges for packaging at a regulated fruit market in Siliguri, West Bengal, India, 2008. **OVERLEAF:** Citrus harvest, Kedarpur village, Maharashtra, India, 2014.

India's citrus production is economically important, but South Asia is probably the home of the Asian citrus psyllid, an insect that nowadays carries the bacterium that causes huanglongbing, or citrus greening disease. In the 1800s, as the international citrus trade spread across the world, so too did the insect – and the disease.

Oranges grow well in the Lamghanat, Bajaur and Sawad. The Lamghanat one is smallish, has a navel, is very agreeable, fragile and juicy…. The Bajaur orange is about as large as a quince, very juicy and more acid than other oranges. It had always been in my mind that the word naranj *was an Arabic form; it would seem to be really so, since everyone in Bajaur and Sawad says* narang.

BABUR ON INDIAN CITRUS, *BABURNAMA* (MEMOIRS OF BABUR), 1519

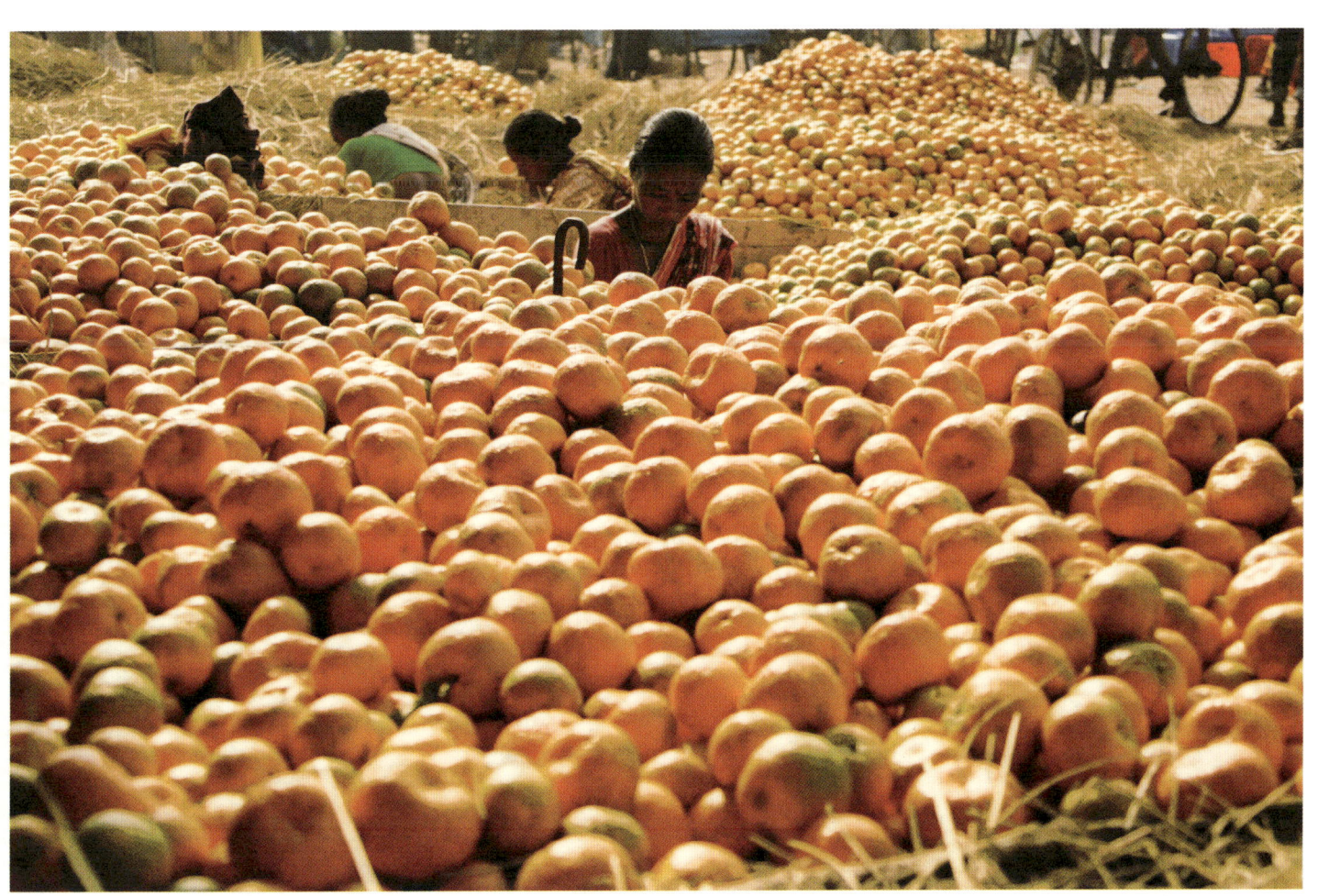

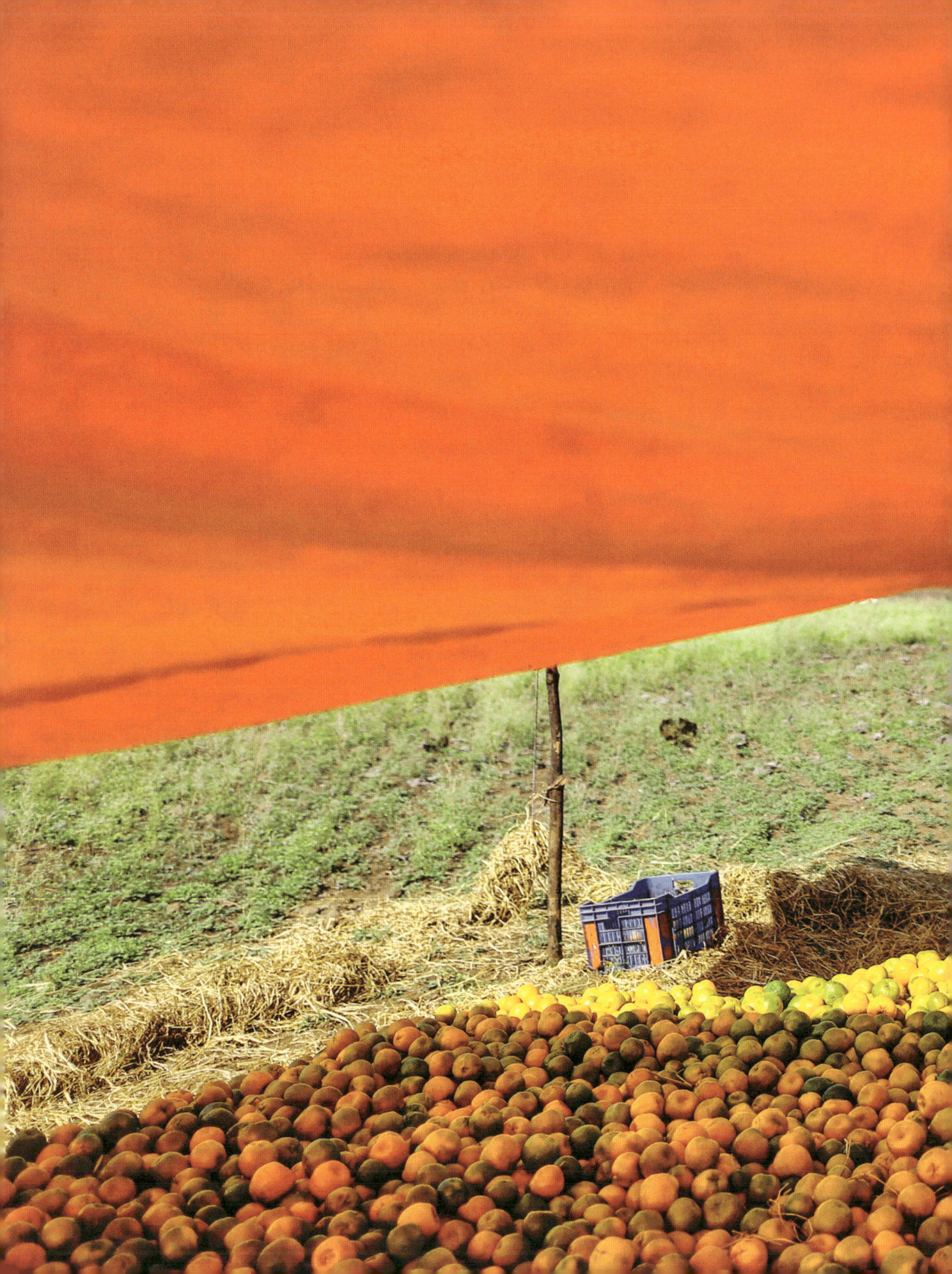

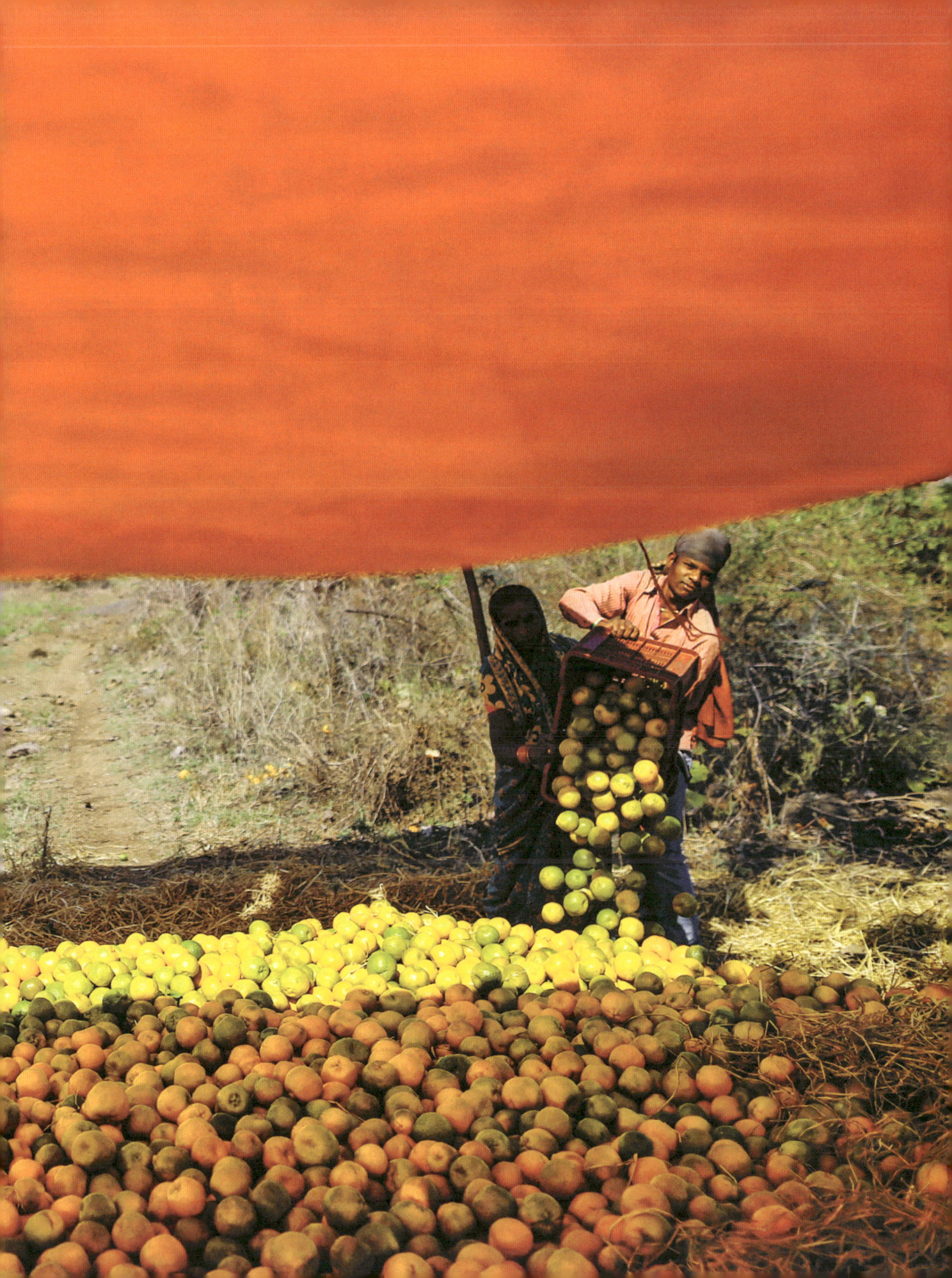

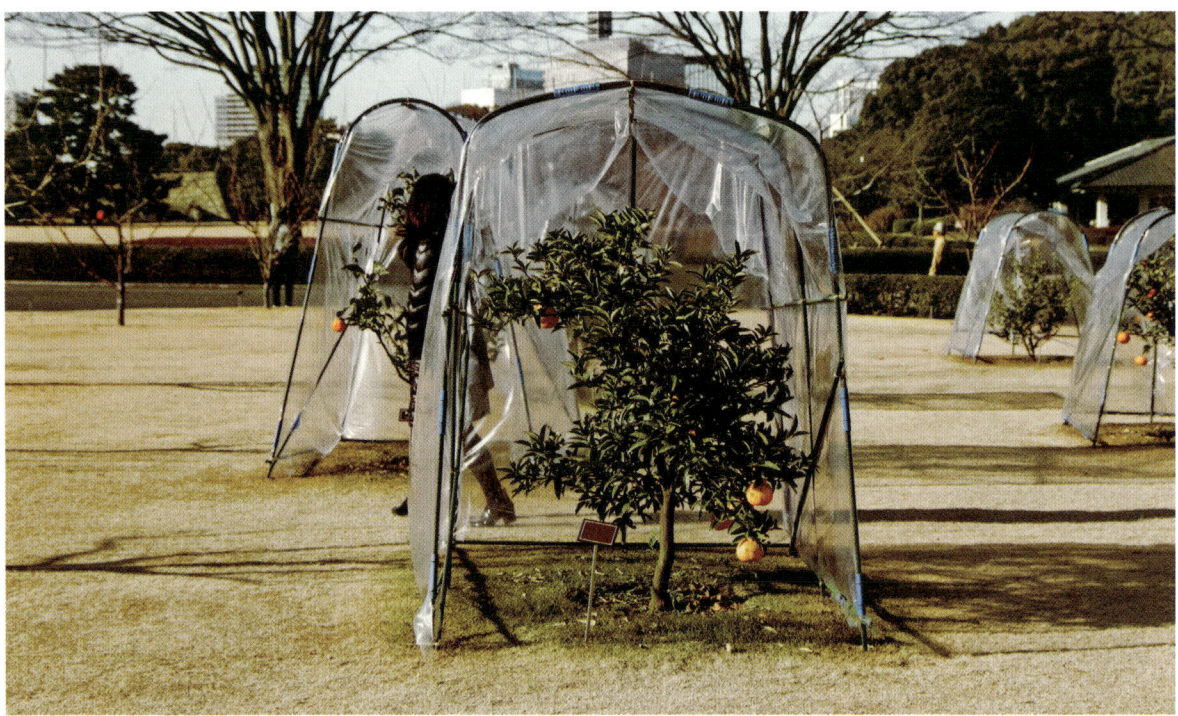

Individual citrus trees covered by frames and plastic sheeting to protect against possible infection, Japan. If trees are infected with huanglongbing, there is no cure.

in navies by spreading citrus material – alas infected – for the establishment of plantations.

This severe form has prevented the establishment of a modern viable citrus industry in much of tropical Asia and, even before 1920, was very destructive of the Pakistan industry. Between 1960 and 1970 three million trees were destroyed in Indonesia alone and, by 1983, commercial groves in most parts of Java and Sumatra were abandoned. In the Philippines, HLB was largely responsible for the 60-per cent reduction in area planted with citrus between 1961 and 1970. In 1971 alone it killed more than one million trees in just one province.

By 1980, up to 95 per cent of the trees in the northern and eastern provinces of Thailand were affected. It was first recorded in Papua New Guinea in 2002. By September 2004 it was found in Brazil and is now rampant in South and Central America. The first infection in the USA was detected in 2005 in Florida City and soon reached Louisiana, Texas, Georgia, Mississippi, South Carolina, Alabama and California.

Today it is found throughout the citrus-growing parts of Florida, where it has spread to affect 80 per cent of the state's orange groves.

The usual control would have been pesticides. Indeed, in Asia their (unsustainable) use can lead to trees surviving for up to fifteen years in well-managed orchards, but such treatments, even with up to seventy-three sprays every year, have failed to prevent spread of the disease. Moreover, intensive insecticide application schedules lead to the selection of insecticide-resistant psyllids, besides devastating natural biological control of the pest by weaver ants. There is some hope, though, in studies, largely by the Western Sydney University, Australia, that indicate that mineral oils used to control other pests and diseases may be as effective as conventional insecticides, without the risk of resistance.

Some biological control of the Asian citrus psyllid is also possible, but introduced parasitoids, first described in India in 1922, have been unable to prevent the spread of the disease in Asia and the Americas. However, in 2023 workers found that tobacco planted in groves could attract and kill the psyllids, while interplanting with guava seems to slow spread of the insect.[64] In desperation, all kinds of other ideas have been put forward, including applications of aspirin-like antibacterials and various nutrients. Antibiotics have also been used but, due to cost, phytotoxicity and potentially harmful residues, their use is not practical. Thermotherapy, based on water-saturated hot air treatments, has even been tried to eliminate the bacteria.

Nowhere in the world is there an effective management regime in terms of treatment of infected plants or prevention of infection in healthy ones. In the USA, the situation is dire, and only in recent years has the industry admitted it has a major problem: the 2014 Farm Bill allotted the first $23 million of *c*. $125 million to HLB research over the subsequent five years. Florida once produced 63 per cent of citrus fruits grown in the USA and supported some 76,000 jobs, but yields are falling year on year. Since the first report of HLB in Florida in 2005, citrus acreage has decreased by 38 per cent and production by 74 per cent. In 2004, Florida had an estimated 7,000 growers; in 2023 there were about 2,000, the projected yield, exacerbated by hurricane damage, for the year being 61 per cent less than in 2022.

More and more citrus, fresh and juiced, has to be imported from other countries, especially Brazil, but the simple fact is that orange juice – and indeed the whole citrus industry worldwide – seems doomed.

The Florida economy has been severely affected as growers attempt to switch to other crops, or abandon the dying citrus to act as reservoirs for healthy trees nearby. The effect on other citrus-growing areas has not received so much attention: for example, what will be the fate of Bhutan, where more than 60 per cent of the rural population is involved in mandarin production, the country's major fruit export?[65]

So far, the psyllid, and therefore the disease, has not become established in the Mediterranean or Australia. This is difficult to explain other than invoking some effect from climatic constraints, particularly in the case of the Mediterranean, where quarantine protocols are less stringent and the African citrus psyllid, an efficient vector of *C*Las, is spreading – now reaching Israel.[66] Quarantine to prevent entry of the bacteria and their vectors is crucial to the continued well-being of crops and growers. But in the long run a more fundamental solution has to be found – one that enables deterrents to the insect or disease resistance to be bred into the crop.

Citrus trees at an orchard in Arcadia, Florida, 2023. Individual citrus trees may be protected only to a certain degree. Infected plants must be destroyed for the protection of others.

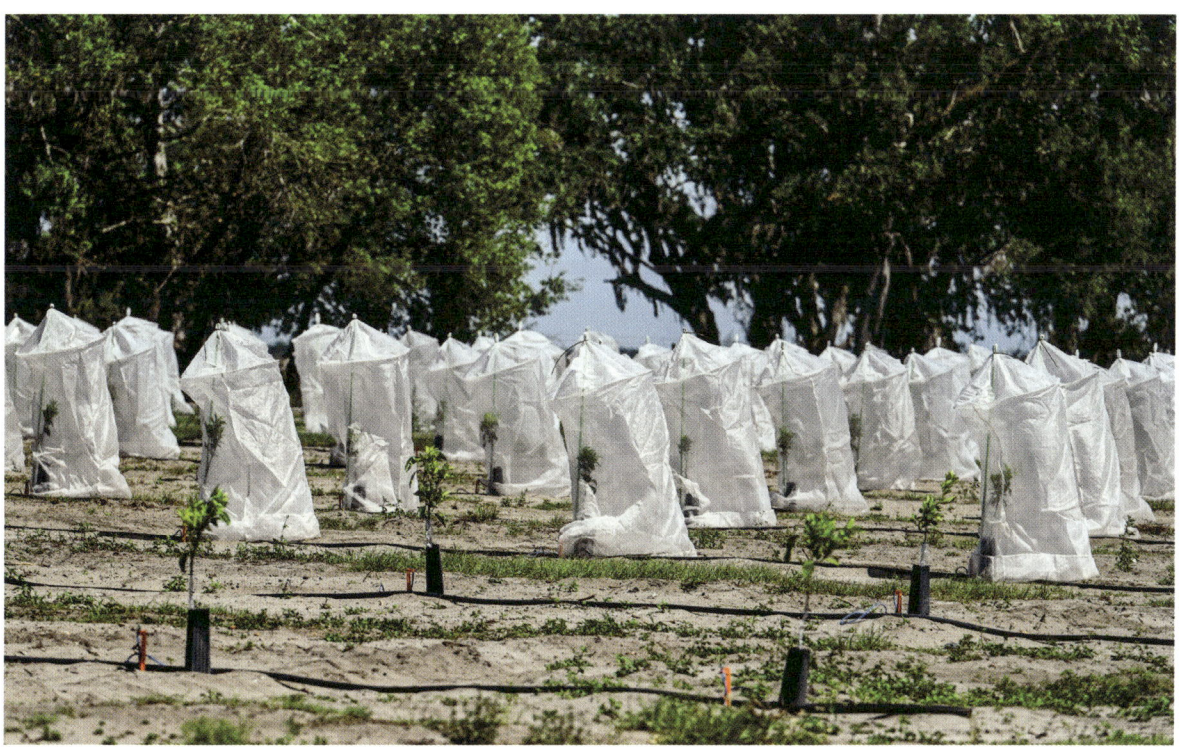

Observations that the psyllid does not thrive above certain altitudes in Asia suggest that UVB light may affect the plants, such that their anatomy or phytochemistry become inimical to the psyllid, and that seeking *Citrus* genes that promote this could be a way forward. Encouragingly, research in the 2010–20s focused on comparative analysis between HLB-sensitive cultivars and tolerant taxa led to the identification of natural defence genes potentially responsible for HLB tolerance. One of these candidate regulators from *Citrus australasica*, named stable antimicrobial peptide (SAMP), has been shown to kill *Liberibacter crescens*, a Liberibacter strain, in culture and to prevent infections of *C*Las in greenhouse trials. These peptides not only reduce disease symptoms but also induce immunity to fresh infections, though these promising results need confirmation in field trials. Although there are already selected cultivars that can live for some productive years before succumbing, and some of the *C*Las-resistant plants could be used as resistant rootstocks, an effective breeding programme incorporating the germane genes could take many years.

One of the problems in citrus breeding is the extended juvenile phase in most kinds of citrus that make breeding programmes very lengthy.

In Japan, for instance, it can take about twenty years for a new cultivar to go from development to being cultivated for commercial shipment. Breeders are often deterred from undertaking programmes for disease-resistance breeding since traditional methods produce undesirable traits such as this, as well as low fruit quality, small fruits, hard rind and long thorns.[67] This problem is common to many fruit-tree crops. In 2001, however, the generation time of citrange was reduced to some eighteen months by introducing genes from other plants and, since then, the technique has been used to incorporate tristeza resistance of *Citrus trifoliata* into commercial cultivars. If genetic modification is the only answer for such a 'clean and green' product as citrus and it succeeds, thereby making GMOs acceptable to those markets so far resistant to their commercial use, it will be a cakewalk for the GM industry with absolutely anything else.

BELOW: Orange harvest, Sicily, 1984. Mount Etna, whose snowy slopes are visible in the background, has given the region mineral-rich volcanic soil. **OVERLEAF:** Destroying a surplus of oranges as directed by the AIMA, the Italian body responsible for regulating agricultural surpluses.

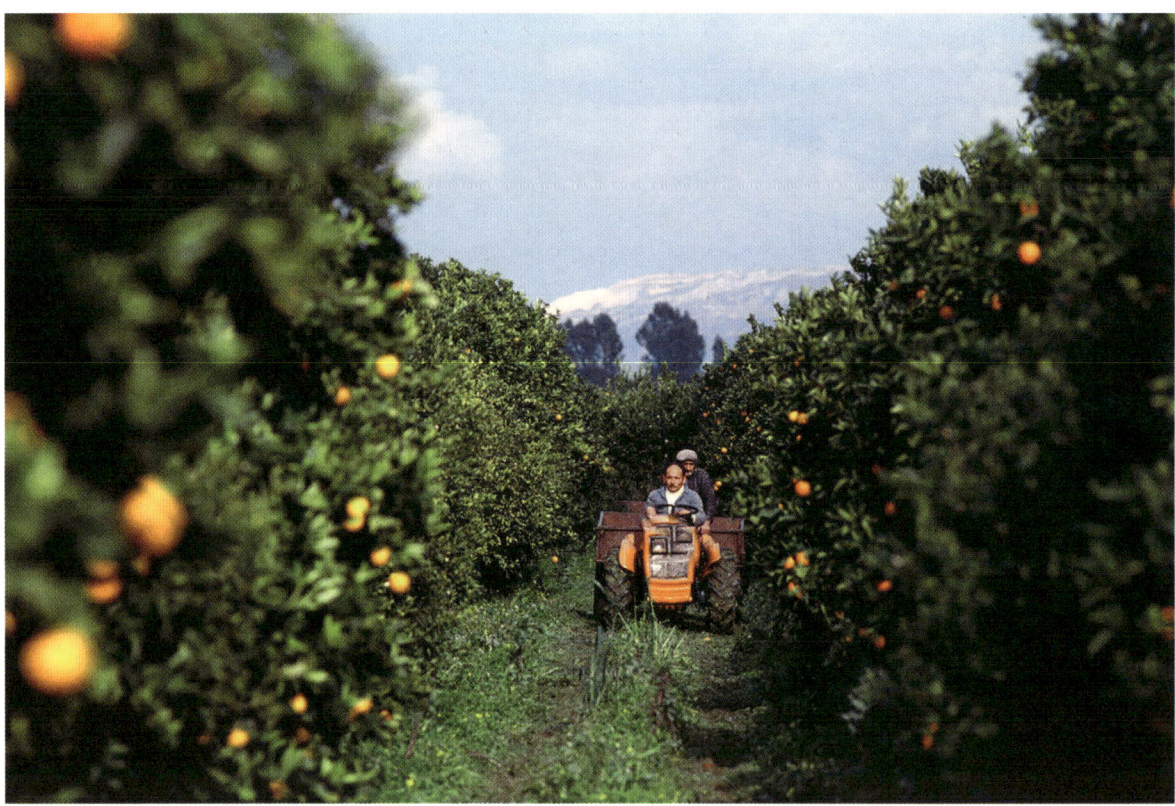

In 2006, Japanese breeders reported successful hybridization between citrus and the African genus *Citropsis*, which has resistance to a number of citrus pests and diseases, something that Walter Swingle had been unable to do, despite his considering them related. These hybrids had not yet flowered by 2013 when Australian breeders[68] successfully crossed *Citropsis gabunensis* from tropical Africa with *Citrus wakonai*, which has the shortest juvenile period of any known citrus as well as continuously flowering and easily crossing with other *Citrus* species (though it is very prone to tristeza). Australian species seem to be the best source of genes for HLB resistance.

The importance of all this is that, given the close-knit interbreeding group that makes up commercial citrus, besides the clonal nature of most of these crop plants, much of the industry is reliant on a very narrow genetic base, derived from sporadic hybrids and mutants. Walter Swingle appreciated this and advocated the establishment of citrus germplasm collections in the USA. From 1929 onwards, such a collection was accumulated at Riverside, California, with a view to improving variation, disease resistance and so forth in the US industry. However most of these, amounting to 325 accessions from some 25 countries brought into California before 1955,[69] seem to have been from plants in cultivation – in other words, from the genetically depleted gene pool. Similar could be said of other, more recent, heritage collections of citrus cultivars in many countries. As important as these are in many respects, this *ex situ* conservation largely misses the point.

It seems all too obvious now, in view of the HLB collapse – and earlier disasters like the comparable Great Famine in nineteenth-century Ireland, where the clonal potato crop rapidly succumbed to potato late blight (*Phytophthora infestans*), or the collapse of the similarly genetically narrow coffee industry in what is now Sri Lanka and Malaysia shortly afterwards – that crop wild relatives need to be identified and viable, variable populations of them conserved. Only recently have the parental species of commercial citrus been identified and a workable taxonomic framework for them and the hybrid clones in cultivation elaborated.[70] So what do we now find?

FUTURE OR FINALE?

We do not know where the commercial citrus hybrids were first accidentally made. It is likely that the forests where the parental species grew have been lost or greatly modified. The parental wild species of the major crops of oranges and lemons include the

original mandarin, now restricted to some mountain ranges in southern China, yet, of the other two, the pomelo and the citron, it is not clear that any truly 'wild' populations survive. There are reserves in India for the citron relation, *Citrus indica*, while populations of the Chinese kumquat and *Citrus mangshanensis*, besides at least one Australian species, the Humpty Doo lime, are also in reserves – but what of the rest?

It may be that it is too late for many, but other avenues are being explored. For example, at least some citrus are amenable to cryopreservation of seeds in seed banks, offering an alternative strategy of *ex situ* conservation for potentially significant germplasm.[71] This has been shown to be effective for not only wild taxa like the Australian *Citrus australasica* (finger lime), *Citrus inodora* and *Citrus garrawayi*, but also for cultivars of lime, lemon and sour orange.

BEFORE THE GREAT southern supercontinent of Gondwana split apart, its constituent pieces colliding with the land of the Northern Hemisphere, the ancestors of citrus plants, their descendants now surviving in Africa and Australasia with a few having reached into Asia, were rainforest trees with pinnate leaves. Those to be domesticated in the China region diverged into species dispersed by birds (kumquats), primates (the original mandarins) and water (pomelos) and were largely geographically isolated. Their closest relations, morphologically and ecologically more diverse, survive in the rainforests as well as dry habitats of the south-west Pacific.

Humans brought the Asian species together and spontaneous hybrids arose – the progenitors of modern oranges and lemons. Easily transported because of their tough rind and battery of chemical defences against insects, fungi and bacteria, their fruits were taken all around the world, infiltrating not only the cuisine but also the art, language and literature of most peoples. Now faced with a calamity due to the narrow genetic basis for these ubiquitous Asian crop plants, perhaps to be saved by their Pacific relations, the question has to be asked: Had the Pacific species, with their higher levels of vitamin C, higher resistance to HLB and greater variation than their Asian counterparts, been domesticated and hybridized, would the modern world of citrus be very different?

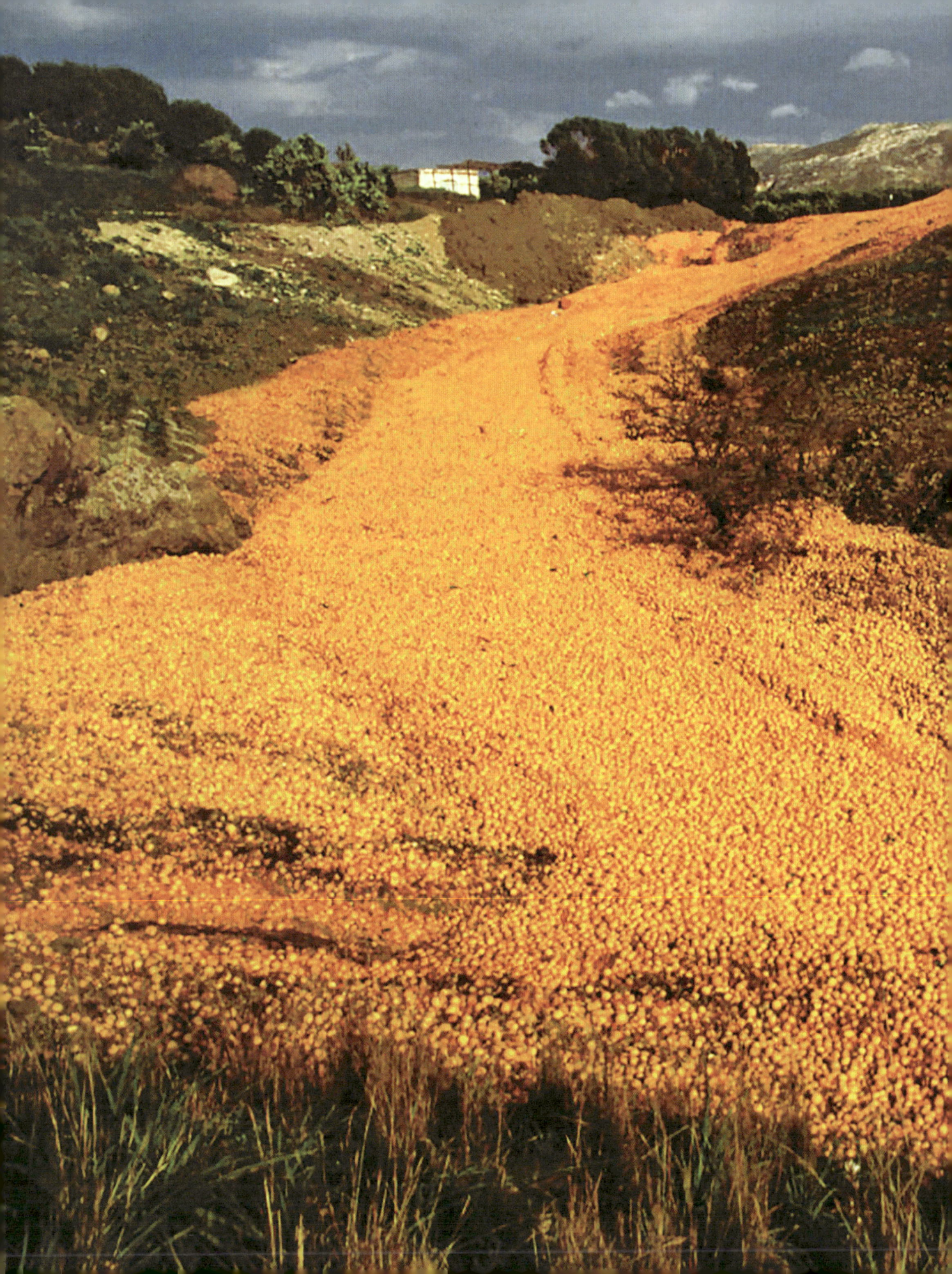

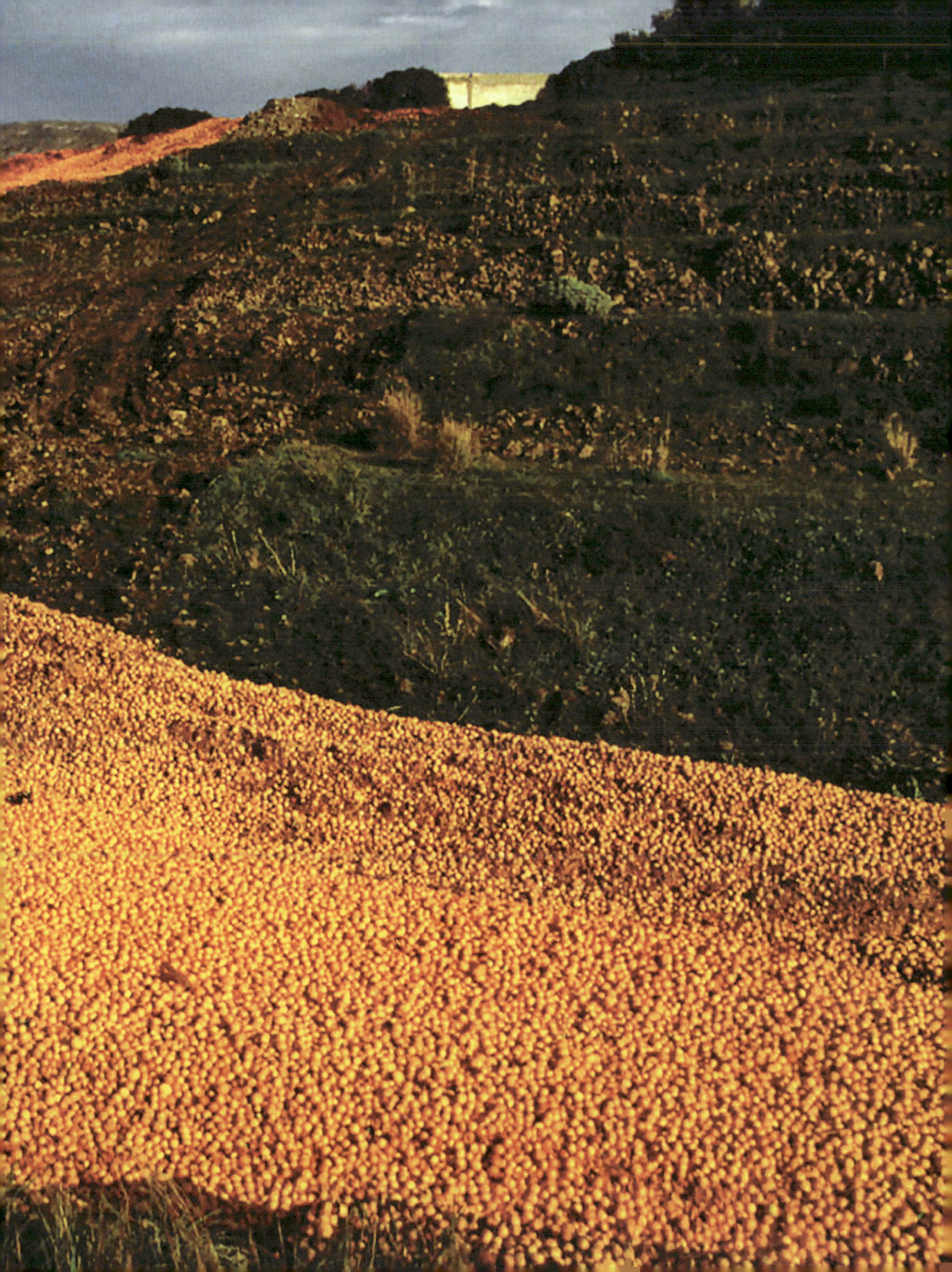

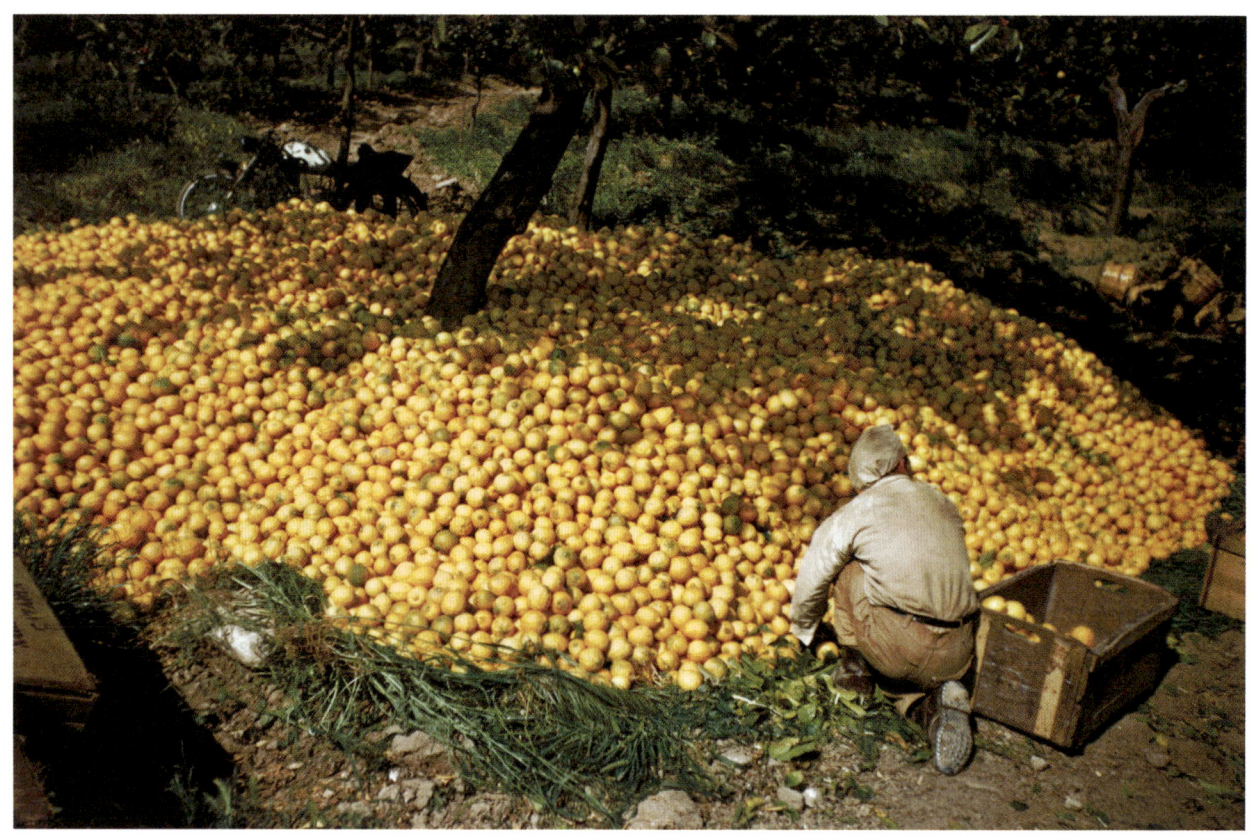

ABOVE: Lemon harvest, Sicily, 1970. **OPPOSITE:** Orange seller, Italy, 1960s.

In the early 19th century, Sicily was a major exporter of lemons, the increase in demand for lemon juice leading to wealth in Sicilian citrus-growing areas. High profits and a rather lax application of the law, combined with low levels of trust and otherwise high levels of poverty, left lemon producers open to exploitation. Neither the Bourbon administration nor the government formed in 1861 after independence was effective in enforcing property rights. The newly wealthy therefore resorted to taking on mafia associates, not only for personal and property protection but also as intermediaries with exporters at ports. No other commodity had anywhere near this effect. It has been cogently argued that the mafia domination of the citrus trade gave rise to the modern Mafia (from the Sicilian *mafiusu*, 'swaggering' or 'bravado'), both in Italy and, subsequently, the US.

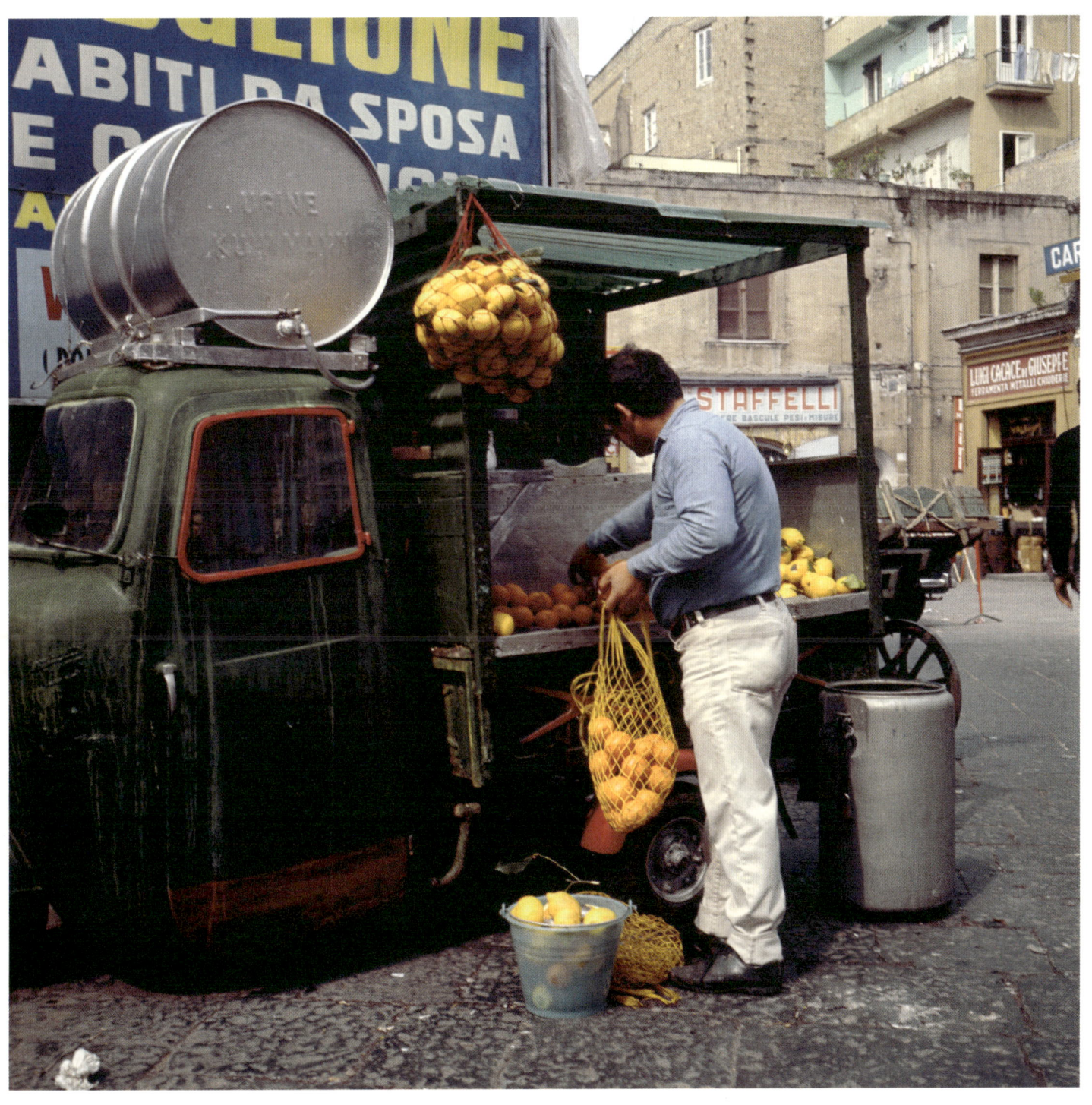

In one half day, I have learned that I am not a citrus picker. If you know of any seasonal farm work in the state that doesn't require the combined agility of a monkey and the stamina of a horse, I would like to hear of it.

LETTER TO THE FLORIDA STATE EMPLOYMENT SERVICE FROM A SEASONAL WORKER

PROGRESS AND PERILS

OPPOSITE: Orange harvest, Florida, 1960s.

Citrus trees and shrubs were introduced into Florida by the Spanish in the late 16th century. The development of the citrus industry in Florida dates from the 1870s and 1880s, coinciding with the construction of the railways, which were suitable for transporting perishable products to market. The citrus industry drew workers from around the USA for the yearly harvests. The Great Freeze of 1894–95 destroyed huge numbers of trees; the Mediterranean fruit fly destroyed many more. In the 1960s, the competition between Florida and California orange producers intensified, and agribusiness corporations took over the small producers.

OVERLEAF: Agricultural workers plant rows of young citrus trees in recently cleared land to create an orange grove near the city of Winter Garden, Florida, *c.* 1950.

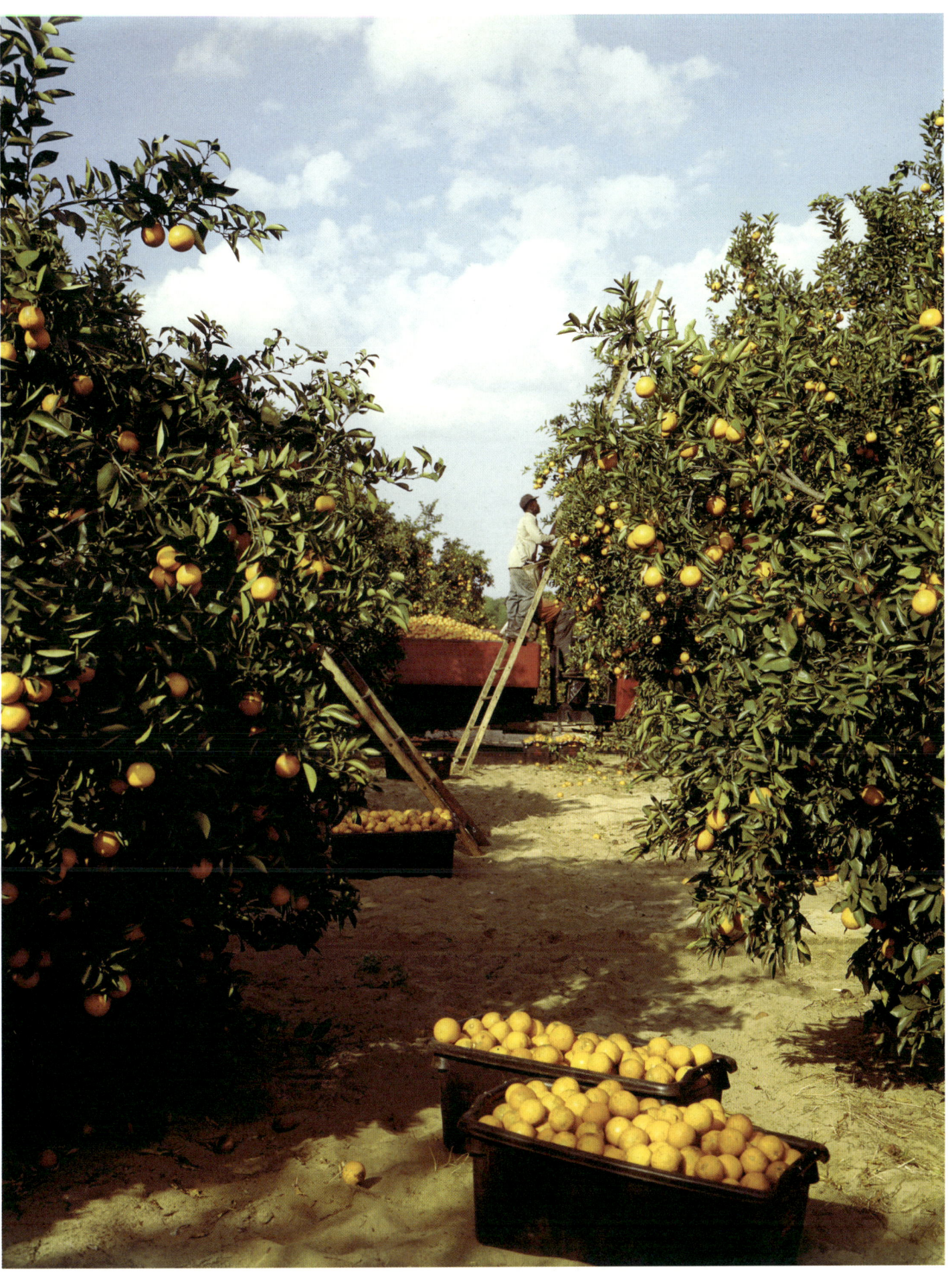

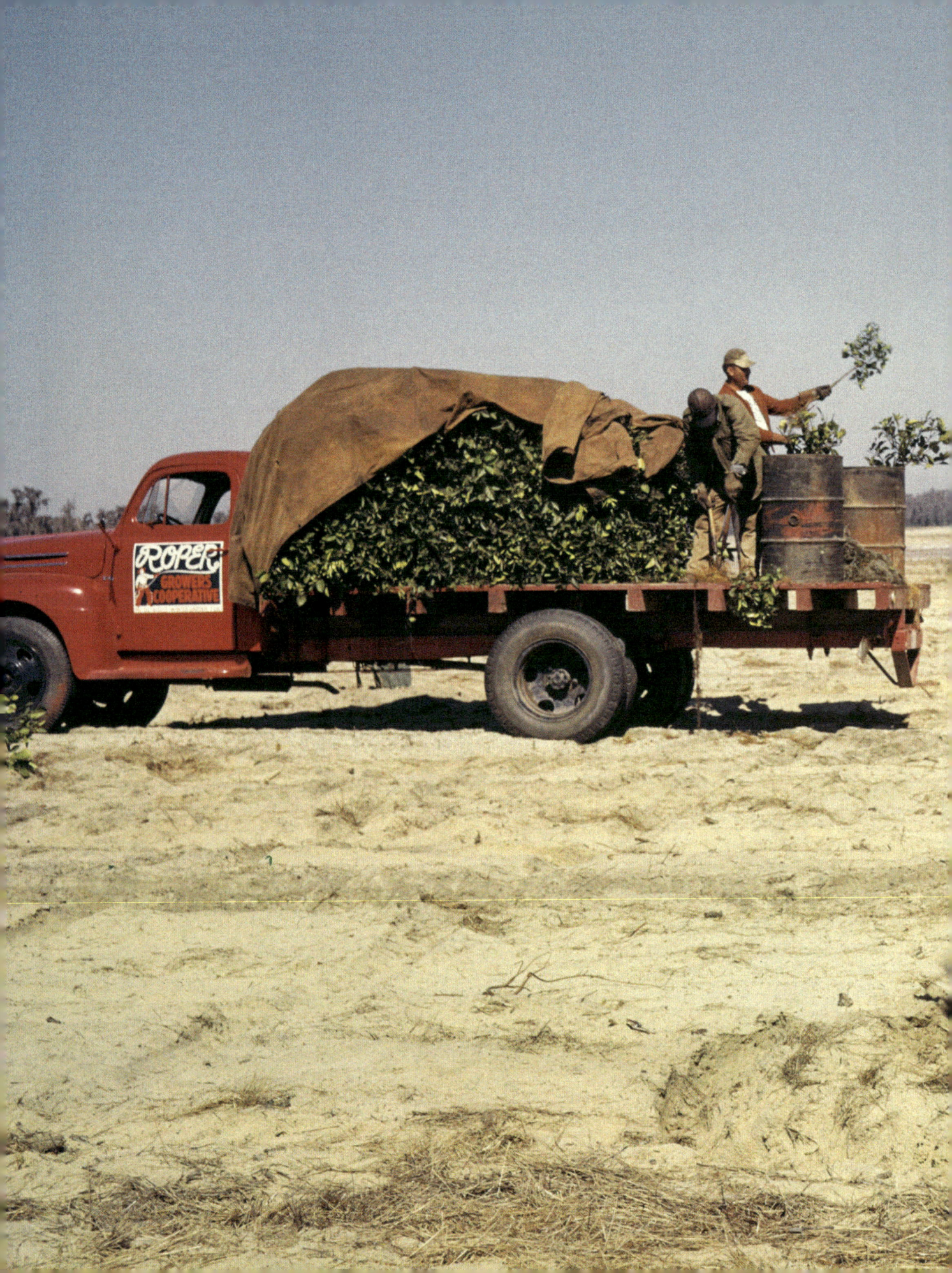

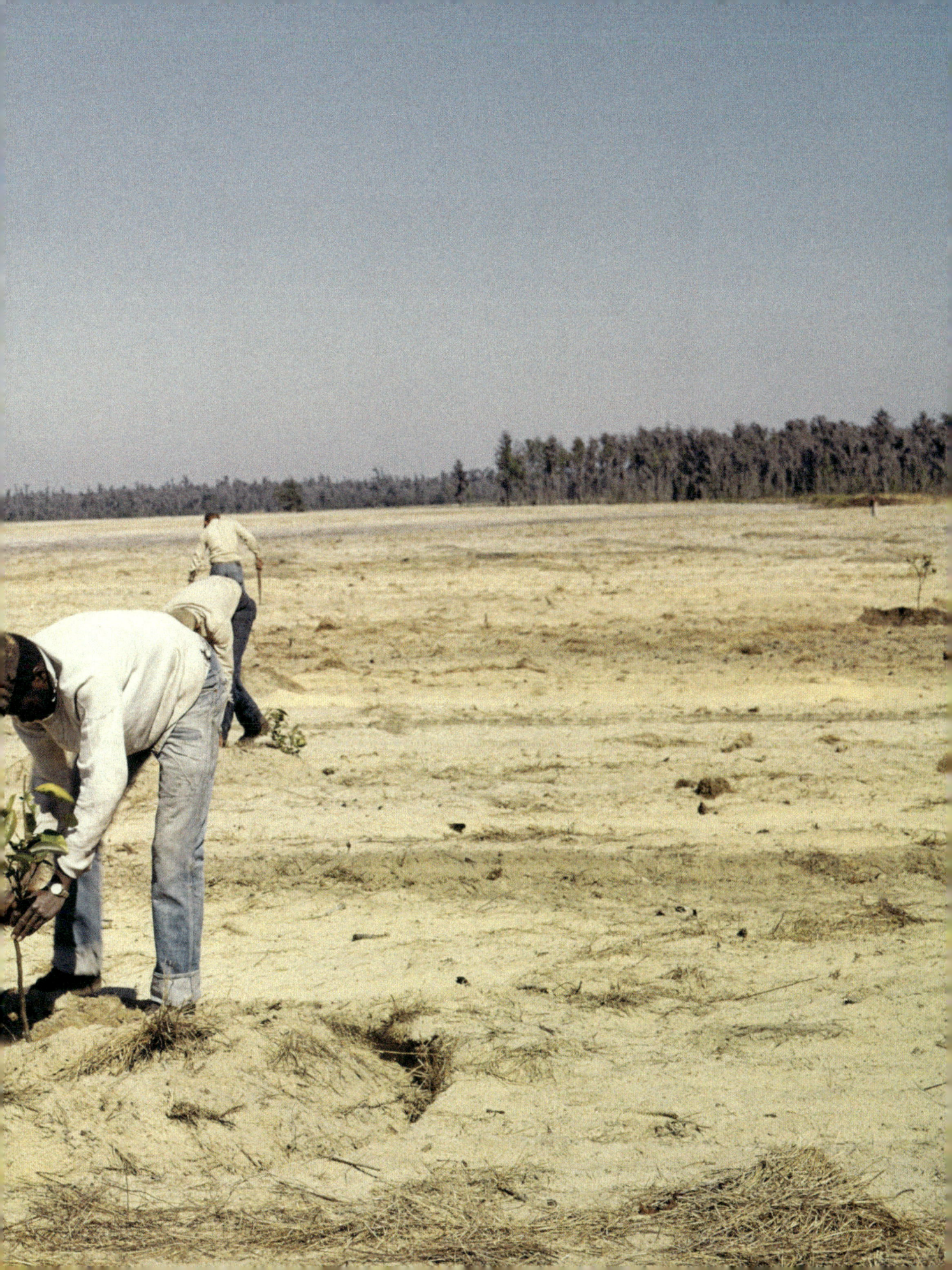

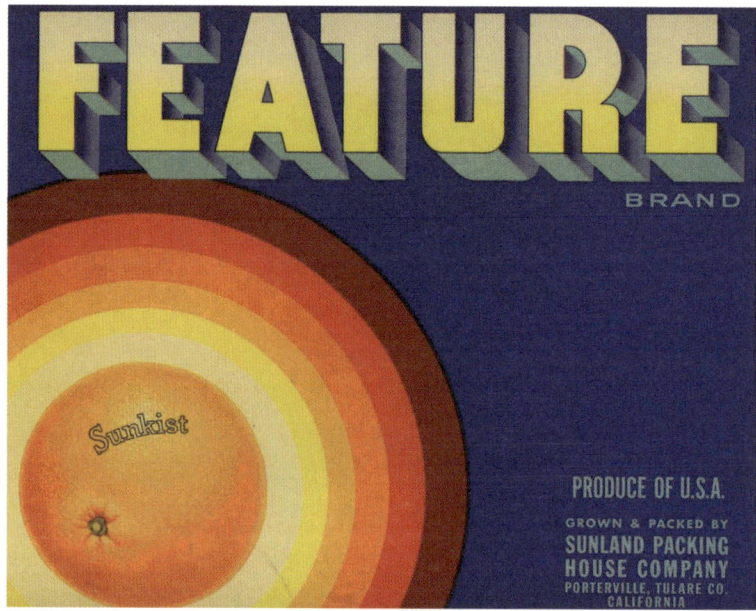

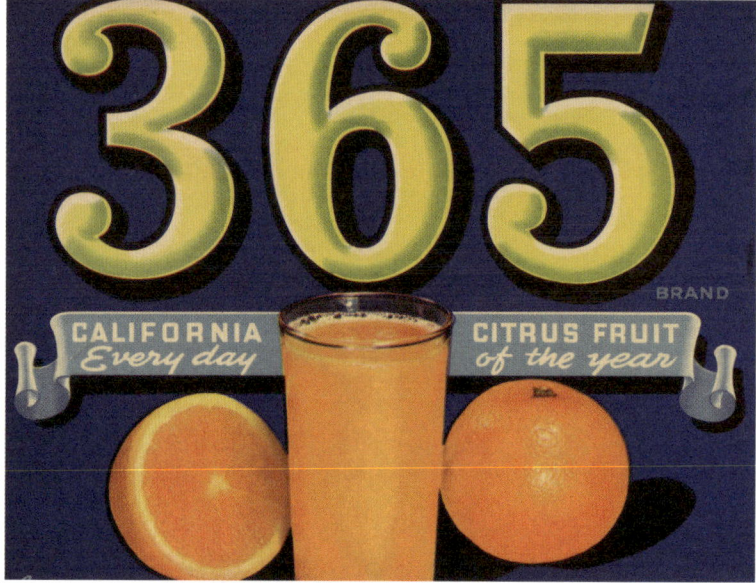

Florida citrus crate labels, early to mid-20th century.

Advances in mass printing and colour reproduction meant that citrus-growers could create eye-catching paper labels to attach to their wooden fruit crates. From 1885 onwards, lithographed labels (*c.* 25.5 × 27 cm / 10 × 11 in.) were soaked overnight and pasted to both ends of the wooden crates that succeeded barrels until, in turn, these were replaced by cardboard boxes with printed designs in the 1950s. By the 1930s, popular artists were producing designs for some 2,000 brands – but, as this was 'commercial' work, they remained unsigned. In their heyday, around 5,000 designs were in use at any one time; 10,000 were produced in California alone.

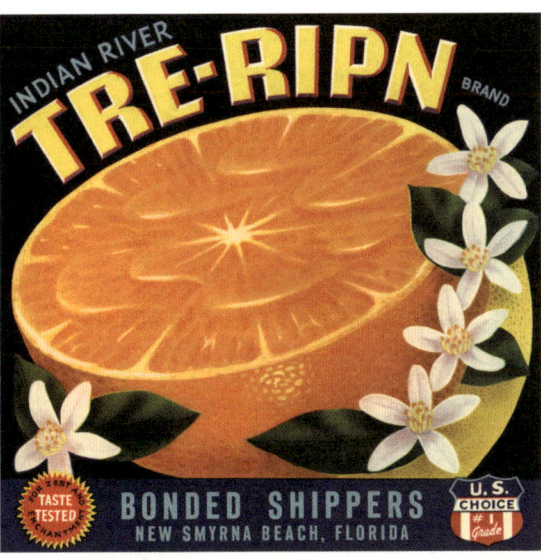

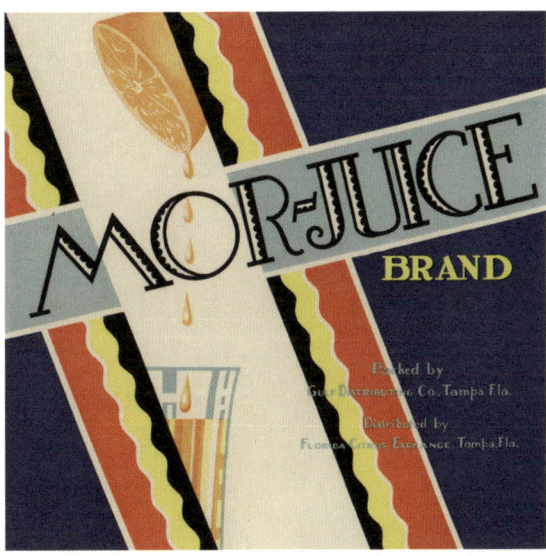

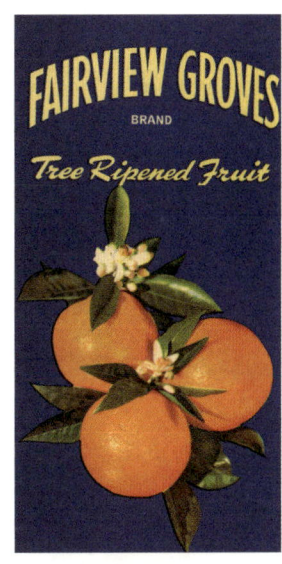

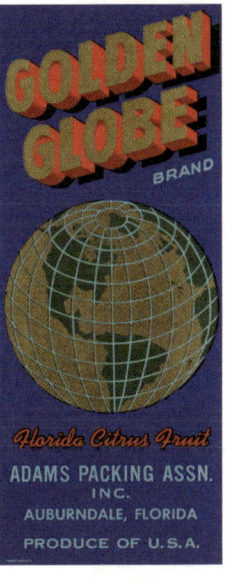

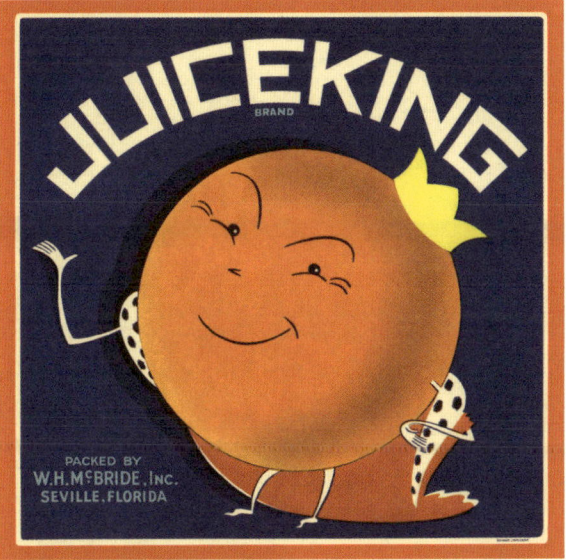

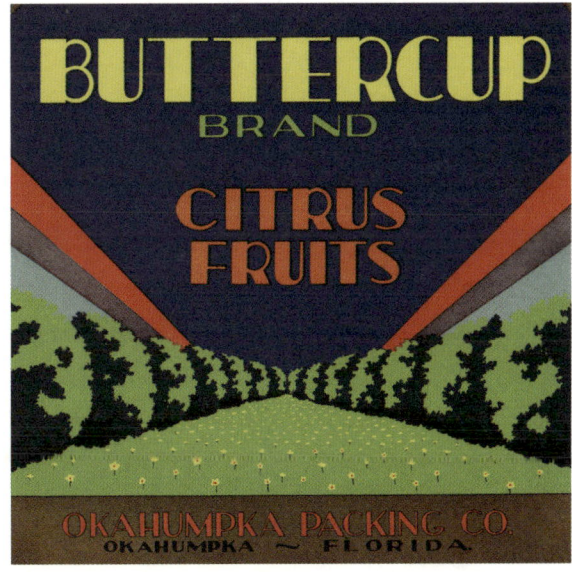

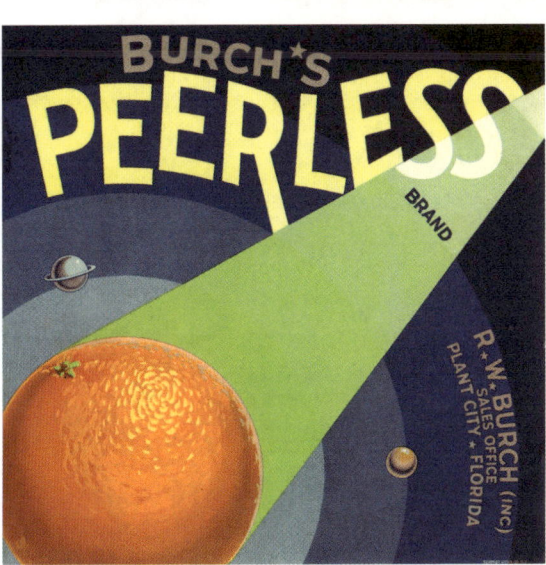

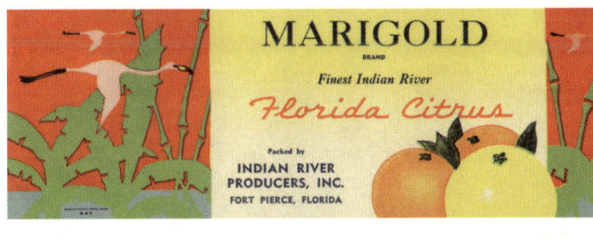

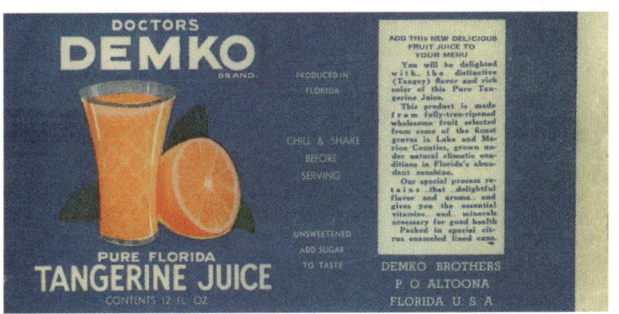

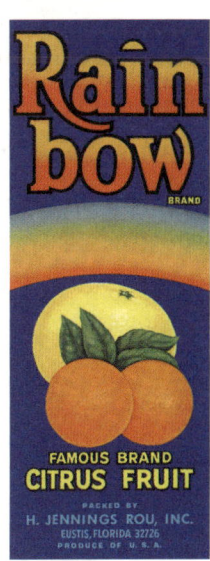

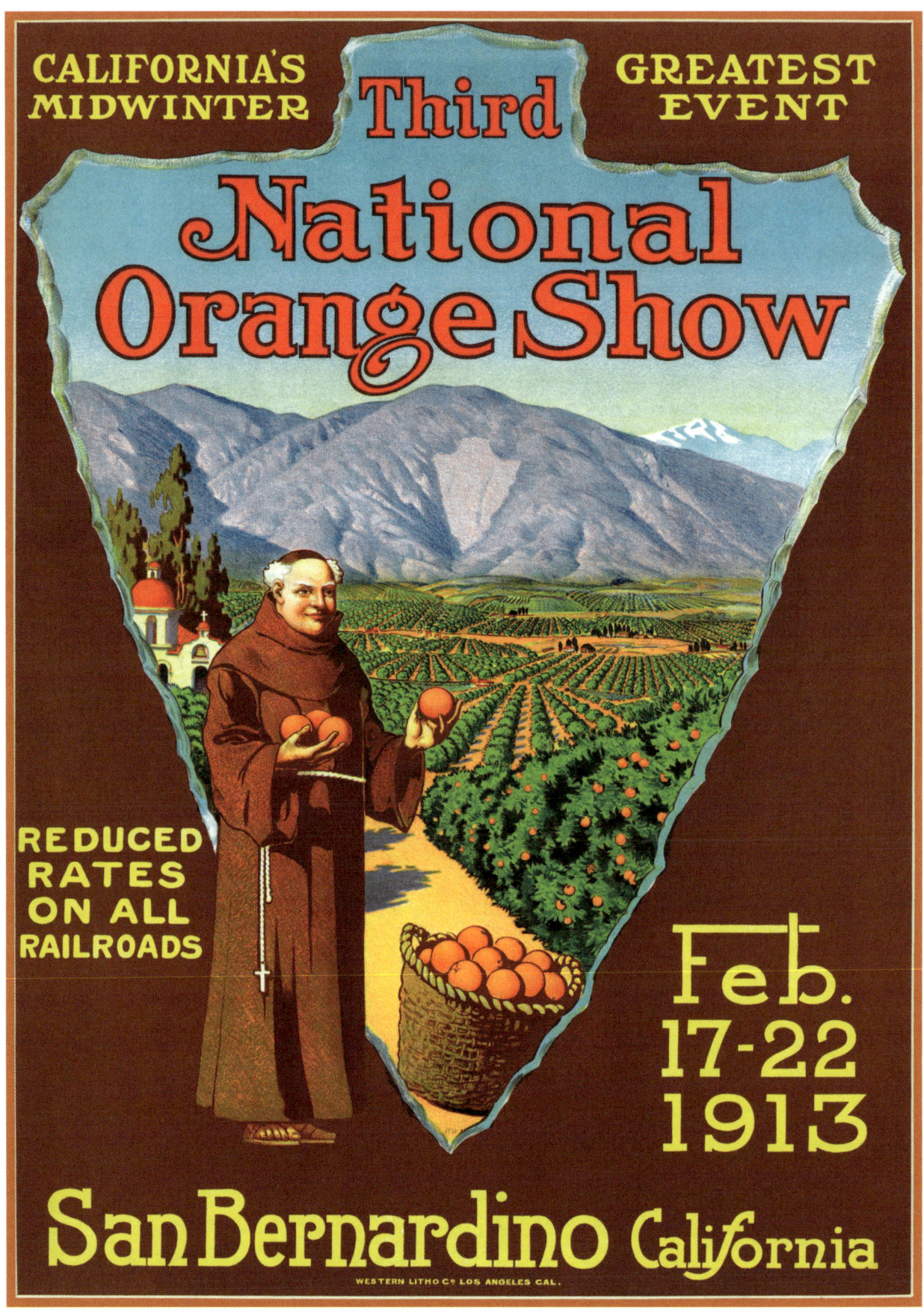

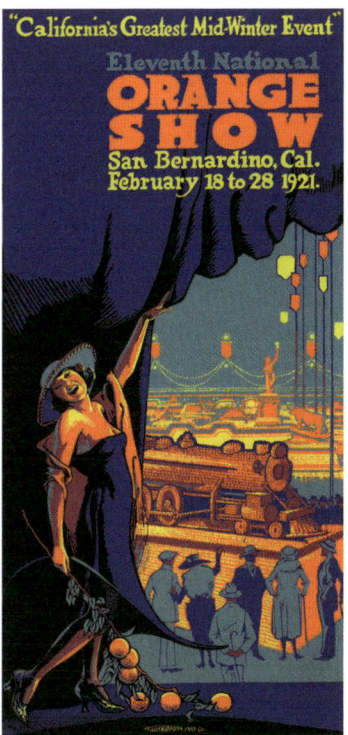

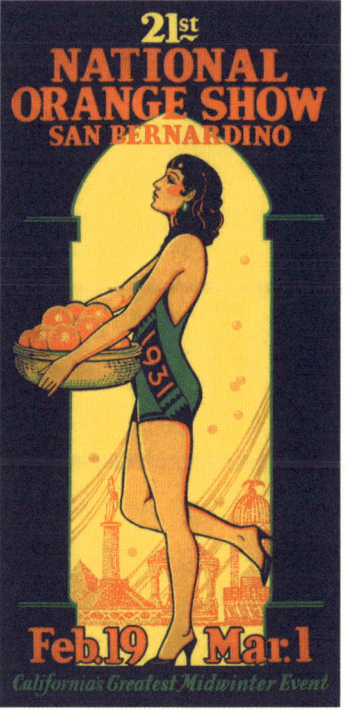

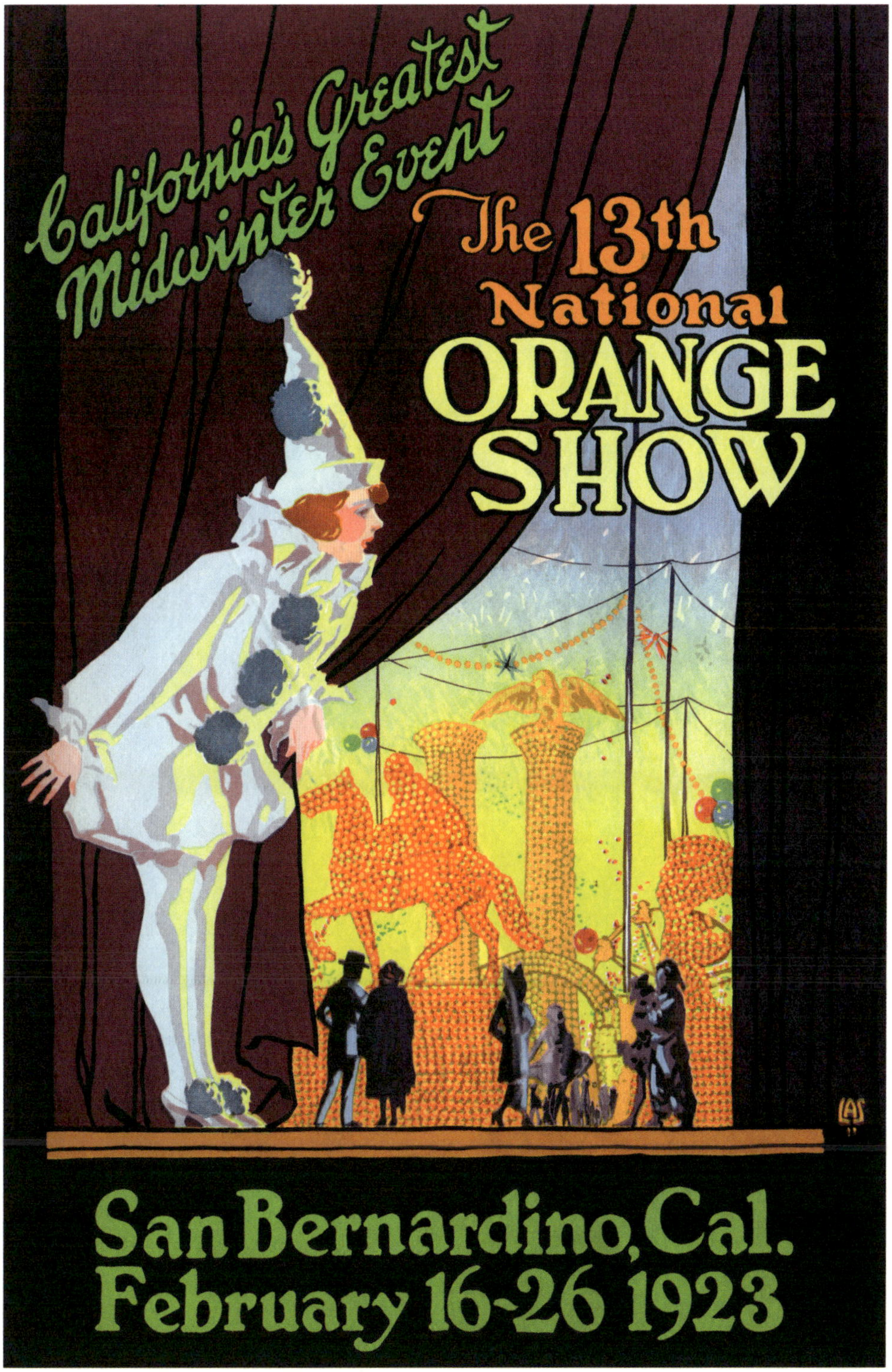

OPPOSITE AND ABOVE: Posters advertising the San Bernardino National Orange Show, 1913–31.

The world's first citrus fair was held in California in 1879. In 1889, the very first 'Orange Show' was staged in San Bernardino, becoming the National Orange Show in 1911. It has taken place every year since, with the exception of four years during World War II and the Covid-19 pandemic 2020–21.

OVERLEAF: Posters advertising the Fête du Citron, Menton, France, 1984–86. The carnival is held at the end of the winter.

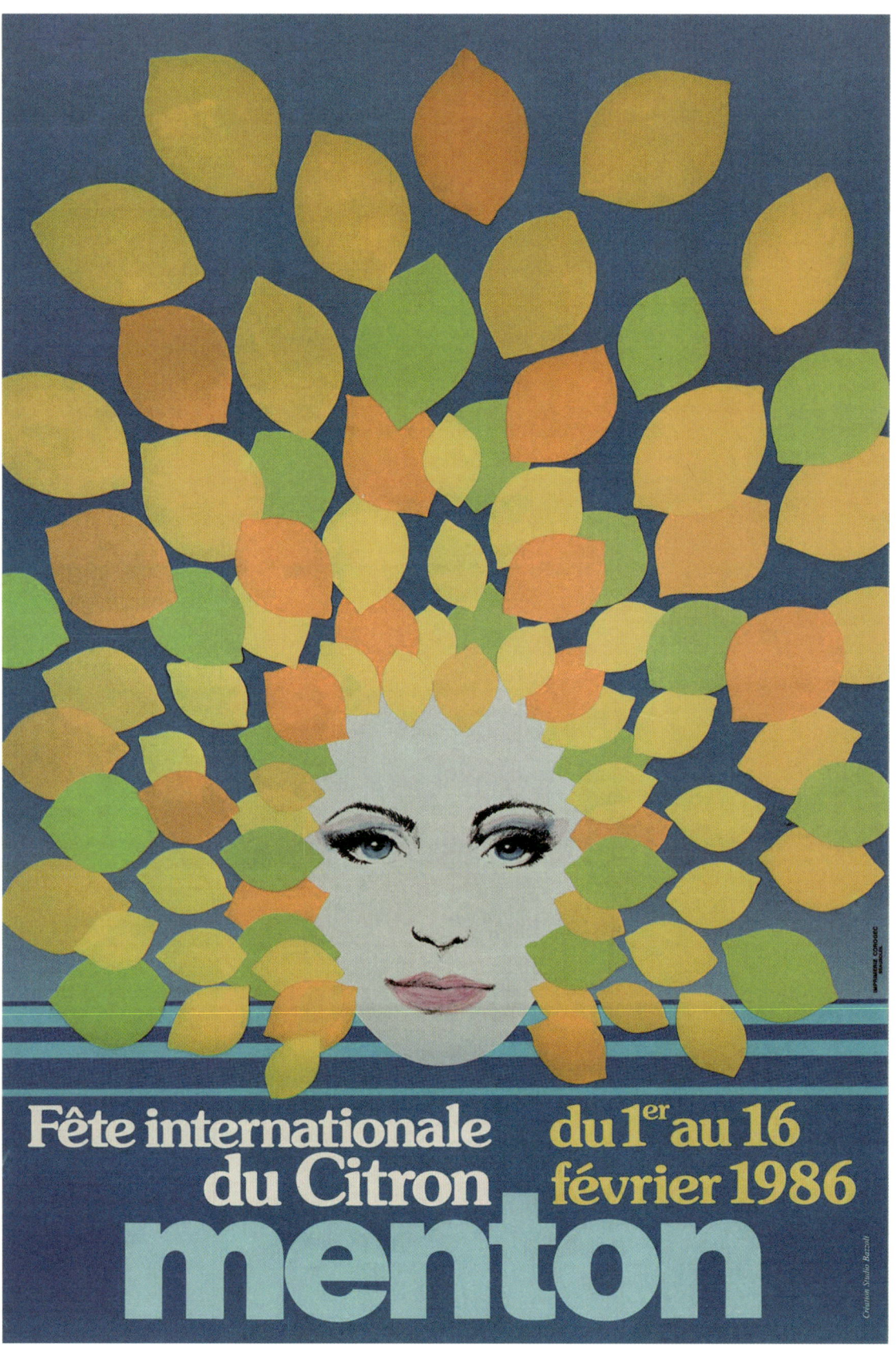

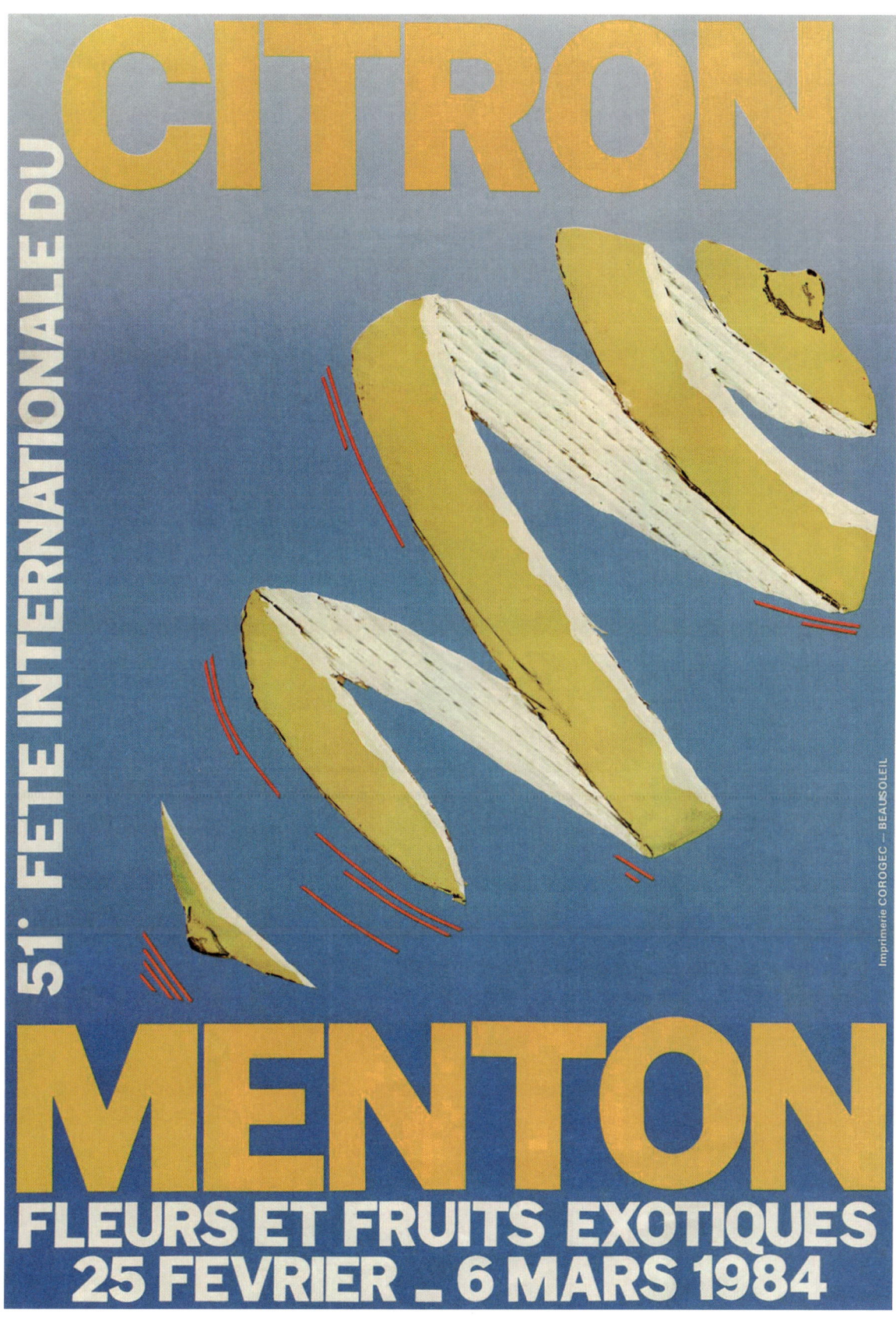

CITRON

51ᵉ FETE INTERNATIONALE DU

MENTON

FLEURS ET FRUITS EXOTIQUES
25 FEVRIER _ 6 MARS 1984

Imprimerie COROGEC – BEAUSOLEIL.

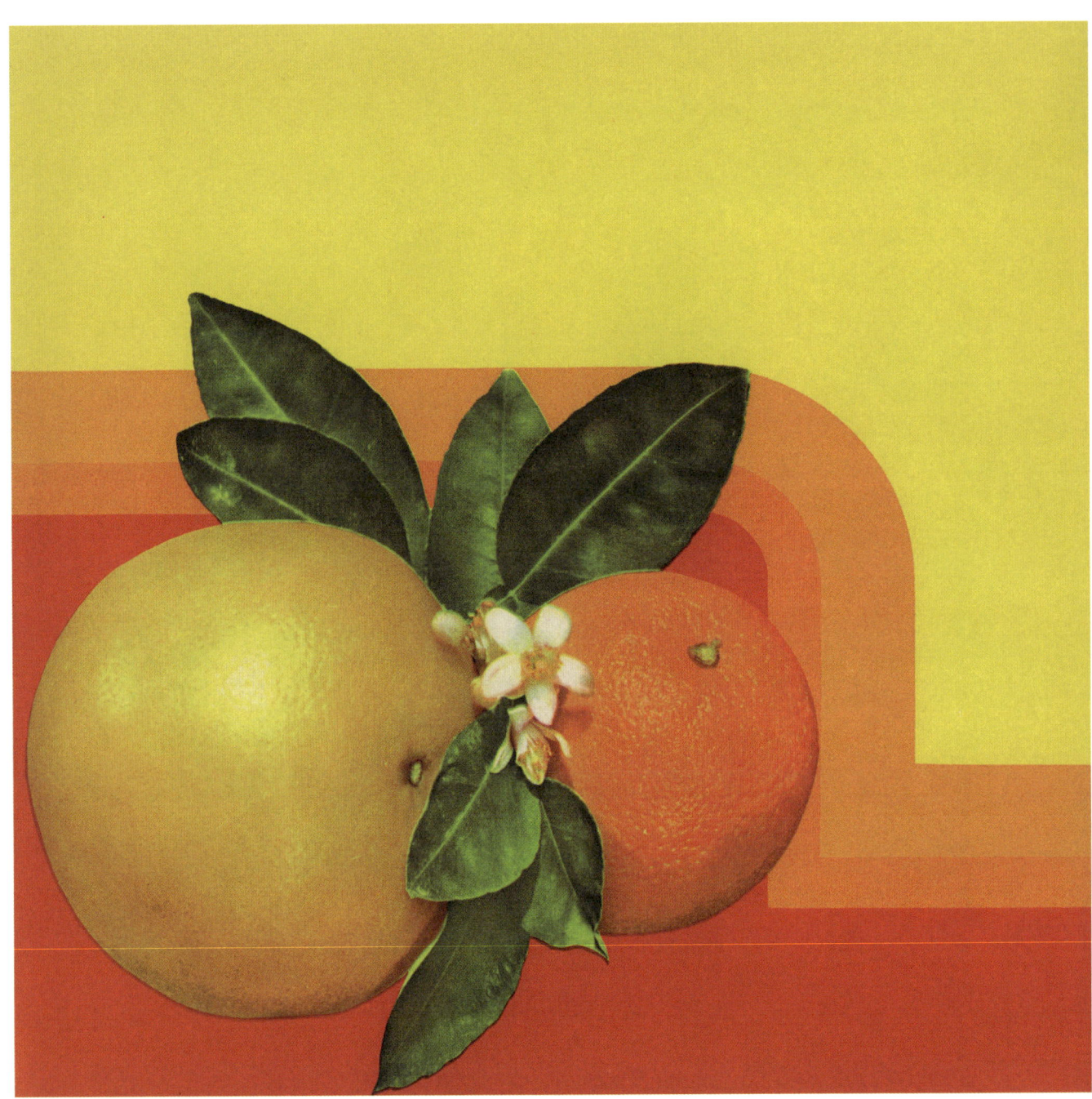

ABOVE AND OPPOSITE: Blank citrus labels, 20th century.

The boom in growers, distributors and shippers of citrus fruit produced an equally diverse boom in graphic design schemes for labels and posters. A business simply had to select a fruit and colour scheme, and fill in the brand name.

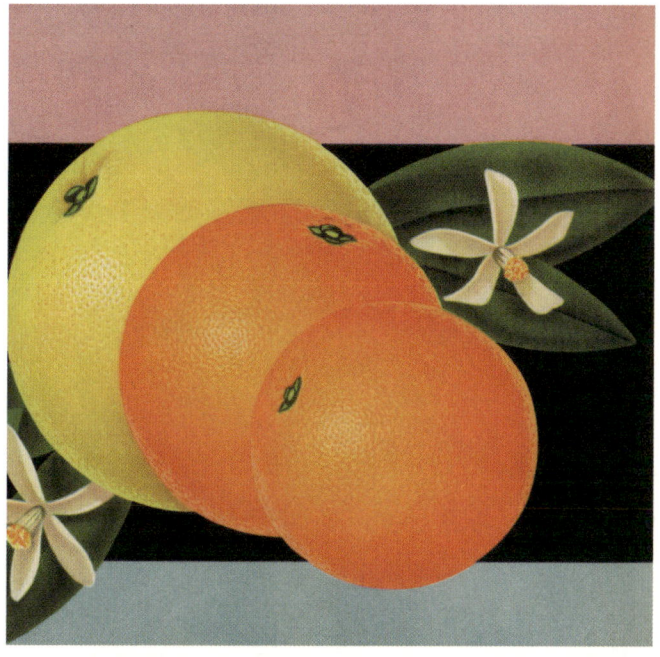

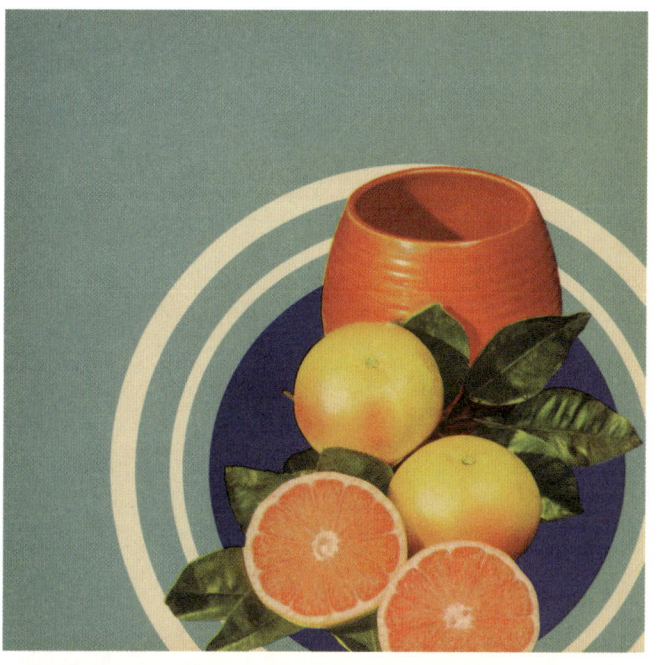

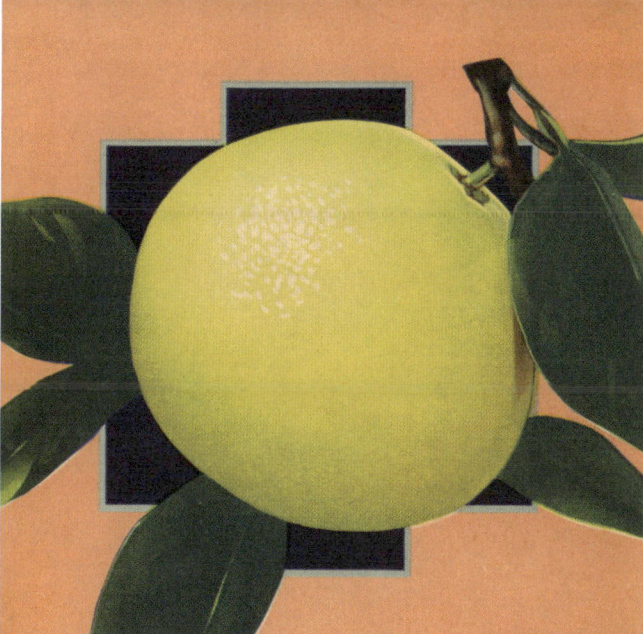

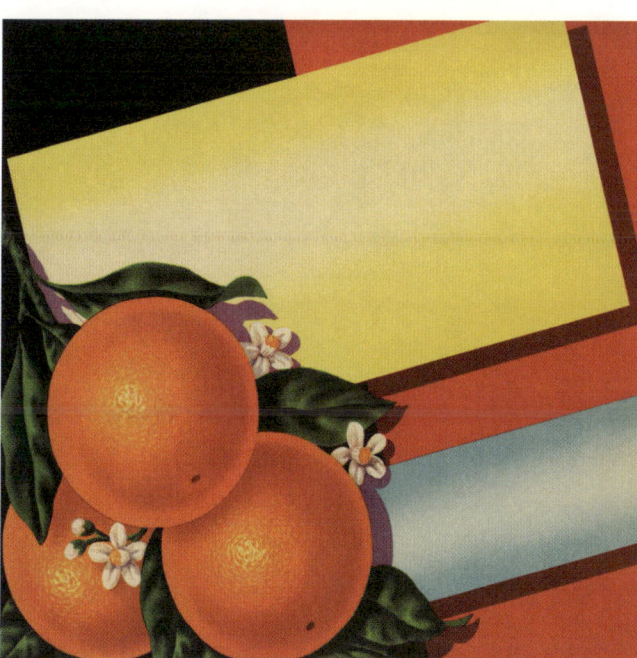

OVERLEAF: Lemon crate labels, California (p. 244). Grapefruit crate labels, Florida and Texas (p. 245). Orange crate labels, California (pp. 246–47). Orange crate labels, Spain (pp. 248–51).

In popular culture and decoration, there were not only the orange-box labels, also shown here, but even the wrappers (often printed with toxic inks) around individual fruits.

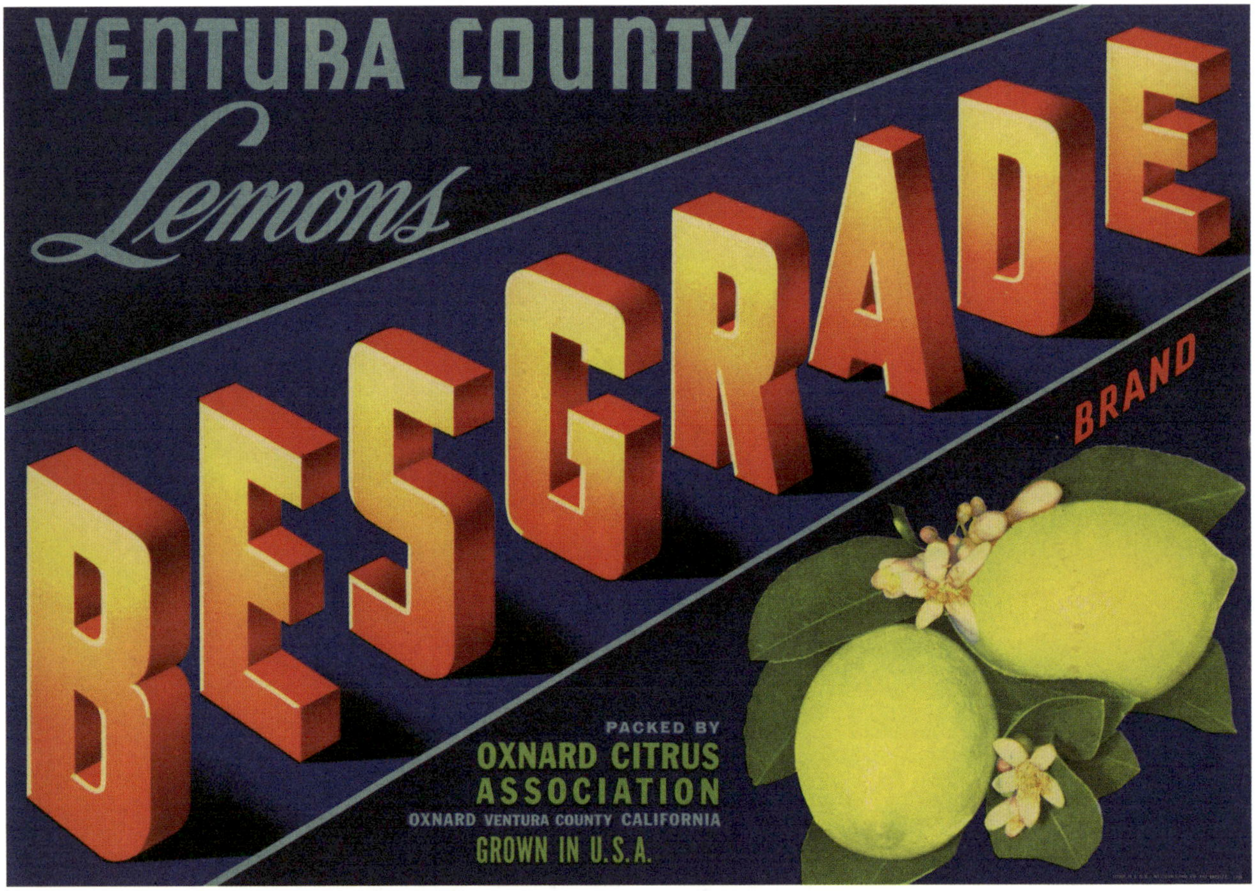

VENTURA COUNTY
Lemons
BESGRADE
BRAND

PACKED BY
**OXNARD CITRUS
ASSOCIATION**
OXNARD VENTURA COUNTY CALIFORNIA

GROWN IN U.S.A.

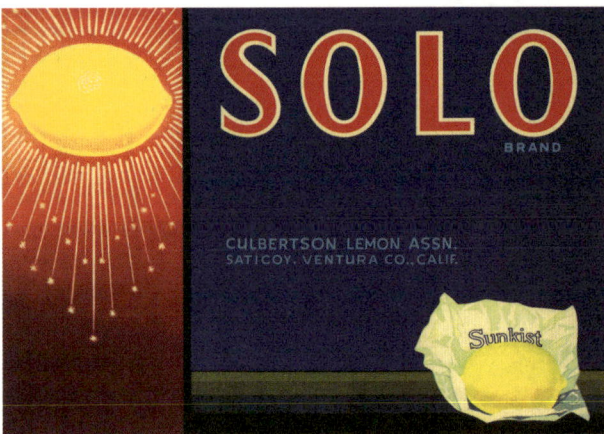

SOLO
BRAND

CULBERTSON LEMON ASSN.
SATICOY, VENTURA CO., CALIF.

Sunkist

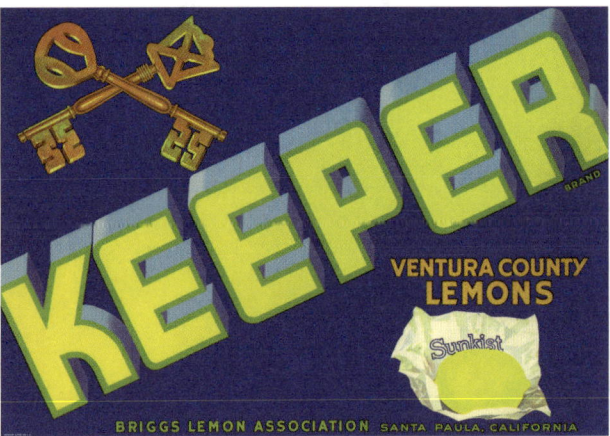

KEEPER
BRAND

**VENTURA COUNTY
LEMONS**

Sunkist

BRIGGS LEMON ASSOCIATION   SANTA PAULA, CALIFORNIA

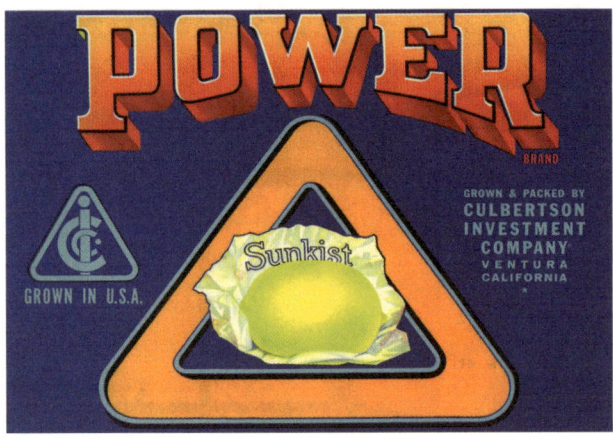

POWER
BRAND

GROWN & PACKED BY
**CULBERTSON
INVESTMENT
COMPANY**
VENTURA
CALIFORNIA

GROWN IN U.S.A.

Sunkist

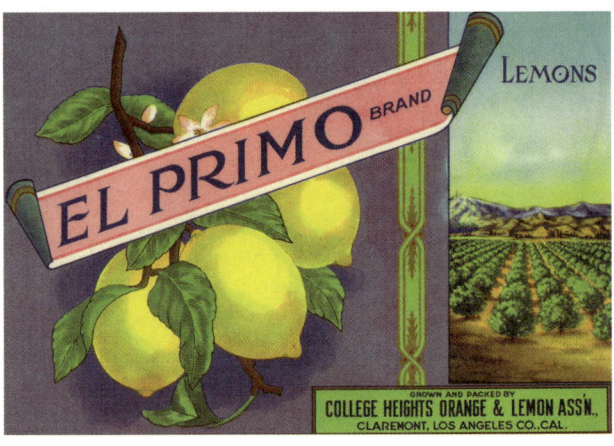

EL PRIMO
BRAND

LEMONS

GROWN AND PACKED BY
COLLEGE HEIGHTS ORANGE & LEMON ASS'N.
CLAREMONT, LOS ANGELES CO., CAL.

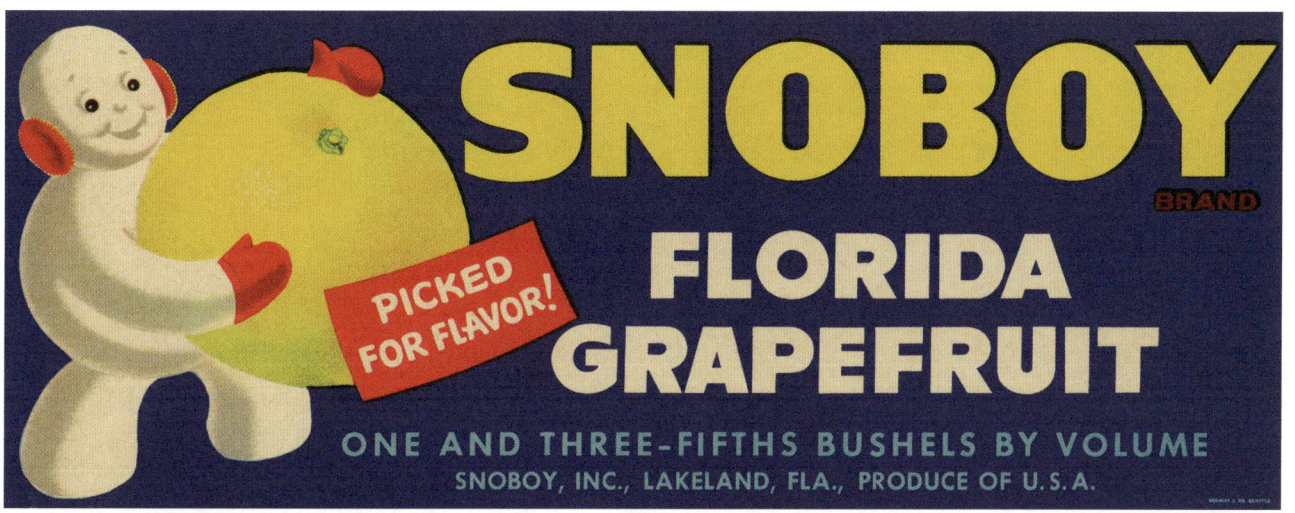

SNOBOY BRAND

FLORIDA GRAPEFRUIT

ONE AND THREE-FIFTHS BUSHELS BY VOLUME
SNOBOY, INC., LAKELAND, FLA., PRODUCE OF U.S.A.

PICKED FOR FLAVOR!

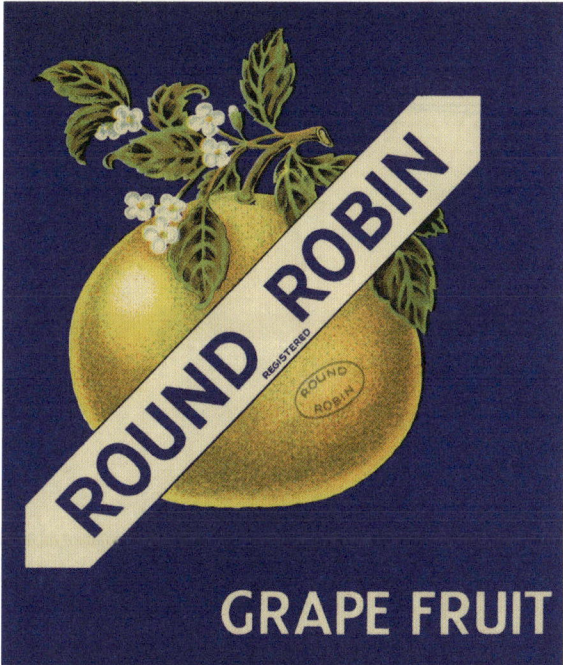

ROUND ROBIN

GRAPE FRUIT

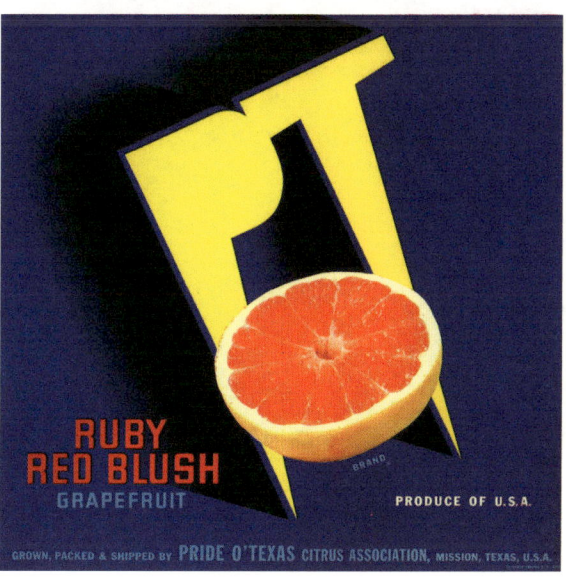

RUBY RED BLUSH GRAPEFRUIT

GROWN, PACKED & SHIPPED BY PRIDE O'TEXAS CITRUS ASSOCIATION, MISSION, TEXAS, U.S.A.

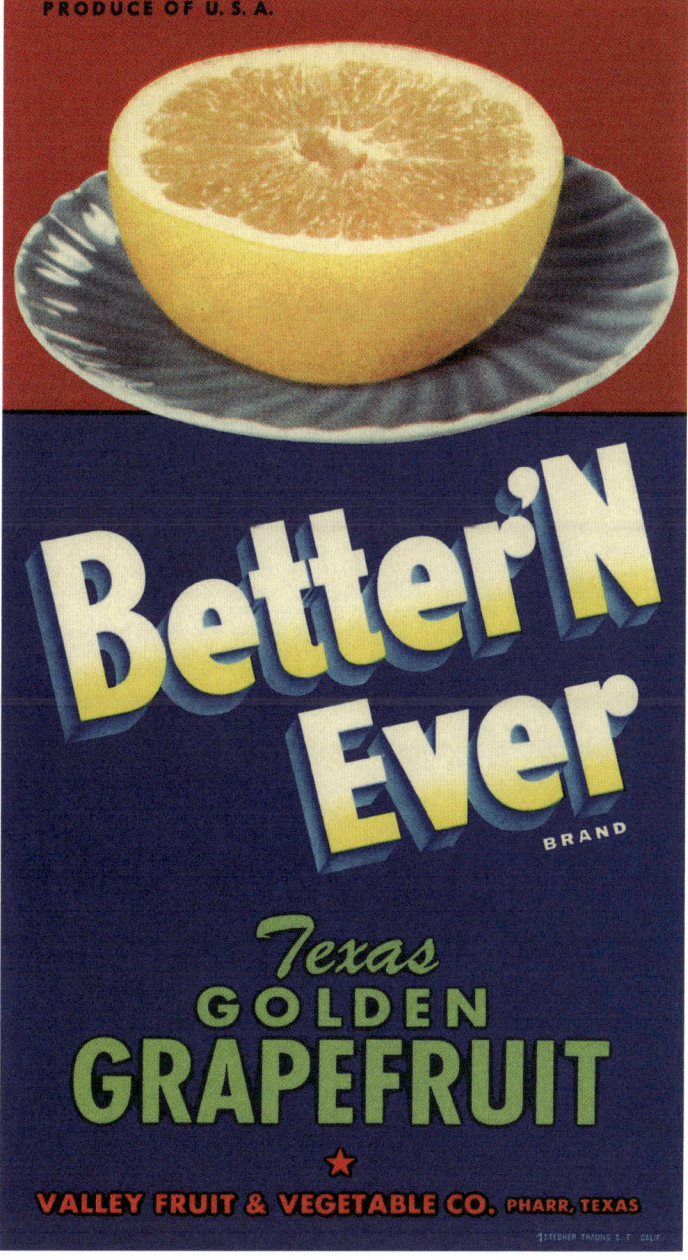

PRODUCE OF U.S.A.

Better'N Ever BRAND

Texas GOLDEN GRAPEFRUIT

VALLEY FRUIT & VEGETABLE CO. PHARR, TEXAS

PRODUCE OF U.S.A.

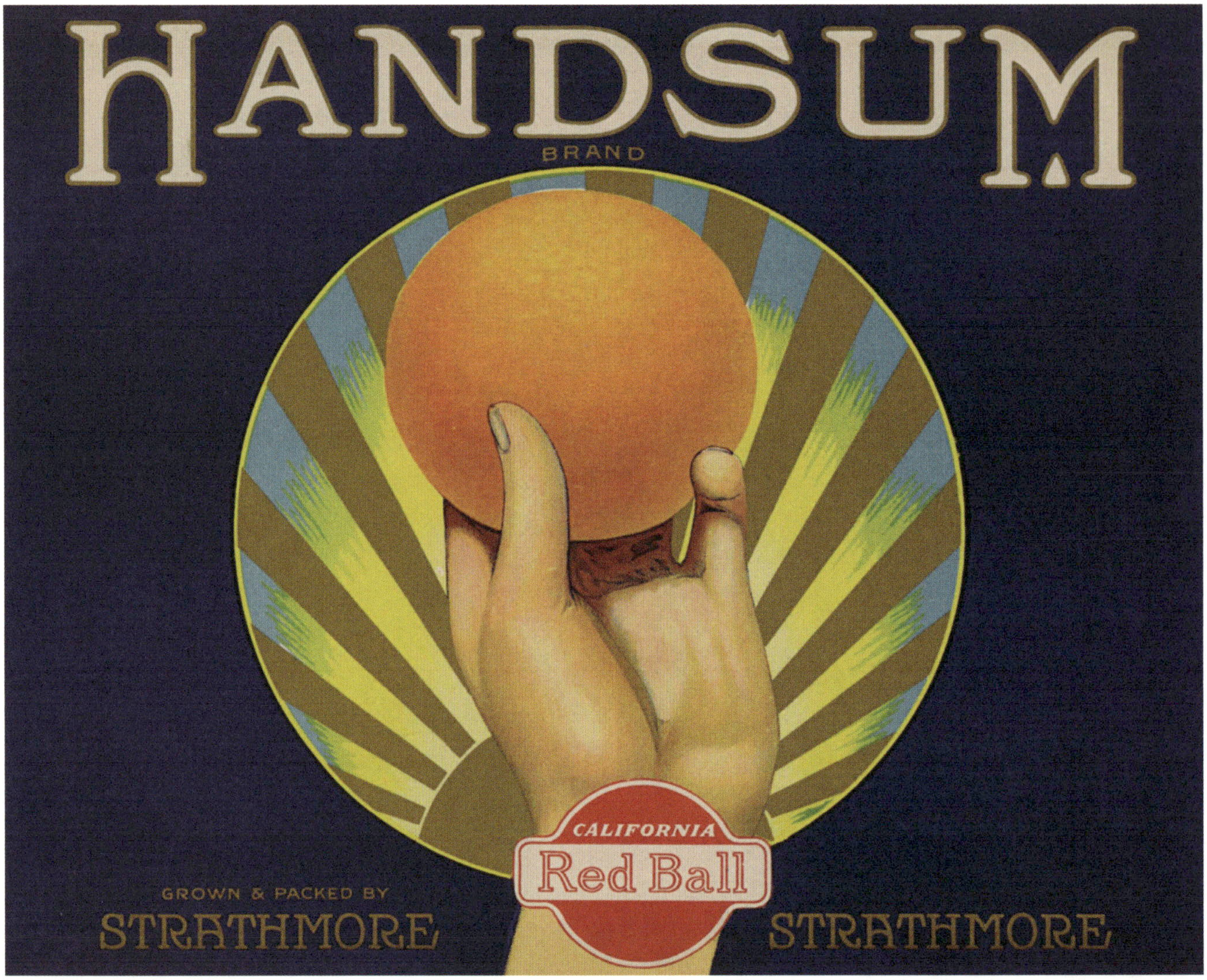

HANDSUM
BRAND

CALIFORNIA
Red Ball

GROWN & PACKED BY
STRATHMORE   STRATHMORE

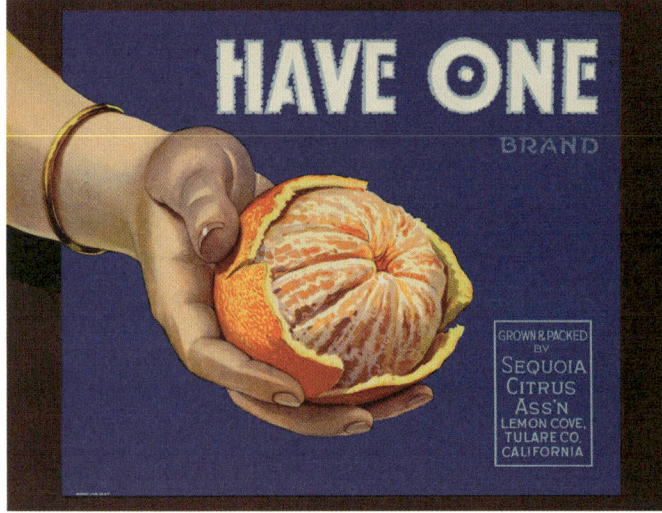

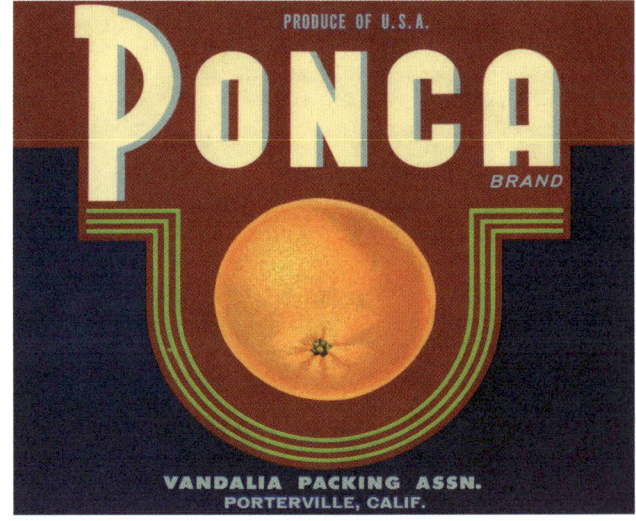

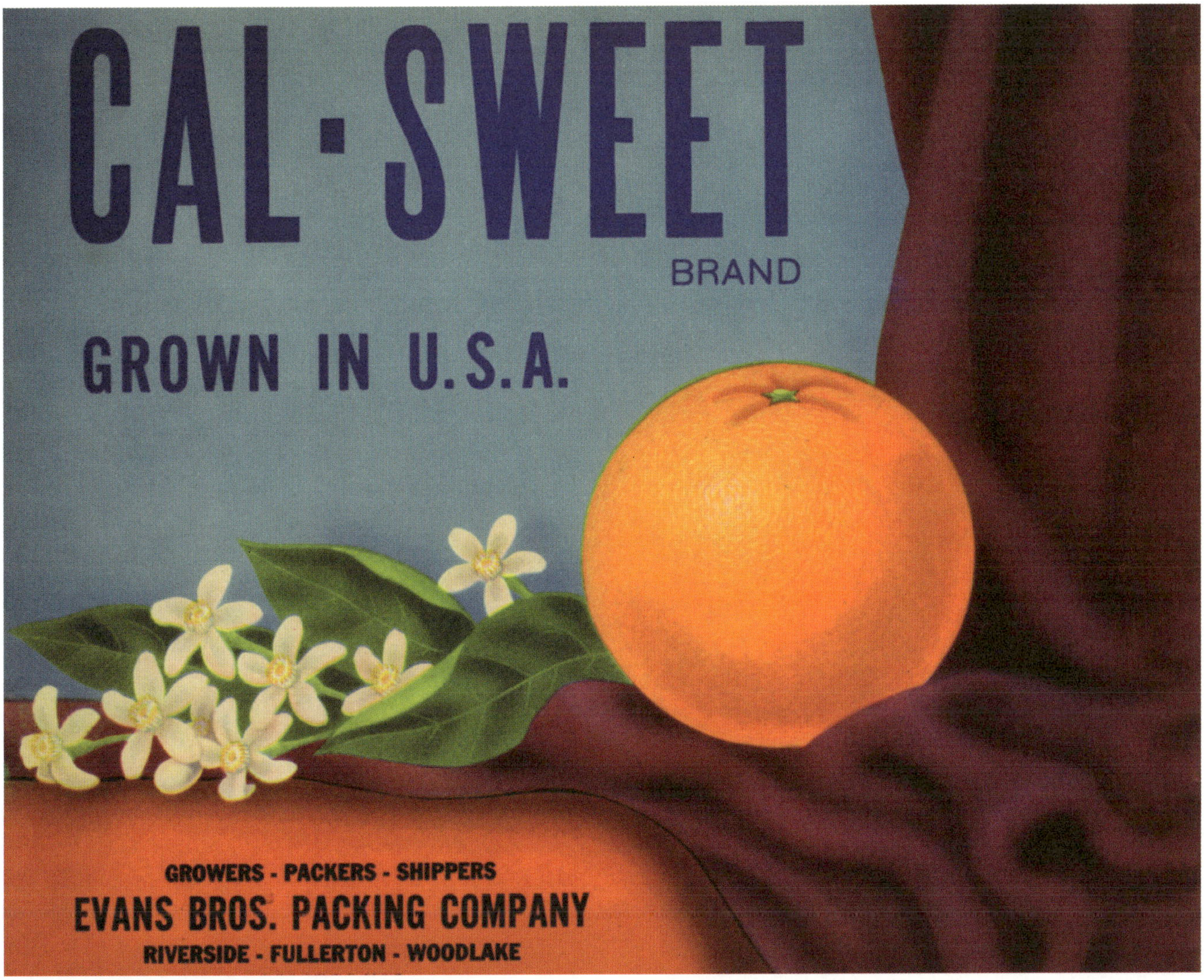

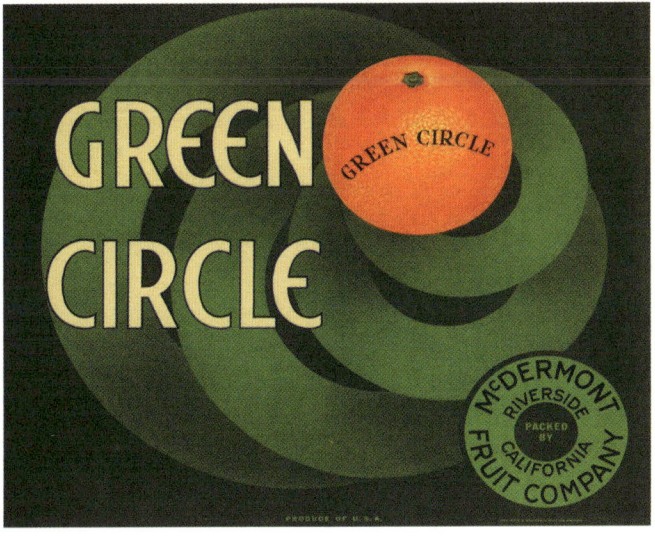

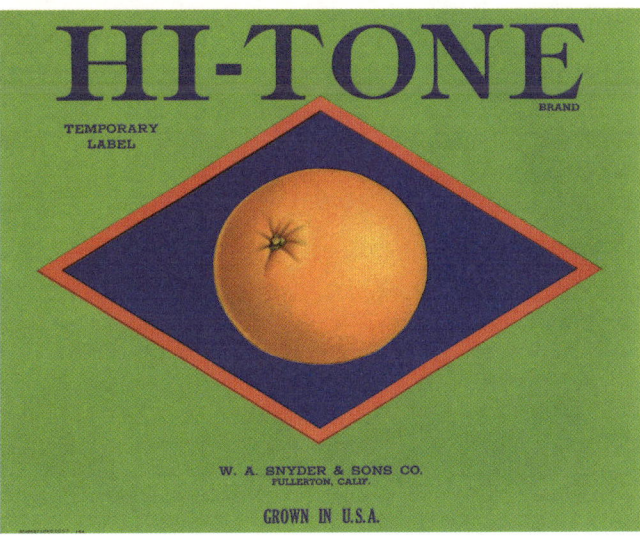

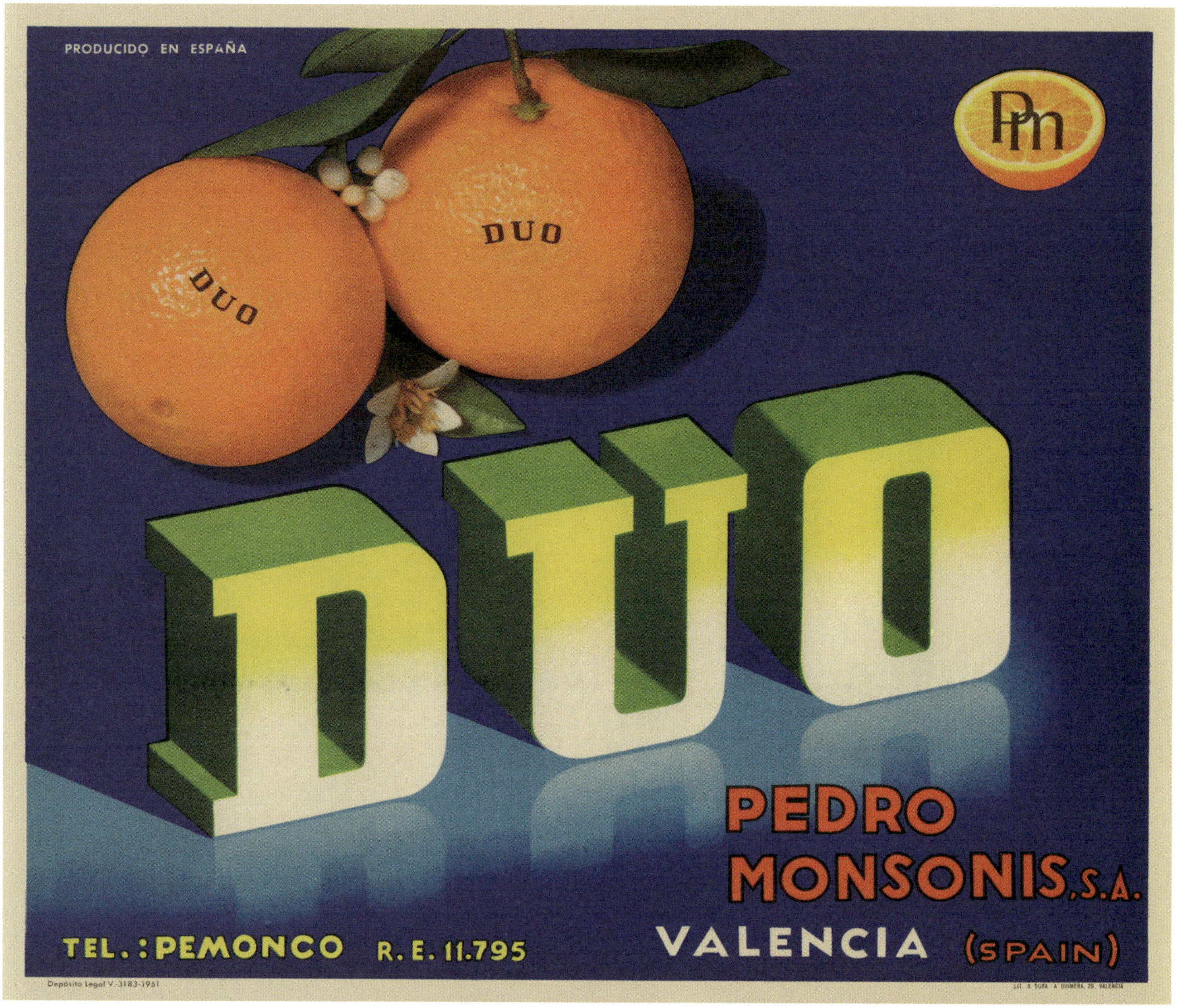

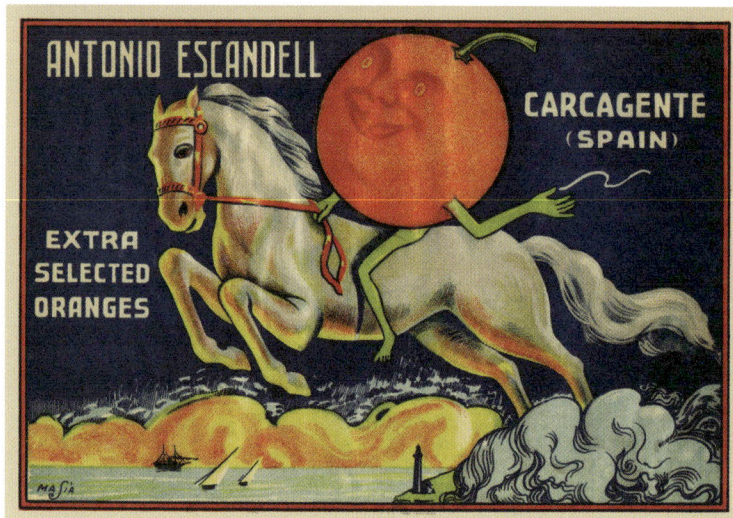

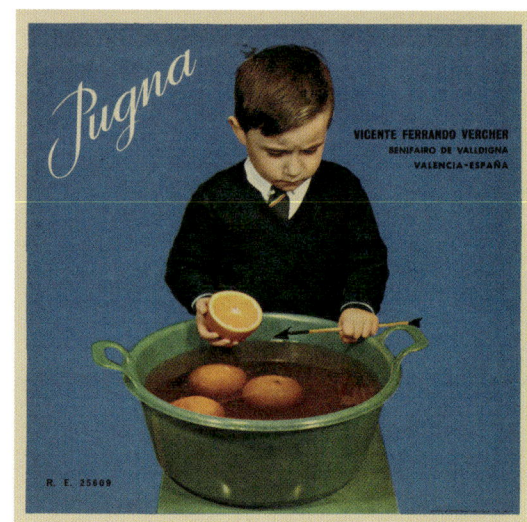

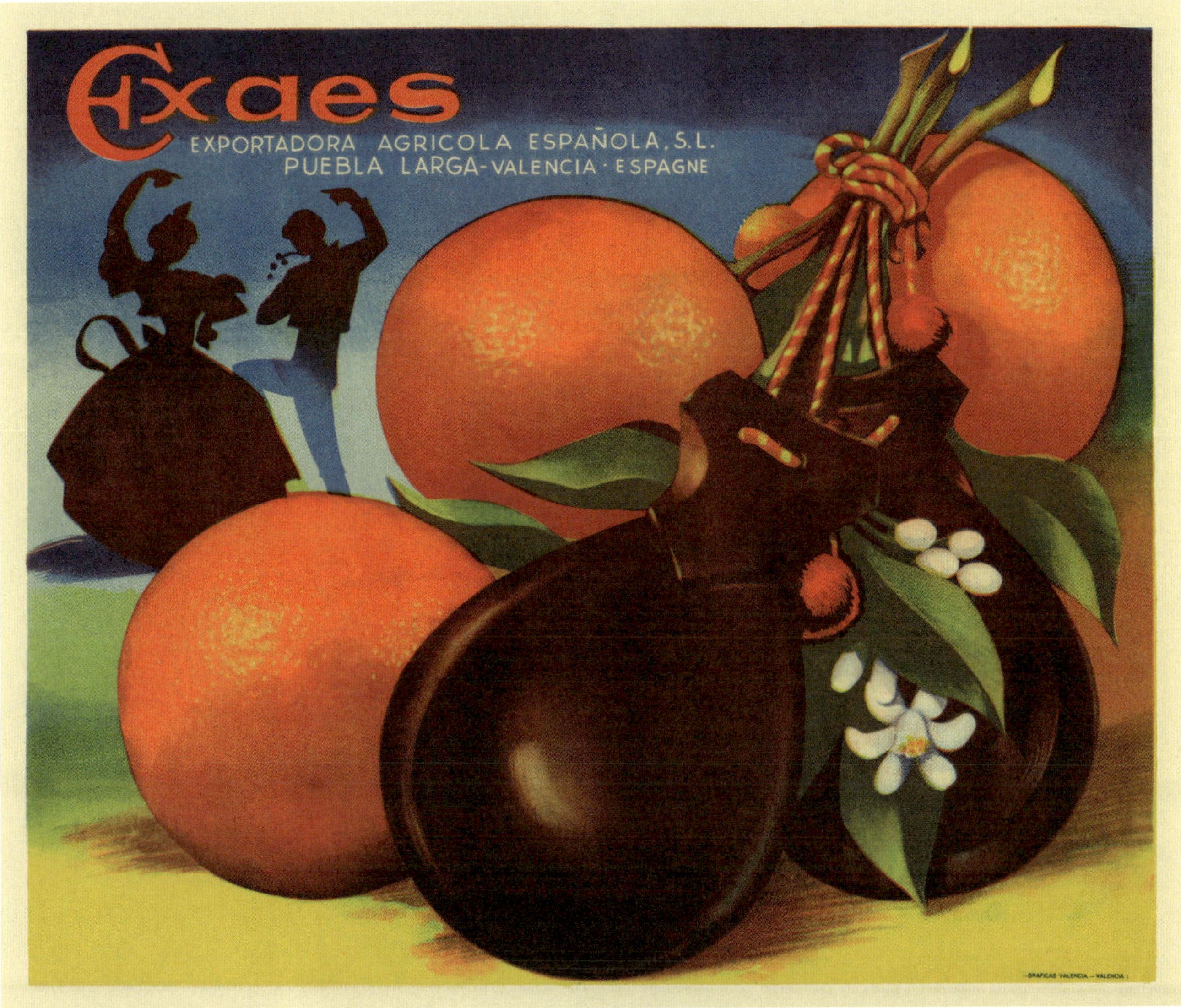

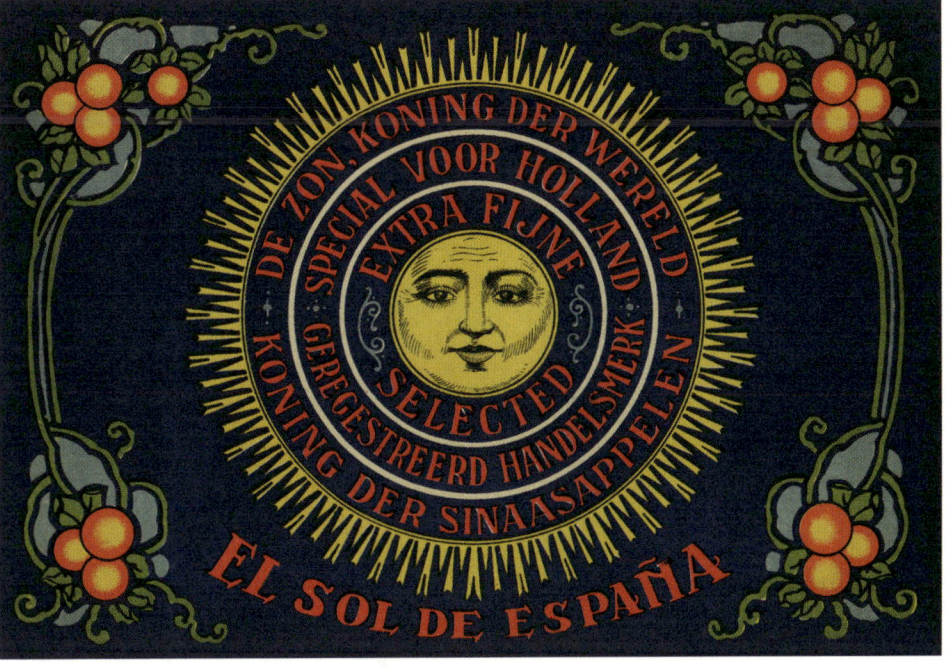

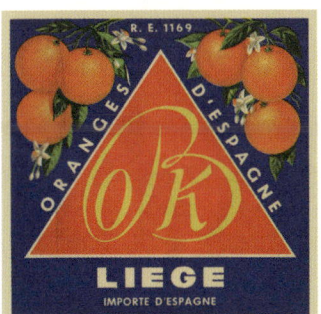

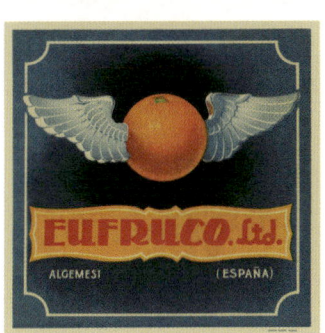

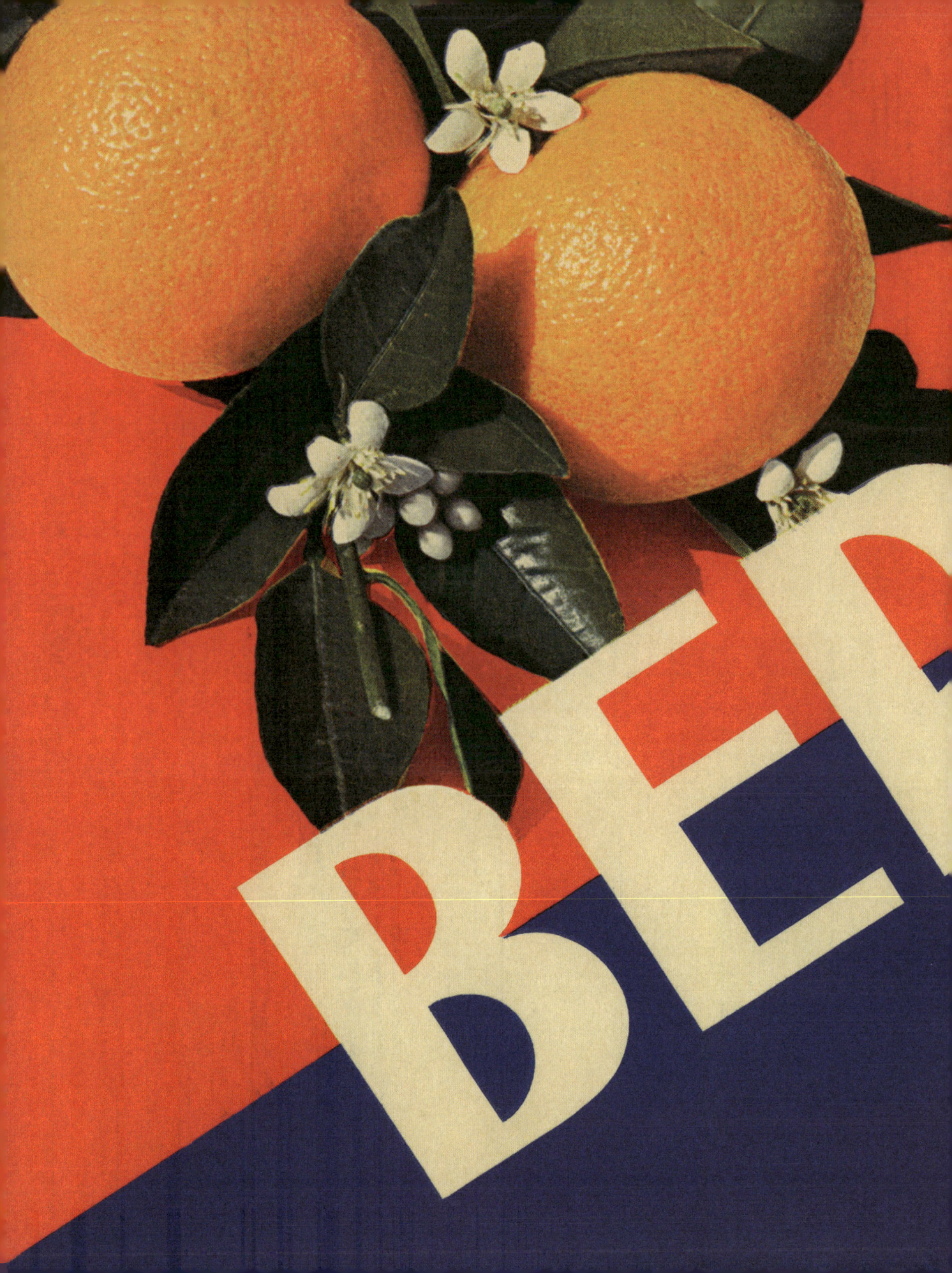

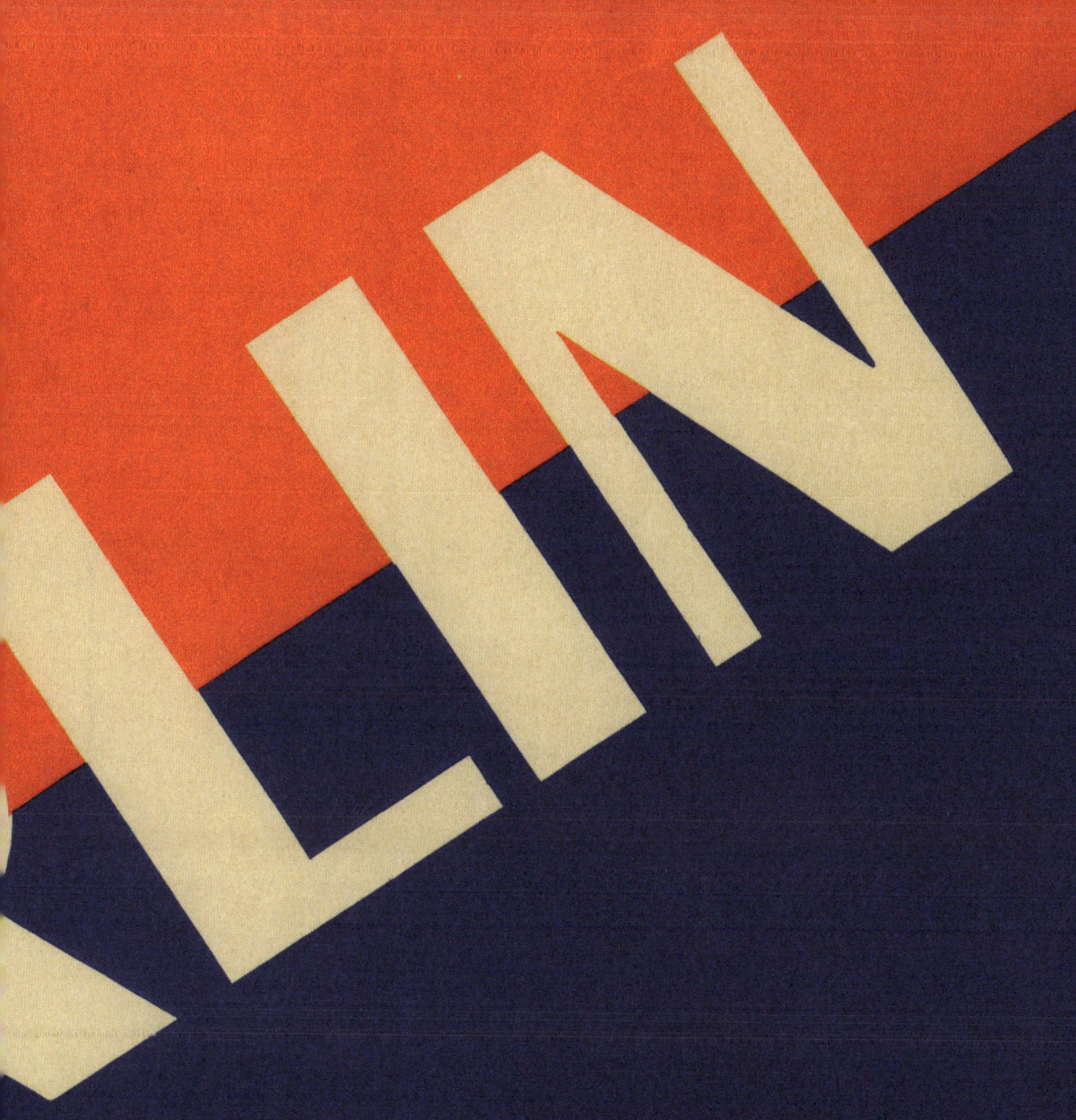

SALA-GOMEZ, S. L.

PEGO - SPANIEN - R. E. 8789

PRODUCIDO EN ESPAÑA

NOTES

INTRODUCTION

1. https://knoema.com/atlas/World/topics/Agriculture/
Crops-Production-Quantity-tonnes/Citrus-fruit-
production, accessed 14 February 2023.
2. https://www.statista.com/statistics/961248/production-
of-apples-worldwide/, accessed 14 February 2023.
3. https://knoema.com/atlas/World/topics/Agriculture/
Crops-Production-Quantity-tonnes/Bananas-
production, accessed 14 February 2023.
4. https://www.producereport.com/article/world-citrus-
organization-announces-202122-statistics,
accessed 14 February 2023.
5. Meiggs, *Trees and Timber*, pp. 286–91.
6. Strasburger, *Rambles*, p. 53.
7. Mabberley, 'Citrus'.
8. Karsten, *Old Company's Garden*, p. 17 f.n.
9. Privé-Gill, 'Eocene'.
10. Fischer & Butzmann, '*Citrus meletensis*'.
11. Xie et al., '*Citrus linczangensis*'.
12. Mabberley, 'Biology of the citron'.
13. Heads, *Molecular panbiogeography* pp. 420–22.
14. Kubitzki, 'Rutaceae'; Mabberley, *Plant-book*, passim;
Appelhans et al., 'New subfamily classification'.
15. Bowett, *Woods in British Furniture-making*, pp. 63,
110, 175–76.
16. Hammond and Scullard, *Oxford Classical Dictionary*,
p. 511.
17. Bartholomew and Reed, 'General morphology', p. 693
18. Kretschmar and Baumann, 'Caffeine'
19. Wright et al., 'Caffeine in floral nectar'; subsequent
findings in other plants show that nicotine has a
similar effect.
20. Muth et al., 'Discovery of octopamine and tyramine'.
21. Ali and Celik, 'Cytotoxic effects'.
22. Kato-Noguchi and Tanaka, 'Potential'.
23. *Common Sense Pest Control* 19(1): 16 (2003).
24. Agusti et al., 'Advances'.
25. Agusti and Primo-Millo, 'Flowering and fruit-set'.
26. Bain, 'Valencia orange'.
27. Bartholomew and Reed, 'General morphology', p. 695.
28. Letter to the author, 21 July 2016.
29. e.g. Lim, 'Citrus and citroids'.
30. Bonavia, *Cultivated Oranges and Lemons*, pp. 40–41,
51, 65, 194.
31. Ibid., pp. 202, 203.
32. Corlett, 'Impact of hunting'.
33. Letter to the author, 21 July 2016.
34. Letter to the author, 12 March 2023.
35. Juniper and Mabberley, *Extraordinary Story of
the Apple*.
36. D. K. Hore letter to S. K. Malik and J. Wearn,
8 March 2010.
37. *Save Our Citrus*, USDA blog, 2012.

THE ANCIENT WORLD

1. Needham, *Science and Civilisation*, p. 104.
2. Scora, 'On the history' p. 369: jambira used for both
citron and lemon.
3. Needham, *Science and Civilisation*, pp. 82–3.
4. Ibid., pp. 89, 104, 363.
5. Ibid., p. 104.
6. Ibid., p. 365.
7. Ibid., pp. 88–89, 363.
8. Ibid., p. 365.
9. Schafer, *The Golden Peaches of Samarkand*, p. 120.
10. *Time*, 29 August 1988, p. 61.
11. Nesom, '*Citrus trifoliata*'.
12. *Poncirus* of the erratic polymath C. S. Rafinesque (in
his *Flora Telluriana*, 1838, p. 143) with no etymology
but apparently derived from 'poncire', a kind of citron,
though perhaps suspiciously like '[*Limon*] *poncinus*',
the *ponzino* lemon grown in southern Italy.
13. Juniper and Mabberley, *Extraordinary Story of the
Apple*, p. 114 etc.
14. Mudge et al., 'History of grafting', p. 447.
15. Spencer, *Oriental Asia* pp. 89, 92.
16. Scora, 'On the history'.
17. It has been argued (see Uma Shaanker and Ganeshaiah,
'Polyembrony in plants') that, according to inclusive
fitness models, it is to the mother plant's advantage
(in terms of production of numbers of offspring) by
comparison with 'sibling rivalry' between offspring.
18. Wang et al., 'Genomic analyses'.
19. Maheshawari, *Introduction to the Embryology*,
pp. 226, 335.
20. Needham, *Science and Civilization*, p.363.
21. Ibid., p.369.
22. Zhang and Mabberley, 'Citrus'.
23. Needham, *Science and Civilization*, p. 371.
24. Wu et al., 'Diversification'; Mabberley and Xu,
'Proposal'.
25. Very often the name 'Citrus sinensis' was used for
cultivated chinotto plants in Europe, though it was
more universally applied to what is now known as
C. × aurantium Sweet Orange Group (Mabberley
2023), yet another of the very many nomenclatural
confusions surrounding citrus.
26. It is unclear that the currently grown 'chinotto' is
exactly the same as that figured in the past, so all

these mutants are included in the Myrtifolia Group, the earliest valid name for one of them being 'Humile' (*Auratium humile* of Miller, 1768).

27. Li, 'Origin and development'.
28. Mabberley and Xu, 'Proposal'.
29. Needham, *Science and Civilization*, p. 106.
30. Hagerty (trans), *Monograph on the Oranges of Wên-chou, Chekiang*.
31. Early recorded in Preyel, *Artificia hominum*, p. 997; Preyel's name appears only as 'A. P. F. B.' at the end of the Praefatio.
32. On the other hand, it is held by www.etymologiebank.nl that it comes directly from the Portuguese *pomo limões* a name, given by Portuguese mariners and taken over (and corrupted) by the Dutch.
33. Barry et al., 'Commercial scion varieties'.
34. Bonavia, *Oranges and Lemons*, p. 40.
35. Kumamoto et al., 'Forbidden fruit'; the *New York Times Magazine*, 16 December 2007, featured 'Captain Shaddock's Pomelo Oil' promoted by the Chop Shop barber in New York City.
36. Rumpf called it 'Decumanum', i.e. large, which name Linnaeus later used to replace his 'var. *grandis*', when assessing it anew as a species in its own right. The accepted Latin name today, *Citrus maxima*, is based on *Aurantium maximum*, which is based on Rumpf's illustration in his *Herbarium Amboinense*.
37. Beekman, *Ambonese Herbal* 2, p. 140.
38. Ibid., p. 142.
39. More likely *Strychnos nux-vomica* (see Church 1981).
* Quote on p. 34 see Needham, *Science and Civilisation*.
40. The cultivated plant collected by Bunge is considered by certain workers to be a cross between different cultivar groups treated as species by some.
41. Coats, *Plant Hunters*, p. 65.
42. Bretschneider, *History of European Plant Discoveries*, pp. 323–30.
43. García-Lor et al., 'Genetic diversity'; Ng, 'Pomelo'.
44. 'Pierre Julien Cavalerie', Wikipedia, https://fr-m-wikipedia-org.translate.goog/wiki/Pierre_Julien_Cavalerie?_x_tr_sl=fr&_x_tr_tl=en&_x_tr_hl=en&_x_tr_pto=sc, accessed 3 March 2023.
45. In modern Chinese 'youzi', as pronounced in contemporary Mandarin, is used for the pomelo, yuzu being called xiangcheng or luohancheng.
46. Tanaka, 'Citrus fruits'.
47. Mahon, 'Asian citrus'.
48. Compton and Thijsse, 'The remarkable'.
49. Ibid.
50. Tanaka, 'Taxonomy'.
51. Shimizu et al., 'Hybrid origins'.
52. Tanaka, 'Citrus fruits'.
53. see Wu et al. 'Diversification'.
54. 'Hinamatsuri', Wikipedia, https://en.wikipedia.org/wiki/Hinamatsuri, accessed 4 March 2023.

55. Andrews, 'Acclimatization'.
56. Ramón-Laca, 'Cultivated citrus'.
57. Espargham, S. 'The Mythical Criticism of the Girl of Narenj o Toranj Tale', *Literary Criticism* vol. 3, no. 12, 2010, ://lcq.modares.ac.ir/article-29-1257-en.html, accessed 26 February 2023.
58. Needham, *Science and Civilisation*, p. 37.
59. Valder, *Garden Plants*, p. 254.
60. Ferrari, *Hesperides*, p. 215; Eric Danell, pers. comm. 11 Nov 2016, 7 March 2023.
61. Lim, *Edible, Medicinal and Non-medicinal*, p. 683.
62. Innes Miller, *Spice Trade*, p. 194.
63. Swingle, 'Botany of citrus'.
64. Gmitter and Hu, 'Yunnan'.
65. Karp and Hu, 'The citron'.
66. Bonavia, *The Cultivated Oranges and Lemons*, p. 70.
67. Ramón-Laca, 'Cultivated citrus'; Wearn and Mabberley, 'Citrus and orangeries'.
68. Lane Fox. *Alexander*, p. 11.
69. Ibid., pp. 25–27.
70. Greene, *Landmarks*, p. 128.
71. Ibid., pp. 129 et seqq.
72. Hort (trans.), *Theophrastus*.
73. Though, bearing in mind its well-known medicinal uses, one cannot rule out some punning wordplay here.
74. cf. Greene, *Landmarks* passim.
75. Lipschits et al., 'Riddle'.
76. Laboury, 'Archaeological and Textual Evidence'.
77. *Enquiry into Plants* I, XIII, 4 (Hort, *Theophrastus* 1, p. 95).
78. Greene, *Landmarks*, p. 129.
79. Stearn, 'Earliest European acquaintance'.
80. Pagnoux et al., 'Introduction'
81. Ibid.
82. Andrews, 'Acclimatization'.
83. Fairclough, *Virgil*.
84. Calabrese, 'Origin and history'.
85. Dioscorides, *De materia medica*, p. 189.
86. Andrews, 'Acclimatization'.
87. Ramón-Laca, 'Cultivated citrus'.
88. Tolkowsky, *Hesperides*, p. 62.
89. Gallesio, *Orange Culture*, p. 39.
90. Andrews, 'Acclimatization'.
91. In *shmitta* years (sabbatical, i.e. the seventh year of the agricultural cycle), fields, orchards etc. must be left unworked, so etrogim have to be imported to Israel from e.g. Morocco (Jeremy Montagu pers. comm. 7 October 2015).
92. Janick, 'Fruits of the Bibles'.
93. Interestingly, the Greek word *kedros* was the source of the Latin *citrus* (see p. 17), see Helmut Baumann. *The Greek Plant World: in Myth Art and Literature*, Portland, OR: Timber Press, 1993 ed. [tr. W. T. and E. R. Stearn], p. 139.
94. Tolkowsky, *Hesperides*, p. 53.

95. Loc. cit.
96. Andrews, 'Acclimatization'.
97. Ramón-Laca, 'Cultivated citrus'.
98. D. Moster, 'Does Peri Etz Hadar Mean Etrog?', *The Lehrhaus*, 20 September 2018, https://thelehrhaus. com/scholarship/does-peri-etz-hadar-mean-etrog/.
99. Gallesio, *Orange Culture*, p. 36.
100. Isaac, 'Influence of religion'.
101. Chancey, Mark A., and Adam Porter. 'The Archaeology of Roman Palestine.' *Near Eastern Archaeology*, vol. 64, no. 4, December 2001, pp. 164–203.
102. Isaac, 'Influence of religion'.
103. Gallesio, *Orange Culture*, p. 39.
104. Tolkowsky, *Hesperides*, p. 171.
105. Goldschmidt, 'New insights'; persistence can be induced in oranges by the use of the synthetic plant hormone Picloram.
106. 'Diamante' is a likely father of the *pompia*, a kind of lemon characteristic of Sardinia (Luro et al., 'Genetic origin of Pompia').
107. Maruca et al., 'Religious and cultural significance'.
108. Mudge et al., 'History of grafting', p. 451.
109. Salmon, 'Corfu Etrog polemic'.
110. Volkamer, *Hesperides*, pp. 202–203.
111. It was later known, again a confusion, as 'C. sinensis', a name formerly used for *C. × aurantium* L. Sweet Orange Group.
112. Mabberley, 'Biology of the citron'.
113. Mabberley, 'Citrus'.
114. Needham, *Science and Civilisation*, p. 370 f.n.
115. Laufer, 'Lemon'.
116. *Oxford English Dictionary*.
117. Needham, *Science and Civilisation*, p. 370 f.n.
118. Needham, *Science and Civilisation*, p. 377.

HERBALS TO HESPERIDES

1. Hibbert, *House of Medici*; Padgett and Ansell, 'Robust action'.
2. Tolkowsky, *Hesperides*, pp. 168 et seqq.; Ruggiero, 'Coat of arms'.
3. Tolkowsky, *loc. cit.*
4. Isaac, 'Influence of religion'.
5. Schottenhammer, 'Transfer'; Chaffee, *Muslim Merchants*.
6. Anon., *Ancient Accounts*, pp. 13 et seqq. and *Remarks* pp. 124 et seqq.
7. Ibid.
8. Hämeen-Anttila, *Last Pagans of Iraq*.
9. Crone and Cook, *Hagarism*.
10. Bonavia, *Cultivated Oranges and Lemons*, pp. 247–48.
11. Webber, 'Cultivated varieties', p. 9.
12. Pérès, 'Abu 'l-Khayr al-Ishbili'.
13. Clément-Mullet, *Le livre de l'agriculture*, p. 300.

14. Ramón-Laca, 'Cultivated citrus'.
15. Tolkowsky, *Hesperides*, pp. 123, 144.
16. Needham, *Science and Civilisation*, p. 367.
17. Gallesio, *Orange Culture*, p. 49.
18. Ibid., p. 54.
19. Ibid.
20. Webber, 'Cultivated varieties', p. 9.
21. Beekman, *Ambonese Herbal 2*, p.168.
22. Ramón-Laca, 'Cultivated citrus'.
23. Gallesio, *Orange Culture*, p. 49.
24. *Oxford English Dictionary*.
25. Spary, *Eating the Enlightenment*, pp. 133–39.
26. Tolkowsky, *Hesperides*, p. 298.
27. Gallesio, *Orange Culture*, p. 46.
28. Mayer, *Crusades*, p. 185.
29. 6, tr. 1 cap xi 54–55.
30. Goody, *Culture of Flowers*, pp. 158, 167, 175, 246, 297.
31. Nelson, 'Victorian wedding flowers'.
32. Ibid.
33. Tolkowsky, *Hesperides*, pp. 153.
34. Zylberberg, 'Oranges and Seders'.
35. Attlee, *Land Where Lemons Grow*, p. 8.
36. Lightbown, *Sandro Botticelli*.
37. Attlee, *Land Where Lemons Grow*, pp. 13-17.
38. Ibid., p. 21.
39. Gallesio, *Orange Culture*, p. 28.
40. Gallesio, *Orange Culture*.
41. Anon., 'Some Hortulan Communications'.
42. Ragionieri, 'Origin'.
43. *Revue Horticole* 1, p. 341, 1831.
44. Uphof, 'Wissenschaftliche Beobachtungen'.
45. Zhou et al., 'Interactions'.
46. Liu, 'Historical and modern genetics'.
47. Juniper and Mabberley, *Extraordinary Story of the Apple*, p. 135.
48. Attlee, *Land Where Lemons Grow*, p. 25.
49. *Howick Hall and Gardens*, website, https://howickhallgardens.com/things-to-see-do-at-howick-hall-gardens/earl-grey-tea-house-at-howick-hall-gardens/, accessed 7 April 2023.
50. *Foods of England*, www.foodsofengland.co.uk/earlgreytea.htm, accessed 7 April 2023.
51. Bonavia, *Oranges and Lemons*, p.142.
52. Attlee, *Land Where Lemons Grow*, p. 131.
53. Gallesio, *Orange Culture*, pp. 49–50.
54. Ibid., p. 50.
55. Paulus, 'Humanistic tradition'.
56. Ibid.
57. Raphael, *Oak Spring Pomona*, pp. 177–80; Freedberg, 'Cassiano, Ferrari'.
58. Freedberg, 'Cassiano, Ferrari'.
59. Haskell and McBurney, 'General Introduction'.
60. Ibid.
61. Freedberg, 'Cassiano, Ferrari'.

62. Ibid.
63. Ibid.
64. Tolkowsky, *Hesperides*, pp. 161–62.
65. Laszlo, *Citrus*, p. 11.
66. Freedberg, 'Cassiano, Ferrari'.
67. Beekman, *Ambonese Herbal* 1, pp. 1–156.
68. Beekman *Ambonese Herbal* 2, p. 163.
69. Ibid., pp.167–68.
70. Henniger 'Dutch contributions'.

THE HOUSE OF ORANGE

1. Wedgwood, *William the Silent*, passim.
2. Green, *Dictionary of Celtic Myth and Legend*, p. 33.
3. Not to be confused with Oranienbaum known by its post-war name of Lomonosov, the estate is the oldest of the imperial palaces around St Petersburg, Russia; it is held that a small greenhouse with orange trees was found on the land of the future estate, each with a label 'Oranienbaum'. Peter the Great therefore wanted the estate to be so named – and an orange tree became the symbol of the palace as well as the town.
4. Vorstand der Kulturstiftung DessauWörlitz (ed.), *Oranien-Orangen-Oranienbaum* passim.
5. Ramsay, *Peel*, pp. 21-48.
6. 'How an old Dutch flag became a racist symbol'. *Economist* 23 June 2015.
7. Van der Meer, 'History of *Citrus*'.
8. Ibid.
9. Dietz, *Port and Trade*, pp. 28–45 (no. 277), pp. 79–97 (no. 533).
10. Dietz, *Port and Trade*, pp. 1–14 (no. 45), pp. 28–45 (no. 212).
11. Wilson, *Marmalade*, p. 43.
12. Mathew, *Keiller's of Dundee*.
13. *Sydney Morning Herald* 27 May 2021.
14. Henniger, 'Dutch contributions'.
15. 'How did carrots become orange?', *The Economist*, https://www.economist.com/the-economist-explains/2018/09/26/how-did-carrots-become-orange, accessed 31 March 2023.
16. Raphael, *Oak Spring Pomona*, p. 183.
17. Wearn and Mabberley, Citrus and orangeries'; in an illuminated manuscript, *Manuel des vertus des animaux et des végétaux* (Ruas et al., 2017) of 1480, '*citrona et naratia*' (lemons and oranges) are shown in wooden planter-boxes, suggesting they were moved inside in winter, perhaps in Venice where the manuscript was illuminated.
18. Paulus, 'Humanstic tradition'.
19. Grove, *Green Imperialism*, p. 77.
20. Tolkowsky, *Hesperides*, pp. 206–207.
21. *Herbier de l'amateur* 7, sub t. 454 (1827).
22. McPhee, *Oranges*, p. 23.
23. Tolkowsky, *Hesperides*, p. 207.
24. Henrey, *British botanical and horticultural literature* 1, p. 182.
25. Raphael, *Oak Spring Pomona*, p. 184.
26. Ibid., p. 183.
27. Ibid., p. 192.
28. Ibid., p. 192-3).
29. Paulus, 'Humanistic tradition'.
30. Le Rougetel, *Chelsea gardener*, pp. 63-64.
31. see Wearn and Mabberley, 'Citrus and Orangeries'.
32. Gallesio, *Orange Culture*, p.14.
33. Ibid., p. 19.
34. Ibid.
35. Ibid., p. 25.
36. Ibid., p. 27.
37. Raphael, *Oak Spring Pomona*, p. 205.
38. Noehden, 'An account'.
39. Ibid.
40. 'Orangery Palace', Wikipedia, https://en.wikipedia.org/wiki/Orangery_Palace, accessed 26 March 2023.
41. Paulus, 'Humanistic tradition'.

EMPIRE, EXPORTS AND VITAMIN C

1. Grove, *Green Imperialism*, p. 29.
2. Pauling, *Vitamin C*, p. 16.
3. Quoted in Bown, *Scurvy*, p. 46.
4. Tickner and Medvei, 'Scurvy'.
5. Bown, *Scurvy*, p. 45.
6. Ibid., pp. 56–57.
7. Ibid., p. 44.
8. Tannahill, *Food in History*, p 227.
9. Bown, *Scurvy*, passim.
10. Ibid., p. 18.
11. Ibid., pp. 361–63.
12. Ibid., p. 35.
13. Ibid., pp. 88–92.
14. Tolkowsky, *Hesperides*, p. 253.
15. Ibid.
16. Ibid., pp. 256–57.
17. Karsten, *Old Company's Garden*, pp. 31, 42.
18. Bown, *Scurvy*, p. 50.
19. Ibid., p. 36.
20. Wilson, *Marmalade*, p. 125.
21. Bartholomew, 'Lind'.
22. Drummond and Wilbraham, *Englishman's Food*, pp. 391–92.
23. Bown, *Scurvy*, p. 121.
24. Ibid., p. 131.
25. Kodicek and Young, 'Cook'.
26. Williams, *Naturalists at Sea*, pp. 79–80, 270.
27. Bown, *Scurvy*, p. 209.
28. Ibid., p. 255.
29. Pauling, *Vitamin C*, p. 15.

30. Drummond and Wilbraham, *Englishman's Food*, p. 396.
31. Bennett, 'Put the lime in the coconut'.
32. Webber, 'Cultivated varieties', p. 14.
33. Knox, *Ceylon*, pp. 126-7.
34. B. M. P. Singhakumara pers. comm. 14 iii 14, ex B. B. Wijeratne, School Teacher, Wegama Maha Vidyalaya, Kivulapitalanda, Bibile (Monaragala district), Sri Lanka.
35. Beekman, *Ambonese Herbal* 2, p. 155.
36. Ibid., p. 157.
37. Honychurch, *A to Z*, Rose, Lime Company.
38. Bonavia, 'Limes'.
39. Mabberley, 'Lime leaves'.
40. Lim, *Edible Medicinal and Non-medicinal* p. 745.
41. *True Secrets of Key West Revealed*, Eden Entertainment, 2011, p. 49 [online edition], https://www.google.com.au/books/edition/True_Secrets_of_Key_West_Revealed/ZvEAAwAAQBAJ, accessed 13 May 2023.
42. Tolkowsky, *Hesperides*, p. 258.
43. Barns, 'America's first citrus grove'.
44. Passos et al., 'Navel orange'.
45. Laszlo, *Citrus*, p. 30.
46. Hume, 'History'.
47. Webber, 'History and development of the citrus industry' in *The Citrus Industry* vol. 1, p. 1.
48. Kumamoto et al., 'Forbidden fruit'.
49. Ibid..
50. Swingle, 'Botany of citrus', p. 27.
51. Webber, 'Cultivated varieties', p. 579.
52. Laszlo, *Citrus*, p. 32.
53. Wästberg, *Sparrman*, p. 136.
54. Bennett, 'Put the lime in the coconut'.
55. Tolkowsky, *Hesperides*, p. 263.
56. *Gardeners' Chronicle* 17 July 1886, p. 77.
57. Tolkowsky, *Hesperides*, p. 264.
58. Passos et al., 'Navel orange'.
59. 'Washington Navel Orange tree', Wikipedia, https://en.wikipedia.org/wiki/Washington_navel_orange_tree_(Riverside,_California), accessed 8 April 2023.
60. Webber, 'Cultivated varieties', p. 523.
61. Laszlo, *Citrus*, p. 34.
62. Tobey and Wetherell, 'Citrus industry'.
63. Klein, The Fruits of Empire.
64. *Gardeners' Chronicle*, 26 Dec 1885, p. 818.
65. Hume, 'History of citrus culture in Florida'.
66. *Gardeners' Chronicle*, 3 July 1886, p. 15; 9 Sept 1886, p. 467.
67. Winmill, *George and Sarah Suttor*, p. 76.
68. Ibid., p. 145.
69. 'Lemon Industry', http://www.livinglibrary.edu.nf/_/Lemon_Industry.html, accessed 9 April 2023.
70. Anon., 'Lane Cove orangery'.
71. Bennett, *Gatherings of a Naturalist*, p. 314.
72. Dowe and Broughton, 'F. M. Bailey's ascent'.
73. Peekel, *Flora*, p. 1.
74. Richards, *The Secret War*, p. 93, 235-36.
75. Dowe and Broughton, 'F. M. Bailey's ascent'.
76. Winters, '*Microcitrus papuana*'.
77. Forster and Smith, *Citrus wakonai*.
* Quote on p. 189 from J. Gasquet, *Cézanne: A Memoir with Conversations*, London, 1991, p. 220.

PROGRESS AND PERILS

1. Himes, *Medical History of Contraception*, pp. 17-18, 181, 195.
2. The late Roger Short (1930-2021), University of Melbourne pers. comm. 6 February 2004.
3. Short et al., 'Lemon and lime juice'.
4. Hort, *Theophrastus* I, pp. 311, 313.
5. Alvarez Arias & Ramón-Laca, 'Pharmacological properties'.
6. Baldini, 'The role'.
7. Imbesi and De Pasquale, 'Citrus species', pp. 580 *et seqq*.
8. Alistair Hay, pers. comm. 20 March 2023.
9. Bisignano and Saija, 'Biological activity'.
10. *New York Times*, 10 January 1984, Sect. C, p. 3.
11. Panara et al., 'Review'.
12. Genovese et al., 'Prenylated coumarins'.
13. Liu et al., 'History'.
14. McPhee, *Oranges*, p. 127.
15. Van der Meer, 'History'.
16. *Independent on Sunday*, 30 March 2008, p. 75.
17. Webber, 'Cultivated varieties', p. 563.
18. Tanaka, 'Revision'.
19. Webber, 'Cultivated varieties', p. 567.
20. Laszlo, *Citrus*, p. 23.
21. Webber, 'Cultivated varieties', p. 558.
22. Biography, The Swingle Plant Anatomy Reference Collection, http://swingle.miami.edu/wtswingle.html, accessed 30 April 2023.
23. Laszlo, *Citrus*, pp. 40-41.
24. Webber, 'Cultivated varieties', p. 643.
25. Webber and Swingle, 'New citrus creations'.
26. Holmes, 'Great Socialist Beyond'.
27. Swingle et al., 'New citrus hybrids'.
28. Swingle, 'Botany of Citrus', p. 389.
29. Lim, *Edible Medicinal and Non-medicinal*, p. 732.
30. Mabberley, 'Classification' .
31. Linnaeus, *Species plantarum*.
32. Swingle, 'Botany of Citrus', p. 388.
33. Ibid., p. 389.
34. Alexander Sennikov, pers. comm. 4 May 2023.
35. Mabberley, 'Classification'.
36. Webber, 'Cultivated varieties', p. 630.
37. Recupero et al., 'New promising citrus hybrids'.
38. Mabberley, 'Sweet ... and sour'.

39. Mabberley and Placito, *Algarve Plants and Landscape*, p. 156.
40. Hensz, 'Mutation breeding'.
41. *New York Times*, 14 March 2007, p. D5.
42. Laszlo, *Citrus*, p. 95.
43. McPhee, *Oranges*, pp. 117–118 .
44. Laszlo, *Citrus*, p. 97.
45. https://www.tropicana.com/our-story/, accessed 1 May 2023.
46. McPhee, *Oranges*, pp. 125–26.
47. Ibid., p. 115.
48. https://www.statista.com/statistics/895621/us-per-capita-consumption-of-orange-juice/, accessed 1 May 2023.
49. https://www.statista.com/statistics/1044896/world-orange-juice-major-producers/, accessed 1 May 2023.
50. Gao et al., 'Consumer preferences'.
51. USDA-ERS 2009a; USDA-ERS 2009b.
52. Gao et al., 'Consumer preferences'.
53. Laszlo, *Citrus*, pp. 106-107.
54. https://www.exchangechambers.co.uk/landmark-high-court-judgment-in-brazilian-orange-cartel-case/, accessed 7 May 2023; R. Brito, 'Brazil prosecutors, Reuters, https://www.reuters.com/world/americas/brazil-prosecutors-seek-25-bln-damages-alleged-orange-juice-cartel-2023-04-06/, accessed 7 May 2023.
55. https://www.bbc.com/future/article/20230424-how-agrivoltaics-helped-save-italys-citron-and-lemon-trees - accessed 7 May 2023.
56. McPhee, *Oranges*, pp. 101–102.
57. Mabberley and Placito, *Algarve Plants and Landscape*, pp. 145–59, 250–57.
58. Smith et al., *Citrus Pests*.
59. Fortey, *Dry Storeroom*, pp.198–99.
60. Navarro et al., 'Improvement of shoot-tip grafting'.
61. https://edis.ifas.ufl.edu/publication/PP323?downloadOpen=true, accessed 8 May 2023.
62. Slightly modified from text in Mabberley, 'Biology of the citron'.
63. George Beattie, pers. comm. March 2023.
64. Zheng et al., '*Nicotiana tabacum*'.
65. Dorji and Yapwattanaphun, 'Morphological identification'.
66. Reynaud et al., 'African citrus psyllid'.
67. Endo et al., 'Fast-track breeding'.
68. Smith et al., 'First fruiting'.
69. Hodgson, 'Citrus introductions'.
70. Mabberley, 'Classification'.
71. Hamilton et al. 'Thermal analysis'.
* Quote on p. 232 from McPhee, *Oranges*.

BIBLIOGRAPHY

Agusti, M. and Primo-Millo, E. 2020. 'Flowering and fruit-set', pp. 219–44 in M. Talon et al. (eds), *The Genus Citrus*. Woodhead, Duxford, United Kingdom.

Agusti, M., Reig, C., Martínez-Fuentes, A. and Mesejo, C. 2022. 'Advances in citrus flowering: a review'. *Frontiers in plant science* 13: https://doi.org/10.3389/fpls.2022.868831.

Ali, O. and Celik, T. A. 2007. Cytotoxic effects of peel extracts from *Citrus limon* and *Citrus sinensis*. *Caryologia* 60: 48–51.

Alvarez Arias, B. and Ramón-Laca, L. 2005. 'Pharmacological properties of citrus and their ancient and medieval uses in the Mediterranean region'. *Journal of Ethnopharmacology* 97: 89–95.

Andrews, A. C. 1961. 'Acclimatization of citrus fruits in the Mediterranean region'. *Agricultural History* 35: 35–46.

Anon. 1666/7. 'Some Hortulan communications about the curious engrafting of Oranges and Lemons, or Citrons, upon one anothers Trees, and of one Individual Fruit, half Orange and half Lemon growing on such Trees, et al'. *Philosophical Transactions* 2: 553–54.

Anon. 1733. *Ancient Accounts of India and China by Two Mohammedan Travellers*. Harding, London.

Anon. 1857. 'The Lane Cove Orangery'. *Maitland Mercury & Hunter River General Advertiser* 20 Oct. 1857: 4.

Appelhans, M. S., Bayly, M. J., Heslewood, M. M., Groppo, M., et al. 2021. 'A new subfamily classification of the *Citrus* family (Rutaceae) based on six nuclear and plastid markers'. *Taxon* 70: 1035–61.

Attlee, H. 2015. *The Land where Lemons Grow*. Penguin, London.

Bain, J. M. 1928. 'Morphological, anatomical, and physiological changes in the developing fruit of the Valencia orange, *Citrus sinensis* (L.) Osbeck'. *Australian Journal of Botany* 6: 1–24 + tt. 4.

Baldini, E. 1997. 'The role of Cassiano dal Pozzo's Paper Museum in citrus taxonomy', pp. 85–99 in D. Freedberg and E. Baldini, *Part One: Citrus Fruit (The Paper Museum of Cassiano del Pozzo Series B - Natural History)*. Harvey Miller, London.

Barns, V. K. 1934. 'America's first citrus grove'. *Citrus Industry* 15: 8, 9, 24.

Barry, G. H., Caruso, M. and Gmitter, F. G. 2020. 'Commercial scion varieties', pp. 83–104 in M. Talon et al. (eds), *The Genus Citrus*. Woodhead, Duxford, United Kingdom.

Bartholomew, E. T. and Reed, H. S. 1943. 'General morphology, histology, and physiology', in H. J. Webber and L. D. Batchelor, *The Citrus Industry* vol. 1. *History, Botany, and Breeding*. University of California Press, Berkeley and Los Angeles.

Bartholomew, M. 2004. 'Lind, James (1716–1794)'. Article 166669 [accessed 5 November 2004]. *Oxford Dictionary of National Biography*. Oxford University Press, Oxford.

Beekman, E. M. 2010, 2011. *The Ambonese Herbal* vols 1, 2. Yale University Press and National Tropical Botanical Garden, New Haven and London.

Bennett, G. 1860. *Gatherings of a Naturalist in Australasia*. Van Voorst, London.

Bennett, J. A. 2021. '"Put the lime in the coconut". The rise and decline of lime juice in the Cook Islands, c. 1850s–1930'. *Locale: the Australasian-Pacific journal of regional food studies* 8: 1–26.

Bhuiyan, M. N. I., Begum, J., Sardar, P. K. and Rahman, M. S. 2009. 'Constituents of peel and leaf essential oils of *Citrus Medica* L.'. *Journal of Scientific Research* 1: 387–92.

Bisignano, G. and Saija, A. 2002. 'The biological activity of citrus oils', pp. 602–30 in G. Dugo and A. Di Giacomo (eds), *Citrus: The Genus Citrus*. CRC Press, London.

Bonavia, E. 1886. 'On the probable wild source of the whole group of cultivated true limes (*Citrus acida*, Roxb., *C. medica*, var. *acida* of Brandis, Hooker, and Alph. De Candolle)'. *Journal of the Linnean Society, Botany* 22: 213–18.

Bonavia, E. 1888. *The Cultivated Oranges and Lemons, etc. of India and Ceylon, with researches into their origin and the derivation of their names, and other useful information* [with Atlas]. Allen, London.

Bowett, A. 2012. *Woods in British Furniture-making 1400-1900*. Oblong Creative, Wetherby.

Bowman, F. T. 1955. *A History of Citrus-growing in Australia 1788-1900* [reprinted from *Citrus News*, July–Dec. 1955]. Melbourne.

Bown, S. 2003. *Scurvy: How a Surgeon, a Mariner and a Gentleman Solved the Greatest Medical Mystery of the Age of Sail.* Viking/Penguin Australia, Camberwell.

Bretschneider, E. 1898. *History of European Botanical Discoveries in China.* Imperial Russian Academy of Sciences, St Petersburg.

Calabrese, F. 2002. 'Origin and history', pp. 1–19 in G. Dugo and A. Di Giacomo (eds), *The Genus Citrus.* CRC Press, London.

Chaffee, J. W. 2018. *The Muslim Merchants of Premodern China, the History of a Maritime Asian Trade Diaspora, 750–1400.* Cambridge University Press.

Church, A. H. 1981†. 'The botany of the garden in Eden', pp. 237–45 in D. J. Mabberley (ed.), *Thalassiophyta and Other Essays of A. H. Church.* Clarendon Press, Oxford.

Clément-Mullet, J.-J. (trans.) 1864. *Le livre de l'agriculture d'Ibn-al-Alam (Kitab-al-felahah).* Herold, Paris.

Coats, A. M. 1969. *The Plant Hunters.* Studio Vista, London.

Commelin, J. 1676. *Nederlantze Hesperides, Dat is, Oeffening en Gebruik Van de Limoen- en Oranje-boomen.* Doornik, Amsterdam.

Commelin J. (trans. N., G. V.). 1683. *The Belgick, or Netherlandish Hesperides.* Holford, London.

Compton, J. A. and Thijsse, G. 2013. 'The remarkable P. F. B. von Siebold, his life in Europe and Japan'. *Curtis's botanical magazine* 30: 275–314.

Corlett, R. T. 2007. 'The impact of hunting on the mammalian fauna of tropical Asian forests'. *Biotropica* 39: 292–303.

Crone, P. and Cook, M. 1977. *Hagarism: the Making of the Islamic World.* Cambridge University Press.

Dietz, B. (ed.)1972. *The Port and Trade of Early Elizabethan London: Documents.* London Record Society, London.

Dioscorides [trans. T. S. Osbaldeston and R. P. A. Wood]. 2000. *De materia medica.* Ibidis Press, Johannesburg.

Dorji, K. and Yapwattanaphun, C. 2011. 'Morphological identification of mandarin (*Citrus reticulata* Blanco) in Bhutan'. *Kasetsart Journal (Natural Science)* 45: 783–802.

Dowe, J. L. and Broughton, A. D. 2007. 'F. M. Bailey's ascent of Mt Bellenden-Ker in 1889, and notes on the publication priority of new vascular plant species from the expedition'. *Austrobaileya* 7: 555–66.

Drummond, J. C. and Wilbraham, A. 1959. *The Englishman's Food. A History of the Customs of English Diet.* Rev. ed., Readers Union and Cape, London.

Endo, T., Fujii, H., Omura, M. and Shimada, T. 2020. 'Fast-track breeding system to introduce CTV resistance of trifoliate orange into citrus germplasm, by integrating early flowering transgenic plants with marker-assisted selection'. *BMC Biology* 2: art. 224.

Faggin, G. T. 1970. 'Catalogue of works', pp. 85–102, *The Complete Paintings of the Van Eycks.* Weidenfeld & Nicolson, London.

Fairclough, H. R. (trans.) 1960. *Virgil: Eclogues, Georgics, Aeneid* 1–6. Heinemann, London.

Ferrari, G. B. 1646. *Hesperides sive De malorum aureorum cultura et usu.* Scheus, Rome.

Fischer, T. and Butzmann, R. 1998. '*Citrus meletensis* (*Rutaceae*), a new species from the Pliocene of Valdarno (Italy)'. *Plant Systematics and Evolution* 210: 51–55.

Forster, P. I. and Smith, M. W. 2010. '*Citrus wakonai* P. I. Forst. and M. W. Sm. (Rutaceae), a new species from Goodenough Island, Papua New Guinea'. *Austrobaileya* 8: 133–38.

Fortey, R. 2008. *Dry Storeroom No. 1: the Secret Life of the Natural History Museum.* Knopf, New York.

Freedberg, D. 1997. 'Series B: Natural History', pp. 33–43; 'Cassiano, Ferrari and their drawings of citrus fruit', 45–83 in D. Freedberg and E. Baldini, *Part One: Citrus Fruit* (*The Paper Museum of Cassiano del Pozzo Series B - Natural History*). Harvey Miller, London.

Gallesio, G.† (trans. S. D. Wilcox and Mrs C. A. Cowgill). 1876. *Orange Culture. A Treatise on the Citrus Family.* Florida Agriculturist, Jacksonville.

Gao, Z., House, L. O., Gmitter, F. G., Valim, M., et al. 2011. 'Consumer preferences for fresh citrus: impacts of demographic and behavioral characteristics'. *International Food and Agribusiness Management Review* 14: 23–40.

García-Lor, A., Luro, F., Ollitrault, P. and Navarro, L. 2015. 'Genetic diversity and population structure analysis of mandarin germplasm by nuclear, chloroplastic and mitochondrial markers'. *Tree Genetics and Genomes* 11: art. 123.

Genovese, S., Taddeo, V. S., Epifano, F. and Fiorito, S. 2018. 'Prenylated coumarins of the genus *Citrus*: an overview of the 2006–2016 literature data'. *Current Medicinal Chemistry* 25: 1186–93.

Gmitter, F. G. and Hu, X. 1990. 'The possible role of Yunnan, China, in the origin of contemporary *Citrus* species (Rutaceae)'. *Economic Botany* 44: 267–77.

Goldschmidt, E. E. 2017. 'New insights in citron (*Citrus medica* L.) genomics and fruit development'. *HortScience* 52 (6): 1–4.

Goody, J. 1993. *The Culture of Flowers*. Cambridge University Press.

Green, M. J. 1997. *Dictionary of Celtic Myth and Legend*. Thames & Hudson, London.

Greene, E. L. 1983. *Landmarks of botanical history*, parts I and II (ed. F. N. Egerton). Stanford University Press, Stanford.

Grove, R.H. 1995. *Green imperialism: Colonial expansion, tropical island Edens and the origins of environmentalism, 1600-1860*. Cambridge University Press, Cambridge.

Hagerty, M. J. (trans.) 1923. *Han Yen-chih's Chü Lu. Monograph on the Oranges of Wên-chou, Chekiang*. Brill, Leiden.

Hämeen-Anttila, J. 2006. *The Last Pagans of Iraq: Ibn Wahshiyya and his Nabatean Agriculture*. Brill, Leiden.

Hamilton, K. N., Ashmore, S. E. and Pritchard, H. W. 2009. 'Thermal analysis and cryopreservation of seeds of Australian wild *Citrus* species (Rutaceae): *Citrus australasica*, *C. inodora* and *C. garrawayi*'. *CryoLetters* 30: 268–79.

Hammond, N. G. L. and Scullard, H. H. 1970. *The Oxford Classical Dictionary*. 2nd edn, Clarendon, Oxford.

Hanks, P. and Hodges, F. 1988. *A Dictionary of Surnames*. Oxford University Press, Oxford and New York.

Haskell, F. and McBurney, H. 1997. 'General introduction to Catalogue Raisonné', pp. 9–26 in D. Freedberg and E. Baldini, *Part One: Citrus fruit* (*The Paper Museum of Cassiano del Pozzo Series B - Natural History*). Harvey Miller, London.

Heads, M. 2012. *Molecular Panbiogeography of the Tropics*. University of California Press, Berkeley and Los Angeles; London.

Henniger, J. 1988. 'Dutch contributions to the study of exotic natural history in the seventeenth and eighteenth centuries', pp. 59–66 in W. Eisler and B. Smith, *Terra Australis. The Furthest Shore*. International Cultural Corporation of Australia.

Henrey, B. 1975. *British Botanical and Horticultural Literature Before 1800*. 3 vols, Oxford University Press, Oxford.

Hensz, R. A. 1977. 'Mutation breeding and the development of the "Star Ruby" grapefruit'. *Proceedings of the International Society of Citriculture* 2: 582–85.

Hibbert, C. 1974. *The House of Medici: Its Rise and Fall*. Allen Lane, London.

Himes, N. E. 1936. *Medical History of Contraception*. Williams and Wilkins, Baltimore.

Hodgson, R. W. 1955. 'Citrus introductions at UCLA'. *California Citrograph* 40: 164–76.

Holmes, L. E. 2022. 'To the Great Socialist Beyond: Fairhope, the Organic School, and Russia (the journey of Willard and Helen Edwards)'. *Journal of Russian American Studies* 6: 88–119.

Honychurch, L. *c.* 2001. *A-Z of Dominica Heritage*. The Chronicle, Roseau, Dominica.

Hort, A. (trans.) 1916. *Theophrastus. Enquiry into Plants and Minor Works on Odours and Weather Signs*. 2 vols, Harvard University Press, Cambridge, Mass., and Heinemann, London.

Hume, H. H. 1935. 'History of citrus culture in Florida'. *Citrus Industry* 16 (10): 5, 8, 20.

Imbesi, A. and de Pasquale, A. 2002. 'Citrus species and their essential oils in traditional medicine', pp. 577–601 in G. Dugo and A. Di Giacomo (eds), *Citrus. The Genus Citrus*. CRC Press, London.

Innes Miller, J. 1969. *The Spice Trade of the Roman Empire*. Clarendon Press, Oxford.

Isaac, E. 1959. 'Influence of religion on the spread of citrus'. *Science* 129: 179–86.

Janick, J. 2007. 'Fruits of the Bibles'. *HortScience* 42: 1072–76.

Juniper, B. E. and Mabberley, D. J. 2019. *The Extraordinary Story of the Apple*. Kew Publishing, Kew.

Karp, D. and Hu, X. 2018. 'The citron (*Citrus medica* L.) in China'. *Horticultural Reviews* 45: 143–96 + tt.

Karsten, M. C. 1951. *The Old Company's Garden at the Cape and Its Superintendents*. Maskew Miller, Cape Town.

Kato-Noguchi, H. and Tanaka, Y. 2006. 'Potential of *Citrus junos* fruit waste from the food processing industry for weed management'. *HortScience* 41: 1516–17.

Klein, S. 2020. *The Fruits of Empire: Art, Food, and the Politics of Race in the Age of American Expansion*. University of California Press, Oakland.

Knox, R. [1681], ed. J. Ryan, 1911. *An Historical Relation of Ceylon Together with Somewhat Concerning Severall Remarkeable Passages of My Life That Hath Hapned since My Deliverance out of My Captivity*. Maclehose, Glasgow [reprint by Gunasena, Colombo 1981].

Kodicek, E. H. and Young, F. G. 1969. 'Captain Cook and scurvy'. *Notes and Records of the Royal Society of London* 24: 43–63.

De Koning, J., van Uffelen, G., Zemanek, A. and Zenanek, B. (eds). 2008. *Drawn After Nature: The Complete Botanical Watercolours of the 16th-century Libri picturati*. KNNV, Zeist.

Kretschmar, J. A. and Baumann, T. W. 1999. 'Caffeine in *Citrus* flowers'. *Phytochemistry* 52: 19–23.

Kubitzki, K. 2011. 'Rutaceae', pp. 276–356 in K. Kubitzki (ed.), *Families and Genera of Vascular Plants* vol. 10. Springer, Koenigstein.

Kumamoto, J., Scora, R. W., Lawton, H. W. and Clerx, W. A. 1987. 'Mystery of the forbidden fruit: historial epilogue on the origin of the grapefruit, *Citrus paradisi* (Rutaceae)'. *Economic Botany* 41: 97–107.

Laboury, D. 2007. 'Archaeological and textual evidence for the function of the "Botanical Garden" of Karnak in the initiation ritual', pp. 27–34 in B. Bryan and P. Dorman (eds), *Sacred Spaces and Sacred Function in Ancient Thebes. Occasional Proceedings of the Theban Workshop*. University of Chicago Press, Chicago.

Lane Fox, R. 1975. *Alexander the Great*. Futura, London.

Laszlo, P. 2007. *Citrus. A History*. University of Chicago Press, Chicago and London.

Laufer, B. 1934. 'The lemon in China and elsewhere'. *Journal of the American Oriental Society* 54: 143–60.

Le Rougetel, H. 1990. *The Chelsea Gardener*. SagaPress/Timber Press, Portland.

Li, W. B. 1992. 'Origin and development of mandarins in China before the Song Dynasty (AD 960–1279)'. *Proceedings of the International Society of Citriculture 1992* 1: 61–66.

Lightbown, R. 1989. *Sandro Botticelli: Life and Work*. Thames & Hudson, London.

Lim, A. K. 2001. 'Citrus and citroids: recent biological findings'. *Malayan Nature Journal* 55: 221–29.

Lim, T. K. 2012. *Edible Medicinal and Non-Medicinal Plants*. Vol. 4, *Fruits*. Springer, Dordrecht.

Lipschits, O., Gadot, Y. and Langgut, D. 2012. 'The riddle of Ramat Rahel; the archaeology of a royal Persian period edifice'. *Transeuphratène* 41: 57–78.

Liu, Y. 2006. 'Historical and modern genetics of plant graft hybridization'. *Advances in Genetics* 56: 101–29.

Liu, Y-Q., Heyring, E. and Tanumihardjo, S. A. 2012. 'History, global distribution, and nutritional importance of citrus fruits'. *Comprehensive Reviews in Food Science and Food Safety* 11: 530–45.

Luro, F., Viglietti, G., Marchi, E., Costantino, G., et al. 2019. 'Genetic, morphological and chemical investigations reveal the genetic origin of Pompia (*Citrus medica tuberosa* Risso and Poiteau) - an old endemic Sardinian citrus fruit'. *Phytochemistry* 168: art. 112083.

Mabberley, D. J. 2002. 'Limau hantu and limau purut: the story of lime-leaves (*Citrus hystrix* DC., Rutaceae)?' *Gardens' Bulletin Singapore* 54: 173–84.

Mabberley, D. J. 2004. '*Citrus* (Rutaceae): a review of recent advances in etymology, systematics and medical applications'. *Blumea* 49: 481–98.

Mabberley, D. J. 2017. *Mabberley's Plant-book: A Portable Dictionary of Plants, their Classification and Uses*. 4th edn, Cambridge University Press, Cambridge.

Mabberley, D. J. 2020. 'Sweet … and sour [review of A. Gentile et al. (eds), *The Citrus Genome*]'. *Australasian Systematic Botany Society Newsletter* 185: 59–60.

Mabberley, D. J. 2022. 'A classification for edible citrus: an update, with a note on *Murraya* (Rutaceae)'. *Telopea* 25: 271–84.

Mabberley, D. J. 2023. 'The biology of the citron (*Citrus medica* L., Rutaceae-Aurantioideae-Aurantieae), its hybrids and allies', pp. 3–40 in E. E. Goldschmidt and M. Bar-Joseph (eds), *The Citron Compendium*. Springer, Cham, Switzerland.

Mabberley, D. J. and Placito, P. J. 1993. *Algarve Plants and Landscape: Passing Tradition and Ecological Change*. Oxford University Press, Oxford.

Mabberley D. J. and Xu, Q. 2022. 'Proposal to conserve the name *Citrus reticulata* (Rutaceae) with a conserved type'. *Taxon* 71: 1123–24.

Maheshwari, P. 1950. *An Introduction to the Embryology of Angiosperms*. McGraw-Hill, New York.

Mahon, S. 2018. 'Asian citrus with a real zest for life'. *Daily Telegraph* 9 Jan.

Maruca, G., Laghetti, G. and Hammer, K. 2015. 'Religious and cultural significance of the citron (*Citrus medica* L.

"Diamante") from Calabria (South Italy): a biblical fruit of the Mediterranean land'. *Journal of Environmental Science and Engineering* A 4: 203–9.

Mathew, W. M. 1999. *Keiller's of Dundee: the Rise of the Marmalade Dynasty 1800–1879*. Abertay Historical Society, Dundee.

Mayer, H. E. [trans. J. Gillingham]. 1988. *The Crusades*. 2nd edn, Oxford University Press, Oxford.

McPhee, J. 2016. *Oranges*. Daunt, London.

Meiggs, R. 1982. *Trees and Timber in the Ancient Mediterranean World*. Clarendon Press, Oxford.

Mudge, K., Janick, J., Scofield, S. and Goldschmidt, E. 2009. 'A history of grafting'. *Horticultural Reviews* 35: 437–93.

Muth, F., Philbin, C. S., Jeffrey, C. S. and Leonard, A. S. 2022. 'Discovery of octopamine and tyramine in nectar and their effects on bumblebee behavior'. *iScience* 25 (8): 104765.

Navarro, L., Roistacher, C. N. and Murashige, T. 1975. 'Improvement of shoot-tip grafting *in vitro* for virus-free citrus'. *Journal of the American Society of Horticultural Science* 100: 471–79.

Needham, J. 1986. *Science and Civilisation in China*, vol. 6 *Biology and Biological Technology*, part I: *Botany*. Cambridge University Press, Cambridge.

Nelson, E. C. 2009. 'Victorian wedding flowers: orange, myrtle, and the apotheosis of white heather'. *Garden History* 37: 231–36.

Nesom, G. L. 2014. '*Citrus trifoliata* (Rutaceae): review of biology and distribution in the USA'. *Phytoneuron* 2014–46: 1–14.

Ng, C. 2015. 'Pomelo – Citrus maxima – the indigenous mega-citrus of South-East Asia'. *UTAR Agriculture Science Journal* 1: 29–34.

Noehden, G. H. 1818/1820. 'Account of the different varieties of the genus Citrus which are cultivated in Italy, according to *Dr. Sickler's* statement'. *Transactions of the Horticultural Society* 3, appendix: 1–19.

Padgett J. F. and C. K. Ansell. 1973. 'Robust action and the rise of the Medici, 1400–1434'. *American Journal of Sociology* 98: 1259–1319.

Pagnoux, C., Celant, A., Coubray, S., Fiorentino, G. and Zech-Matterne, V. 2013. 'The introduction of *Citrus* to Italy, with reference to the identification problems of seed remains'. *Vegetation History and Archaeobotany* 122: 421–38.

Panara, K., Joshi, K. and Nishteswar, K. 2012. 'A review on phytochemical and pharmacological properties of *Citrus medica* Linn.' *International Journal of Pharmaceutical & Biological Archives* 3: 1292–97.

Passos, O. S., Coelho, Y. S. and Cunha Sobrinho, A. P. 1977. 'The history of the navel orange and its behavior in the state of Bahia, Brazil'. *Proceedings of the International Society of Citriculture* 2: 645–47.

Pauling, L. 1970. *Vitamin C and the Common Cold*. Freeman, San Francisco.

Paulus, H.-E. 2017. 'The humanistic tradition of citrus culture in Central Europe from the 15th to the 18th century', in V. Zech-Matterne and G. Fiorentino (eds), *Agrumed: Archaeology and History of Citrus Fruit in the Mediterranean. Acclimatization, Diversifications, Uses*. Centre Jean Bérard, Naples. https://books.openedition.org/pcjb/2107.

Peekel, [Father, 'P.'] G. (1984, trans. E. E. Henty). *Flora of the Bismarck Archipelago for Naturalists*. Office of Forests, Lae, Papua New Guinea.

Pérès, H. 1999. 'Abu 'l-Khayr al-Ishbili', in *Encyclopaedia of Islam* [CD-ROM edition, v. 1.0]. Brill, Leiden.

[Preyel, A.] 1655. *Artificia hominum miranda naturae, in Sina and Europa*. Serlini and Fickwirth, Frankfurt.

Privé-Gill, C. 1981. 'Eocene upper Lutetian dicotyledonous woods from the Paris basin France'. *Palaeontographica Abt. B Palaeophytologie* 6: 119–35.

Ragionieri, A. 1927. 'Origin of the Florentine bizzarria'. *Journal of Heredity* 18: 527–28.

Ramón-Laca, L. 2003. 'The introduction of cultivated citrus to Europe via northern Africa and the Iberian Peninsula'. *Economic Botany* 57: 502–14.

Ramsay, A. A. W. 1928. *Sir Robert Peel*. Constable, London.

Raphael, S. 1990. *An Oak Spring Pomona: a Selection of the Rare Books on Fruit in the Oak Spring Garden Library*. Oak Spring Garden Library, Upperville.

Recupero, S., Russo, G. and Reforgiato-Recupero, G. 2015. 'New promising citrus hybrids for ornamental use'. *Acta Horticulturae* 1065: 223–28.

Reynaud, B., Turpin, P., Molinari, F. M., Grondin, M., et al. 2022. 'The African citrus psyllid *Trioza erytreae*: An efficient vector of *Candidatus* Liberibacter asiaticus'. *Frontiers in Plant Science* 13: https://doi.org/10.3389/fpls.2022.1089762.

Richards, J. 2008. *The Secret War: a True History of Queensland's Native Police*. University of Queensland Press, St Lucia.

Risso, A. 1813. *Essai sur l'histoire naturelle des orangers, bigaradadiers, limettiers, cedratiers, limoniers ou citronniers, cultivés dans le départment des Alpes Maritimes*. Dufour, Paris.

Risso, A. and Poiteau, A. 1818–22. *Histoire naturelle des Orangers*. Le Doux, Paris.

Ruas, M.-P., Mane, P., Hallavant, C. and Lemoine, M. 2017. 'Citrus fruit in historical France: written sources, iconographic and plant remains', in V. Zech-Matterne and G. Fiorentino (eds), *Agrumed: Archaeology and History of Citrus Fruit in the Mediterranean. Acclimatization, Diversifications, Uses*. Centre Jean Bérard, Naples. https://books.openedition.org/pcjb/2107.

Ruggiero, R. 2021. 'Medici coat of arms'. https://www.youtube.com/watch?v=b-W48LJMTrk (accessed 14 Jan. 2022).

Salmon, Y. 2018. 'The Corfu etrog polemic through the middle of the 19th century', pp. 315–67 in E. E. Goldschmidt and M. Bar-Joseph, *The Etrog Citron (Citrus medica L.): Tradition and Research*. Mosad Harav Kook, Jerusalem.

Schafer, E. H. 1963. *The Golden Peaches of Samarkand. A Study of T'ang Exotics*. University of California Press, Berkeley and Los Angeles.

Schottenhammer, A. 2010. 'Transfer of *Xiangyao* from Iran and Arabia to China: a reinvestigation of *Youyang zazu* (863)', pp. 117–49 in R. Kauz (ed.), *Aspects of the Maritime Silk Road: from the Persian Gulf to the East China Sea*. vol. 10 of *East Asian Economic and Socio-cultural Studies - East Asian Maritime History*. Harrassowitz, Wiesbaden.

Scora, R. W. 1975. 'On the history and origin of citrus'. *Bulletin of the Torrey Botanical Club* 102: 369–75.

Shimizu, T., Kitajima, A., Nonak, K. and Yoshioka, T. 2016. 'Hybrid origins of citrus varieties inferred from DNA marker analysis of nuclear and organelle genomes', *PLoS One*, 11 (11): e0166969.

Short, R. V., McCoombe, S. G., Maslin, C., Naim, E. and Crowe, S. 2004. 'Lemon and lime juice as potent natural microbicides'. *Abstracts of Papers for World AIDS Conference Bangkok 8 -16 July 2004*.

Sickler, F. 1815. *Der vollkommene Orangerie-Gärtner, oder Vollständige Beschreibung der Limonen, citronen und Pomeranzen, oder fer Agrumi in Italien, und ihrer Cultur*. Weimar.

Siebold, P. F. von. 1830. 'Synopsis plantarum oeconomicarum universi regni Japonici'. *Verhandelingen van het Bataviaasch Genootschap van Kunsten en Wetenschappen* 12: 1–74.

Smith, D., Beattie, G. A. C. and Broadley, R. 1997. *Citrus pests and their natural enemies: integrated pest management in Australia*. Department of Primary Industries, and Horticultural Research and Development Corporation, Brisbane.

Smith, M. W., Gultzow, D. L. and Newman, T. K. 2013. 'First fruiting intergeneric hybrids between *Citrus* and *Citropsis*'. *Journal of the American Society for Horticultural Science* 138: 57–63.

Spary, E. C. 2012. *Eating the Enlightenment: Food and the Sciences in Paris, 1670-1760*. University of Chicago Press, Chicago.

Spencer, J. E. 1973. *Oriental Asia: Themes toward a Geography*. Prentice-Hall, Englewood Cliffs.

Stearn, W. T. 1977. 'The earliest European acquaintance with tropical vegetation'. *Gardens' Bulletin Singapore* 29: 13–18.

Sterbeeck, F. van. 1682. *Citricultura oft Regeringhe der uythemsche Boomen te weten Oanien, Citroenen, Limoenen, Granatum, Laurieren en andere*. Jacops, Antwerp.

Strasburger, E. [trans. O and B. Comerford Casey]. 1906. *Rambles on the Riviera*. Fisher Unwin, London.

Swingle, W. T. 1943. 'The botany of Citrus and its wild relatives of the orange subfamily (family Rutaceae, subfamily Aurantioideae)', pp. 129–474 in H. J. Webber and L. D. Batchelor, *The Citrus Industry* vol. 1. *History, Botany, and Breeding*. University of California Press, Berkeley and Los Angeles.

Swingle, W. T., Robinson, T. R and Savage, E. M. 1931. 'New citrus hybrids'. *United Sates Department of Agriculture Circular* 181. Superintendent of Documents, Washington, DC.

Tanaka, T. 1922. 'Citrus fruits of Japan with notes on their history and the origin of varieties through bud variation'. *Journal of Heredity* 13: 243–53.

Tanaka, T. 1927. 'Taxonomy of the citrus fruits of the Pacific region'. *Memoirs of the Tanaka Citrus Experiment Station* 1: 15–36.

Tanaka, T. 1959. 'A revision of Assam Citrus: Revisio Aurantiacearum XI'. *Bulletin of the University of Osaka Prefecture*, Series B, *Agriculture and Biology* 9: 29–39.

Tickner, F. J. and Medvei, V. C. 1958. 'Scurvy and the health of European crews in the Indian Ocean in the seventeenth century'. *Medical history* 2: 36–46.

Tobey, R. and Wetherell, C. 1995. 'The citrus industry and the revolution of corporate capitalism in southern California, 1887–1944'. *California History* 74: 6–21.

Tolkowsky, S. 1938. *Hesperides, a History of the Culture and Use of Citrus Fruits*. Bale, London.

Uma Shaanker, R. and Ganeshaiah, K. N. 1996. 'Polyembryony in plants: a weapon in the war over offspring numbers?' *Trends in Ecology and Evolution* 11: 26–27.

Uphof, J. C. T. 1935. 'Wissenschaftliche Beobachtungen und Versuche an Agrumen'. *Gartenbauswissenschaften* 9: 421–26. Valder, P. 1999. *The Garden Plants of China*. Florilegium, Glebe, New South Wales.

Van der Meer, W. 'The history of *Citrus* in the Low Countries during the Middle Ages and the Early Modern Age', in V. Zech-Matterne and G. Fiorentino (eds), *Agrumed: Archaeology and History of Citrus Fruit in the Mediterranean. Acclimatization, Diversifications, Uses*. Centre Jean Bérard, Naples. https://books.openedition.org/pcjb/2197.

Volkamer, J. C. 1708. *Nürnbergische Hesperides, Oder Gruendliche Beschreibung Der Edlen Citronat- Citroen- und Pomeranzenfruechte*. Nuremberg.

Volkamer, J. C. 1714. *Continuation der Nürnbergische Hesperidum*. Nuremberg.

Vorstand der Kulturstiftung Dessau-Wörlitz (ed.). 1999. *Oranien-Orangen-Oranienbaum*. Deutscher Kunstverlag, Munich and Berlin.

Wästberg, P. 2010. *The Journey of Anders Sparrman*. Granta, London.

Wang, X., Xu, Y., Zhang, S., Cao, L., et al. 2017. 'Genomic analyses of primitive, wild and cultivated citrus provide insights into asexual reproduction.' *Nature Genetics* 49: 765–72.

Wearn, J. and Mabberley, D. J. 2016. 'Citrus and orangeries in northern Europe.' *Curtis's Botanical Magazine* n.s. 33: 94–107.

Webber, H. J. 1943. 'Cultivated varieties of citrus', pp. 475–668 in H. J. Webber and L. D. Batchelor (eds), *The Citrus Industry I. History, Botany, and Breeding*. University of California Press, Berkeley and Los Angeles.

Webber, H. J. and Swingle, W. T. 1905. 'New citrus creations of the Department of Agriculture'. *United States Department of Agriculture Yearbook* 1904: 221–40.

Wedgwood, C. V. 1944. *William the Silent: William of Nassau, Prince of Orange, 1533–1584*. Cape, London.

Williams, G. 2013. *Naturalists at Sea. Scientific Travellers from Dampier to Darwin*. Yale University Press, Newhaven and London.

Wilson, C. A. 2010. *The Book of Marmalade*. Rev. ed., Prospect, Totnes.

Winmill, M. 2015. *George and Sarah Suttor: Pioneers of Early Australian Horticulture*. Legion Office Works, Castlemaine, Victoria.

Winters, H. F. 1976. '*Microcitrus papuana*, a new species from Papua New Guinea (Rutaceae)'. *Baileya* 20: 19–24.

Wright, G. A., Baker, D. D., Palmer M. J., Stabler, D., et al. 2013. 'Caffeine in floral nectar enhances a pollinator's memory of reward'. *Science* 339: 1202–4.

Wu, G. A., Sigimoto, C., Kinjo, H., Azama, C., et al. 2021. 'Diversification of mandarin citrus by hybrid speciation and apomixis'. *Nature Communications* 12: 4,377.

Xie, S., Manchester, S. R., Liu, K., Wang, Y., Sun, B. 2013. '*Citrus linczangensis* sp. n., a leaf fossil of Rutaceae from the late Miocene of Yunnan China'. *International Journal of Plant Science* 174: 1201–07.

Zhang, D. and Mabberley, D. J. 2008. '*Citrus* Linnaeus, *Sp. Pl.* 2: 782. 1753', pp. 90–96 in Z. Y. Wu et al. (eds), *Flora of China* 11 (Oxalidaceae through Aceraceae). Science Press, Beijing/Missouri Botanical Garden Press, St Louis.

Zheng, L., Xu, Q., Gong, G., Liao, Y., et al. 2023. '*Nicotiana tabacum* as a dead-end trap for adult *Diaphorina citri*: a potential biological tactic for protecting citrus orchards.' *Frontiers in Plant Science* 13: https://doi.org/10.3389/fpls.2022.1081663.

Zhou, J., Hirata, Y., Nou, I.-S., Shiotani, H. and Ito, T. 2002. 'Interactions between different genotypic tissues in citrus graft chimeras'. *Euphytica* 126: 355–64.

Zylberberg, S. 2002. 'Oranges and Seders: symbols of Jewish women's wrestlings'. *Nashim* 5: 148–71.

SOURCES OF ILLUSTRATIONS

a: above; b: below; r: right; l: left; c: centre

2–3, 81 Uffizi, Florence. Inv. no. 1890 n. 8360; **4–5** Harn Museum of Art, University of Florida, Gainesville. The Florida Art Collection, Gift of Samuel H. and Roberta T. Vickers. Obj. no 2020.18.1095. Image courtesy Harn Museum of Art. Photo Randy Batista; **6** British Library Archive/Bridgeman Images; **7** Wellcome Collection, London. 25392i; **8–9, 162–163** Botanischer Garten und Botanisches Museum Berlin; **10–11** Printed by Louis Roesch Co., Litho. P. K. Yonge Library of Florida History, The Jerry Chicone Jr Citrus Crate Label Collection; **14a, 40** British Library Archive/Bridgeman Images; **14c, 63** BnF, Dist. GrandPalaisRmn/image BnF; **14b, 114, 132** Kupferstichkabinett Berlin, Staatliche Museen zu Berlin. Photo Jörg P. Anders; **15a, 150** History of Advertising Trust/Heritage Images/Getty Images; **15b, 212** DaTo Images/Bridgeman Images. © Estate of Franz Krausz; **17** British Library Archive/Bridgeman Images; **18, 20** BnF, Paris, département Arsenal, FOL-S-659; **21** National Agricultural Library, Baltimore. Henry G. Gilbert Nursery and Seed Trade Catalog Collection; **23** Peter H. Raven Library, Missouri Botanical Garden; **24** History and Art Collection/Alamy Stock Photo; **26** DeAgostini Picture Library/Scala, Florence; **28** Philadelphia Museum of Art. Purchased with funds contributed by Edward B. Robinette from the Simkhovitch Collection, 1929. Acc. no. 1929-40-213o; **29** Wellcome Collection, London; **31** Peter H. Raven Library, Missouri Botanical Garden; **32** Dumbarton Oaks Research Library, Washington DC. RARE-FOLIO QK357.5 A85 1798z; **35** Wellcome Collection, London. 28084i; **36** Wellcome Collection, London. 23244i; **37** Naturalis Biodiversity Center, Siebold collection, Leiden. RMNH.ART.635; **38** British Museum, London. 1930,0319,0.1; **42** Minneapolis Institute of Art. The Minnich Collection, The Ethel Morrison Van Derlip Fund, 1966. P.14,202; **43** National Library of Poland. SD XVII.4.10505; **44** Universitäts- und Landesbibliothek Sachsen-Anhalt; **45** Peter Barritt/Alamy Stock Photo; **47** Photo collection of Rijksmuseum, Amsterdam. RP-F-F01148-CY; **48l** Scala, Florence; **48r** Luisa Ricciarini/Bridgeman Images; **49** Su concessione del Ministero della Cultura – Parco Archeologico di Pompei; **50** Rijksmuseum, Amsterdam. RP-P-OB-51.709; **51** Heritage Image Partnership Ltd/Alamy Stock Photo; **52a** The Jewish Museum, New York. Gift of Dr Harry G. Friedman. F 4715; **52bl** akg-images/Bible Land Pictures; **52br** The Jewish Museum, New York. Gift of Henry L. Moses in memory of Mr and Mrs Henry P. Goldschmidt. JM 5-50; **53** Duby Tal/Albatross/Alamy Stock Photo; **55** BnF, Paris, département Arsenal, FOL-S-659; **56** Victoria and Albert Museum, London. Acc. no. IM.276-1913; **59** Yale Medical Historical Library, Cushing/Whitney Medical Library, New Haven. Call no. Manuscript Persian 23; **60** The Metropolitan Museum of Art, New York. Rogers Fund, 1913. Acc. no. 13.152.6/Art Resource/Scala, Florence; **61l** BnF, Paris. Département des manuscrits. Arabe 2771; **61r** New York Public Library. Spencer Collection. Persian MS. 39; **63a** Österreichische Nationalbibliothek, Vienna. Cod. Ser. n. 2644; **63c** BnF, Dist. GrandPalaisRmn/image BnF; **63b** Österreichische Nationalbibliothek, Vienna. Cod. 2396; **64** Biblioteca Pública de Évora; **66** British Library Archive/Bridgeman Images; **67** The British Library, London. Or. 3714; **68** BnF, Paris. Département des Manuscrits. Arabe 2221; **69** Bodleian Libraries, University of Oxford. MS. Pococke 375; **71** Scala, Florence; **72** akg-images; **73** Wallraf-Richartz-Museum, Cologne/Rheinisches Bildarchiv Cologne, rba_c002372; **74** Kunsthis Artefact/Alamy Stock Photo; **75** Bayerische Staatsgemäldesammlungen – Alte Pinakothek München. Inv. no. 8709; **76–77** Scala, Florence; **78–79** Bayerische Staatsbibliothek München. BSB Cod. icon. 277; **80** Cambridge University Library. Classmark Inc.3.A.1.8[37]; **82** BnF, Paris. Département des Manuscrits. Latin 9474; **83–84** British

Library Archive/Bridgeman Images; **86–87** Jagiellonian Library, Krakow University; **88** Private collection; **89a** Royal Collection Trust, London. RCIN 919330; **89bl** Österreichische Nationalbibliothek, Vienna. Pk 485b, 1; **89br** BnF, Paris, département Arsenal, FOL-S-659; **91** Historical image collection by Bildagentur-online/Alamy Stock Photo; **93** The Metropolitan Museum of Art, New York. The Elisha Whittelsey Collection, The Elisha Whittelsey Fund, 1967. Acc. no. 67.828; **94–97** Biblioteka Narodowa, Warsaw. SD XVII.4.10505; **99** Private collection; **100–103** All images Biblioteka Narodowa, Warsaw. SD XVII.4.10505, except **102cl, 102bc, 103ac** Heritage Auctions; **103cc** Christie's Images/Bridgeman Images; **104, 105ar, 105bl, 105br** Minneapolis Institute of Art. The Minnich Collection, The Ethel Morrison Van Derlip Fund, 1966; **105al** Heritage Auctions; **107** Private collection; **108–109** akg-images/Rabatti & Domingie; **110–111, 112–113** Villa Medicea di Poggio a Caiano – Museo della Natura Morta, Poggio a Caiano. Photos Scala, Florence – courtesy of the Ministero Beni e Att. Culturali e del Turismo; **116** Tekniska museet, Gösta Bodmans arkiv, Stockholm. GBN-Ö1-12; **117** Rijksmuseum, Amsterdam. Obj. no. SK-A-3889; **118** The National Gallery, London. Inv. no. NG186; **119** GrandPalaisRmn/René-Gabriel Ojeda; **120** The J. Paul Getty Museum, Los Angeles. Obj. no. 86.PB.538; **121** The J. Paul Getty Museum, Los Angeles. Obj. no. 2001.29; **122** Minneapolis Institute of Art. The Minnich Collection The Ethel Morrison Van Derlip Fund, 1966. Acc. no. P.14,813; **124** Bridgeman Images; **125** Château de Versailles, Dist. GrandPalaisRmn/Christophe Fouin; **126–127** Peter H. Raven Library, Missouri Botanical Garden; **128** University of California Libraries; **129a** The Metropolitan Museum of Art, New York. Harris Brisbane Dick Fund, 1937. Acc. no. 37.48(7); **129b** Rijksmuseum, Amsterdam. Obj. no. RP-P-1899-A-21596; **130–131** Beaux-Arts de Paris, Dist. RMN-Grand Palais/image INHA; **133a** British Library, London. Maps.C.9.e.7.(75); **133b** Universal History Archive/UIG/Bridgeman Images; **134** Peter H. Raven Library, Missouri Botanical Garden; **135** Private collection; **136l** The Cleveland Museum of Art. Gift of Kennedy & Company, NY, for The Donald Gray Memorial Collection. Acc. no. 1946.177; **136–137, 137r** Sunny Celeste/imageBROKER/Shutterstock; **139** The Picture Art Collection/Alamy Stock Photo; **140** Musée Carnavalet, Histoire de Paris. Inv. no. PH76362; **142** Leiden University Library. Acc. no. KITLV 37A29; **143** Leiden University Library. Acc. no. KITLV 37A28; **144l** Royal Collection Trust, London. RCIN 921148; **144ar** Royal Collection Trust, London. RCIN 921209; **144br** Royal Collection Trust, London. RCIN 921184; **145** Peter H. Raven Library, Missouri Botanical Garden; **146** Statens Museum for Kunst, Copenhagen. Inv. no. KKSgb2948/63. Photo SMK/Jacob Schou-Hansen; **147** Statens Museum for Kunst, Copenhagen. Inv. no. KKSgb2948/62. Photo SMK/Jacob Schou-Hansen; **148** BnF, Paris. Département Réserve des livres rares, S-670; **152** British Library, London; **153** National Maritime Museum, Greenwich, London. BHC4185; **154** Private collection; **155** Real Jardín Botánico, Madrid; **157** Look and Learn/Illustrated Papers Collection/Bridgeman Images; **159** Superstock; **160** Wellcome Collection, London. 16337i; **164** DeGolyer Library, Dallas. Doris A. and Lawrence H. Budner Theodore Roosevelt Photograph Collection; **165** Library of Congress, Prints and Photographs Division, Washington DC. Call. No. LC-D4-3604; **167** National Library of Australia, Canberra. 7790045; **168** Smithsonian Libraries, Washington DC.; **170–171** New York Public Library, Rare Book Division. b14485031; **172–173** BnF, Paris, département Arsenal, FOL-S-659; **174** The LuEsther T. Mertz Library, New York Botanical Garden; **175** Smithsonian Libraries, Washington DC.; **176–177** Peter H. Raven Library, Missouri Botanical Garden; **178–179** Biblioteka Narodowa, Warsaw. 1.407.053 A;

SOURCES OF ILLUSTRATIONS

180–181 Real Jardín Botánico, Madrid. Ff(914)BLA; **183** Courtesy Arader Galleries, Philadelphia; **184–187** Marianne North Gallery, Royal Botanic Gardens, Kew; **188a** GrandPalaisRmn (musée d'Orsay)/ Hervé Lewandowski; **188b** Barnes Foundation, Philadelphia. Acc. no. BF17; **189** Fitzwilliam Museum, University of Cambridge/Bridgeman Images; **190** The Museum of Modern Art, New York. Martha Jackson Gallery. Acc. no. SC206.1982. Photo The Museum of Modern Art, New York/Scala, Florence; **191a** Statens Museum for Kunst. Inv. no. KMS7070; **191b** Harry Ransom Center, The University of Texas at Austin. © Banco de México Diego Rivera Frida Kahlo Museums Trust, Mexico, D. F./DACS 2024; **192–193** National Agricultural Library, U.S. Department of Agriculture Pomological Watercolor Collection; **194** BnF, Paris. ENT DN-1 (CAPPIELLO, Leonetto/2)-GRAND ROUL; **196** Library of Congress Prints and Photographs Division, Washington DC. G. Eric and Edith Matson Photograph Collection. Call no. LC-M33- 10636; **197** Joanne Levesque/Getty Images; **198al** Hera Vintage Ads/Alamy Stock Photo. Mercier © ADAGP, Paris and DACS, London 2024; **198ar** Private collection. Loupot © ADAGP, Paris and DACS, London 2024; **198b** Collection Dupondt/ akg-images. Mercier © ADAGP, Paris and DACS, London 2024; **199** Collection Galerie 1 2 3, Geneva. www.galerie123.com. Mercier © ADAGP, Paris and DACS, London 2024; **200** incamerastock/Alamy Stock Photo; **202** Fairchild Tropical Botanic Garden Archive; **203** University of Miami Library, Walter Tennyson Swingle Collection. Object no. asm0188000007; **204al** Library of Congress, Prints and Photographs Division, Washington DC. Photo Truman Ward Ingersoll. Call no. LOT 12418, no.18; **204ar** Henry Miller News Picture Service/FPG/Archive Photos/Getty Images; **204b** Library of Congress, Prints and Photographs Division Washington DC. California Citrus Heritage Recording Project, Riverside. Call no. HAER CAL,33-RIVSI,6—11; **205al** Felix H. Man/ullstein bild/ TopFoto; **205cl** Edward Malindine/Daily Herald Archive/National Science & Media Museum/SSPL/Getty Images; **205bl** Library of Congress. Prints and Photographs Division, Washington DC. Photo Alfred S. Campbell. Call NO. STEREO FOREIGN GEOG FILE - Egypt— Cairo; **205r** Library of Congress, Prints and Photographs Division Washington DC. G. Eric and Edith Matson Photograph Collection. Call no. LC-M31- A-21; **206–207** University of Massachusetts Amherst Libraries; **208** Library of Congress, Prints and Photographs Division, Washington DC. U.S. Farm Security Administration – Office of War Information Photograph Collection. Call no. LC-USF33-003648-M4; **209al** Library of Congress Prints and Photographs Division Washington DC. Historic American Engineering Record. Call no. HAER CAL,33-RIVSI,4; **209ar** Library of Congress, Prints and Photographs Division, Washington DC. U.S. Farm Security Administration – Office of War Information Photograph Collection. Photo Arthur Rothstein. Call no. LC-USF33- 002367-M3; **209b** New York Public Library. The Miriam and Ira D. Wallach Division of Art, Prints and Photographs, Photography Collection. Photo Arthur Rothstein (Farm Security Administration). Object no.002368-M1 c.2; **210** Library of Congress, Prints and Photographs Division, Washington DC. Farm Security Administration – Office of War Information. Photo Jack Delano. Call. no. LC-USW36-882; **211** Apic/ Getty Images; **213** Swim Ink 2, LLC/Corbis/Getty Images. Ansieau © ADAGP, Paris and DACS, London 2024; **214** Transcendental Graphics/Getty Images; **215al** Bibliothèque Forney, Paris. AF 221619 T 846. © Toni; **215ar** Bibliothèque Forney, Paris. AF 220321. Villemot © ADAGP, Paris and DACS, London 2024; **215cr** Christie's Images/Bridgeman Images. Villemot © ADAGP, Paris and DACS, London 2024; **215b** Francis Elzingre/Alamy Stock Photo; **218** The Asahi Shimbun/Getty Images; **219a** Kazuki Yamada/Adobe Stock; **219bl** The Yomiuri Shimbun/Associated Press/Alamy Stock Photo; **219br** The Asahi Shimbun/Getty Images; **220, 222–223** Dhiraj Singh/Bloomberg/Getty Images; **221** Diptendu Dutta/AFP/Getty Images; **224** Elena Vyrypayeva/Adobe Stock; **225** Chandan Khanna/ AFP/Getty Images; **226, 228–229** © Ferdinando Scianna/Magnum Photos; **230** Garry Hogg/Stringer/Getty Images; **231** Allan Cash Picture Library/Alamy Stock Photo; **233** State Library and Archives of Florida. M99-26, Charles Barron photographs, 1951–77; Box 1, FF32; **234–235** Paul Popper/Popperfoto/Getty Images; **236a** Feature Brand by Sunland Packing House Company, Porterville. Special Collections & University Archives, University of California, Riverside. Collection of citrus labels (MS 038), box 3, item 83; **236b** 365 Brand by W. A. Snyder & Sons Company, printed by Schmidt Litho. Co. Special Collections & University Archives, University of California, Riverside. Collection of citrus labels (MS 038), box 2, item 3; **237 from top to bottom, left to right** Indian River Tre-Ripn Brand by Bonded Shipers, New Smyrna Beach, Florida. P. K. Yonge Library of Florida History. The Jerry Chicone Jr Citrus Crate Label Collection; Mor-Juice Brand, packed by Gulf Distributing Co., Tampa, Florida, distributed by Florida Citrus Exchange, Tampa, Florida. P. K. Yonge Library of Florida History, The Jerry Chicone Jr Citrus Crate Label Collection; Fairview Groves Brand. P. K. Yonge Library of Florida History. The Jerry Chicone Jr Citrus Crate Label Collection; Golden Globe Brand by Adams Packing Association, Inc., printed by Schmidt Litho. Co. P. K. Yonge Library of Florida History, The Jerry Chicone Jr Citrus Crate Label Collection; Juiceking Brand by W. H. McBride, Inc., printed by Schmidt Litho. Co. P. K. Yonge Library of Florida History. The Jerry Chicone Jr Citrus Crate Label Collection; Buttercup Brand by Okahumpka Packing Co., Florida, printed by Florida Grower Press, Tampa. P. K. Yonge Library of Florida History. The Jerry Chicone Jr Citrus Crate Label Collection; Burch's Peerless Brand by R. W. Burch, Inc., printed by Schmidt Litho. Co. P. K. Yonge Library of Florida History. The Jerry Chicone Jr Citrus Crate Label Collection; Marigold Brand by Indian River Producers, Inc., Tampa. P. K. Yonge Library of Florida History. The Jerry Chicone Jr Citrus Crate Label Collection; Doctors Demko Brand by Demko Brothers, Altoona, Florida. P. K. Yonge Library of Florida History. The Jerry Chicone Jr Citrus Crate Label Collection; Rainbow Brand by by H. Jennings Rou Inc. P. K. Yonge Library of Florida History. The Jerry Chicone Jr Citrus Crate Label Collection; **238** The Protected Art Archive/Alamy Stock Photo; **239al, 239r** Pierce Archive LLC/Buyenlarge/Getty Images; **239bl** Vintage Travel and Advertising Archive/Alamy Stock Photo; **240–241** Archives municipales de Menton; **242, 243ar 243bl** Printed by Schmidt Litho. Co. Jerry Chicone Jr Citrus Crate Label Collection, George A. Smathers Library, University of Florida; **243al, 245br** Printed by Louis Roesch Co., Litho. P. K. Yonge Library of Florida History. The Jerry Chicone Jr Citrus Crate Label Collection; **244a** Besgrade Brand by Oxnard Citrus Association, California, printed by Western Litho. Co. Special Collections & University Archives, University of California, Riverside. Collection of citrus labels, MS 038, Box 3, Item 32; **244cl** Solo Brand by Culbertson Lemon Assn. Saticoy, Ventura County, California. Special Collections & University Archives, University of California, Riverside. Collection of citrus labels. MS 038 Box 2, Item 54; **244cr** Keeper Brand by Briggs Lemon Association, printed by Schmidt Litho. Co. Special Collections & University Archives, University of California, Riverside. Collection of citrus labels. MS 038, Box 3, Item 106; **244bl** Power Brand by Culbertson Investment Company, Ventura, California, printed by Schmidt Litho. Co. Special Collections & University Archives, University of California, Riverside. Collection of citrus labels. (MS 038), box 2, item 138; **244br** El Primo Brand by College Heights Orange & Lemon Association, Claremont, Los Angeles, printed by Schmidt Litho. Co. Special Collections & University Archives, University of California, Riverside. Collection of citrus labels, MS 038, Box 3, Item 55-A; **245a** Snoboy Brand by Snoboy Inc., printed by Schmidt Litho. Co. P. K. Yonge Library of Florida History. The Jerry Chicone Jr Citrus Crate Label Collection; **245cl** Round Robin Brand. P. K. Yonge Library of Florida History. The Jerry Chicone Jr Citrus Crate Label Collection; **245bl** Ruby Red Blush Grapefruit Brand by Pride O'Texas Citrus Association, Mission, Hidalgo, Texas. UC Davis Library, Archives and Special Collections, Davis; **245br** Better N'Ever Brand, Valley Fruit & Vegetable Co., Pharr, Texas, printed by Stecher-Traung Lithograph Co. P. K. Yonge

Library of Florida History. The Jerry Chicone Jr Citrus Crate Label Collection; **246a** Handsum Brand, Strathmore. Photo GraphicaArtis/Getty Images; **246bl** Have One Brand by Sequoia Citrus Association, Lemon Cove, Tulare, California. Boston Public Library; **246br** Ponca Brand by Vandalia Packing Association, Porterville, California, printed by Schmidt Litho. Co. Special Collections & University Archives, University of California, Riverside. Collection of citrus labels (MS 038), box 2, item 139-A; **247a** Cal-Sweet Brand by Evans Brothers Packing Company, printed by Louis Roesch Co. Special Collections & University Archives, University of California, Riverside. Collection of citrus labels. (MS 038), box 1, item 69-A; **247bl** Green Circle Brand by McDermont Fruit Company, Riverside, California, printed by Western Litho. Co. Special Collections & University Archives, University of California, Riverside. Collection of citrus labels (MS 038), box 1, item 53-A; **247br** Hi-Tone Brand by W. A. Snyder & Sons Company, Fullerton, California, printed by Schmidt Litho. Co. Special Collections & University Archives, University of California, Riverside. Collection of citrus labels (MS 038), box 3,

item 123; **248a** Duo Brand by Pedro Monsonis, S. A., Valencia, printed by Lit. S. Dura A. Guimera, 29. P. K. Yonge Library of Florida History. The Jerry Chicone Jr Citrus Crate Label Collection; **248bl** Antonio Escandell, Carcaixent, printed by Masia. P. K. Yonge Library of Florida History. The Jerry Chicone Jr Citrus Crate Label Collection; **248br** Pugna Brand by Vicente Ferrando Vercher, Valencia, printed by Simat de Valldigna: Gráficas Artísticas Valldigna. Biblioteca Valenciana Nicolau Primitiu, Colección: BV Fondo gráfico, Ubicación: BV Colección Valenciana, Signatura: Etiqueta/155; **249a** Exaes Brand by Exportadora Agricola Espanola, S. L., Valencia, printed by Lit. Graficas Valencia. P. K. Yonge Library of Florida History. The Jerry Chicone Jr Citrus Crate Label Collection; **249bl** Liege Brand. P. K. Yonge Library of Florida History. The Jerry Chicone Jr Citrus Crate Label Collection; **249br** El Sol de España Brand. P. K. Yonge Library of Florida History. The Jerry Chicone Jr Citrus Crate Label Collection **250–251** Berlin Brand by Sala-Gomez, S. L., Pego, Alicante, printed by Lit. S. Dura A. Guimera, 29. P. K. Yonge Library of Florida History. The Jerry Chicone Jr Citrus Crate Label Collection.

ACKNOWLEDGMENTS

In writing this book, I have been helped, over many years, by many people, who answered my questions, chased up obscure literature, took me into the field, or provided facts and pointers for both text and illustrations. I am indebted to them all, but, in particular, to Peter Arthur, Alistair Hay, John McPhee, Ryan Moore and Jim Sait (Sydney), the late Pieter Baas, Anja Henselaar, Gerard Thijsse and Gerda van Uffelen (Leiden), Andrew Barron (Macquarie University, Sydney), Randy Bayer (University of Memphis, Tennessee), George (Andrew) Beattie and Philip Nobis (Western Sydney University, Australia), Andrew Brown and Helen Pickering (London), Julia Buckley, Lydia Elstone, Rafael Govaerts, Elizabeth Howard, Marie Humphries, Mark Nesbitt and James Wearn (Royal Botanic Gardens Kew, England), Roger Butler (Canberra), Michael and Barbara Christ (Woltersdorf, Germany), John Clarkson (Mareeba, Queensland), Joel Cohen, Lesley Elkan, Miguel Garcia, Simon Goodwin, Anna Hallett, Kristina McColl, Hannah McPherson, John Siemon and Ifeanna Tooth (Botanic Gardens of Sydney), Eric Danell (Åkarp, Sweden), Micheal Do (Sydney Opera House), Stefan Dressler (Senckenberg Museum, Frankfurt), Edgardo Etxeberria and Fred Gmitter (Citrus Research and Education Center, Lake Alfred, Florida), Kazumi Fujikawa (Makino Botanic Garden, Japan), Rabbi Philip Graubart (Washington DC), Mark Griffiths, Yoko Otsuki and Anne Sing (Oxford), Andrea Hart and Jacek Wajer (Natural History Museum, London), D.K. Hore (India), Shih-Chung Hsu (Institute of Botany, Taipei), Hiroyuki Iketani (Okayama University of Science, Japan); Dale Johnson (Washington, Missouri), Ruth Kiew (Forest Research Instiute Malaysia), Phillip Kodela (ABRS, Canberra), Mark Large (Auckland), Aaron Liston

(Oregon State University), Laura Mabberley (Madrid), John McNeill and Henry Noltie (Royal Botanic Garden Edinburgh), Alison Mitchell (Riverton, South Australia), the late Jeremy Montagu (Wadham College, Oxford), Quentin Phillipps (Borneo Research Consultants Ltd.), Karen Preuss (Montgomery, Alabama), Luis Ramón-Laca Menéndez de Luarca (Universidad de Alcalá, Spain), Peter and Pat Raven (Missouri Botanical Garden, Saint Louis), the late Sarah Reichard (University of Washington Botanic Gardens, Seattle), Michael Saalfeld (Denham, Buckinghamshire, England), Alexander Sennikov (University of Helsinki), Philip Short (State Herbarium of South Australia, Adelaide), the late Roger Short (University of Melbourne), Balangoda Singhakumara and B.B. Wijeratne (Sri Lanka), Daniel Stashower (Bethesda, Maryland), Lena Struwe (Rutgers University, New Jersey), Steve Sykes (Adelaide), Bernhard Voss (Jork, Hamburg), Akira Wakana (Kyushu University, Japan), Ian Warrington (Palmerston North, New Zealand), the late Harold Winters (Beltsville, Maryland), Qiang Xu (Huazhong Agricultural University, Wuhan), John Wiersema (Smithsonian Institution, Washington) and Dianxiang Zhang (South China Botanical Garden, Guangzhou).

Philip Watson at Thames & Hudson has been a constant source of encouragement and guidance, while Emma Barton has more than ably chivvied my text towards the acceptable. I am grateful, as ever, to Robert Fernandez for keeping my physical frame up to the task and, above all, to my long-suffering partner, Andrew Drummond, who has patiently endured the gestation of *yet another* book.

INDEX

Page numbers in *italics* refer to illustrations and their captions.

Adam's apple 33, 65, 70, *73*, 141
Agathosma spp. 19–21
agrivoltaics 216
Alberti, Leandro 65
Alberti, Leon Battista 98
Albertus Magnus 70
Alexander the Great 39, 46, 152
Alfonso, King 92–8
America 161–6 (*see also* California; Florida)
Amyris balsamifera 21
Ansieau, Roland *213*
anuga 156
apomixis 30, 106
archaeological remains 46–7
Arcimboldo, Giuseppe (*Winter*) 74
Aristotle 41
Armit, William Edington de Margrat 168
Arshile Gorky (*Still Life with Lemons*) *190*
Australasia 166–9
Australia 116, 166–9

Babur (*Baburnama*) 58–64, *66*, *67*, 221
'Bahia' sweet orange 64–5
Bailey, Frederick Manson *168*, 168–9
Banks, Joseph 154
Bassano, Jacopo (*Last Supper*) 71, *75–7*
Bauhin, Johann and Gaspard 33, 41
bears 25
Belvedere gardens, Vienna 135
bergamot 90–2, *90*
Bergera koenigii (curry leaf) 19
Bernat, Toni *215*
Bessa, Pancrace *183*
Bible 50
Bimbi, Bartolomeo 80, *110–13*
Binche 99
biological pest control 216, 224
bitter oranges (*Citrus trifoliata*) 28–9, *34*, 226
bizzaria 81–90, *88–9*, *137*
Blane, Gilbert 155
Bligh, Captain William 154
Bloemaert, Cornelis 98–9
Bonavia, Emanuel 22–4, 39, 156, 182
Bordone, Paris (*Bathsheba Bathing*) *73*
Botticelli, Sandro (*Primavera*) 80, *81*
Bourdichon, Jean 80, *82*
Bourette, Charlotte-Jacquéline Reynier 65
breeding 226–7
British Navy 154
Brown River finger lime (*Citrus wintersii*) 169
Buc'hoz, Pierre-Joseph *148*, *149*
Buddha's hand citrons *38*, 39, 149
Bunge, Alexander von 36

calamondins (*Citrus ×microcarpa*) 30, 36, *180*
California 161–6, *164*, 204
Calycophyllum candidissimum 21
Campari *200*

candied peel 39
Candolle, Augustin Pyramus de 157
Cantimpré, Thomas de 65–70
Carew, Sir Francis 119
Cariani, Giovanni (*Sacred Conversation*) *72*
Caribbean 161
Casanova, Giacomo Girolamo 196
Cavalerie, Pierre Julien 36
Cézanne, Paul (*Still Life with Apples and Oranges*) *189*
Chaddock, Captain Philip 33, 161
Charles VIII of France 98
chemical compounds 22
chimaeras 81–90
China 19, 28–31, 36, 39, 54
Cima da Conegliano, Giovanni Battista (*Madonna of the Orange Tree*) 70
citrangedins 202
citrangequats (*Citrus ×georgiana*) 202, *207*
citranges (*Citrus ×insitorum*) 196, *206–7*
citrons *see Citrus medica*
Citropsis 227
Citrus (genus) origins 19
Citrus ×aurantiifolia (Key lime) 65, 155–7, *155*, 161
Citrus ×aurantium 17, 28, 31, 64–5
   'Chinotto' 31
   'Citrus unshiu' 219
   mandarin × pomelo hybrids 201
   Seville oranges 92, 118, 166
   sour oranges *82*, 138–41, *176*
   sweet oranges 64–5, *134*, 161–6
Citrus ×floridana (limequat) 202, 207
Citrus ×georgiana (citrangequat) 202, *207*
Citrus ×insitorum (citrange) 196, *206–7*
Citrus ×junos (yuzu) 36–7, *36*
Citrus ×latifolia (Tahiti lime) 157–61
Citrus ×limon (lemon) 19, 54, *55*, 65, 141, *177*
Citrus ×microcarpa (calamondin) 30, 36, *180*
Citrus ×otaitensis 21, 54, 65, 210
Citrus ×sudachi 37, 219
Citrus ×tachibana 22, 37, *37*
Citrus australasica (finger lime) 168, 226
Citrus australis (dooja) 166
citrus canker 217
Citrus cavaleriei (Ichang papeda) 36
citrus fights 99
Citrus garrawayi (Mount White lime) 169
Citrus glauca (desert lime/lime bush/ desert kumquat) 168
Citrus gracilis (Humpty Doo lime) 169, 227
citrus greening disease see huanglongbing
Citrus halimii 25
Citrus 'Hormish' 90
Citrus hystrix (makrut lime) 25, 30, 156–61
Citrus indica 25, 227
Citrus inodora (Russell River lime) 169
Citrus japonica (kumquat) 22, 34, *35*, 36
Citrus linczangensis 19
Citrus macroptera 25
Citrus mangshanensis 227
Citrus maxima (pomelo) 22, 28, 30, *32*, 33–6, 161, *181*
Citrus medica (citron) 17, 30, 37–9, 46, 51, *86–7*, 138–41, *139*, 210
   uses 39

Citrus meletensis 19
Citrus nigra 19
Citrus oxanthera 168
Citrus polyandra 169
Citrus reticulata (mandarin) 30, 31, *31*, 54
Citrus trifoliata (bitter orange) 28-9, *34*, 226
Citrus wakonai 169, 217
Citrus warburgiana 168
Citrus wintersii (Brown River finger lime) 169
citrusmania 134
Citrusoxylon 19
cleaning products 201
clementines 201
Cointreau 197, *198-9*
collections, plant 80, 92
Columbus, Christopher 161
Commelin, Jan (*Nederlantze Hesperides*) *126-7*, 134-5, *134*
contraceptives 196
Cook, Captain James 154, 161
cooking 117
cordial, lime 156, *156*, *157*
crate labels *236-7*, *242-3*
Crimean War 155-6
Crusades 65, 70
Cutrale family 216
Cydonia oblonga (quince) 21, 118

da Gama, Vasco 65, 153
dal Pozzo, Cassiano 98
Darwin, Charles 90
De Caus, Salomon 124-34
'Diamante' citron 51
Dioscorides, Pedanius 17, 47, 58, 61, *61*, 80
diseases 25, 29, 216-27
dooja (*Citrus australis*) 166
Dossi, Dosso (*Saint Sebastian*) 70-1
drinks 156, *156*, *157*, 197-201
drugs, controlled 22
Duccio 70
Dundee 118
Dutch East India Company 118, 153
Dutch royal family 116
Dutch traders and settlers 116, 118-19

Earl Grey tea 92
eau de Cologne 92
Egypt 16, *205*
Eleanor of Castile 70
elephants 24-5
Engelbrecht, Martin *123*
essential oils 21-2
etrogim 51, *53*
Evelyn, John 138
evolution 22-4

fairs *238-41*
fairy tales 38-9
Fan Chengda 54
Farina, Giovanni Maria 92
Ferdinand I, Holy Roman Emperor 124
Ferrari, Giovanni Battista (*Hesperides*) *42*, *43*, 54, *92*, *94-7*, 98-106, *100-5*
finger limes (*Citrus australasica*) 168, 226
fingered citrons *38*, *39*, 106, 149, *149*
flags 117
Florida 161, 165-6, *165*, 216, 225, *232-5*
flowering 22

Foppens van Es, Jacob (*Still Life with Pitcher*) 119
fossils 19
fragrance 21-2, 138-41
frost 216
fungicidal qualities 22

Galeotti, Paolo 90
Galipea spp. (angostura) 19
Gallesio, Giorgio 65, 81, 138-41
Gargnano 92
Garraway, Roland Walter 169
Garzoni, Giovanna (*Still Life with Bowl of Citrons*) *121*
genetic modification 226-7
germplasm 227
Gibeau Orange Julep restaurant *197*
Gilbertus Anglicus 153
Giovanni Agostino da Lodi (*Madonna and Child with an Angel*) 70
golden apples of the Hesperides 21, *42-5*, 47, *47*
graft hybrids 81-90
grafting 29, *34*, 51, 106, 202
grapefruits 161, 211
Gwyn, Eleanor 'Nell' 65

hadar 50-1
Han dynasty 28
Han Yanzhi 33, 34
Harmonillus 106
Hawaii 161
Hawkins, Sir Richard 153, *153*, 158
hedging 29
Henriette Catherine of the House of Orange-Nassau 116
herbals 80, *80-5*
Herbert, Sir Thomas 156
herbicides 22
Hesperides (nymphs) 44, 92, 101
Hesperides, golden apples of the 21, *42-5*, 47, *47*
hesperidium 21
Het Loo Palace, Apeldoorn *128*, 134
Hibat-Allah Zayd Ibn Jami 54
Hill, Richard 166
Hippocrates 41
HIV 196
Holtzbecker, Hans Simon 132, *146*, *147*
Hore, D. K. 25
Hortus Palatinus 124-34
huanglongbing (HLB) 25, 217-27, *224*, *225*
humanism, Renaissance 98
Humbert II de la Tour-du-Pin 92
Humpty Doo lime (*Citrus gracilis*) 169, 227
hybridization 29-30, 54, 90, 138-41

Ibn al-Awwam 64
Ibn Butlan 62, *62*
Ibn Wahshiyya (*Kitab al-Filaha al-Nabatiyya*) 58
Ichang papeda (*Citrus cavaleriei*) 36
al-Idrisi, Muhammad (*Nuzhat al-mushtaq fi ikhtiraq al-afaq*) *68*, *69*, 70
India 28, 39, 46, *220*, *221*, *222-3*
Indonesia 106
insects 21-2, 196, 217-27
Ivrea 99

Jacques de Vitry 65
Jaffa *205*
James Keiller and Son 118

Japan 37, *218*, *219*
Jardin de Tuileries, Paris *140*, 141
Jardin des Pamplemousses, Mauritius 33–6
Ju Lu (Register of Citrus) 29, 33, *34*
Juanes, Juan de (*The Last Supper*) 71
Judaism 50–1, *50*, *51*, *52*, *53*

kaffir lime 157
Kahlo, Frida (*Untitled*) *190*
Kew, Royal Botanic Gardens 138
Key limes (*Citrus ×aurantiifolia*) 65, 155–7, *155*, 161
Knox, Robert 156
Köhler, Hermann *176–7*
Krausz, Franz *212*
kumquats (*Citrus japonica*) 22, 34, *35*, 36
Kuo I-Kung 30

La Quintinie, Jean-Baptiste de *124*
La Trémoille, Marie Anne de 92
Lancaster, Sir James 153
leech limes (*Citrus hystrix*) 157
Leighton, Frederic (*The Garden of the Hesperides*) *45*
lemon houses (*limonaie*) 92–8, *99*
lemon juice 196, 201
lemon sherbet 54
lemonade 65
lemons (*Citrus ×limon*) 19, 54, *55*, 65, 141, *177*
lemonwood 21
Leonardi, Vincenzo *89*, *99*, *99*, *102–5*
Leviticus 50, 53
Li Shizhen *29*
Liguria 65
lime juice 155–6
limequats (*Citrus ×floridana*) 202, 207
limes 65, 155–61, *155*
limonadiers 65
Limonia crenulata 23
limonoids 22
Lind, James (*Treatise of the Scurvy*) *154*, *154*
Linnaeus, Carl 17, 41, 210
liqueurs 197–201
longevity of fruit 22, 29, 156
Loret, Victor 46
Louis XIV 22, 134
Louise Henriette of Orange-Nassau 116
Lundstrøm, Vilhelm (*Still Life with White Jar, Orange and Book*) *190*

Mafia *230*
Magellan, Ferdinand 153
makrut limes (*Citrus hystrix*) 25, 30, 156–61
Malus domestica (orchard apple) 25
Malus Medica 25, 41
mandarins (*Citrus reticulata*) 30, 31, *31*, 54, 201
Mantegna, Andrea (*Agony in the Garden*) 70, *71*
Markée, Cornelis *142*, *143*, *144*
marmalade 118
Martínez, Bartolomé 161
Mattioli, Pietro Andrea (*Comementarii*) 80
Maupassant, Guy de 140
Medici family 58, *78*, *79*, 80–1, *110–13*
medicinal uses 117, 196–7
Menton Fête du Citron *240–1*
Menzies, Archibald 161
Merian, Maria Sibylla *144*, *147*
Meydenbach, Jacob (*Ortus Sanitatis*) 80
Michel, Étienne (*Traité du citronier*) 141
Millais, John Everett (*The Bridesmaid*) *189*

Minute Maid 216
Mitchell, Major Thomas 116, 168
Mombasa *64*, 65
Mount White lime (*Citrus garrawayi*) 169
Munting, Abraham (*Naauwkeurige Beschryving der Aardgewassen*) 135
Musée de l'Orangerie, Paris 141
Muslim traders, medieval 58

Nati, Pietro 81
Needham, Joseph 30, 39
neroli oil 92
Netherlands, the 116–17
New York 116
Newton, Amanda *192–3*
North America 116
North, Marianne 182, *184*, *185*, *186–7*
nutritional values 197

Oliver, Daniel 210
orange blossom 22, 70, *189*
orange juice 211, *211*, *212–15*
Orange Order, Protestant 116–17
orangeries 92–8, *94–7*, 119–35, *126–33*, 135–8, *140*
oranges *see Citrus ×aurantium*
orangewood 21
Orangina *211*, 215
Oranienbaum Palace 116
Oranienburg 116
ornamental shrubs 21, 210

Pacific islands 161
Palladius 47–50
Papua New Guinea 166–9
Paris 141
parrots 22–4
Passmore, Deborah Griscom *192–3*
Pauling, Linus 155
Paulus, Helmut-Eberhard 98
Peekel, Father Gerhard 169
peel 22
Peel, Robert 117
Peiresc, Nicolas-Claude Fabri de 98
pekoe tea 118
Pepys, Samuel 120
pests 216
Philippe, Odet 161
phytophthora 202, 216–17
place names 116
plague doctors *158*
Pliny the Elder 41, 48
Poiteau, Pierre-Antoine *172–3*
polyembryony 30
pomelos *see Citrus maxima*
Pomology, USDA Department of *192–3*
Pompeii 47
Poncirus 29
Portugal 216
Portugal oranges 65
Potsdam 141
Prague 124
psyllids 217–27

quinces (*Cydonia oblonga*) 21, 118

Ramat Rahel, Jerusalem 46
Rangpur lime (Canton lemon) 54

Redouté, Pierre-Joseph 141, *170-1*
Reggio Calabria 99
Renaissance paintings 70-1
Renoir, Pierre-Auguste (*Apples, Oranges and Lemons*) *189*
rhinos 24-5
ripening 22
Risso, Antoine (*Histoire naturelle des orangers*) *18, 20, 55, 89*, 141, *172-3*
Romans, Ancient 47, *48*, 49
rootstocks 202
Rose, Lauchlan 156, *156, 157*
Rossi, Anthony 211
Rothstein, Arthur *209*
rough lemons 54, 65
Rumpf, Georg Eberhard (*Herbarium Amboinense*) 33, 106, *107, 155*, 156
Russell River lime (*Citrus inodora*) 169
Ruta graveolens (rue) 19
Rutaceae family 19-21

San Bernardino National Orange Show *238-9*
satsumas 201, 219
Schönbrunn, Austria 135-8
Schutt, Ellen Isham *192-3*
Schweinfurth, Georg *162-3*
Scora, Reinhard 30
Scott, Captain Robert 154
scurvy 152-5
sea voyages 152-3
seed dispersal 22-5
seedless cultivars 210-11
Seville oranges 92, 118, 166
shaddocks 33
Shakespeare, William 118
Sicily 68, 70, 165-6, *226, 230*
Sickler, Friedrich (*Der vollkommene Orangerie-Gärtner*) 141
Siebold, Philipp Franz von 36-7
Sloane, Hans 33
sour oranges *82*, 138-41, *176*
South Africa 116
'Star Ruby' grapefruits 211
Steadman, Royal Charles *192-3*
still lifes *119, 121, 189-91*
Strasburger, Eduard 30
Su Shi *24*
Sucocítrico Cutrale 216
sudachi 219
Sukkot 46, 50-1, *50, 53*, 65
surpluses *226*
sweet oranges 64-5, *134*, 161-6
Swingle, Walter Tennyson 19, 201-10, *202, 206-7*, 227
Szent-Györgyi, Albert 155

Tabula Rogeriana 68, 69, 70
tachibana (*Citrus ×tachibana*) 22, 37, *37*
Tahiti limes (*Citrus ×latifolia*) 157-61
Tahiti oranges 21, 54
Tajimamori 37
al-Tajir, Sulaiman 58
tangelos 202, *202*
tangerines 28, 201
tangors 201, 202
taxonomy 17-19, 30, *34, 44*, 138, 210
tea 118
Tetraclinis articulata 17
Theophrastus 25, *40, 41-6*, 196
Thunberg, Carl Pehr 36

Tibbets, Eliza Lovell 164-5
timbers 21
Tolkowsky, Shemuel 50
traders 58, 65, 117-18
Traini, Francesco (*Triumph of Death*) 70
Treviso 99
tristeza virus 29, 217
Tropicana 211
Trotter, Thomas 152
Troyer, Albert Melville 202
tryptamine 22
Tura, Cosmè (*Madonna and Child in a Garden*) 70

van Cleve, Joos (*The Holy Family*) 71
van Eyck, Jan 71, 117, *118*
van Hulsdonck, Jacob (*Still Life with Lemons, Oranges and a Pomegranate*) *120*
van Leeuwenhoek, Antonie 30
van Sterbeeck, Frans (*Citricultura*) 134-5, *135*
Vavilov, Nikolai 19
Veronese, Paolo (*Annunciation*) 70
Versailles *124, 133*, 134
vesicles 21, 22
Victoria, Queen 70
Villa di Castello, Florence 80, 90, *106, 108-9*
Villemot, Bernard *215*
Virgil 47
Virgin Mary 22, 70, *73*
viruses 216-17
Vitamin C 33, 152, 155
Volkamer, Johann Christoph (*Nürnbergische Hesperides*) 44, *90*, 99, 123, 135, *136-7, 139*
Volkamer lemon 54

water dispersal of seeds 22, 25
Webber, Herbert 210
weddings 70
Weinmann, Johann Wilhelm *144*
Whitton Park 138
wild species 30, *31*, 227
William III 116-17, *117*
William of Orange (the Silent) 116, 117
Wolfskill, William 164, 165
Woodall, John (*The Surgeon's Mate*) 153
wrappers, citrus *243-50*

Yu Gong (Tribute of Yu) 28, *28*
yuzu (*Citrus ×junos*) 36-7, *36*

Zanthoxylum 21
Zhao Lingrang *24*
Zhou Qufei 54
Zwinger Palace, Dresden 134

For Andrew Drummond

ON THE COVER: *Citrons and Lemons*, Bartolomeo Bimbi, 1715, Villa Medicea di Poggio a Caiano, Museo della Natura Morta, Poggio a Caiano. Photo Scala, Florence, courtesy of the Ministero Beni e Att. Culturali e del Turismo (main image); xsense/Shutterstock (decorative pattern).

PAGES 2–3: Sandro Botticelli, see p. 81. **PAGES 4–5:** *Florida Lemons*, William Joseph McCloskey, 1889. **PAGE 6:** Grapefruit, unknown artist, *c.* 1849–55. **PAGE 7:** Citron (*Citrus medica*): fruiting branch, coloured etching by J. Pass after J. Ihle, *c.* 1800. **PAGES 8–9:** Georg Schweinfurth, see pp. 162–63. **PAGES 10–11:** Citrus crate label, San Francisco, California, early 20th century. **PAGE 26:** Detail from a fresco by Carlo Maratta and Ciro Ferri, Villa Falconieri La Rufina, Frascati, Italy, 17th century. **PAGE 56:** Babur supervising the laying out of the Garden of Fidelity, from a copy of the *Baburnama*, *c.* 1590. **PAGE 114:** Hans Simon Holtzbecker, see p. 132. **PAGE 150:** Detail of an advertisement for Pink's marmalade, *c.* 1890. **PAGE 194:** Advertisement for Zeste liqueur by Leonetto Cappiello, 1906.

First published in the United Kingdom in 2024 by
Thames & Hudson Ltd, 181A High Holborn, London WC1V 7QX

First published in the United States of America in 2024 by
Thames & Hudson Inc., 500 Fifth Avenue, New York, New York 10110

Citrus: A World History © 2024 Thames & Hudson Ltd, London
Text © 2024 David J. Mabberley

All Rights Reserved. No part of this publication may be reproduced or transmitted in any form or by any means, electronic or mechanical, including photocopy, recording or any other information storage and retrieval system, without prior permission in writing from the publisher.

British Library Cataloguing-in-Publication Data
A catalogue record for this book is available from the British Library

Library of Congress Control Number 2024937217

ISBN 978-0-500-02636-6

Printed and bound in China by C&C Offset Printing Co. Ltd

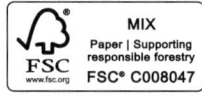

FSC
www.fsc.org
MIX
Paper | Supporting
responsible forestry
FSC® C008047

Be the first to know about our new releases,
exclusive content and author events by visiting
thamesandhudson.com
thamesandhudsonusa.com
thamesandhudson.com.au